T0305820

Theory and Computation of Hydrodynamic Stability

The study of hydrodynamic stability is fundamental to many subjects, ranging from geophysics and meteorology through to engineering design. This treatise covers both classical and modern aspects of the subject, systematically developing it from the simplest physical problems, then progressing to the most complex, considering linear and nonlinear situations, and analyzing temporal and spatial stability. The authors examine each problem both analytically and numerically. Many relevant fluid flows are treated, including those where the fluid may be compressible, or those from geophysics, or those that require salient geometries for description. Details of initial-value problems are explored equally with those of stability.

The text includes copious illustrations and an extensive bibliography, making it suitable for courses on hydrodynamic stability or as an authoritative reference for researchers. In this second edition the opportunity has been taken to update the text and, most importantly, provide solutions to the numerous extended exercises.

W. O. CRIMINALE is Professor Emeritus in the Department of Applied Mathematics at the University of Washington. His research focuses on the areas of initial-value problems in shear flows, large-scale oscillations in turbulent flows, mixing and nonlinear mechanics. Professor Criminale is the recipient of many honors, which include: the Alexander von Humboldt Senior Research Award; Fellow, American Physical Society; Guest Scientist, Stanford–Ames Turbulence Research Center; the Faculty Research Award, Battelle Pacific Northwest Laboratories; and Royal Society Fellow in the United Kingdom.

T. L. JACKSON is a research professor in the Department of Mechanical and Aerospace Engineering at the University of Florida, Gainesville. He is currently a fellow of the American Physical Society (APS), a fellow of the American Society of Mechanical Engineers (ASME), an associate fellow of the American Institute of Aeronautics and Astronautics (AIAA) and a member of the Combustion Institute. He was previously an associate editor for the *AIAA Journal,* and he currently serves on the editorial advisory board for the AIAA *Journal of Propulsion and Power.* His expertise is in the area of basic fluid mechanics, combustion, stability, solid propellant combustion, energetic material modeling and the large-scale simulation thereof.

R. D. JOSLIN is a permanent program director in the Engineering Directorate at the US National Science Foundation in Alexandria, Virginia. He manages the Fluid Dynamics Program and other cross-Foundation programs. Dr. Joslin was previously a program manager at the Office of Naval Research, where he managed the Turbulence, Stratified Wakes, Submarine Maneuvering, Ocean Energy, Multi-Platform Interactions and Supercavitation Programs. He is a member of the APS, ASME and is an associate fellow in the AIAA. His areas of expertise include fundamental fluid mechanics, turbulence and transition, stability theory, DNS/LES/CFD, supercavitation and renewable energy.

Established in 1952, the *Cambridge Monographs on Mechanics* series has maintained a reputation for the publication of outstanding monographs, a number of which have been re-issued in paperback. The series covers such areas as wave propagation, fluid dynamics, theoretical geophysics, combustion, and the mechanics of solids. Authors are encouraged to write for a wide audience, and to balance mathematical analysis with physical interpretation and experimental data, where appropriate. Whilst the research literature is expected to be a major source for the content of the book, authors should aim to synthesize new results rather than just survey them.

A complete list of books in the series can be found at www.cambridge.org/mathematics.

RECENT TITLES IN THIS SERIES

Magnetoconvection
N. O. WEISS & M. R. E. PROCTOR

Waves and Mean Flows (Second Edition)
OLIVER BÜHLER

Turbulence, Coherent Structures, Dynamical Systems and Symmetry (Second Edition)
PHILIP HOLMES, JOHN L. LUMLEY,
GAHL BERKOOZ & CLARENCE W. ROWLEY

Elastic Waves at High Frequencies
JOHN G. HARRIS

Gravity-Capillary Free-Surface Flows
JEAN-MARC VANDEN-BROECK

Lagrangian Fluid Dynamics
ANDREW F. BENNETT

Theory and Computation of Hydrodynamic Stability

SECOND EDITION

W. O. CRIMINALE
University of Washington

T. L. JACKSON
University of Florida

R. D. JOSLIN
National Science Foundation

CAMBRIDGE
UNIVERSITY PRESS

Shaftesbury Road, Cambridge CB2 8EA, United Kingdom

One Liberty Plaza, 20th Floor, New York, NY 10006, USA

477 Williamstown Road, Port Melbourne, VIC 3207, Australia

314–321, 3rd Floor, Plot 3, Splendor Forum, Jasola District Centre, New Delhi – 110025, India

103 Penang Road, #05–06/07, Visioncrest Commercial, Singapore 238467

Cambridge University Press is part of Cambridge University Press & Assessment,
a department of the University of Cambridge.

We share the University's mission to contribute to society through the pursuit of
education, learning and research at the highest international levels of excellence.

www.cambridge.org
Information on this title: www.cambridge.org/9781108475334

DOI: 10.1017/9781108566834

First edition © Cambridge University Press & Assessment 2003
Second edition © W. O. Criminale, T. L. Jackson, and R. D. Joslin 2019

First published 2003
Second edition 2019

A catalogue record for this publication is available from the British Library

Library of Congress Cataloging-in-Publication data
Names: Criminale, William O., author.
Title: Theory and computation of hydrodynamic stability / W.O. Criminale,
University of Washington, T.L. Jackson, University of Florida, R.D.
Joslin, National Science Foundation, Alexandria, Virginia.
Description: Second edition. | Cambridge, United Kingdom ; New York, NY :
Cambridge University Press, 2019. | Series: Cambridge monographs on
mechanics | Includes bibliographical references and index.
Identifiers: LCCN 2018024498| ISBN 9781108475334 (hardback : alk. paper) |
ISBN 9781108466721 (pbk. : alk. paper)
Subjects: LCSH: Hydrodynamics. | Stability.
Classification: LCC QA911 .C75 2019 | DDC 532/.5–dc23
LC record available at https://lccn.loc.gov/2018024498

ISBN   978-1-108-47533-4   Hardback
ISBN   978-1-108-46672-1   Paperback

Contents

viii   *Contents*

Preface to the Second Edition

This second edition started soon after the first printing because it was recognized that this specialty field involves significant mathematics and the exercises at the end of each chapter may prove challenging to many. Therefore, it was determined that providing solutions to the exercises may be more useful to the learning process, and a number of solutions to the exercise problems are provided in Appendix C. We have also taken the opportunity to move the discussion of all of the numerical algorithms to one place, namely Appendix B. Also, many thanks to M. R. Malik and H. C. Kuhlmann for the published reviews of the first edition. Although the authors caught most of the issues presented in the reviews in the preprint process, the final printed version failed to address some typos and some poor quality images. It is our hope that this second edition will be more useful to the readers, especially with the addition of solutions to the exercises.

Preface

The subject of hydrodynamic stability, or stability of fluid flow, is one that is most important in the fields of aerodynamics, hydromechanics, combustion, oceanography, atmospheric sciences, astrophysics and biology. Laminar or organized flow is the exception rather than the rule to fluid motion. As a result, exactly what may be the reasons or causes for the breakdown of laminar flow has been a central issue in fluid mechanics for well over a hundred years. And, even with progress, it remains a salient question, for there is yet to be a definitive means for prediction. The needs for such understanding are sought in a wide and diverse list of fluid motions because the stability or instability mechanisms determine, to a great extent, the performance of a system. For example, the underprediction of the laminar to turbulent transitional region on aircraft – which is due to hydrodynamic instabilities – would lead to an underestimation of a vehicle's propulsion system and ultimately result in an infeasible engineering design. There are numerous such examples.

The seeds for the writing of this book were sown when one of us (WOC) was contacted by two friends, Philip Drazin and David Crighton, with the suggestion that it was perhaps time for a new treatise devoted to the subject of stability of fluid motion. A subsequent review was taken by asking many colleagues as to their assessment of this thought and, if this was positive, what should a new writing of this subject entail? The response was enthusiastic and revealed three major requirements: (i) a complete updating of all aspects of the field; (ii) the presentation should provide both analytical and numerical means for solution of any problem posed; (iii) the scope of the treatment should cover the full range of the dynamics, ranging from the transient to asymptotic behavior as well as linear and nonlinear formulations. Then, since the computer is now a major tool, a final need suggested that direct numerical simulation (DNS) must be included as well.

This challenge was accepted and with intensive collaboration, we have attempted to meet these goals. All prototype flows are considered whether confined (Chapter 3), semi-confined (Chapter 3), in the absence of boundaries (Chapter 2) and both parallel, almost parallel or flows with curved streamlines (Chapter 6). In addition, the topics of spatial versus temporal stability (Chapter 4), compressible (Chapter 5) as well as incompressible fluids geophysical flows (Chapter 7), transition and receptivity (Chapter 10) and optimization and control of flows (Chapter 12) are given full attention. Also, specific initial-value problems (Chapter 8) would be examined as well as the question of stability. In every case, the basics are developed with the physics and the mathematical needs (Chapters 1, 2) with emphasis on numerical methods for solution. To this end, in formulating the organization of the book, it was decided that it would be beneficial if, at the end of each chapter that dealt with a specific topic, in addition to exercises for illustration, an appendix, when appropriate, would be attached that provided a numerical basis for that particular area of need. The reader would then be able to develop their own code. Nonlinear stability (Chapter 9) and direct numerical simulation, i.e., DNS (Chapter 11) are supplemented with a cursory review of what is known from experiments (Chapter 13).

The book can easily be used as a text for either an upper-level undergraduate or graduate course for this subject. For those who are already knowledgeable, we hope that the book will be a welcome and useful reference.

There are many friends who have helped us with the formulation and writing. Indeed, all have given us both criticism and advice when needed. Particular recognition should be given to Richard DiPrima, who was the mentor for one of us (TLJ) and was a person who provided more than a rationale to be engaged in the field of hydrodynamic stability with his teaching, expertise and major contributions to the subject. In a similar manner, Robert Betchov provided the initial impetus for another (WOC). More recently, M. Gaster, C. E. Grosch, F. Hu, D. G. Lasseigne, L. Massa and P. J. Schmid have made their time available so that our writing would benefit and the contents be made to fit our goal. To each, we extend our sincere thanks. And, to the late Robert Betchov, Dick DiPrima, David Crighton and Philip Drazin, a firm note of gratitude. The passing of our colleagues is a loss. Finally, we have had assistance from some who have helped with technical needs. In particular, Frances Chen, Michael Campbell and Peter Blossey should be cited.

1

Introduction and Problem Formulation

Mathematics is the key and door to the sciences

– Galileo Galilei (1564–1642)

1.1 History, Background and Rationale

In examining the dynamics of any physical system, the concept of stability becomes relevant only after first establishing the possibility of equilibrium. Once this step has been taken, the concept of stability becomes pervasive, regardless of the actual system being probed. As expressed by Betchov & Criminale (1967), stability can be defined as the ability of a dynamical system to be immune to small disturbances. It is clear that the disturbances need not necessarily be small in magnitude and therefore may become amplified. As such, there is a departure from the state of equilibrium. Should no equilibrium be possible, then it can already be concluded that the particular system in question is statically unstable and the dynamics is a moot point.

Such tests for stability can be and are made in any field, such as mechanics, astronomy, electronics and biology, for example. In each case from this list, there is a common thread in that only a finite number of discrete degrees of freedom are required to describe the motion and there is only one independent variable. Like tests can be made for problems in continuous media but the number of degrees of freedom becomes infinite and the governing equations are now partial differential equations instead of the ordinary variety. Thus, conclusions are harder to obtain in any general manner, but it is not impossible. In fact, successful analysis of many such systems has been made and this has been particularly true in fluid mechanics. This premise is even more so today because there are far more advanced means of computation available to supplement analytical techniques. Likewise the means for experimentation has improved in profound ways and will be highlighted throughout the text in validation of the theoretical and computational results.

Fundamentally, there is no difficulty in presenting the problem of stability in fluid mechanics. The governing Navier–Stokes continuum equations for the conservation of momentum and mass that is often expressed by constraints, such as incompressibility that requires the fluid velocity to be solenoidal in a somewhat general sense, are the tools of the science. A specific flow is then fully determined by satisfying the boundary conditions that must be met for that flow. Other considerations involve the importance of the choice of the co-ordinate system that is best to describe the flow envisioned and whether or not there is any body force, say. Then, the important first step is to identify a flow that is in equilibrium. For this purpose, a flow that is in equilibrium need not necessarily be time independent, but the system is no longer accelerated due to the balance of all forces. For such flows meeting these conditions very few, if any, remain that have not been theoretically evaluated using this approach, but, because the governing equations of motion are a set of nonlinear partial differential equations, the results are most often the result of approximations. Nevertheless these flows are well established, many have been experimentally confirmed, and they are all laminar. In addition, a few exact solutions of the governing equations are known. In such cases, where more complex physics is entailed, such as compressibility or electrical conductivity of the fluid, similar arguments can be made and results have been equally obtainable.

Essentially there are three major categories of base mean flows, namely: (a) flows that are parallel or almost parallel; (b) flows with curved streamlines and; (c) flows where the mean flow has a zero value. Examples of the parallel variety are channel flows, such as plane Couette and Poiseuille flows where the flows are confined by two solid boundaries. There is one mean component for the mean velocity and it is a function of the coordinate that defines the locations of the boundaries. In a polar coordinate system, pipe flow is another example of note. Almost parallel flows are of two main categories: (i) free shear flows, such as the jet, wake and mixing layer where there are no solid boundaries in the flow and (ii) the flat-plate boundary layer where there is but one solid boundary. In these terms, (i) and (ii) have two components for the mean velocity, and they are both functions of the coordinate in the direction of the flow as well as the one that defines the extent of the flow. In Cartesian terms, if U and V are the mean velocity components in the x and y spatial directions, respectively, then almost parallel assumes that $V \ll U$ and that the variation of U with respect to the downstream variable x is weak. Group (b) has flows such as that between concentric circular cylinders (Taylor problem) or flow on concave walls (Görtler problem). The cases where there is no mean flow (e.g., Rayleigh problem, Bénard cells) are simply special cases of the more general picture. Whether from the point of view of the physics or the mathematics needed to make analyses, each of these prototypes has its own unique features and it is the stability of the system that is the question to be answered. It should be clear that the actual causes of any resulting instability will vary as well.

It should be again stressed that, regardless of the methods required for obtaining any mean flow, they are laminar and are in equilibrium or near equilibrium. But, unfortunately, just as the adage states, "turbulence is the rule and not the exception to fluid motion." In other words, laminar flows are extremely hard to maintain; transition to turbulence will occur in the short or the long time. One need only to observe the flow over the wings of an airplane, the meandering of a river, the outflow from the garden hose or the resulting flow behind bluff bodies in both the laboratory and in nature to witness this predominance first hand. Laminar flow is orderly, can be well predicted and is most generally desired. The illustrations of Figs. 1.1, 1.2 and 1.3 vividly demonstrate the more-than-subtle differences for these two flows in the boundary layer setting. Benefits of laminar flow include less drag and reduced acoustics when compared to the turbulent state. Figure 1.1 shows the clean streamline pattern over a flat plate, reminiscent of laminar flow, whereas Fig. 1.2 shows the random turbulent boundary layer over a segment of the same flat plate. Although transition occurs via a different mechanism on a rotating cone, Fig. 1.3 shows the entire set of fluid states whereby the flow is laminar at the apex of the cone. The focus of this text becomes clear as the flow is disturbed and "transitions" to a state between laminar and turbulent. Finally, the flow is fully random, chaotic or what is called turbulent. Contrary to the benefits of laminar flow, a case where a benefit from turbulent flow would be desired over laminar is mixing, for example. The goal of predicting or even approximating the process of transition has been a stated goal throughout the history of fluid mechanics and, it was once thought, stability analysis would be able to do this. Any success has been limited but stability analysis can explain – for almost all of the major cases – why a basic flow cannot be maintained indefinitely.

Although the main focus of the text is on the mathematics of predicting flow instabilities, the classical experiments of Reynolds (1883) are introduced in Fig. 1.4, which shows the circular pipe flow experiment. Note the very raw experimental setup of the era compared with modern-day more advanced laboratory systems. Figure 1.5 shows the classical experiment due for flow in a circular pipe whereby dye was inserted and the mean flow run at different values through a number of pipe diameters. This was an extremely important series of experiments to modern day fluid mechanics, so it is worth revisiting

Figure 1.1 Laminar boundary layer on a flat plate (Werlé, 1974).

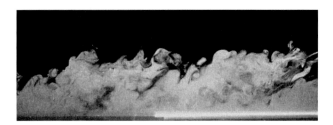

Figure 1.2 Turbulent boundary layer on a flat plate (Reprinted from Falco, 1977 with the permission of AIP Publishing).

these results briefly, as well as the thoughts of Osborne Reynolds. At the beginning of his paper, Reynolds stated the following:

> There appeared to be two ways of proceeding – the one theoretical, the other practical. The theoretical method involved the integration of the equations for unsteady motion in a way that had not been accomplished and which, considering the general intractability of the equations, was not promising. The practical method was to test the relation between U, μ/ρ, and c.

The first way of proceeding – theory – is the primary focus of this text and clearly shows the progress made over time and, with the advent of computers, the equations have become tractable. The second way of proceeding – namely, experimentation – was important to the contemporary scientist because the variation of velocity U, kinematic viscosity μ/ρ, and pipe radius c was the advent of the Reynolds number $Re = \rho U c/\mu$.

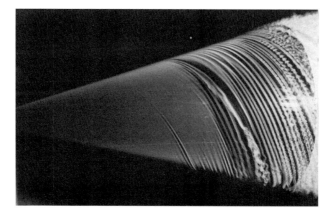

Figure 1.3 Spiral vortices on a cone in rotation with freestream (Kobayashi, Kohama & Kurosawa, 1983, reproduced with permission).

In returning to the discussion of Reynolds' main observations in Fig. 1.4, the original organized parallel laminar flow is seen at several stages with the ultimate breakdown and fully random three-dimensional motion transpiring. At low "Reynolds number," the dye is transported through the pipe evident as a straight line at the top of the image. As the Reynolds number increases, or a critical velocity is reached, Reynolds noted:

> And it was a matter of surprise to me to see the sudden force with which the eddies sprang into existence, showing a highly unstable condition to have existed at the time the steady motion broke down.

As the critical velocity increases, the dye image clearly shows a more random or turbulent pattern. Ironically, this problem is one where stability theory has not been able to make any conclusions whatsoever and remains an enigma in the field. In short, linear theory has been used to investigate this flow in many ways and no solutions that predict instability have been found. This has been found to be true regardless of any added complexities that might be envisioned – for example, axisymmetric versus non-axisymmetric disturbances. Still, it is clear that this flow is unstable.

Drawings of vortices can be traced as far back as those of Leonardo da Vinci that were made in the fifteenth century. The first significant contribution to the theory of hydrodynamic stability is that due to Helmholtz (1868). The principal initial experiments are due to Hagan (1855). Later a major list of contributions can be cited. Reynolds (1883), Kelvin (1880, 1887a,b) and Rayleigh (1879, 1880, 1887, 1892a,b,c, 1895, 1911, 1913, 1914, 1915, 1916a,b) were all ac-

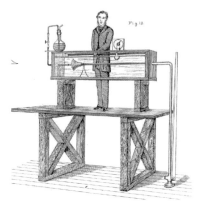

Figure 1.4 Sketch of the Reynolds pipe flow experiment (Reynolds, 1883).

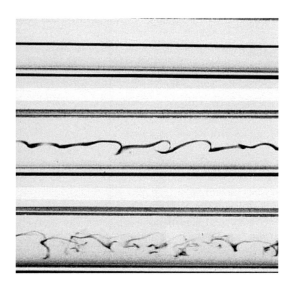

Figure 1.5 Repetition of Reynolds' dye pipe experiment (van Dyke, 1982).

tive in this period. Here, the birth of the Reynolds number as well as the first theorems due to Rayleigh appeared. As has been noted before, Lord Rayleigh was thirty-six when he considered the stability of flames and then published his work on jets. At seventy-two he began to do work in nonlinear stability theory! Unlike Reynolds' pipe experiment, which was intrinsically viscous, the exceptional theoretical work of Kelvin and Rayleigh was done using the inviscid approximation in the analysis.

Independently, Orr (1907a,b) and Sommerfeld (1908) framed the viscous stability problem. Both were attempting to investigate channel flow, with Orr considering plane Couette flow, and Sommerfeld plane Poiseuille flow. Of course one case is the limit of the other and the combination has led to the Orr–Sommerfeld equation that has become the essential basis in the theory of hydrodynamic stability. But, even here, it should be remembered that it was not until twenty-two years after the derivation of this equation that any solution at all could be produced. Tollmien (1929) calculated the first neutral eigenvalues for plane Poiseuille flow and showed that there was a critical value for the Reynolds number. This work was made possible by the development of Tietjens' functions (Tietjens, 1925) and analysis of Heisenberg (1924), connected with the topic of resistive instability. Romanov (1973) proved theoretically that plane Couette flow is stable. Unlike pipe flow, there is no experimental controversy here. Plane Poiseuille flow, on the other hand, is unstable.

Schlichting (1932a,b, 1933a,b,c, 1934, 1935) continued the work of Tollmien and extended it even further. The combination of these efforts have led to the designation for the oscillations that are now the salient results for the stability of parallel or nearly parallel flows, namely Tollmien–Schlichting waves. It should be noted that such waves correspond to those waves where friction is critical and do not exist for any problem that does not include viscosity and are known to be present only in flows where a solid boundary is present in the flow. Also, in the limit of infinite Reynolds number, the flow is stabilized.

Prandtl (1921–1926, 1930, 1935) was active in problems related to stability in the hopes that the theory might lead to the prediction of transition and the onset of turbulence. As mentioned, to date no such success has been achieved but the effort continues as the understanding makes progress. But, for the first time during this period, a major boost to stability analysis was given by the work of Taylor (1923) where theory was confirmed by his experiment for the case of rotating concentric cylinders. Taylor himself was responsible for this, and the work continues to be a model for understanding the stability of mean flows with curved stream lines.

The advent of matched asymptotic expansions and singular perturbation analysis brought new vigor to the theory. Lin (1944, 1945) made use of these tools and re did all previous calculations, thereby confirming the earlier results that had been obtained by less sophisticated means. Experiments also gained momentum with the work of Schubauer & Skramstad (1943) in the investigation of the flat-plate boundary layer setting the standard. Here, a vibrating ribbon was employed to simulate a controlled disturbance, that is a Tollmien–Schlichting wave, at the boundary. This method is still employed by many today. Theoretical calculations were confirmed and, equally important, for the first time it became apparent that the value of the critical Reynolds number meant the stability boundary for the onset of unstable Tollmien–Schlichting waves and not the threshold for the onset of turbulence. Figure 1.6, depicting the results of this experiment, is a hallmark in this field. This conclusion has been further substantiated today. For example, Schubauer & Klebanoff (1955, 1956), Klebanoff, Tidstrom & Sargent (1962) and Gaster & Grant (1975) performed even more extensive experiments for the boundary layer.

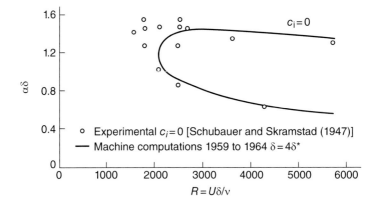

Figure 1.6 Experimental and theoretical stability results for neutral oscillations of the Blasius boundary layer (after Betchov & Criminale, 1967).

Investigating the stability of compressible flows was not done until much later with the theoretical work of Landau (1944), Lees (1947) and Dunn & Lin (1955) being the principal contributors at this time. Physically and mathematically, this is a far more complex problem and, in view of the time span it took to resolve the theory in an incompressible medium, this was understandable. A wide range of problems have been investigated here, including different prototypes and Mach numbers up to hypersonic in value. Likewise, there are experiments that have been done for these flows (see Kendall, 1966).

The use of numerical computation for stability calculations was made with the work of Brown (1959, 1961a,b, 1962, 1965), Mack (1960, 1965a,b) and Kaplan (1964) being the principal contributions. Neutral curves that were previously obtained by asymptotic theory and hand calculations are now routinely determined by numerical treatment of the governing stability equations. Such numerical evaluation has proven to be more efficient and far more accurate than any of the methods employed heretofore. Furthermore, the complete and unsteady nonlinear Navier–Stokes equations are evaluated by the use of high-order numerical methods in tandem with machines that range from the personal computer to supercomputers and the parallel class of machines, which are the standard tool for solving fluid mechanics problems today. By numerical calculations, one of the earliest results for the full Navier–Stokes calculations was obtained by Fromm & Harlow (1963), where the problem of vortex shedding from a vertical flat plate was investigated. Since this time, the complete Navier–Stokes equations are routinely used to study the vortex shedding process. Among others, Lecointe & Piquet (1984), Karniadakis & Triantafyllou (1989) and Mittal & Balachandar (1995), for example, have all numerically solved the full equations in order to investigate instability and vortex shedding from cylinders. A summary of this vortex shedding problem is provided in a review by Williamson (1996).

Effort has been made to assess nonlinearity in stability theory. Meksyn & Stuart (1951), Benney (1961, 1964) and Eckhaus (1962a,b, 1963, 1965) were all early contributors to what is now known as weakly nonlinear theory. Each effort was directed to different aspects of the problems. For example, the nonlinear critical layer, development of longitudinal or streamwise vortices in the boundary layer and the possibility of a limiting amplitude for an amplifying disturbance were examined. The role of streamwise vorticity in the breakdown from laminar to turbulent flow has recently been explored using the complete Navier–Stokes equations. For this purpose, Fasel (1990), Fasel & Thumm (1991), Schmid & Henningson (1992a,b) and Joslin, Streett & Chang (1993) have introduced oblique wave pairs at amplitudes ranging from very small to finite values. The interaction of such oblique waves leads to dominant streamwise vortex structure. When the waves have small amplitudes, the disturbances first amplify but then decay at some further downstream location. When finite, the nonlinear interactions of the vortex and the oblique waves result in breakdown.

Since the experimental setting for probing in this field is almost unequivocally one where any disturbance changes in space and only oscillates in time, thought has been given to the question of spatial instability so that theory may be more compatible with experimental data. The problem can be posed in very much the same way as the temporal one, but the equations must be adapted for this purpose (e.g., see Section 1.8 for the discussion based on Gaster (1965a,b)). This is true even if the problem is governed by the linear equations. Direct numerical simulation also has major complexities when computations are made in this way. Nevertheless, this is done. For this purpose, reference to the summaries of Kleiser & Zang (1991) and Liu (1998) can be made where the use of direct numerical simulation for many instability problems has been given. More specifically, among this vast group, Wray & Hussaini (1984) and Spalart & Yang (1987) both investigated the breakdown of the flat-plate boundary layer by use of a temporal numerical code. In other words, an initial value problem was prescribed at time $t = 0$, and the disturbance developed for later times. By contrast, when a spatial code is employed, and initial values are given at a fixed location and then the development thereafter downstream, the work of Fasel (1976), Murdock (1977), Spalart (1989), Kloker & Fasel (1990), Rai & Moin (1991a,b) and Joslin, Streett & Chang (1992, 1993) should be noted. For three-dimensional mean flows, where cross flow disturbances are present, Spalart (1990), Joslin & Streett (1994) and Joslin (1995a) studied the breakdown process by means of direct numerical simulation.

Stability theory uses perturbation analysis in order to test whether or not the equilibrium flow is unstable. Consider the flows that are incompressible, time independent and parallel or almost parallel by defining the mean state as

$$\underline{U} = (U(y),0,0); \quad P$$

in Cartesian coordinates where $U(y)$ is in the x-direction with y the coordinate that defines the variation of the mean flow, z is in the transverse direction and P is the mean pressure. For some flows, such as that of channel flow, this result is exact; for the case of the boundary layer or one of the free shear flows, then this is only approximate but, as already mentioned, the U-component of the velocity, $U \gg V$ and $U \gg W$, as well as U varying only weakly with x, and hence the designation of almost parallel flow. In this configuration, both x and z range from minus to plus infinity with y giving the location of the solid boundaries, if there are any. P is the mean pressure and the density is taken as constant.

Now assume that there are disturbances to this flow that are fully three-dimensional and hence

$$\underline{u} = (U(y)+\tilde{u},\tilde{v},\tilde{w}); \quad p = P+\tilde{p}$$

can be written for the velocity and pressure of the instantaneous flow. By assuming that the products of the amplitudes (defined nondimensionally with the measure in terms of the mean flow) of the perturbations, as well as the products of the perturbations with the spatial derivatives of the perturbations, are small, then, by subtracting the mean value terms from the combined flow, a set of linear equations can be found and are dimensionally

$$\frac{\partial \tilde{u}}{\partial x} + \frac{\partial \tilde{v}}{\partial y} + \frac{\partial \tilde{w}}{\partial z} = 0, \tag{1.1}$$

for incompressibility, and

$$\frac{\partial \tilde{u}}{\partial t} + U\frac{\partial \tilde{u}}{\partial x} + \frac{dU}{dy}\tilde{v} = -\frac{1}{\rho}\frac{\partial \tilde{p}}{\partial x} + \nu\nabla^2\tilde{u}, \tag{1.2}$$

$$\frac{\partial \tilde{v}}{\partial t} + U \frac{\partial \tilde{v}}{\partial x} = -\frac{1}{\rho} \frac{\partial \tilde{p}}{\partial y} + \nu \nabla^2 \tilde{v}, \tag{1.3}$$

$$\frac{\partial \tilde{w}}{\partial t} + U \frac{\partial \tilde{w}}{\partial x} = -\frac{1}{\rho} \frac{\partial \tilde{p}}{\partial z} + \nu \nabla^2 \tilde{w}, \tag{1.4}$$

for the momenta, where ρ is the density of the fluid, ν is the kinematic viscosity and the three-dimensional Laplace operator is given by

$$\nabla^2 = \frac{\partial^2}{\partial x^2} + \frac{\partial^2}{\partial y^2} + \frac{\partial^2}{\partial z^2}.$$

Hereafter, $()' = d/dy$, $()'' = d^2/dy^2$, etc. It is more prudent to nondimensionalize the equations and this will be done eventually but, for the purposes of the discussion of the basic concepts, they will here be considered dimensionally throughout this chapter. When nondimensionalization has been done in this case, all quantities are redefined and the coefficient of viscosity is replaced with the reciprocal of the Reynolds number, defined in terms of the chosen length and velocity scales of the particular flow.

1.2 Initial-Value Concepts and Stability Bases

At this stage a temporal initial-value, spatial boundary-value problem has been prescribed and must be solved in order to determine whether or not the given flow is unstable. In this respect, it is well defined but, as will be seen, there are many difficulties in actually performing this task. There is, of course, more than one definition for stability that can be used, but the major concern is whether or not the behavior of the disturbances causes an irreversible alteration in the mean flow. In short, if, as time advances from the initial instant, there is a return to the basic state, then the flow is considered stable. There are various ways that instability can occur but it is first essential to understand what means are possible for solving these problems in order that any decision can be made. At the outset it can already be seen that the order of the system is higher than the traditional second-order boundary value problems of mathematical physics. As a result, some of the classic methods of exploration are of limited value; others that may be used require extensions or alterations in order to be employed here.

Any velocity vector field can be decomposed into its solenoidal, rotational and harmonic components. For the problems being discussed here there is no solenoidal part due to the fact that the fluid is incompressible and $\nabla \cdot \underline{u} = 0$. On physical grounds the rotational part of the velocity corresponds to the perturbation vorticity with the harmonic portion related to the pressure. This analogy makes for better interpretation of the physics for, even though the boundary conditions must be cast in terms of the velocity, the initial specification can be considered as that of vorticity. In this respect, each of the mean flows that has been cited, when the governing equations are written in terms of the vorticity, the vorticity is essentially a quantity that is diffused or advected from what it was initially and the velocity profile is the result of this action. The same reasoning can be made for the perturbation field.

The reasoning for the decomposition of the velocity can be best understood by actually using the definitions for the divergence and the curl. First, operate on (1.2) to (1.4) by taking the divergence, and use (1.1) to give

$$\nabla^2 \tilde{p} = -2\rho U' \frac{\partial \tilde{v}}{\partial x}. \tag{1.5}$$

The relation (1.5) is an equation for the perturbation pressure and has an inhomogeneous term that is effectively a source for the pressure, due to the interaction of the fluctuating and mean strain rates. When neither is strained, then the pressure is harmonic. If the velocity had not been solenoidal, then factors relating to the compressibility of the fluid would come into play.

Now, the definitions of the perturbation vorticity components are

$$\tilde{\omega}_x = \frac{\partial \tilde{w}}{\partial y} - \frac{\partial \tilde{v}}{\partial z}, \tag{1.6}$$

$$\tilde{\omega}_y = \frac{\partial \tilde{u}}{\partial z} - \frac{\partial \tilde{w}}{\partial x}, \tag{1.7}$$

and

$$\tilde{\omega}_z = \frac{\partial \tilde{v}}{\partial x} - \frac{\partial \tilde{u}}{\partial y}, \tag{1.8}$$

respectively, since $\underline{\omega} = \nabla \times \underline{u}$.

By using these definitions and the operation of the curl on the same set of equations for the momenta, the following are obtained:

$$\left(\frac{\partial}{\partial t} + U \frac{\partial}{\partial x}\right) \tilde{\omega}_x - \nu \nabla^2 \tilde{\omega}_x = -U' \frac{\partial \tilde{w}}{\partial x} = \Omega_z \frac{\partial \tilde{w}}{\partial x}, \tag{1.9}$$

$$\left(\frac{\partial}{\partial t} + U \frac{\partial}{\partial x}\right) \tilde{\omega}_y - \nu \nabla^2 \tilde{\omega}_y = -U' \frac{\partial \tilde{v}}{\partial z} = \Omega_z \frac{\partial \tilde{v}}{\partial z}, \tag{1.10}$$

$$\left(\frac{\partial}{\partial t} + U \frac{\partial}{\partial x}\right) \tilde{\omega}_z - \nu \nabla^2 \tilde{\omega}_z = -U' \frac{\partial \tilde{w}}{\partial z} + U'' \tilde{v} = \Omega_z \frac{\partial \tilde{w}}{\partial z} - \Omega'_z \tilde{v}, \tag{1.11}$$

where $\Omega_z = -dU/dy$ is the single component of the mean vorticity and is in the z-direction. Each of these equations has the expected transport by the mean velocity and diffusion but, in case there is also an inhomogeneous term that is due to the interaction of the fluctuating strain and the mean vorticity. Just as in the pressure relation, these interactions are needed for any generation of the respective fluctuating component. But, it is important to note, such generation here is due to three-dimensionality for, if there was neither the $\tilde{w}$-component of the velocity nor the spatial dependence in the transverse z-direction, as it would be for the two-dimensional problem, then the fluctuating vorticity components, except for $\tilde{\omega}_z$, could only be advected and diffused regardless of any initial input.

In order to seek a solution for this problem, the number of equations needs to be reduced. There are several ways of doing this but one in particular is more than efficient. From kinematics it can be shown that

$$\nabla^2 \tilde{v} = \frac{\partial \tilde{\omega}_z}{\partial x} - \frac{\partial \tilde{\omega}_x}{\partial z}. \tag{1.12}$$

Thus, by differentiating (1.9) by z and (1.11) by x and combining equations using (1.12), then

$$\left(\frac{\partial}{\partial t} + U \frac{\partial}{\partial x} \right) \nabla^2 \tilde{v} - \nu \nabla^4 \tilde{v} = U'' \frac{\partial \tilde{v}}{\partial x} = -\Omega_z' \frac{\partial \tilde{v}}{\partial x} \tag{1.13}$$

can be obtained and, although still in a partial differential equation form, it is the Orr–Sommerfeld equation of stability theory. It is fortuitous that this equation uncouples in such a way as to only be fourth order and fully homogeneous in the $\tilde{v}$ dependent variable. The solution of (1.13) is then to be used in (1.10) for the solution of $\tilde{\omega}_y$. In like manner, the results found for $\tilde{\omega}_y$ are combined with $\tilde{v}$ and the problem is complete when these are used in (1.7) together with (1.1) to determine $\tilde{u}$ and $\tilde{w}$. Finally, $\tilde{p}$ can be evaluated from one of the momenta, (1.2) to (1.4). If the initial data and boundary conditions are satisfied, the problem is complete and the query as to stability can now be answered.

One last observation should be noted here. Equation (1.10) is actually the Squire equation that is known to accompany that of Orr–Sommerfeld. In this form, however, the dependent variable is the component of the vorticity that is perpendicular to the x–z plane and is only of interest in the full three-dimensional perturbation problem, strictly speaking. The importance of this cannot be stressed enough, for it leads to the understanding of the physics of the problem and details of the flow. It is not necessary if only the stability of the flow is the requirement. This equation also provides the other two orders of the anticipated sixth-order system. Unlike (1.13) though, it is not homogeneous.

1.3 Classical Treatment: Modal Expansions

The traditional classical method for solving (1.13) for $\check{v}$ is by modal expansion (normal modes). First, it is recognized that the coefficients in (1.13) are functions of y only. Therefore, since the extent of the planes perpendicular to y defined by the x, z spatial variables is doubly infinite, $\check{v}$ can be Fourier transformed in these two variables. Accordingly, define

$$\check{v}(\alpha, y, \gamma, t) = \int_{-\infty}^{+\infty} \int_{-\infty}^{+\infty} \tilde{v}(x, y, z, t) e^{i(\alpha x + \gamma z)} \, dx \, dz. \qquad (1.14)$$

With this step, the governing equation remains a partial differential equation in terms of the variables y and t, but the far-field boundary conditions in x and z, namely boundedness as $x, z \to \pm\infty$, are satisfied by the rigid conditions for Fourier transforms with α and γ, both real and which are the streamwise and spanwise wavenumbers. At this point, since the problem is linear, it would be natural to reduce the equation even further by employing a Laplace transform in time so that an ordinary differential equation for $\check{v}$ results. This procedure will be reserved until later for it deserves its own treatment. Suffice it for the moment to note that this has been done by Gustavsson (1979). Instead, the classical method for solution has been made by assuming that the time dependence can be separated from that of y. Thus,

$$\check{v}(\alpha, y, \gamma, t) = \sum_{n=0}^{\infty} \hat{v}_n(\alpha, y, \gamma) e^{-i\omega_n t} \qquad (1.15)$$

is taken and, as noted, (1.15) should be the infinite sum of all such model solutions. Moreover, ω_n is taken as a complex frequency with a positive imaginary part indicating an unstable mode. The substitution of (1.15) into (1.13) after the Fourier decomposition that is prescribed by (1.14) has been done, then the Orr–Sommerfeld equation is reduced to that of an ordinary differential equation for $\hat{v}$. Solutions are then required to meet the boundary conditions at the respective locations marked in terms of the y-variable; at y_1 and y_2, say. First, this means that $\hat{v}$ must satisfy conditions at y_1 and y_2. In Fourier space, the equivalent of (1.1) is $i(\alpha \hat{u} + \gamma \hat{w}) = \hat{v}'$, where $\hat{u}$ and $\hat{w}$ are defined in exactly the same manner as was done in (1.14) for $\check{v}$ as well as the solution form of (1.15). Thus, the conditions for $\hat{u}$ and $\hat{w}$ are now in terms of the first derivative of $\hat{v}$. The combination leads to the result that ω_n is a function of α, γ and kinematic viscosity (equivalently, the Reynolds number) of the flow for all n. From the point of view of the Laplace transform method, such solutions would be tantamount to finding poles in the complex Laplace space. But, in this way, the determination using (1.15) is more direct. Provided there are homogeneous boundary conditions in y, then the problem is that of the eigenvalue, eigenfunction variety with ω_n the eigenvalue. But, here the analogy to classical homogeneous eigenvalue–eigenfunction problems ends. First, as already noted, this differential equation is fourth order rather than second. Also, it is not self-adjoint, has a small parameter (reciprocal of the Reynolds number that is large compared to one) multiplying the highest derivative, thereby constituting a singular perturbation problem analytically or a stiff problem numerically. In short, neither the analysis nor the numerics are straightforward. Both of these topics will be treated in more detail. On the other hand, if only the question of stability is to be answered, then only *one* unstable eigenvalue need be found. But, no details of any specific initial-value problem or a determination of the full dynamics of any disturbance will follow in this way and it is not necessary for such a stability decision. But, if the modal expansions are to be used for this purpose, then all modes must be known and this includes those that are damped. The transient period becomes critical and it cannot be evaluated without this information. This topic will be presented in detail in Chapter 8 but, for now, it must be indicated that, among other things, it relates in part to the boundary conditions of the problem. Doubly bounded flows such as the channels have a different foundation in terms of the mathematics than those that have only one boundary (the boundary layer), or those without boundaries whatsoever as the jet, wake and mixing layer.

An alternative to using Fourier transformations is to use normal modes. Because the disturbances are described by linear equations, solutions can be sought in the following form

$$\tilde{v}(x,y,z,t) = \hat{v}(y)e^{i(\alpha x + \gamma z - \omega t)}. \tag{1.16}$$

This method is commonly referred to as the normal mode approach, and is tantamount to using Fourier transformations.

After making the substitution given by (1.16), the familiar Orr–Sommerfeld equation is found and is, for each mode,

$$(\alpha U - \omega)\Delta\hat{v} - \alpha U''\hat{v} = -i\nu\Delta^2\hat{v}, \tag{1.17}$$

where

$$\Delta = \frac{d^2}{dy^2} - \tilde{\alpha}^2, \tag{1.18}$$

and

$$\tilde{\alpha}^2 = \alpha^2 + \gamma^2, \tag{1.19}$$

with $\tilde{\alpha}$ the scalar polar wavenumber in the α-γ plane of Fourier space. This form of the equation offers some interesting properties. This is best seen by returning to the original set of equations (1.1) to (1.4) and making the transformation on all of the dependent variables, and then using the modal form of solution (1.16).

The ordinary differential equations that are obtained by the prescribed operations are

$$i(\alpha\hat{u} + \gamma\hat{w}) + \hat{v}' = 0, \tag{1.20}$$

$$i(\alpha U - \omega)\hat{u} + U'\hat{v} = -\frac{i\alpha}{\rho}\hat{p} + \nu\Delta\hat{u}, \tag{1.21}$$

$$i(\alpha U - \omega)\hat{v} = -\frac{1}{\rho}\hat{p}' + \nu\Delta\hat{v}, \tag{1.22}$$

and

$$i(\alpha U - \omega)\hat{w} = -\frac{i\gamma}{\rho}\hat{p} + \nu\Delta\hat{w}. \tag{1.23}$$

Squire (1933) introduced what should be properly called an equivalent transformation. And, once this is done, the very useful Squire theorem in stability theory emerges. For this purpose, let

$$\tilde{\alpha}\bar{u} = \alpha\hat{u} + \gamma\hat{w}, \tag{1.24}$$

$$\tilde{\alpha}\bar{w} = -\gamma\hat{u} + \alpha\hat{w}. \tag{1.25}$$

Just as $\tilde{\alpha}$ was the polar variable in the α-γ plane, it should be clear that $\bar{u}$ in (1.24) is the fluctuating component parallel to the wavenumber vector and $\bar{w}$ of (1.25) is in the angular direction. With the polar angle $\varphi = \tan^{-1}(\gamma/\alpha)$ defined in the plane, we see that $\bar{u}, \bar{w}$ are therefore the polar components of the velocity in the α-γ plane. With the use of these definitions the set of equations (1.20) to (1.23) can be combined to give

$$i\tilde{\alpha}\bar{u} + \hat{v}' = 0, \tag{1.26}$$

$$i\tilde{\alpha}(\alpha U - \omega)\bar{u} + \alpha U'\hat{v} = -\frac{i\tilde{\alpha}^2}{\rho}\hat{p} + \nu\tilde{\alpha}\Delta\bar{u}, \tag{1.27}$$

and

$$i(\alpha U - \omega)\hat{v} = -\frac{1}{\rho}\hat{p}' + \nu\Delta\hat{v}. \tag{1.28}$$

Thus, if the additional changes of variables,

$$\bar{v} = \hat{v}, \tag{1.29}$$

$$\frac{\bar{p}}{\tilde{\alpha}} = \frac{\hat{p}}{\alpha} \tag{1.30}$$

are used along with

$$\omega = \alpha c, \tag{1.31}$$

$$\widetilde{c} = c, \tag{1.32}$$

where $\widetilde{c}$ or c is the phase speed, then equations (1.27) and (1.28) read

$$i\widetilde{\alpha}(U - \widetilde{c})\bar{u} + U'\bar{v} = -\frac{i\widetilde{\alpha}}{\rho}\bar{p} + \frac{v\widetilde{\alpha}}{\alpha}\Delta\bar{u} \tag{1.33}$$

and

$$i\widetilde{\alpha}(U - \widetilde{c})\bar{v} = -\frac{1}{\rho}\bar{p}' + \frac{v\widetilde{\alpha}}{\alpha}\Delta\bar{v}. \tag{1.34}$$

Clearly these equations are analogous to those of a purely two-dimensional system except, that is, for the factor that multiplies the coefficient of viscosity. When the pressure is eliminated between (1.33) and (1.34) then

$$(U - \widetilde{c})\Delta\bar{v} - U''\bar{v} = -i\frac{\widetilde{v}}{\widetilde{\alpha}}\Delta^2\bar{v} \tag{1.35}$$

with

$$\widetilde{v} = \frac{v\widetilde{\alpha}}{\alpha} = \frac{v}{\cos\varphi} \tag{1.36}$$

becomes the Orr–Sommerfeld equation in this notation. From (1.35), the well-known Squire theorem can now be identified. Except for the viscosity, the equations governing a three-dimensional and a two-dimensional perturbation are the same. The relation (1.36), when written in terms of the nondimensional Reynolds number, Re, is simply $\widetilde{Re} = Re\cos\varphi$; $Re = \rho U_0 L/\mu = U_0 L/v$ with U_0 and L as characteristic scales of the mean flow. Now, as was shown in Fig. 1.5 or 1.6, there is a minimum Reynolds number for the onset of instability. Although this result is for the flat-plate boundary layer, it is also true for plane Poiseuille flow. Consequently, by use of the Squire transformation, the Squire theorem can be noted. The minimum Reynolds number for instability will be higher for an oblique three-dimensional wave than for a purely two-dimensional wave. Note that this statement does not rule out the possibility that, for a high enough values of the Reynolds number, an unstable oblique oscillation is possible even though the purely two-dimensional one that has the same value of α is damped. This last point is one referred to by Watson (1960) as well as Betchov & Criminale (1967) but has not been exploited to date.

The equation for the other component of the polar wave velocity is found directly by combining the definition (1.25) with operations on the appropriate equation and is

$$i(\alpha U - \omega)\bar{w} - v\Delta\bar{w} = \sin\varphi U'\bar{v}, \tag{1.37}$$

which is nothing more than

$$i(\alpha U - \omega)\hat{\omega}_y - \nu \Delta \hat{\omega}_y = -i\gamma U'\bar{v} = -i\tilde{\alpha}\sin\varphi U'\bar{v} \qquad (1.38)$$

when written using normal modes, as can be seen by taking normal modes of (1.7) and using (1.21) and (1.23). Regardless of the choice, this equation has become known as the Squire equation in stability theory. It is important to notice that the inhomogeneous term depends upon the solution for $\bar{v}$ and has the factor that is a measure of the obliquity of the wave. This term has been referred to as "lift up" by several authors and is attributed to Landahl (1980). When the angle of obliquity is perpendicular to the direction of the flow ($\varphi = \pi/2$), then the mean flow no longer has any influence and the equation, in this limit, can be solved exactly for $\bar{v}$ even without the assumption of modes.

The completion of the problem from this basis can be made by (i) solving for $\bar{v}$ from (1.35); (ii) determining $\bar{u}$ from the condition of incompressibility,

$$i\tilde{\alpha}\bar{u} = -\bar{v}';$$

(iii) solving for $\bar{w}$ from (1.37) and then inverting the transformations given by (1.24) and (1.25), namely

$$\hat{u} = \cos\varphi\bar{u} - \sin\varphi\bar{w}, \qquad (1.39)$$

and

$$\hat{w} = \sin\varphi\bar{u} + \cos\varphi\bar{w} \qquad (1.40)$$

to obtain the original Cartesian velocity components. And, as has already been shown, the pressure can be subsequently determined. From this summary, it can be seen that the central part of the analysis clearly rests with the success of solving both the Orr–Sommerfeld and the Squire equations if a full examination is desired. This is far different than merely determining whether or not the flow is stable.

1.4 Transient Dynamics

By comparison, the transient portion of the dynamics of perturbations has only relatively recently become a topic of some importance in stability theory. On the one hand, because of the many complexities in the mathematics and the lack of adequate computing in the early stages of the development, it was practically impossible to actually accomplish this task. At the same time, traditional thought on this matter did not indicate that this aspect could have any bearing on the ultimate behavior, and was simply ignored. Today, it is now quite clear that the results of stability calculations in the modal form are really more for the purpose of predicting the asymptotic fate of any disturbance, and the transient dynamics can have and do lead to events that make this part of the problem even more of interest than it ever was.

It can be recalled that the leading equations to be used in the stability analysis have different properties than those that are more common in initial-value, boundary value problems. For iteration, the principal ones, namely the Orr–Sommerfeld equation, is fourth order and is not self-adjoint. Thus, for a specific initial-value designation, there is the question of exactly how to express arbitrary functions or even what set of functions are to be used for expansion of these given functions. The Orr–Sommerfeld equation does not have a set of known functions. Of course, there are means to form inner products (cf. Drazin & Reid, 1984) in this case and therefore all constants needed can be evaluated. But, it is only the channel flows that have a complete set of eigenfunctions (cf. DiPrima & Habetler, 1969) so long as the problem is viscous. Inviscidly, there is only a continuous spectrum (Case, 1960a, 1961; Criminale, Long & Zhu, 1991). The boundary layer (Mack, 1976) and the free shear flows have been shown to have only a finite number of such modes. But, regardless, the fact that there must be a continuous spectrum to make the problem complete is already a recognition to the salient fact that there can be temporal behavior that is algebraic rather than just exponential.

The use of the Laplace transform in time to transform the partial differential equations to ones that are but ordinary has been made by Gustavsson (1979) as an alternative to modal expansions for initial-value problems. In this way the problem is completely specified and, in principle, can be made tractable. Unfortunately, only general properties can actually be found using this approach, since the ordinary differential equation that must be solved is the same as the Orr–Sommerfeld. However, the important algebraic behavior is shown to exist along with the exponential modes, and is due to the existence of a continuous spectrum, because there must be branch cuts as well as poles when the inversion to real time is to be made. This method also closes the gap for those flows where there is the lack of modes for the arbitrary initial-value problem, and the continuous spectrum, together with the discrete modes, allows for arbitrary expansions. This approach, where both the discrete and continuous spectra are used, has been well described by Grosch & Salwen (1978) and Salwen & Grosch (1981). There is no continuous spectrum for viscous channel flows since, as stated, the modes form a complete set with the problem confined to one of finiteness where there is normally only a discrete spectrum.

Then, there is yet another way in which algebraic behavior can arise. This can be seen by referring to the Squire equation where there is the one inhomogeneous term that is proportional to the normal velocity component, that is, the term attributed to lift-up. This equation, unlike that of the homogeneous Orr–Sommerfeld equation, can be resonant if there is a matching of the frequencies of the respective modes of the normal velocity with the dependent variable of this equation. This phenomenon has been shown to be possible for plane channel flow by Benney & Gustavsson (1981) but, it was concluded, resonance is not possible for the boundary layer. The case for resonance in the free shear flows is yet to be determined.

Exactly how dominant the algebraic behavior might be depends upon the particular problem and, to some extent, whether or not the problem is treated with or without viscosity. For any of the cases where there is the existence of a continuous spectrum it should be noted that perturbations can increase algebraically to quite large amplitudes before any exponentially growing mode supersedes its progress. The algebraic growth is ultimately damped by viscous action if viscosity is included in the problem. Otherwise, for some problems, the portion that grows algebraically can do so without bound and thus the assumption of linearity is overcome long before the dominance of any exponential growth. Thus, the concept of stability needs to be put in the proper context and it would be better to ask such questions as the existence of (a) optimum or maximum growth of disturbances or (b) behavior of the relative components of the perturbation velocity or vorticity, for example. Such undertakings have been and are continuing to be made.

1.5 Asymptotic Behavior

As has been stated, one answer to the question of whether or not a given flow is stable is to determine whether or not there is at least one eigenmode that results in exponential growth. Then, regardless of the time scale, there will eventually be an unlimited increase of the perturbation amplitude and the flow cannot in any way be stable. And, this may be possible with or without any early transient algebraic development. In short, it is the long time limit that must now be found. For this purpose there are numerous numerical schemes that can be used to make the determination in a reasonably efficient manner. The question of the many or infinite number of modes does not actually need to be answered for only one growing mode is required to answer the question. Typical results of this strategy results in an eigenvalue expression that has the complex frequency as a function of the polar wavenumber, angle of obliquity and the Reynolds number if viscous forces are included. Or, because of the Squire transformation from the Cartesian to the polar wavenumber variables, the determination of these values can be made without resorting to the oblique angle value. If the behavior for three-dimensionality is desired, it can be inferred from the equivalent two-dimensional data by use of the transformation as was shown by Watson (1960) or Betchov & Criminale (1967), for example.

There are other means of assessing the asymptotic fate of a particular initial input in more detail. For example, in order to predict the complete spatial behavior of the initial distribution, then the Fourier transforms that were made in the x and z variables must be inverted for this purpose. In the asymptotic state any transient response has long been exceeded by the exponential modal behavior and thus the leading behavior of these double integrals is exponential in time and can be evaluated by the method of steepest descent. In this way, the general features of the evolving disturbance can be predicted as well as the maximum amplitude. Such features include the location and distribution of the maximum part of the evolving disturbance or, as is better known, the description of the wave packet. A very early attempt for this kind of analysis for a localized disturbance in the laminar boundary layer was made by Criminale & Kovasznay (1962) and it was found that a wave packet ultimately was formed with the wave fronts swept back (three-dimensional) and the wavenumbers and frequencies those of the band of amplified Tollmien–Schlichting waves. The relative widths of the packet could also be determined by this method. The important point here is that modal expansions can and do provide the critical information required for the asymptotic behavior.

1.6 Role of Viscosity

The role of viscosity in the stability of parallel or almost parallel flows has two parts and is both the cause of the instability and has the role of damping as well. This scenario is, in many ways, unique in fluid mechanics but the phenomenon is known to exist in other fields. It is best explained by analogy. First, as in many other physical problems, viscous forces do ultimately act as damping but not necessarily at all times or in all situations in certain flows. An unstable Tollmien–Schlichting wave not only requires viscosity to be unstable but have only been shown to exist only in the presence of solid boundaries.

As suggested, an explanation as to why viscosity is destabilizing can best be illustrated by analogy. Such an analogy was suggested by Betchov & Criminale (1967) and it remains valid today. For an oscillator with mass m and a linear restoring force proportional to k but with a time delay τ, the equation of motion can be written as

$$m\frac{d^2x(t)}{dt^2} + kx(t-\tau) = 0. \tag{1.41}$$

Then, for small values of the delay, τ, (1.41) takes the form

$$m\frac{d^2x(t)}{dt^2} - \tau k\frac{dx(t)}{dt} + kx(t) = 0. \tag{1.42}$$

Thus, it is clear from this result that such action is destabilizing and it is essentially a question of phasing. Although the conclusions to be drawn from (1.42) may appear simple minded, it expresses the elements required. In the more subtle arguments that will be used to demonstrate this point more precisely, it will be expressed in terms of Reynolds stress and interaction with the mean flow but it is the certain phasing that must be correct in order for there to be an instability in the flow.

Still, there are many problems that can be investigated without viscosity and, historically, this is exactly what was done. Most notably the contributions of Rayleigh (1879, 1880, 1887, 1892a,b,c) were all made by inviscid analysis. And, interestingly enough, when investigating these problems, Rayleigh only examined two-dimensional perturbations. The Squire transformation and theorem that demonstrates that the two-dimensional problem is all that need be done in order to determine the stability came much later. Except for the much earlier work that is now referenced as Kelvin–Helmholtz (Helmholtz, 1868; Kelvin, 1871), this work provided much of the important bases that remains even to this day in the field of hydrodynamic stability. There are several theorems due to Rayleigh that are important both for the mathematics and to the understanding of the physics of such flows and this is true even if viscous effects are retained. However, the flows that were extensively examined in this manner by Rayleigh were those of the jet, wake and the mixing layer.

One need only to return to the fundamental equations, (1.33)–(1.35), which were expressed as an equivalent two-dimensional system, as $\gamma = 0$ for no z-variation for true two-dimensionality, and neglect the viscous terms to derive the Rayleigh equation. This is straightforward where the pressure is eliminated and it is found that

$$(U-c)\left(\hat{v}'' - \alpha^2\hat{v}\right) - U''\hat{v} = 0. \tag{1.43}$$

Unlike the Orr–Sommerfeld equation, (1.43) is second order and, although not self-adjoint, it can easily be so constructed and the more conventional rules for boundary-value problems can be used. Unfortunately, there is no set of known functions for this equation save for some special $U(y)$ distributions and, for the initial-value part of the problem, a continuous spectrum must be added since there are only a finite number of discrete eigenmodes. If one is interested in the full three-dimensional problem, then the equivalent Squire equation must be included.

Comparison of (1.43)–(1.35), say, tacitly reveals another point that is well known in the theory. Just as the ignoring of viscous effects is tantamount to lowering the order of the governing equation and thereby making it singular, the Rayleigh equation can also be singular if $(U - c) = 0$ somewhere in the flow. For this to be true, then c must be purely real and thus the interpretation is that the phase speed for the mode, c_r, is equal to the value of the mean flow at some y-location in the flow. Likewise, this implies that the flow is neutrally stable in this case. Exploitation of this fact is the basis for many of the theorems due to Rayleigh. It is also part of the reasoning for the emergence of a continuous spectrum of eigenvalues, as demonstrated by Case (1960a, 1961). Chapter 2 is devoted to inviscid problems.

1.7 Geometries of Relevance

The mean flows envisioned and described have been at least tacitly assumed to be those that are two-dimensional and unidirectional, whether or not they are the true or approximate solutions to the Navier–Stokes equations. *Under these restrictions, the channel, flat-plate boundary layer and free shear flows of the jet, wake and mixing layer are classical problems studied under this approximation.* Cartesian coordinates describe these situations quite well. But there is no reason that the parallel flow that exists in a round pipe or the wake or jet of a round nozzle cannot be explored in the same manner. Flow along curved walls is another similar analogy. Then, there is the flow that can exist between concentric cylinders. All of these are important and can be investigated but the respective governing equations in these cases are better cast in terms of polar or other coordinates. Such action has been taken and these problems generate still more surprises for the chain of logical thought. In short, a great deal of the physics can be transposed to the new geometries but the results are not nearly so satisfying. Some of the failures can be explained but some are still enigmas; others yield even more salient conclusions. Examples for each of such flows will be examined in detail.

1.8 Spatial Stability Bases

As has been suggested, the major experiments that have been done in the investigation of stability in flat-plate boundary layers for example, does not, strictly speaking, correspond to the theory defined by a temporal initial-value problem. Instead, exploration was made by introducing a disturbance at an initial x-location upstream in the flow. Then, subsequent measurements are made downstream from this location in order to determine the resulting flow. By definition, this is a spatial initial-value problem. The behavior in time is simply periodic and neither decreases nor increases. From the definition given by (1.16), ω must be purely real. As a result, some alteration in the formulation must be made so that this formulation has merit.

The set of equations, (1.20)–(1.23), are still valid. However, these were developed with the understanding that the respective wavenumbers, α and γ, were real. For the spatial problem, these parameters are to be complex. But, if an initial value for the relevant quantity is to be given as a function of y and z at $x = x_0$, γ must be real in order to satisfy the far-field boundary conditions as $z \to \pm\infty$. In like manner, the integral that defines the limits for the x-variable in (1.14) would only be for $x > x_0$, much in the manner of a Laplace transform in time with $t > 0$. The net result leads to an amended Orr–Sommerfeld equation in the sense that α is the eigenvalue with γ, ω and the Reynolds number as parameters. Also, in this case, $\alpha_i < 0$ implies instability. The Squire transformation is still permitted and the theorem is valid since it implies the neutral locus where $\alpha_i = \omega_i = 0$.

A general problem can be constructed that combines both the temporal and the spatial bases. The boundary value requirements remain the same but now the resulting dispersion relation can be seen to be one where there are two complex variables, namely ω and α. The wavenumber γ remains real and the Reynolds number is again a parameter if the system is taken as viscous. It is for this reason that Gaster (1965a,b) offered an alternative when the question of spatial stability was originally proposed. In like manner, Briggs (1964) used this formulation in studying instabilities in plasmas. Here, solutions were sought by use of the normal modes decomposition, and the dispersion relation that is developed once the boundary conditions have been met is taken as a function of the two complex variables ω and α. For many problems as the boundary layer, the amplification rates are small and this allowed Gaster to make a local Laurent expansion and use the Cauchy–Riemann relations of complex variable theory to establish a correspondence between the temporal values that were already computed and the spatial quantities that were unknown. These relations have proven to be of major importance. When amplification rates are large, however, care must be taken and the complex eigenvalues must be computed directly, as shown by Betchov & Criminale (1966) for jet and wake problems. Mattingly & Criminale (1972) extended this work and added experimental confirmation results as well. Direct numerical calculations have also been made for the boundary layer by Kaplan (1964), Raetz (1964) and Wazzan, Okamura & Smith (1966) and the agreement with the analytical continuation method of Gaster was shown to be quite accurate.

Once the spatial-temporal problem is established, then other issues must be considered. Briefly stated, means of instability – now known as convective and absolute instabilities – have been identified when viewed from the spatial initial-value construction. These concepts relate to the fact that, for convective instability, at a fixed spatial location, amplification can occur and then pass as it is convected downstream. Absolute instability is one that, when amplification has begun, it does not cease and local break down is inevitable. Chapter 4 will discuss such problems in detail.

So to move from this historical perspective of hydrodynamic instabilities, the discussion will move from the simpler problems and mathematical approximations to solving the full Navier–Stokes equations in later chapters. Chapter 2 will begin by introducing dimensionless equations and moving directly into the temporal instability formulation for inviscid incompressible equations, followed in Chapter 3 by viscous incompressible flows. Chapters 4 and 5 will discuss the spatial stability of incompressible and compressible flows, respectively. Chapters 6, 7 and 8 will introduce centrifugal, geophysical and transient dynamics, respectively. The remaining chapters move beyond classical linear hydrodynamic instability to nonlinear model interactions. Chapter 9 will primarily focus on nonlinear modes pertaining to the flat plate. Chapter 10 will look at how instabilities are initiated (i.e., receptivity) and methods of predicting the transition from laminar to turbulent flow. Chapters 11 and 12 will briefly discuss direct numerical simulation and control within the framework of the unsteady Navier–Stokes equations. Finally, the text will close with the challenging discussion of how experiments have evolved from the early days of observations by Reynolds (1833) to our ability to make measurements of these "infinitesimal" disturbances – called hydrodynamic instabilities.

2

Temporal Stability of Inviscid Incompressible Flows

Simplification of modes of proof is not merely an indication of advance in our knowledge of a subject, but is also the surest guarantee of readiness for further progress.

– Baron William Thomson (Lord Kelvin) (1824–1907)

2.1 General Equations

The essentials for a good fundamental understanding of stability theory are presented in this chapter. The discussion is limited to two-dimensional incompressible flows, and the general equations are derived related to specific physics of the flow. As was discussed in Chapter 1, for three-dimensional flows there exists a theorem that allows the discussion to be reduced to an equivalent two-dimensional problem. In this way, the bases are provided for a discussion of the oscillations of uniform flows, shear flow away from walls (mixing layers, jets and wakes) and of shear flows along one or two walls. A shear flow along a single wall is, generally, called a boundary layer and is of special interest to aeronautical engineers. The oscillations of a boundary layer have played a large role in the historical development of stability theory because it lends itself relatively conveniently to experimental observation and measurements. The oscillations of boundary layers will be analyzed in detail in the next chapter.

The two-dimensional Navier–Stokes equations for an incompressible flow are

$$\frac{\partial u^\dagger}{\partial x^\dagger} + \frac{\partial v^\dagger}{\partial y^\dagger} = 0, \tag{2.1}$$

$$\rho^\dagger \left(\frac{\partial u^\dagger}{\partial t^\dagger} + u^\dagger \frac{\partial u^\dagger}{\partial x^\dagger} + v^\dagger \frac{\partial u^\dagger}{\partial y^\dagger} \right) + \frac{\partial p^\dagger}{\partial x^\dagger} = \mu^\dagger \nabla^2 u^\dagger, \tag{2.2}$$

and

$$\rho^\dagger \left(\frac{\partial v^\dagger}{\partial t^\dagger} + u^\dagger \frac{\partial v^\dagger}{\partial x^\dagger} + v^\dagger \frac{\partial v^\dagger}{\partial y^\dagger} \right) + \frac{\partial p^\dagger}{\partial y^\dagger} = \mu^\dagger \nabla^2 y^\dagger, \tag{2.3}$$

where $\nabla^2 = (\;)_{x^\dagger x^\dagger} + (\;)_{y^\dagger y^\dagger}$ is the two-dimensional Laplace operator. Note that $\dagger$ denotes a dimensional quantity. Also, $u^\dagger(x^\dagger, y^\dagger, t^\dagger)$ is the fluid velocity component parallel to the $x^\dagger$-axis and $v^\dagger(x^\dagger, y^\dagger, t^\dagger)$ is the fluid velocity component parallel to the $y^\dagger$-axis, and are typically expressed in units of *m/s* or *ft/s* for the velocity and *m* or *ft* for the axial directions. More precisely, $u^\dagger$ and $v^\dagger$ are the Eulerian velocity components. It is convenient to refer to the $x^\dagger$-axis as the "horizontal" axis and to the region $y^\dagger > 0$ as being "above" the region $y^\dagger < 0$ (or normal to the wall for boundary layers). This reference is a convention for drawing the axes and does not imply the existence of gravity forces. The function $p^\dagger(x^\dagger, y^\dagger, t^\dagger)$ is the scalar pressure, while $\rho^\dagger$ is the density assumed constant throughout the entire flow field. Equation (2.1) insures the conservation of mass for an incompressible fluid. Equation (2.2) states that any horizontal acceleration is produced by a combination of a pressure gradient and a viscous force proportional to $\mu^\dagger$, the coefficient of viscosity. Equation (2.3) plays the same role for the vertical acceleration.

It is assumed that $\mu^\dagger$ is a constant and, in those cases in which $\mu^\dagger$ varies with the temperature, it may become necessary to add additional terms. Some possibilities for the variable $\mu^\dagger$ will be considered in Chapter 5.

The coordinates $x^\dagger$, $y^\dagger$ and the time, $t^\dagger$, are independent variables and the functions $u^\dagger$, $v^\dagger$ and $p^\dagger$ are the dependent variables and functions of $x^\dagger$, $y^\dagger$, $t^\dagger$. Note that (2.2) and (2.3) are nonlinear, thereby making the solution of these equations nontrivial and discussed later in Chapters 9 and 11. As a result, theoretical assumptions or simplifications are made so that the nonlinear partial differential equations can be reduced to solving a problem of either linear partial differential equations or linear ordinary differential equations.

The equations are now derived that control the small oscillations of a parallel and steady mean flow. By parallel, the dependent variables for the mean (or base) flow are at most a function of only one independent variable, while steady denotes that the mean flow does not change in time. This derivation is done in three steps: (i) separation of fluctuations, (ii) linearization and (iii) recourse to complex functions. But first the system will be nondimensionalized so that a rational means of approximation can be made.

2.1.1 Nondimensionalization

The following scalings are introduced to nondimensionalize the governing system of equations.

$$u = u^\dagger / U_c, \quad v = v^\dagger / U_c, \quad p = p^\dagger / (\rho^\dagger U_c^2), \qquad (2.4)$$

for the velocities and the pressure. U_c is some characteristic flow velocity of the problem (e.g., wind tunnel speed, etc.). The length and time scales are chosen so that

$$x = x^\dagger/L, \quad y = y^\dagger/L, \quad t = t^\dagger/(L/U_c), \tag{2.5}$$

where L is some characteristic length scale of the problem (e.g., boundary layer thickness, momentum thickness, body diameter, etc.). Substitution of these variables into (2.1)–(2.3) produces the following nondimensional system:

$$u_x + v_y = 0, \tag{2.6}$$

$$u_t + uu_x + vu_y + p_x = Re^{-1}\nabla^2 u, \tag{2.7}$$

and

$$v_t + uv_x + vv_y + p_y = Re^{-1}\nabla^2 v, \tag{2.8}$$

where the subscript denotes a partial derivative; i.e., $u_x = \partial u/\partial x$. Henceforth, all variables will be assumed to be nondimensional unless otherwise noted. The parameter, Re, is the Reynolds number, defined as

$$Re = \frac{\rho^\dagger U_c L}{\mu^\dagger}. \tag{2.9}$$

It is extremely important to ensure that the units are consistent in all nondimensional parameters. When consistency in units is not maintained, then severe repercussions can result, as drawn out in the failure of the MARS Climate Orbiter.

**MARS CLIMATE ORBITER TEAM FINDS
LIKELY CAUSE OF LOSS[a]**

A failure to recognize and correct an error in a transfer of information between the Mars Climate Orbiter spacecraft team in Colorado and the mission navigation team in California led to the loss of the spacecraft last week, preliminary findings by NASA's Jet Propulsion Laboratory internal peer review indicate.

The peer review preliminary findings indicate that one team used English units (e.g., inches, feet and pounds) while the other used metric units for a key spacecraft operation. This information was critical to the maneuvers required to place the spacecraft in the proper Mars orbit.

[a]NASA Press Release: 99–113, September 30, 1999

2.1.2 *Mean Plus Fluctuating Components*

Assume that the flow can be decomposed into a laminar basic flow and a fluctuating component that oscillates about the basic flow

$$u = U(y) + \tilde{u}(x,y,t),$$

$$v = \tilde{v}(x,y,t),$$

$$p = P(x) + \tilde{p}(x,y,t), \tag{2.10}$$

where $(\tilde{u}, \tilde{v}, \tilde{p})$ are the fluctuation components and (U,P) is the basic flow. If these relations are substituted into (2.6)–(2.8) and the mean flow equations are subtracted from the resultant equations, the substitution yields

$$\tilde{u}_x + \tilde{v}_y = 0, \tag{2.11}$$

$$\tilde{u}_t + U\tilde{u}_x + U'\tilde{v} + \tilde{p}_x + \underline{(\tilde{u}\tilde{u}_x + \tilde{v}\tilde{u}_y)} = Re^{-1}\nabla^2\tilde{u}, \tag{2.12}$$

and

$$\tilde{v}_t + U\tilde{v}_x + \tilde{p}_y + \underline{(\tilde{u}\tilde{v}_x + \tilde{v}\tilde{v}_y)} = Re^{-1}\nabla^2\tilde{v}. \tag{2.13}$$

For simplicity, note that $U' = dU/dy$. This system of equations describes the nonlinear evolution of disturbances about the base flow. The underlined terms denote the nonlinear terms.

2.1.3 *Linearized Disturbance Equations*

The nonlinear terms in (2.12) and (2.13) are products of the fluctuating velocities and their derivatives and correspond to an effect of a fluctuation on another fluctuation. If the fluctuation has a frequency ω, then the coupled terms will have frequency 0 or 2ω. Therefore this interaction will either modify the non-fluctuating basic flow (often referred to as mean-flow distortion) and feedback to the fluctuating components or introduce higher harmonics (e.g., $3\omega, 4\omega$, etc.). Such difficulties are removed by assuming that the products of the fluctuations and their derivatives have small amplitudes. The terms that are underlined can then be neglected (in comparison to the magnitude of the other terms) because a small fluctuation multiplied by a small fluctuation results in an order of magnitude smaller term and no longer influences the equations to this order of approximation. As such, the following set of linear equations result

$$\tilde{u}_x + \tilde{v}_y = 0, \tag{2.14}$$

$$\tilde{u}_t + U\tilde{u}_x + U'\tilde{v} + \tilde{p}_x = Re^{-1}\nabla^2\tilde{u}, \tag{2.15}$$

and

$$\tilde{v}_t + U\tilde{v}_x + \tilde{p}_y = Re^{-1}\nabla^2\tilde{v}. \tag{2.16}$$

An interesting property of linear equations is that if $(\tilde{u}_1, \tilde{v}_1, \tilde{p}_1)$ and $(\tilde{u}_2, \tilde{v}_2, \tilde{p}_2)$ are solutions of the equations, then the sum $(\tilde{u}_1 + \tilde{u}_2, \tilde{v}_1 + \tilde{v}_2, \tilde{p}_1 + \tilde{p}_2)$ will be a solution as well. The same fundamental property of linearity occurs in acoustics, electromagnetics and ordinary quantum mechanics, in which it is guaranteed that the simultaneous oscillations will evolve independently because the nonlinear terms that would permit interaction have been neglected in (2.14)–(2.16). In thermodynamics, ferromagnetism, electronics and fluid dynamics, nonlinear equations must often be retained to capture sufficient flow physics for engineering applications. Fortunately, the solution of the linear system is often sufficient for problems such as when very small disturbances are found to be fluctuating in a laminar basic flow. The amplitude of these disturbances in this case is much smaller than that of the basic flow. Later, as these disturbances grow in energy, the nonlinear disturbance equations (2.11)–(2.13) are required to compute the subsequent disturbance evolution. (See Chapter 9 for a discussion on nonlinear disturbances.)

Special relations are now derived from the basic equations of motion that will yield insights toward understanding the physical processes occurring in such oscillatory flows. More specifically, equations for the streamfunction, the pressure and the vorticity will be developed.

(a) Velocity Disturbance Equation

A single equation for the tangential velocity disturbance $\tilde{v}$ can be obtained by taking the curl of the momentum equations (2.15)–(2.16) and substituting the resultant equations into the continuity equation (2.14). Or, step-by-step, one can differentiate (2.15) with respect to y and (2.16) with respect to x and subtract the resulting equations to eliminate the pressure. This results in two equations with two unknowns – the continuity equation (2.14) and the following equation

$$\frac{\partial}{\partial t}(\tilde{u}_y - \tilde{v}_x) + U\frac{\partial}{\partial x}(\tilde{u}_y - \tilde{v}_x) + U''\tilde{v} = Re^{-1}\nabla^2(\tilde{u}_y - \tilde{v}_x). \tag{2.17}$$

The next step is to differentiate (2.17) with respect to x and eliminate $\tilde{u}$ by means of the continuity equation (2.14), yielding

$$\left(\frac{\partial}{\partial t} + U\frac{\partial}{\partial x}\right)\nabla^2\tilde{v} - U''\tilde{v}_x = Re^{-1}\nabla^4\tilde{v}, \tag{2.18}$$

where $\nabla^4 = \nabla^2 \cdot \nabla^2 = (\ )_{xxxx} + 2(\ )_{xxyy} + (\ )_{yyyy}$ is called the biharmonic operator. Equation (2.18) is a single partial differential equation for the dependent variable $\tilde{v}$ and, in principle, can be solved given appropriate boundary and initial conditions. This equation is commonly referred to as the *Orr–Sommerfeld equation*.

(b) Streamfunction Disturbance Equation

An alternate form of the linear disturbance equation (2.18) can be derived using a streamfunction formulation, which exactly satisfies the continuity equation (2.14) by definition. Define the perturbation streamfunction, $\tilde{\psi}$, in the usual manner with

$$\tilde{u} = \tilde{\psi}_y \qquad \text{and} \qquad \tilde{v} = -\tilde{\psi}_x. \tag{2.19}$$

Take the curl of the momentum equations, which leads to (2.17), and substitute the velocities from (2.19) into (2.17). The single partial differential equation for $\tilde{\psi}$ results

$$\left(\frac{\partial}{\partial t} + U \frac{\partial}{\partial x} \right) \nabla^2 \tilde{\psi} - U'' \tilde{\psi}_x = Re^{-1} \nabla^4 \tilde{\psi}. \tag{2.20}$$

(c) Pressure Disturbance Equation

The pressure appears in (2.7) and (2.8) as a scalar and, for incompressible flow, is decoupled from velocity. One might attempt to arbitrarily select or impose a pressure solution as a sort of potential. However, an arbitrary pressure field might create a velocity field in the momentum equations that would violate the continuity equation (2.6). The fluid would accumulate in certain places and the density would be unphysically forced to be nonconstant. In general, it takes great energy to alter the density. As long as the fluid velocity is much smaller than the speed of sound, the density physically should remain constant and (2.6) is always satisfied (for the moment this assumes that we do not have a density stratified flow, such as would occur in the ocean or in the atmosphere). This continuity equation and constant density impose a restriction on the pressure fluctuations, which can be formulated in the following sense. Consider the linearized equations (2.14)–(2.16). Differentiate (2.15) with respect to x and (2.16) with respect to y (or the divergence of the equations), add the two equations and simplify using (2.14). The end product is a Poisson equation given by

$$\nabla^2 \tilde{p} = -2U' \tilde{v}_x. \tag{2.21}$$

Drawing a parallel to engineering mechanics, this equation is similar to that of an elastic membrane loaded with some external force $\phi(x,y)$. If the deflection of the membrane is η, the basic equation is

$$\nabla^2 \eta = \kappa \phi.$$

Thus, just as the external force causes a deflection of the membrane, the product $U'\tilde{v}_x$ is the source of pressure fluctuations. In the absence of any "source," the pressure obeys the Laplace equation and is determined solely by the boundary conditions.

Another convenient form of the pressure equation can be obtained by applying the linear operator

$$\frac{\partial}{\partial t} + U\frac{\partial}{\partial x}$$

to (2.21) and, using the momentum equation (2.16) to eliminate the time derivative of $\tilde{v}$, the following equation results

$$\left(\frac{\partial}{\partial t} + U\frac{\partial}{\partial x}\right)\nabla^2 \tilde{p} - 2U'\tilde{p}_{xy} = -2U'Re^{-1}\nabla^2\tilde{v}_x. \qquad (2.22)$$

For an inviscid flow ($Re \to \infty$), the right-hand side vanishes and the equation reduces to a single equation for the disturbance pressure. For the viscous problem, note that, although an equation for the pressure is valuable to understand the physics, it is not useful for solving the boundary value problem. This is not necessarily true for the inviscid problem where the right-hand side of (2.22) vanishes and a partial differential equation emerges that is homogeneous in the pressure.

(d) Vorticity Disturbance Equation

Finally, a single equation can be derived for the spanwise vorticity component. In general, the vorticity is a vector that indicates the rotation of a small mass of fluid with respect to the chosen coordinates. More precisely, consider a small surface element immersed in the fluid and define the circulation as the integral of the velocity along the perimeter. The circulation is equal to the flux of the vorticity through the surface; no vorticity implies no circulation. In a two-dimensional flow, the vorticity vector is always perpendicular (i.e., out of plane) to the (x,y)-plane, and, because its orientation is fixed, the vorticity can be treated like a scalar quantity because there is only one component for two-dimensional flows. Thus for the disturbance vorticity, $\tilde{\omega}_z$, and the mean flow, Ω_z, we have

$$\tilde{\omega}_z = \tilde{v}_x - \tilde{u}_y , \qquad \Omega_z = -U'. \qquad (2.23)$$

Note, for linear disturbance equations, the basic flow state (U, P, Ω) is very often referred to as mean, basic and base flow, and as such may be used in any of these terms throughout this discussion. The vorticity fluctuations obey an important equation, which can be obtained by taking the curl of the momentum equations (2.15) and (2.16) to eliminate the pressure, resulting in

$$\left(\frac{\partial}{\partial t} + U\frac{\partial}{\partial x}\right)\tilde{\omega}_z - U''\tilde{v} = Re^{-1}\nabla^2\tilde{\omega}_z. \tag{2.24}$$

This equation is analogous to that of the heat conduction-diffusion equation in the presence of sources and sinks of heat. Indeed, in a copper sheet of constant thickness and unit-specific heat, the temperature T obeys the following basic equation:

$$T_t = \kappa\nabla^2 T + Q,$$

where κ is the coefficient of thermal diffusivity and $Q(x,y,t)$ is proportional to the rate of production or withdrawal of heat. If the copper is replaced by a layer of mercury moving with velocity $U(y)$, another term must be included to account for the convection of energy. The equation in this case becomes

$$T_t + UT_x = \kappa\nabla^2 T + Q.$$

In the absence of conductivity and production, the equation has the solution $T(x - Ut, y)$, which shows that each particle of mercury retains the same temperature. In a comparison between the energy equation (in terms of temperature) and the vorticity equation (2.24), the term $U''\tilde{v}$ plays a comparable role as Q. However, unlike Q, the disturbance velocity $\tilde{v}$ is linked to $\tilde{\omega}_z$ (see (2.23)).

2.1.4 Recourse to Complex Functions

Because the disturbances are described by linear equations, solutions can readily be obtained by making use of complex functions and reducing the system of partial differential equations (2.14)–(2.16) to ordinary differential equations. Thus, one can hope to find normal mode solutions of the type

$$\tilde{u}(x,y,t) = \tfrac{1}{2}(\check{u} + \check{u}^*) = \frac{1}{2}\left[\hat{u}(y)e^{i\alpha(x-ct)} + \hat{u}^*(y)e^{-i\alpha^*(x-c^*t)}\right],$$

$$\tilde{v}(x,y,t) = \tfrac{1}{2}(\check{v} + \check{v}^*) = \frac{1}{2}\left[\hat{v}(y)e^{i\alpha(x-ct)} + \hat{v}^*(y)e^{-i\alpha^*(x-c^*t)}\right],$$

$$\tilde{p}(x,y,t) = \tfrac{1}{2}(\check{p} + \check{p}^*) = \frac{1}{2}\left[\hat{p}(y)e^{i\alpha(x-ct)} + \hat{p}^*(y)e^{-i\alpha^*(x-c^*t)}\right], \tag{2.25}$$

where $(\check{u}, \check{v}, \check{p})$ are complex normal mode forms and $(\check{u}^*, \check{v}^*, \check{p}^*)$ are the complex conjugates. Hence, the sum of the normal mode and its complex conjugate is the real disturbance quantity. In principle, since the complex conjugate values can easily be obtained from the quantities themselves, one need only solve for the complex quantities $(\breve{u}, \breve{v}, \breve{p})$. In the relations in (2.25), $\alpha = \alpha_r + i\alpha_i$ is the nondimensional wavenumber in the x-direction and $c = c_r + ic_i$ is the wave velocity. The nondimensional frequency of the disturbance is given by $\omega = \alpha c$, and the nondimensional wavelength of the disturbance is given by $\lambda = 2\pi/\alpha_r$. In general, α, c and ω are complex numbers and form the bases for a generalized temporal-spatial problem. The symbols $(\hat{u}, \hat{v}, \hat{p})$ are used to indicate a complex function of y only and $c^* = c_r - ic_i$ and $\alpha^* = \alpha_r - i\alpha_i$ as complex conjugate quantities. To be a physical solution for the disturbances, the normal mode form (2.25) must obey the continuity and momentum equations (2.14)–(2.16). After substituting (2.25) into these equations, we find that the resulting equations are no longer functions of x or t. Thus, the partial differential equations have been reduced to a system of ordinary differential equations in y.

The advantage of using complex quantities should now be evident; in general, an oscillation has an amplitude and a phase. This means that two numbers must be specified, and we may as well give the amplitude of the cosine and the amplitude of the sine components. In a single complex quantity, the amplitude and the phase can be expressed for a fluctuation. The reader is reminded, however, that although complex quantities are used, the solutions to the original system are real and this fact must be borne in mind when describing the behavior of the original system. This particular approach of using complex quantities is usually called the *normal mode* approach, and the solutions are called *normal modes*. These points were also discussed in Chapter 1 from a somewhat different perspective using Fourier transformations.

To better explain the information contained in a complex normal mode analysis, take an example solution of the form

$$\breve{p} = \hat{p}(y)e^{4i[x - (0.7 + 0.2i)t]}, \qquad \hat{p}(y) = 2y + 3iy^2$$

and assume $\alpha = 4$ and $c = 0.7 + 0.2i$. Recall the relations $e^{i\theta} = \cos\theta + i\sin\theta$, $i = \sqrt{-1}$, and $i^2 = -1$. This leads to the real disturbance solution

$$\tilde{p} = \left\{2y\cos[4(x - 0.7t)] - 3y^2\sin[4(x - 0.7t)]\right\}e^{0.8t}.$$

The imaginary part of c leads to an exponential growth in time, the real part of $\breve{p}$ gives the amplitude of the cosine component and the imaginary part of $\breve{p}$ gives the amplitude of the sine component except for a minus sign. The function grows exponentially in time with growth rate of 0.8. The wavenumber of this function is given by $\alpha = 4$, and thus the wavelength is $\lambda = 2\pi/\alpha = \pi/2$. The phase speed is given by $c_r = 0.7$, which means that the solution grows in time along the particle paths $x = 0.7t + x_0$, where x_0 is any constant along the real axis at time $t = 0$.

In our original problem, if $\alpha = \alpha_r + i\alpha_i$ and $c = c_r + ic_i$, then the amplitudes of the disturbance functions are proportional to

$$\tilde{v} \approx e^{-\alpha_i x + \omega_i t}, \tag{2.26}$$

and similarly for the other disturbance functions, with the complex frequency being given by

$$\omega = \omega_r + i\omega_i, \quad \text{where} \quad \omega_r = \alpha_r c_r - \alpha_i c_i, \quad \text{and} \quad \omega_i = \alpha_r c_i + \alpha_i c_r. \tag{2.27}$$

From (2.26), note that the disturbance can grow exponentially in the positive x-direction if $\alpha_i < 0$, and increase exponentially in time if $\omega_i > 0$.

The use of the normal mode relationship for disturbances (2.25) substituted into the linearized disturbance equations (2.14)–(2.16) transforms the partial differential equations into ordinary differential equations. However, this transformation does not come without its complications. Namely, with this substitution, the complex eigenfunctions $(\hat{u}, \hat{v}, \hat{p})$ are unknown functions of y. The complex wavenumber $\alpha_r + i\alpha_i$ and frequency $\omega_r + i\omega_i$ introduce four additional unknowns, resulting in more unknowns than equations. Hence, one must make assumptions concerning these unknowns in order to obtain any solution. In one case, one may assume that the disturbance amplifies in space and not time with a fixed frequency. As such α becomes the unknown eigenvalue, $\omega_i = 0$, and ω_r is specified. As suggested in Chapter 1, this is referred to as spatial stability theory and will be discussed in some detail in Chapter 4. A second case could assume that the disturbances amplify in time and not space. Hence, ω becomes the unknown eigenvalue, $\alpha_i = 0$, and α_r is specified. This is referred to as "temporal stability theory" and will be discussed in detail in the remainder of this chapter and in Chapter 3. The intersection of temporal and spatial theories occurs at the neutral locations where disturbances neither amplify nor decay in space and time. The neutral location, $\alpha_i = \omega_i = 0$, is where both theories yield the same normal mode solution. The following are the salient limiting cases:

Temporal Stability Theory: $\alpha_i = 0$ and ω complex;
Spatial Stability Theory: $\omega_i = 0$ and α complex.

To restate, temporal stability theory considers disturbances that grow in time, while spatial stability theory considers disturbances that grow in the space. In temporal theory, the wavenumber α is taken as real while the frequency ω is assumed complex; in spatial theory the wavenumber is assumed complex while the frequency is taken to be real. These two limiting cases dominate the subject of hydrodynamic stability theory, and most publications and research fall into one of these two categories.

Historically, temporal theory can be traced to the time of Helmholtz (1868),

Kelvin (1871) and Rayleigh (1879, 1880, 1887, 1892a) in the late nineteenth century, while spatial theory originated in the 1950s. During this later time period, spatial stability was recognized to be important for the case of parallel laminar flows, where various experimentalists observed that instability and transition to turbulence occurred by the growth of disturbances in the downstream x-direction (i.e., spatial instability; see the review article by Dunn, 1960). The method was restricted to first computing temporal modes and then relating those to spatial modes by means of the phase velocity (i.e., $\omega_i = -c_r \alpha_i$). Such a simple relation is not valid in general, and very little progress was made as a result. Gaster put the concept of spatial stability on a firm theoretical foundation, and carried out the first (unpublished) spatial calculations as part of his doctoral thesis. Early spatial calculations confirmed the validity of the approach by comparing the growth rates obtained by numerics to those obtained by experiments. The more general case of space-time growth of disturbances, first published by Gaster (1962, 1965a, 1968), will be considered as a separate topic in Chapter 4.

To obtain an equation that describes the spatial-temporal amplification of disturbances, substitute the normal mode expression for (2.25) into the linear disturbance equation (2.18) to get

$$(U - c)(\hat{v}'' - \alpha^2 \hat{v}) - U'' \hat{v} = (i\alpha Re)^{-1} (\hat{v}^{iv} - 2\alpha^2 \hat{v}'' + \alpha^4 \hat{v}). \qquad (2.28)$$

This now-famous equation was first derived independently by Orr (1907a,b) and Sommerfeld (1908) and is known as the Orr–Sommerfeld equation.[1] Note that the partial differential equation has become a fourth-order ordinary differential equation, which requires four boundary conditions for closure. If the fluid is taken as inviscid ($Re \to \infty$), then the corresponding equation becomes

$$(U - c)(\hat{v}'' - \alpha^2 \hat{v}) - U'' \hat{v} = 0. \qquad (2.29)$$

Equation (2.29) was first derived by Kelvin (1880) and is known as the Rayleigh equation.[2] The Rayleigh equation is a second-order ordinary differential equation, requiring two boundary conditions for closure. The Orr–Sommerfeld equation governs the stability of parallel *viscous* flows, while Rayleigh's equation governs the stability of parallel *inviscid* flows. These two equations mark the cornerstone of all hydrodynamic stability analysis and will be of primary importance for the remainder of this text.

---

[1] Historically, neither Orr nor Sommerfeld actually wrote down this familiar form, which is instead due to Noether (1921).

[2] Rayleigh's equation is sometimes mistakenly referred to as the "inviscid Orr–Sommerfeld" equation; since Rayleigh derived his equation more than 25 years before Orr and Sommerfeld, we shall not use this terminology here.

The Orr–Sommerfeld and the Rayleigh equations can also be derived directly from the streamfunction equation (2.20). Let the perturbation streamfunction $\tilde{\psi}$ be represented by a normal mode form

$$\tilde{\psi}(x,y,t) = \frac{1}{2}(\check{\psi} + \check{\psi}^*) = \frac{1}{2}\left[\phi(y)e^{i\alpha(x-ct)} + \phi^*(y)e^{i\alpha^*(x-c^*t)}\right]. \qquad (2.30)$$

By substituting the normal mode form (2.30) into the linear disturbance equations (2.20), an alternative form of the Orr–Sommerfeld equation results, namely

$$(U-c)(\phi'' - \alpha^2\phi) - U''\phi = (i\alpha Re)^{-1}(\phi^{iv} - 2\alpha^2\phi'' + \alpha^4\phi). \qquad (2.31)$$

Assuming inviscid, equation (2.31) becomes

$$(U-c)(\phi'' - \alpha^2\phi) - U''\phi = 0. \qquad (2.32)$$

This formulation is the one originally used by Rayleigh. Note that, because of the relationship between $\hat{v}$ and ϕ in equation (2.19), the Orr–Sommerfeld equation (2.28) governing the velocity disturbance $\hat{v}$ has the same form as the equivalent equation governing the streamfunction disturbance ϕ.

The proper boundary conditions for the Orr–Sommerfeld and Rayleigh equations depend on the configuration. For bounded flows, the boundary conditions require that ϕ and ϕ' vanish at the walls for the Orr–Sommerfeld fourth-order equation, while ϕ vanishes at the walls for the second-order Rayleigh equation. In this latter case, the normal derivative cannot be required to vanish at the boundary and the flow may slip along the walls since $\hat{u} = \phi'$ may be nonzero. For unbounded flows, the perturbation solution is required to be bounded.

An alternative form of Rayleigh's equation can be derived by substituting the pressure normal mode form (2.25) into the linear disturbance equation (2.22). In the inviscid limit, the pressure disturbance equation becomes

$$\hat{p}'' - \frac{2U'}{U-c}\hat{p}' - \alpha^2\hat{p} = 0. \qquad (2.33)$$

This equation often appears in the literature as an alternative means for solving the inviscid stability problem. The proper boundary conditions require that, in the case of unbounded domains, either $\hat{p}$ vanish at the walls or be bounded.

2.1.5 Three-Dimensionality

The normal mode analysis can be extended from two to three dimensions in a straightforward manner. Let $u(x,y,z,t)$, $v(x,y,z,t)$ and $w(x,y,z,t)$ be the fluid velocities in the x, y and z-direction, respectively, and $p(x,y,z,t)$ be the scalar

pressure. The general nondimensional equations for a three-dimensional incompressible flow are

$$u_x + v_y + w_z = 0, \tag{2.34}$$

$$u_t + u u_x + v u_y + w u_z + p_x = Re_{3d}^{-1} \nabla^2 u, \tag{2.35}$$

$$v_t + u v_x + v v_y + w v_z + p_y = Re_{3d}^{-1} \nabla^2 v, \tag{2.36}$$

$$w_t + u w_x + v w_y + w w_z + p_z = Re_{3d}^{-1} \nabla^2 w, \tag{2.37}$$

where $\nabla^2 = (\ )_{xx} + (\ )_{yy} + (\ )_{zz}$. Here, Re_{3d} is the Reynolds number associated
with the three-dimensional problem, and we use the subscript to distinguish it
from the Reynolds number associated with the two-dimensional problem. This
notation is for convenience only, as will become apparent in the following discussion. Equation (2.34) is the continuity equation in three dimensions, (2.35)
is the mass momentum equation in the x-direction, (2.36) is the mass momentum equation in the y-direction and (2.37) is the mass momentum equation in
the z-direction.

Again, assume the flow oscillates (or fluctuates) about mean values and use
the tilde superscript to indicate a fluctuation. Thus we have the instantaneous
quantities

$$u = U(y) + \tilde{u}(x,y,z,t),$$

$$v = \tilde{v}(x,y,z,t),$$

$$w = \tilde{w}(x,y,z,t),$$

$$p = P(x) + \tilde{p}(x,y,z,t). \tag{2.38}$$

When these relations are introduced into the Navier–Stokes equations (2.34)–
(2.37), the nonlinear disturbance equations result

$$\tilde{u}_x + \tilde{v}_y + \tilde{w}_z = 0, \tag{2.39}$$

$$\tilde{u}_t + U\tilde{u}_x + U'\tilde{v} + \tilde{p}_x + \underline{(\tilde{u}\tilde{u}_x + \tilde{v}\tilde{u}_y + \tilde{w}\tilde{u}_z)} = Re_{3d}^{-1} \nabla^2 \tilde{u}, \tag{2.40}$$

$$\tilde{v}_t + U\tilde{v}_x + \tilde{p}_y + \underline{(\tilde{u}\tilde{v}_x + \tilde{v}\tilde{v}_y + \tilde{w}\tilde{v}_z)} = Re_{3d}^{-1} \nabla^2 \tilde{v}, \tag{2.41}$$

$$\tilde{w}_t + U\tilde{w}_x + \tilde{p}_z + \underline{(\tilde{u}\tilde{w}_x + \tilde{v}\tilde{w}_y + \tilde{w}\tilde{w}_z)} = Re_{3d}^{-1} \nabla^2 \tilde{w}. \tag{2.42}$$

The normal mode approach first linearizes the above system by neglecting the
nonlinear terms (underlined), and then introduces the complex quantities

$$\tilde{u} = \frac{1}{2}(\breve{u} + \breve{u}^*), \quad \tilde{v} = \frac{1}{2}(\breve{v} + \breve{v}^*), \quad \tilde{w} = \frac{1}{2}(\breve{w} + \breve{w}^*), \quad \tilde{p} = \frac{1}{2}(\breve{p} + \breve{p}^*). \tag{2.43}$$

As before, let

$$\breve{u}(x,y,z,t) = \hat{u}(y)e^{i(\alpha x + \gamma z - \omega t)},$$

$$\breve{v}(x,y,z,t) = \hat{v}(y)e^{i(\alpha x + \gamma z - \omega t)},$$

$$\breve{w}(x,y,z,t) = \hat{w}(y)e^{i(\alpha x + \gamma z - \omega t)},$$

$$\breve{p}(x,y,z,t) = \hat{p}(y)e^{i(\alpha x + \gamma z - \omega t)}, \tag{2.44}$$

where α and γ are the complex wavenumbers in the x- and z-directions, respectively, and $\omega = \alpha c$ is the complex frequency. With this notation, the real part of (2.44) is the physical quantity. Substitution of (2.43) and (2.44) into the linearized version of (2.39)–(2.42) results in the following linear system of equations

$$i\alpha\hat{u} + \hat{v}' + i\gamma\hat{w} = 0, \tag{2.45}$$

$$i\alpha(U-c)\hat{u} + U'\hat{v} + i\alpha\hat{p} = Re_{3d}^{-1}\left[\hat{u}'' - (\alpha^2 + \gamma^2)\hat{u}\right], \tag{2.46}$$

$$i\alpha(U-c)\hat{v} + \hat{p}' = Re_{3d}^{-1}\left[\hat{v}'' - (\alpha^2 + \gamma^2)\hat{v}\right], \tag{2.47}$$

$$i\alpha(U-c)\hat{w} + i\gamma\hat{p} = Re_{3d}^{-1}\left[\hat{w}'' - (\alpha^2 + \gamma^2)\hat{w}\right]. \tag{2.48}$$

A single equation for $\hat{v}$ can now be obtained in a straightforward manner. The $\hat{u}$ momentum equation (2.46) is multiplied by $i\alpha$ and the $\hat{w}$ momentum equation (2.48) is multiplied by $i\gamma$. The resulting equations are summed and the continuity equation (2.45) is used to replace the expressions $i\alpha\hat{u} + i\gamma\hat{w}$ with $-\hat{v}'$, resulting in the equation

$$i\alpha(U-c)\hat{v}' - i\alpha U'\hat{v} + (\alpha^2 + \gamma^2)\hat{p} = Re_{3d}^{-1}\left[\hat{v}''' - (\alpha^2 + \gamma^2)\hat{v}'\right]. \tag{2.49}$$

The pressure is eliminated by differentiating (2.49) by y and then use the $\hat{v}$ momentum equation (2.47), resulting in a single equation for $\hat{v}$, namely

$$(U-c)\left[\hat{v}'' - (\alpha^2 + \gamma^2)\hat{v}\right] - U''\hat{v}$$
$$= (i\alpha Re_{3d})^{-1}\left[\hat{v}^{iv} - 2(\alpha^2 + \gamma^2)\hat{v}'' + (\alpha^2 + \gamma^2)^2\hat{v}\right]. \tag{2.50}$$

The above equation is the three-dimensional Orr–Sommerfeld equation, which is a fourth-order ordinary differential equation. The number of unknowns has now increased to six, namely: $\alpha_r, \alpha_i, \gamma_r, \gamma_i$ and ω_r, ω_i.

2.1.6 Squire Transformation

Squire (1933) recognized that, with a simple transformation, equation (2.50) can be reduced to a form equivalent to the two-dimensional Orr–Sommerfeld equation. Define the polar wavenumber, $\tilde{\alpha}$, as

$$\tilde{\alpha} = \sqrt{\alpha^2 + \gamma^2}\,,\qquad(2.51)$$

and a reduced Reynolds number as

$$Re_{2d} = \frac{\alpha Re_{3d}}{\sqrt{\alpha^2 + \gamma^2}} \equiv \frac{\alpha}{\tilde{\alpha}} Re_{3d} = Re_{3d}\cos\varphi,\qquad(2.52)$$

where $\tan\varphi = \gamma/\alpha$, the polar angle in wave space (see Fig. 2.1).

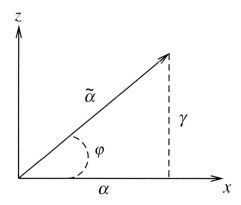

Figure 2.1 Sketch of the polar representation.

Substituting (2.51) and (2.52) into the three-dimensional Orr–Sommerfeld equation (2.50) results in

$$(U - c)(\hat{v}'' - \tilde{\alpha}^2\hat{v}) - U''\hat{v} = (i\tilde{\alpha}Re_{2d})^{-1}\left(\hat{v}^{iv} - 2\tilde{\alpha}^2\hat{v}'' + \tilde{\alpha}^4\hat{v}\right),\qquad(2.53)$$

which has exactly the same form as the two-dimensional Orr–Sommerfeld equation (2.28). Several remarks are in order here. First, the transformation that takes the three-dimensional problem and transforms it into an equivalent two-dimensional problem is now called the "Squire transformation," after Squire (1933) for his important contribution to stability theory. Second, for parallel flows, one needs only study the two-dimensional problem for determining stability with the linear assumption. Once $\tilde{\alpha}$ and Re_{2d} are determined from the two-dimensional problem, we can determine the true wavenumber α and Reynolds number Re_{3d} by inverting the transformation, for a given value of the polar angle φ. Third, since $\alpha \le \tilde{\alpha}$, we see that the three-dimensional and the two-dimensional equations are the same, except that the two-dimensional problem has a lower value of the Reynolds number. Finally, the phase speed, c, is unscaled and hence both the three-dimensional and the two-dimensional linear stability problems have exactly the same phase speed definition. These remarks were originally made by Squire (1933), and we recast them in the form of the following theorem:

> **Squire's Theorem** If an exact two-dimensional parallel flow admits an unstable three-dimensional disturbance for a certain value of the Reynolds number, it also admits a two-dimensional disturbance at a lower value of the Reynolds number.

The theorem could also be restated as, *To each unstable three-dimensional disturbance there corresponds a more unstable two-dimensional disturbance.* Or, *To obtain the minimum critical Reynolds number it is sufficient to consider only two-dimensional disturbances.* Since the Squire transformation relates two- and three-dimensional disturbances, henceforth only two-dimensional disturbances are considered in this chapter. However, we caution that the theorem only applies to parallel flows; for more complicated mean flows (e.g., three-dimensional mean flows or curved flows), three-dimensional disturbances are of utmost importance. Three-dimensional disturbances take on a significant role when nonlinear disturbances are considered in later chapters.

As a final comment, the stream function approach to deriving the Orr–Sommerfeld equation is no longer useful in three dimensions, but the vorticity formulation can be used because the vorticity now has three components and these can be used to derive the counterpart of the Orr–Sommerfeld equation in three dimensions. These points were also outlined in Chapter 1.

2.2 Kelvin–Helmholtz Theory

Before embarking on the solutions to the Rayleigh or Orr–Sommerfeld equation for the general case of parallel mean flows, it is instructive to depart somewhat at this point and consider the stability of piecewise constant flows. Helmholtz and Kelvin (see also Lamb, 1945) gave the first description of such flows, and the theory now bears their names in honor of their contributions. Consider an incompressible inviscid flow of two fluids with different velocities and different densities, as shown in Fig. 2.2. In this section, all quantities will be considered dimensional because the results are easier to understand conceptually than the alternate nondimensional analysis; thus, we drop the † notation for a dimensional quantity in this section. The dimensional mean variables that describe the flow are given by

$$U = \begin{cases} U_1 \\ U_2 \end{cases} \quad \text{and} \quad \rho = \begin{cases} \rho_1 & y > 0, \\ \rho_2 & y < 0. \end{cases} \tag{2.54}$$

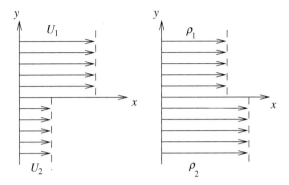

Figure 2.2 Sketch of the piecewise constant approximation to a shear layer; the interface is located at $y = 0$.

As stated, Rayleigh's equation (2.29) governs the stability of the flow on either side of the interface $y = 0$, and has the general solution

$$\hat{v} = \begin{cases} Ae^{-\alpha y} & y > 0, \\ Be^{\alpha y} & y < 0, \end{cases} \tag{2.55}$$

which satisfies the $y \to \pm\infty$ boundary conditions. These solutions are those of a harmonic field. In order to complete the solution, two conditions are needed at the interface $y = 0$ to determine A and B.[3] These are of the form of jump conditions, which are derived in the next section.

2.2.1 Interface Conditions

Two conditions are needed at the interface to complete the solution given above. The first condition comes from the requirement that the jump in the normal stress must be continuous. This condition is equivalent to saying that, for an inviscid fluid, the pressure is continuous at the interface, thus

$$\|p\| = 0. \tag{2.56}$$

Here, the notation $\| \ \| = (\ )|_{y=0^+} - (\ )|_{y=0^-}$ is used to denote the jump across the interface located at $y = 0$. The interface condition can easily be obtained by first linearizing the jump condition (2.56), applying the normal mode technique, and then eliminating the pressure disturbance in terms of the vertical velocity component by combining the linearized continuity and x-momentum equations, or

$$\rho \left[(\omega - \alpha U)\hat{v}' + \alpha U'\hat{v} \right] + i\alpha^2 \hat{p} = 0. \tag{2.57}$$

---

[3] Alternatively, one can set $A = 1$ since the problem is homogeneous, but two conditions are still needed to determine B and the eigenvalue ω.

Equation (2.57) is valid on either side of the interface and, after substituting into (2.56), leads to the jump condition

$$\left\| \rho \left[(\alpha U - \omega)\hat{v}' - \alpha U' \hat{v} \right] \right\| = 0 \quad \text{at} \quad y = 0. \tag{2.58}$$

The second interface condition can be found by appealing to the motion of the interface. Let

$$F(x, y, t) = y - f(x, t) \tag{2.59}$$

describe the position of the interface.[4] The equation for the free surface is given by the material derivative

$$\frac{DF}{Dt} \equiv F_t + u F_x + v F_y = 0. \tag{2.60}$$

This equation states that the change in the quantity F along the particle path must be zero and is also referred to as the kinematic condition at the free surface. After substituting for the definition of F from (2.59), equation (2.60) becomes

$$- f_t - u f_x + v = 0, \tag{2.61}$$

or, in terms of the normal component velocity v,

$$v = f_t + u f_x. \tag{2.62}$$

Assume the shape of the surface can be written in terms of normal modes or

$$f(x, t) = a e^{i(\alpha x - \omega t)}, \tag{2.63}$$

with a the amplitude of the displacement of the interface from its mean position $y = 0$. By substituting (2.63) into (2.62), linearizing and using normal modes, the following results

$$\hat{v} = \begin{cases} i(\alpha U_1 - \omega)a & y = 0^+, \\ i(\alpha U_2 - \omega)a & y = 0^-. \end{cases} \tag{2.64}$$

Since a is a constant, the jump condition becomes

$$\left\| \frac{\hat{v}}{\alpha U - \omega} \right\| = 0 \quad \text{at} \quad y = 0. \tag{2.65}$$

The two interface conditions are now applied to the solution obtained above. Substitute (2.55) into (2.58) and (2.65) to find

$$- \rho_1 (\alpha U_1 - \omega) A \alpha = \rho_2 (\alpha U_2 - \omega) B \alpha, \tag{2.66}$$

---

[4] This equation is only valid if the shape of the interface is single-valued, as would be expected for small disturbances.

and

$$\frac{A}{\alpha U_1 - \omega} = \frac{B}{\alpha U_2 - \omega}. \tag{2.67}$$

In matrix form, equations (2.66) and (2.67) become

$$\begin{bmatrix} \rho_1(\alpha U_1 - \omega) & \rho_2(\alpha U_2 - \omega) \\ (\alpha U_2 - \omega) & -(\alpha U_1 - \omega) \end{bmatrix} \begin{bmatrix} A \\ B \end{bmatrix} = \begin{bmatrix} 0 \\ 0 \end{bmatrix}. \tag{2.68}$$

Since the system is homogeneous, a nontrivial solution exists if and only if the determinant vanishes, which gives the quadratic equation

$$\rho_1(\alpha U_1 - \omega)^2 + \rho_2(\alpha U_2 - \omega)^2 = 0. \tag{2.69}$$

This equation determines the eigenvalue ω and is called the dispersion relation for ω. The solution to the dispersion relation given by (2.69) is

$$\omega = \alpha \left\{ \frac{\rho_1 U_1 + \rho_2 U_2}{\rho_1 + \rho_2} \pm i \sqrt{\frac{\rho_1 \rho_2 (U_2 - U_1)^2}{(\rho_1 + \rho_2)^2}} \right\}, \tag{2.70}$$

which is a linear function of α. Since one root of the imaginary part of ω is positive for $U_1 \neq U_2$, the shear flow is always temporally unstable in an inviscid fluid, even if $\rho_1 = \rho_2$. The root corresponding to $\omega < 0$ is mathematically correct but cannot exist physically. This instability is a direct result of the fact that the dynamic pressure, $[p + \rho|u|^2/2]$, is not equal at the interface. Additional effects, such as buoyancy and surface tension, can be added. These effects will be considered in the exercise section at the end of this chapter and in Chapter 7.

The dispersion relation (2.70) also reveals another important point. In this discontinuous mean model, there is no characteristic length scale and therefore, as $\alpha \to \infty$, $\omega \to \infty$. Large α means small scales and this limit is physically not possible. This point will be made clear in the next section.

2.3 Piecewise Linear Profile

2.3.1 Unconfined Shear Layer

As an example of a piecewise linear profile, consider the unconfined shear layer with nondimensional mean velocity defined by

$$U(y) = \begin{cases} 1 & y > 1, \\ y & -1 < y < 1, \\ -1 & y < -1. \end{cases} \tag{2.71}$$

A sketch of the velocity profile is given in Fig. 2.3. Note there is now a length scale. The length scale L is taken to be half the shear layer thickness and the velocity scale is taken to be the freestream value.

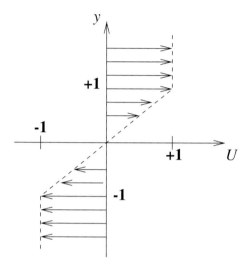

Figure 2.3 Sketch of the unconfined piecewise linear shear layer.

In each region the equation for the disturbance is governed by Rayleigh's equation (2.29) with $U'' = 0$ from (2.71), namely

$$\hat{v}'' - \alpha^2 \hat{v} = 0. \tag{2.72}$$

Again, $\hat{v}$ is a harmonic function. The general solution in each region can easily be determined in such a way to satisfy the far-field boundary conditions where $y \to \pm\infty$. A convenient form is given by

$$\hat{v}(y) = \begin{cases} Ae^{-\alpha(y-1)} & y > 1, \\ Be^{-\alpha(y-1)} + Ce^{\alpha(y+1)} & -1 < y < 1, \\ De^{\alpha(y+1)} & y < -1. \end{cases} \tag{2.73}$$

The jump conditions (2.58) and (2.65) are now applied at $y = \pm 1$. First, the requirement that the pressure disturbance be continuous at the interfaces results in the following two equations

$$A = B\left(\frac{\omega - \alpha - 1}{\omega - \alpha}\right) + Ce^{2\alpha}\left(\frac{\alpha - \omega - 1}{\omega - \alpha}\right), \tag{2.74}$$

and

$$D = Be^{2\alpha}\left(\frac{1-\omega-\alpha}{\omega+\alpha}\right) + C\left(\frac{\alpha+\omega+1}{\omega+\alpha}\right). \tag{2.75}$$

The second interface condition implies that the normal disturbance velocity be continuous across the interfaces, which results in the following two equations

$$A = B + Ce^{2\alpha} \tag{2.76}$$

and

$$D = Be^{2\alpha} + C. \tag{2.77}$$

Satisfying the interface conditions yields four equations for the four unknown constants: A, B, C and D. Rewrite the four equations (2.74)–(2.77) in matrix notation, or

$$\mathbf{J} \cdot \mathbf{x} = 0, \tag{2.78}$$

where

$$\mathbf{x} = (A, B, C, D)^{\mathrm{T}} \tag{2.79}$$

and

$$\mathbf{J} = \begin{bmatrix} 1 & -\left(\frac{\omega-\alpha-1}{\omega-\alpha}\right) & -e^{2\alpha}\left(\frac{\alpha-\omega-1}{\omega-\alpha}\right) & 0 \\ 0 & -e^{2\alpha}\left(\frac{1-\omega-\alpha}{\omega+\alpha}\right) & -\left(\frac{\alpha+\omega+1}{\omega+\alpha}\right) & 1 \\ 1 & -1 & -e^{2\alpha} & 0 \\ 0 & -e^{2\alpha} & -1 & 1 \end{bmatrix}. \tag{2.80}$$

As before, a nontrivial solution will exist if and only if the determinant of $\mathbf{J}$ vanishes. The determinant calculation leads to

$$|\mathbf{J}| = \frac{e^{4\alpha}}{\omega^2 - \alpha^2}\left\{1 - 4\alpha + 4\alpha^2 - 4\omega^2 - e^{-4\alpha}\right\} \tag{2.81}$$

and, by setting (2.81) to zero, the eigenvalue relation results.

$$\omega = \pm\frac{1}{2}\sqrt{(1-2\alpha)^2 - e^{-4\alpha}}. \tag{2.82}$$

This equation is the dispersion relation that relates the wavenumber α to the frequency ω, and was first obtained by Rayleigh (1894). Since α is real, ω is either purely real (stable) or purely imaginary (unstable), depending on the sign of the square root term. The neutral mode can thus be found by setting the square root term to zero, resulting in the neutral mode

$$\omega_N = 0, \quad \text{for} \quad \alpha_N = 0 \quad \text{and} \quad \alpha_N \approx 0.639232. \tag{2.83}$$

The unstable region is given by $0 < \alpha < \alpha_N$, and a graph of the unstable region is shown in Fig. 2.4. Contrast this with the previous example of Kelvin–Helmholtz where the unstable region is infinite with $\alpha > 0$. As mentioned, this is due to the fact that the Kelvin–Helmholtz model has no length scale and therefore is unstable at all values of α. However, as α gets larger, the scale decreases and viscous effects must be eventually considered. From this figure we see that the maximum growth rate, $\omega_{i,max} \approx 0.201186$, occurs for $\alpha \approx 0.39837$. The wavelength $\lambda = 2\pi/\alpha$ associated with the maximum growth rate is

$$\lambda_{max} \approx \frac{2\pi}{0.39837} \approx 15.772. \tag{2.84}$$

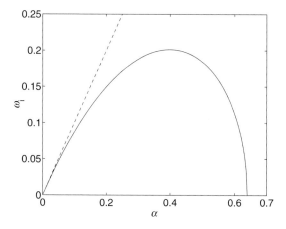

Figure 2.4 Growth rate ω_i for the unconfined piecewise linear shear layer. The dotted line shows the growth rate from the Kelvin–Helmholtz theory with $\rho_1 = \rho_2$ and $U_1 = 1$, $U_2 = -1$.

In an experiment, all modes will be excited, but only the mode that has the largest growth rate will generally be observed. Thus, at large times one would expect to see a wave of length λ_{max} emerge that will continue to grow in amplitude in time, at least until nonlinear effects become important.

Extensions of the analysis we just concluded Esch (1957), who included viscosity by solving the Orr–Sommerfeld equation using the piecewise linear profile given here. Esch showed that the inclusion of viscosity had a damping effect on the growth rate.

2.3.2 *Confined Shear Layer*

A similar analysis can be made for the confined shear layer where it is assumed that there is a wall at $y = \pm h$. The nondimensional mean velocity is defined by

$$U(y) = \begin{cases} 1 & 1 < y < h, \\ y & -1 < y < 1, \\ -1 & -h < y < -1. \end{cases} \qquad (2.85)$$

The length scale L is again taken to be half of the shear layer thickness. This scaling yields nondimensional quantities $y = y^{\dagger}/L$, $h = H/L$, where H is the location of the wall in dimensional units. The velocity scale is, as before, taken as the freestream value. A sketch of the configuration is shown in Fig. 2.5.

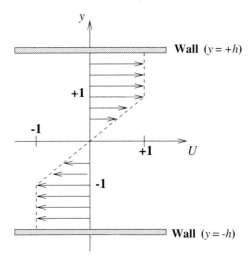

Figure 2.5 Sketch of the confined shear layer.

The solution to the reduced Rayleigh equation that satisfies the zero-velocity boundary conditions at the walls is given by

$$\hat{v}(y) = \begin{cases} A\sinh[\alpha(h-y)] & 1 < y < h, \\ B\sinh(\alpha y) + C\cosh(\alpha y) & -1 < y < 1, \\ D\sinh[\alpha(y+h)] & -h < y < -1. \end{cases} \qquad (2.86)$$

The two interface conditions, applied at $y = \pm 1$, lead to the following four equations:

$$(\alpha - \omega)\left[\alpha B\cosh\alpha + \alpha C\sinh\alpha\right] - \alpha\left[B\sinh\alpha + C\cosh\alpha\right]$$
$$= \alpha(\alpha - \omega)A\cosh[\alpha(h-1)], \quad (2.87)$$

$$(\alpha + \omega)\left[\alpha B \cosh \alpha - \alpha C \sinh \alpha\right] - \alpha\left[B \sinh \alpha - C \cosh \alpha\right]$$

$$= \alpha(\alpha + \omega)D\cosh[\alpha(h-1)], \quad (2.88)$$

$$A\sinh[\alpha(h-1)] = B\sinh \alpha + C \cosh \alpha, \quad (2.89)$$

and

$$-B\sinh \alpha + C \cosh \alpha = D\sinh[\alpha(h-1)]. \quad (2.90)$$

This leads to four equations for four unknowns, and a nontrivial solution exists only if the determinant vanishes. After expanding the determinant, and equating it to zero, the dispersion relation is

$$\omega^2 = \alpha^2 - \frac{\alpha(1+X^2)Y^2 + 2\alpha XY - XY^2}{(1+X^2)Y + X(1+Y^2)} \equiv G(\alpha, \alpha h), \quad (2.91)$$

where

$$X = \tanh \alpha, \quad Y = \tanh[\alpha(h-1)].$$

The flow is stable if G is positive, corresponding to ω real, and is unstable if G is negative, corresponding to ω imaginary. A graph of the contours of G is plotted in Fig. 2.6 in the $(\alpha, \alpha h)$-plane. Note that as $\alpha h \to \infty$, the zero contour, which corresponds to a neutral mode, shows that $\alpha \approx 0.639232$, which is the neutral mode for the shear layer in the unbounded domain.

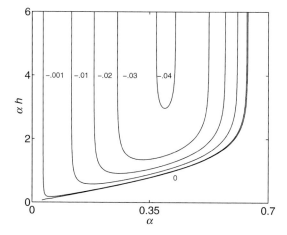

Figure 2.6 Contours of $G(\alpha, \alpha h)$ showing the unstable region for the confined shear layer.

Similar analysis can be applied to other algebraic profiles, and some of these can be found in the exercise section at the end of this chapter.

2.4 Inviscid Temporal Theory

In terms of the stream function, recall that Rayleigh's equation for an inviscid fluid is given by

$$(U - c)(\phi'' - \alpha^2 \phi) - U'' \phi = 0, \tag{2.92}$$

where $c = c_r + i c_i$ is the complex phase speed and α is the real wavenumber. This equation is to be solved subject to the homogeneous boundary conditions

$$\phi = 0 \quad \text{at} \quad y = [a, b], \tag{2.93}$$

which is appropriate for bounded flows. Presented in this section are several important results concerning Rayleigh's equation. It is important to remember that throughout this section we are concerned only with temporal stability; spatial stability is presented in Chapter 4. As a consequence, α is taken as real. Rayleigh used these facts to determine some very important results regarding such flows. Even more, he was able to establish theorems relating to the stability of the flows. Some of these theorems are given in the following proofs, along with important contributions from other authors. In what follows, the notation $\square$ is used to denote the end of a proof. To avoid the use of mathematics beyond the scope of this text, only proofs that are constructive in nature will be presented here. We also switch notation to denote a complex quantity: the superscript $*$ has been replaced by an overbar.

Result 2.1 (Rayleigh's Inflection Point Theorem [1880]) *A necessary condition for instability is that the mean velocity profile $U(y)$ has an inflection point somewhere in the domain of the flow.*

Proof Assume $c_i > 0$. Then, multiply (2.92) by the complex conjugate $\bar{\phi}$ and integrate over the domain defined as $a \leq y \leq b$ to give:

$$\int_a^b \left[\bar{\phi} \phi'' - \alpha^2 \phi \bar{\phi} - \frac{U''}{U - c} \phi \bar{\phi} \right] dy = 0. \tag{2.94}$$

The first term can be integrated by parts, giving

$$\int_a^b \bar{\phi} \phi'' dy = \bar{\phi} \phi' \Big|_a^b - \int_a^b \bar{\phi}' \phi' dy. \tag{2.95}$$

After applying the boundary conditions and using (2.95), equation (2.94) becomes

$$\int_a^b \left[|\phi'|^2 + \alpha^2 |\phi|^2 \right] dy + \int_a^b \frac{U''(U - \bar{c})}{|U - c|^2} |\phi|^2 dy = 0. \qquad (2.96)$$

By separating the real and imaginary parts of (2.96), two equations result:

$$\int_a^b \left[|\phi'|^2 + \alpha^2 |\phi|^2 \right] dy + \int_a^b \frac{U''(U - c_r)}{|U - c|^2} |\phi|^2 dy = 0 \qquad (2.97)$$

and

$$c_i \int_a^b \frac{U''}{|U - c|^2} |\phi|^2 dy = 0. \qquad (2.98)$$

From this last expression, either c_i is zero or the integral must vanish. If c_i is zero, some exceptional neutral solution might be expected, but the very existence of the integral often becomes questionable. When c_i is not zero, the only way for the integral to vanish is for U'' to change signs somewhere in the interval (a, b). This implies that the velocity profile U must have a point of inflection; i.e., $U''(y_s) = 0$ for $a < y_s < b$. $\qquad \square$

Recall that Rayleigh's Theorem states that if an inviscid flow is unstable, then the mean velocity profile U must have a point of inflection somewhere in the bounded domain. The converse is not necessarily true: *If the mean velocity profile has a point of inflection, then the flow is unstable.* The negation of the theorem is, however, true: *If the mean velocity does not have an inflection point, then the flow is stable.* The reader is cautioned against misapplying Rayleigh's Theorem.

A stronger version of Rayleigh's Theorem on the consequence of the inflection point in the mean velocity profile was noted by Fjørtoft (1950) (see also Høiland, 1953). This result is stated as follows.

Result 2.2 (Fjørtoft's Theorem [1950]) *A necessary condition for instability is that $U''(U - U_s) < 0$ somewhere in the flow field, where y_s is the point at which the mean profile has an inflection point, $U''(y_s) = 0$, and $U_s = U(y_s)$.*

Proof Assume $c_i > 0$. Multiply the second condition (2.98) of Rayleigh's Theorem by $(c_r - U_s)/c_i$, where y_s is the point at which $U'' = 0$, and $U_s = U(y_s)$. Substitution into the first condition (2.97) results in the expression

$$\int_a^b \left[|\phi'|^2 + \alpha^2 |\phi|^2 \right] dy + \int_a^b \frac{U''(U - U_s)}{|U - c|^2} |\phi|^2 dy = 0. \qquad (2.99)$$

Since the first integral is strictly positive, $U''(U - U_s) < 0$ over some part of the flow field. $\qquad \square$

As in Rayleigh's Theorem, the converse is not necessarily true: *If $U''(U - U_s) < 0$ somewhere in the flow, then the flow is unstable.* The negation of the theorem is, however, true: *If $U''(U - U_s)$ is not negative somewhere in the flow, then the flow is stable.* The reader is cautioned against misapplying Fjørtoft's Theorem.

Note here again, the mean vorticity for parallel flows is

$$\Omega_z = V_x - U_y = -U'(y). \tag{2.100}$$

Rayleigh's Theorem shows that the mean vorticity must have a local maximum or minimum, while Fjørtoft's Theorem states a stronger condition in that the base vorticity must have a local maximum.

Some examples include plane Couette flow, where the mean velocity profile is given by

$$U = U_o \left(\frac{y}{L}\right) \quad -L < y < L. \tag{2.101}$$

Note that $U'' = 0$ and hence $U''(U - U_s) = 0$, so the flow is inviscidly stable by applying the negation argument of Fjørtoft's Theorem.

For plane Poiseuille flow, the mean velocity is

$$U = U_o(1 - (y/L)^2) \quad -L < y < L. \tag{2.102}$$

Note that $U'' = -2U_o/L^2 < 0$ and the flow is inviscidly stable by the negation argument of Rayleigh's Theorem. Free shear flows, such as the jet, wake and mixing layer, all have points of inflection and thus these flows *may be* unstable to inviscid disturbances. The Blasius boundary layer flow, which does not have an inflection point on the semi-infinite domain $0 < y < \infty$, is also inviscidly stable. On the other hand, the compressible boundary layer will develop an inflection point at high Mach number due to viscous heating, and thus the flow may be susceptible to inviscid disturbances. We shall examine these examples in greater detail later.

Result 2.3 *The phase velocity c_r of an amplified disturbance must lie between the minimum and the maximum values of the mean velocity profile $U(y)$.*

Proof Assume $c_i > 0$. We begin by defining the following function

$$f(y) = \frac{\phi(y)}{U - c}. \tag{2.103}$$

Substitution of (2.103) into Rayleigh's equation (2.92) gives

$$\left[(U - c)^2 f'\right]' - \alpha^2(U - c)^2 f = 0 \tag{2.104}$$

with $f = 0$ at $y = a, b$. We note that this equation is in standard Sturm–Liouville form. Multiplying by the complex conjugate $\bar{f}$ and integrating over the domain (a,b), we get

$$\int_a^b \left[(U-c)^2 f'\right]' \bar{f} dy - \int_a^b \alpha^2 (U-c)^2 f\bar{f}\, dy = 0. \tag{2.105}$$

The first term can be integrated by parts, or

$$\int_a^b \left[(U-c)^2 f'\right]' \bar{f} dy = (U-c)^2 \bar{f} f' \Big|_a^b - \int_a^b (U-c)^2 f' \bar{f}' dy. \tag{2.106}$$

The boundary term is zero from the boundary conditions. For simplicity, let

$$Q^2 = |f'|^2 + \alpha^2 |f|^2, \tag{2.107}$$

and (2.105) becomes

$$\int_a^b (U-c)^2 Q^2 dy = 0. \tag{2.108}$$

Again, separating (2.108) into the real and imaginary parts, we get

$$\int_a^b \left[(U-c_r)^2 - c_i^2\right] Q^2 dy = 0 \tag{2.109}$$

and

$$\int_a^b c_i(U-c_r)Q^2 dy = 0. \tag{2.110}$$

Since $c_i > 0$, we see that the only way the last integral (2.110) can vanish is that $U - c_r$ change sign somewhere in the interval (a,b). Hence $U_{min} < c_r < U_{max}$. $\qquad\square$

Rayleigh (and later Goldstein, 1983) only stated this result for neutral disturbances $c_i = 0$. Lin (1955) gave the extension of the theorem for excited flows (which is stated in Result 2.3), and our proof follows his demonstration.

Result 2.4 (Howard's Semicircle Theorem I [1961]) *If $c_i > 0$, then*

$$U_{min} < \sqrt{c_r^2 + c_i^2} < U_{max}.$$

Proof Assume $c_i > 0$. We begin by rewriting the expression (2.110) of Result 2.3 to get

$$\int_a^b U Q^2 dy = c_r \int_a^b Q^2 dy. \tag{2.111}$$

After substituting (2.111) into expression (2.109) we find that

$$\int_a^b \left[U^2 - (c_r^2 + c_i^2) \right] Q^2 dy = 0. \tag{2.112}$$

For this last expression to be true the term in the square bracket must change sign somewhere in the domain (a,b). That is,

$$U_{min}^2 < c_r^2 + c_i^2 < U_{max}^2 \tag{2.113}$$

or, assuming U to be positive in the domain (a,b),

$$U_{min} < \sqrt{c_r^2 + c_i^2} < U_{max}. \tag{2.114}$$

This is an equation of an annulus in the (c_r, c_i)-plane, with c_i positive.　　□

Result 2.5 (Howard's Semicircle Theorem II [1961])　*If $c_i > 0$, then*

$$\left(c_r - \frac{U_{min} + U_{max}}{2} \right)^2 + c_i^2 \leq \left(\frac{U_{max} - U_{min}}{2} \right)^2.$$

Proof　Begin by noting that $U_{min} \leq U \leq U_{max}$ so that the following inequality is true

$$(U - U_{min})(U - U_{max}) \leq 0, \tag{2.115}$$

where the first term is strictly nonnegative and the second term is strictly non-positive, the product being negative or zero. Multiply equation (2.115) by Q^2 and integrate to get

$$\int_a^b (U - U_{min})(U - U_{max}) Q^2 dy \leq 0, \tag{2.116}$$

or,

$$\int_a^b \left[U^2 - (U_{min} + U_{max})U + U_{min}U_{max} \right] Q^2 dy \leq 0. \tag{2.117}$$

Now assume $c_i > 0$, and recall (2.112)

$$\int_a^b U^2 Q^2 dy = \int_a^b (c_r^2 + c_i^2) Q^2 dy,$$

and (2.111)

$$\int_a^b U Q^2 dy = \int_a^b c_r Q^2 dy. \tag{2.118}$$

Substitute (2.118) into the inequality (2.117) to find

$$\int_a^b \left[c_r^2 + c_i^2 - (U_{min} + U_{max})c_r + U_{min}U_{max} \right] Q^2 dy \leq 0. \tag{2.119}$$

Since Q^2 is positive, the integrand must be negative or

$$c_r^2 + c_i^2 - (U_{min} + U_{max})c_r + U_{min}U_{max} \leq 0. \qquad (2.120)$$

After rearranging terms, (2.120) can be written as

$$\left(c_r - \frac{U_{min} + U_{max}}{2}\right)^2 + c_i^2 \leq \left(\frac{U_{max} - U_{min}}{2}\right)^2. \qquad (2.121)$$

This is the equation of a semicircle with radius $(U_{max} - U_{min})/2$ and coordinates $(c_r - (U_{min} + U_{max})/2, c_i)$. $\qquad \square$

The last four results can be summarized by referring to the following table. Let $A = U_{min}$ and $B = U_{max}$, then the above results can be restated as follows:

Unstable eigenvalue: $\quad c_i > 0$

Result 2.3: $\quad A < c_r < B$

Result 2.4: $\quad A < |c| < B$

Result 2.5: $\quad \left(c_r - \frac{A+B}{2}\right)^2 + c_i^2 \leq \left(\frac{B-A}{2}\right)^2.$

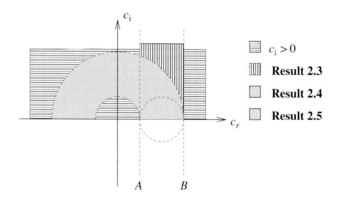

Figure 2.7 Sketch of the regions with $A = U_{min}$ and $B = U_{max}$.

Each of the above results is graphed separately in Fig. 2.7. The left-slanted hatched region gives the exact region where all the discrete temporal eigenvalues of the inviscid, incompressible problem can be found. These are indeed remarkable results, and should be kept in mind when numerically searching for the unstable eigenvalues for any inviscid problem. The reader is reminded, however, that these results only apply to the temporal eigenvalues and cannot be extended to the spatial problem.

As mentioned earlier, neither Rayleigh's nor Fjørtoft's Theorem give a sufficient condition for the instability of a general inviscid flow. A simple counterexample was given by Tollmien for the mean velocity profile $U = \sin(y)$. However, Tollmien did prove sufficiency for a symmetrical channel profile and for monotone profiles and is given next.

Result 2.6 (Tollmien (1935))   *If the mean velocity profile $U(y)$ is either symmetric or monotone in y, then:*

(1) *there is always a neutral disturbance given by $c_N = 0$, $\alpha_N = 0$, $\phi_N = U(y)$;*
(2) *if $U''(y_s) = 0$ for $a < y_s < b$, then there exists a neutral disturbance with $\alpha = \alpha_N > 0$ and $c_N = U_s = U(y_s)$; and*
(3) *for α slightly less than α_N there exists solutions with $c_i > 0$, and for α slightly greater than α_N there are no solutions with $c_i > 0$.*

Tollmien's result is remarkable because it gives us the exact value of the neutral phase speed. The corresponding neutral wave number α_N must, in general, be determined numerically; there are very few examples where α_N can be determined by solving Rayleigh's equation analytically.

Tollmien proved the first two parts but was only able to heuristically prove the third part by adding the further restriction $U'''(y_s) \neq 0$, and then by constructing a series solution about the neutral point to show amplification for $\alpha < \alpha_N$. An alternative to Tollmien's proof for the second part was given by Friedrichs (Von Mises & Friedrichs, 1971), while the third part was formally proved by Lin (1945, Part II). Lin's proof, which removed the restriction on the third derivative of U, is rather involved, and the interested reader can consult the original paper for details. While the proof of part one is trivial, the proof of the second part of Tollmien's theorem as given by Friedrichs is outlined next.

Proof   Begin the proof of Tollmein's second part by considering the variational form of Rayleigh's equation, which is

$$f'' + K(y)f + \lambda f = 0, \qquad f(a) = f(b) = 0, \qquad (2.122)$$

where f is any trial function that satisfies the homogeneous boundary conditions at $y = a$ and at $y = b$; $K(y) = -U''/(U - U_s)$ and $\lambda = -\alpha^2$. We assume that the function $K(y)$ exists and is integrable over the domain (a, b). Since K has no singularities, this equation is in standard Sturm–Liouville form. Thus, there is an infinite sequence of eigenvalues with limit point at $+\infty$, and any eigenvalue can be related to its eigenfunction by Rayleigh's quotient (e.g., Haberman, 1987)

$$\lambda = \int_a^b \left[(f')^2 - Kf^2 \right] dy \Big/ \int_a^b f^2 dy. \qquad (2.123)$$

Equation (2.123) can easily be obtained by multiplying (2.122) by f, integrating the resulting equation over (a,b), integrating by parts the term involving the second derivative and applying the boundary conditions, and finally solving for the eigenvalue λ. The associated variational principle gives the least eigenvalue $\lambda_1 = \min \lambda$, where the minimum is taken over the entire set of trial functions. Now, a neutral mode exists if and only if $\lambda_1 < 0$. Assume that $K(y)$ is nonnegative and the mean velocity U vanishes at the boundaries but not between. This is equivalent to saying that the mean velocity profile is symmetric or is monotonic in y. For the trial function $f = U$, which lies in the space of admissible functions, then

$$
\begin{aligned}
\lambda_1 &= \left. \int_a^b \left[(U')^2 - KU^2 \right] dy \middle/ \int_a^b U^2 dy \right. \\
&= \left. - \int_a^b \left[UU'' + KU^2 \right] dy \middle/ \int_a^b U^2 dy \right. \\
&= \left. - \int_a^b \left[-UK(U-c) + KU^2 \right] dy \middle/ \int_a^b U^2 dy \right. \\
&= \left. - c \int_a^b KU dy \middle/ \int_a^b U^2 dy. \right.
\end{aligned}
$$

Since $K(y) > 0$, $U(y) > 0$ and $c = U(y_s) > 0$, it follows that $\lambda_1 < 0$, and so the existence of a neutral mode has been proven. $\qquad \square$

Result 2.7 (Upper Bound On The Growth Rate) *If there exists a solution with $c_i > 0$, then an upper bound exists and is given by*

$$
\alpha c_i \leq \frac{1}{2} \max |U'(y)|,
$$

where the maximum is taken over the open interval (a,b).

This result is due to Høiland (1953). The reader is asked to prove this last result in the exercise section at the end of this chapter. The outline of the proof will follow that given by Howard (1961).

2.5 Critical Layer Concept

If $(U - c)$ vanishes somewhere in the flow domain, Rayleigh's equation (2.29), or equivalently (2.32), becomes singular in that the term multiplying the highest derivative vanishes, unless $U''(y_c) = 0$. This occurs at a point y_c, say, when the phase speed c is real. Thus, y_c is a singular point and defines the location known as the *critical layer*. Note that when the phase speed is complex, the expression $(U - c)$ is no longer zero and Rayleigh's equation is no longer singular.

To examine the solutions in a neighborhood about the critical layer, we first rewrite Rayleigh's equation as

$$\phi'' + p(y)\phi' + q(y)\phi = 0, \tag{2.124}$$

where

$$p(y) = 0, \qquad q(y) = -\alpha^2 - \frac{U''}{U - c}. \tag{2.125}$$

Assume an extended series solution of the form

$$\phi = \sum_0^\infty a_n (y - y_c)^{n+r}. \tag{2.126}$$

Here, r satisfies the indicial equation

$$r(r - 1) + rp_0 + q_0 = 0, \tag{2.127}$$

where

$$p_0 = \lim_{y \to y_c} (y - y_c)p(y), \qquad q_0 = \lim_{y \to y_c} (y - y_c)^2 q(y). \tag{2.128}$$

Note that $p_0 = 0$ by definition, and that $q_0 = 0$ only if $U'(y_c) \neq 0$. To see this, expand U in a Taylor series about y_c and note that the second term of q gives

$$\lim_{y \to y_c} \frac{(y - y_c)^2 U''}{(U - c)} = \lim_{y \to y_c} \frac{(y - y_c)U''(y_c)}{U'(y_c)} \to 0.$$

Since both p_0 and q_0 vanish in the limit as $y \to y_c$, the roots to (2.127) are $r = 0$ and $r = 1$. Therefore, y_c is a regular singular point with exponents of 0 and 1 (see, for example, Boyce & DiPrima, 1986). Consequently, there exist two linearly independent solutions, valid in some neighborhood about y_c, which are given by

$$\phi_1 = (y - y_c)P_1(y), \tag{2.129}$$

and

$$\phi_2 = P_2(y) + \left(\frac{U_c''}{U_c'}\right)\phi_1 \ln(y - y_c), \tag{2.130}$$

where

$$P_1(y) = 1 + \sum_1^\infty a_n (y - y_c)^n \quad \text{and} \quad P_2(y) = 1 + \sum_1^\infty b_n (y - y_c)^n. \tag{2.131}$$

Here, ϕ_1 is the regular part of the inviscid solution, ϕ_2 is the singular part of the inviscid solution, and P_2 is the regular part of the singular inviscid solution.

Substitution of these results into Rayleigh's equation (2.124) shows that the first few terms of P_1 and P_2 are

$$P_1 = 1 + \frac{U_c''}{2U_c'}(y - y_c) + \frac{1}{6}\left(\frac{U_c'''}{U_c'} + \alpha^2\right)(y - y_c)^2 + \cdots \qquad (2.132)$$

and

$$P_2 = 1 + \left(\frac{U_c'''}{2U_c'} - \frac{(U_c'')^2}{(U_c')^2} + \frac{\alpha^2}{2}\right)(y - y_c)^2 + \cdots . \qquad (2.133)$$

Note that the singular part of the inviscid solution, ϕ_2, has a branch point singularity at $y = y_c$, and hence is multivalued. The correct branch must be chosen, and is given by

$$\text{for } y > y_c, \qquad \ln(y - y_c) = \ln|y - y_c|,$$

$$\text{for } y < y_c, \qquad \ln(y - y_c) = \ln|y - y_c| - \pi i.$$

Also, the path of integration in the y-plane must pass below the singular point y_c.

Now suppose the mean velocity profile U has an inflection point at the critical layer y_c, that is $U_c'' = U''(y_c) = 0$. In this case, both ϕ_1 and ϕ_2 are regular and we can write the general solution as

$$\phi = A\phi_1(y) + P_2(y), \qquad (2.134)$$

where we have normalized ϕ so that $\phi(y_c) = 1$. The two boundary conditions determines A and α. Since $U''(y_c) = 0$ was assumed, according to Tollmien's result (1935) this corresponds to a neutral mode where

$$c_N = U(y_c), \qquad \alpha_N, \quad \text{and} \quad \omega_N = \alpha_N c_N, \qquad (2.135)$$

with eigenfunction $\phi_N = A\phi_1 + P_2$. Recall from Tollmien's result that, if the mean velocity profile is symmetric or is monotone in y of the boundary layer type, and if $U''(y_s) = 0$, then a neutral mode exists with $\alpha_N > 0$ and $c_N = U(y_s)$. Thus, the location of the inflection point and the location of the critical layer coincide, i.e., $y_s = y_c$.

2.5.1 Reynolds Shear Stress

The Reynolds shear stress defined in terms of an average over a wavelength, is

$$\tau = -\overline{\tilde{u}\tilde{v}} = -\frac{\alpha}{2\pi}\int_0^{2\pi/\alpha} \tilde{u}\tilde{v}\,dx, \qquad (2.136)$$

where u and v are the perturbation velocities. Now recall that for the perturbation quantities, the normal mode approach gives

$$\tilde{u}(x, y, t) = \phi'(y)e^{i\alpha(x-ct)} + cc,$$

$$\tilde{v}(x,y,t) = -i\alpha\phi(y)e^{i\alpha(x-ct)} + cc, \qquad (2.137)$$

or, equivalently,

$$\tilde{u} = \left[\phi_r' \cos\{\alpha(x - c_r t)\} - \phi_i' \sin\{\alpha(x - c_r t)\}\right] e^{\alpha c_i t},$$

$$\tilde{v} = \alpha\left[\phi_r \sin\{\alpha(x - c_r t)\} + \phi_i \cos\{\alpha(x - c_r t)\}\right] e^{\alpha c_i t}, \qquad (2.138)$$

when using the streamfunction. Substituting (2.138) into the integral (2.136) for the shear stress leads to

$$\tau = -\frac{\alpha}{2}\left[\phi_r'\phi_i - \phi_r\phi_i'\right]e^{2\alpha c_i t} = \frac{i\alpha}{4}\left[\phi\bar{\phi}' - \bar{\phi}\phi'\right]e^{2\alpha c_i t}, \qquad (2.139)$$

where use of the trigonometric identities,

$$\left.\begin{array}{rcl} \int_0^{2\pi/\alpha} \cos\left[\alpha(x - c_r t)\right]\sin\left[\alpha(x - c_r t)\right]dx &=& 0, \\ \int_0^{2\pi/\alpha} \cos^2\left[\alpha(x - c_r t)\right]dx &=& \frac{\pi}{\alpha}, \\ \int_0^{2\pi/\alpha} \sin^2\left[\alpha(x - c_r t)\right]dx &=& \frac{\pi}{\alpha}, \end{array}\right\} \qquad (2.140)$$

has been made. Now differentiate (2.139) with respect to y and eliminate ϕ'' with the help of Rayleigh's equation or its conjugate. This leads to

$$\frac{d\tau}{dy} = \frac{\alpha c_i}{2}\frac{U''}{|U - c|^2}|\phi|^2 e^{2\alpha c_i t}. \qquad (2.141)$$

Now let $c_i \to 0$, $d\tau/dy$ now vanishes except near the critical layer where it diverges.

Several remarks are in order. First, for $c_i > 0$, integration of (2.141) gives

$$\tau\Big|_a^b = \frac{\alpha c_i}{2}\int_a^b \frac{U''}{|U - c|^2}|\phi|^2 e^{2\alpha c_i t}dy. \qquad (2.142)$$

Since $\tilde{v}$ vanishes at the boundaries, and therefore so does τ, the integral vanishes only if U'' changes sign somewhere in the interval (a,b); this is in fact Rayleigh's Inflection Point Theorem. The second remark concerns the behavior of τ as $c_i \to 0$; that is, as a neutral mode is approached from the unstable side. With a small c_i, consider that ϕ, U' and U'' are nearly constant through the critical layer, and an integration gives

$$\|\tau\| = \frac{\alpha\pi}{2}\frac{U_c''}{U_c'}|\phi_c|^2, \qquad (2.143)$$

where $\|\tau\|$ denotes the jump $\tau(y_c^+) - \tau(y_c^-)$. The easiest way to derive this expression is to start with the definition of τ directly, and take the limit as $y \to y_c$ from above and from below. First recall that the eigenfunction ϕ can be written as a linear combination of ϕ_1 and ϕ_2 as

$$\phi = \phi_1 + \phi_c\phi_2, \qquad (2.144)$$

where ϕ_c is the value of the eigenfunction at the critical layer and is, in general, complex. Now, as $y \to y_c$, we have

$$
\left.
\begin{aligned}
\phi_1 &\to 0, \\
\phi_1' &\to 1, \\
\phi_2 &\to 1, \\
\phi_2'|_+ &\to \tfrac{U_c''}{U_c'}[1 + \ln(y - y_c)], \\
\phi_2'|_- &\to \tfrac{U_c''}{U_c'}[1 + \ln|y - y_c| - \pi i],
\end{aligned}
\right\}
\tag{2.145}
$$

and thus the expression in the shear stress becomes

$$
\phi_r'\phi_i - \phi_r\phi_i' = -|\phi_c|^2 \frac{U_c''}{U_c'}\pi,
\tag{2.146}
$$

and the result for the jump in the shear stress across the critical layer follows immediately. As discussed in Chapter 1, it is this critical phasing that leads to instability.

2.6 Continuous Profiles

Continuous profiles are classified as those mean profiles which are infinitely differentiable. Examples include the class of free shear flows and boundary layer type flows. Free shear flows consists of jets, wakes and mixing layers in bounded or unbounded domains. This class of flows has inflection points and hence is susceptible to inviscid disturbances. Two different approximations to the mixing layer, the example of Kelvin–Helmholtz in Section 2.2 and the piecewise linear profiles of Section 2.3, have already been studied. In this section, two additional approximations, the hyperbolic-tangent profile and the laminar mixing-layer profile, will be studied. Each succeeding approximation, from the Kelvin–Helmholtz to the laminar profile, is more complex than the preceding one and is a more realistic representation to the actual flow of a mixing layer. Other continuous profiles will be considered in the exercise section at the end of the chapter.

2.6.1 Hyperbolic Tangent Profile

Consider the mixing layer given by

$$
U(y) = \frac{1}{2}[1 + \tanh y], \qquad -\infty < y < +\infty,
\tag{2.147}
$$

where the length scale L is taken to be the half-width of the mixing layer and the velocity scale is taken to be the value of the freestream velocity at infinity.

The temporal stability analysis of this profile was first investigated by Betchov & Szewczyk (1963) and Michalke (1964); much of this section draws from the latter reference.

The inflection point is found by setting $U'' = 0$, resulting in $y_s = 0$. This is also the location of the critical layer. The neutral phase speed is thus given by $c_N = U(y_s) = 0.5$. Note that $U''(U - c_N) = -0.5\,\text{sech}^2 y \tanh^2 y < 0$, and so the profile satisfies Fjørtoft's Theorem, and thus the flow *may* be unstable. To find the neutral wavenumber α_N, substitute $c_N = 0.5$ into Rayleigh's equation (2.29) to give

$$\hat{v}'' - (\alpha_N^2 - 2\,\text{sech}^2 y)\hat{v} = 0, \qquad (2.148)$$

which has the solutions

$$\hat{v}_1 = \alpha_N \cosh(\alpha_N y) - \sinh(\alpha_N y)\tanh y \qquad (2.149)$$

and

$$\hat{v}_2 = \alpha_N \sinh(\alpha_N y) - \cosh(\alpha_N y)\tanh y. \qquad (2.150)$$

The $\hat{v}_1$ solution is called the symmetric solution since it is an even function in y, while the $\hat{v}_2$ solution is antisymmetric. The general solution is a linear combination of the symmetric and the antisymmetric solutions, and the only combination that satisfies the boundary conditions $\hat{v} = 0$ as $y \to \pm\infty$ is

$$\alpha_N = 1, \qquad \hat{v}_N = \text{sech}\,y. \qquad (2.151)$$

The corresponding neutral frequency is $\omega_N = \alpha_N c_N = 0.5$. For values in the unstable region away from the neutral mode, Rayleigh's equation must be solved numerically. Table 2.1 shows the results of the numerical solution, and Fig. 2.8 plots the growth rate as a function of wavenumber.

It is of interest to also examine the vorticity distribution since it gives a means to visualize the dynamics of the mixing layer. Define the spanwise vorticity in the usual fashion as

$$\omega_z = v_x - u_y. \qquad (2.152)$$

In terms of the normal mode approach, the spanwise vorticity equation becomes (after substituting the basic and fluctuating velocities)

$$\begin{aligned}
\omega_z &= -\left[U'(y) + \varepsilon \tilde{u}_y(x,y,t)\right] + \varepsilon \tilde{v}_x(x,y,t) \\
&= -U' + \varepsilon\left(-\tilde{\psi}_{yy} - \tilde{\psi}_{xx}\right) \\
&= -U' + \varepsilon\left(-\phi'' + \alpha^2 \phi\right)e^{i\alpha(x-ct)},
\end{aligned} \qquad (2.153)$$

Table 2.1 *Phase speeds and frequencies as a function of the wavenumber* α
for the hyperbolic tangent profile.

α	c_r	c_i	$\omega_r = \alpha c_r$	$\omega_i = \alpha c_i$
1.0	0.5	0.0	0.5	0.0
0.9		0.032693	0.45	0.029424
0.8		0.067334	0.40	0.053867
0.7		0.104321	0.35	0.073025
0.6		0.144162	0.30	0.086497
0.5		0.187511	0.25	0.093755
0.4		0.235225	0.20	0.094090
0.3		0.288447	0.15	0.086534
0.2		0.348728	0.10	0.069746
0.1		0.418227	0.05	0.041823
0.0		0.5	0.0	0.0

where ε is the initial amplitude of the disturbance and $\tilde{\psi}$ is decomposed according to (2.30). Define

$$\breve{\omega}_{z_r}(y) = \mathrm{Re}(-\phi'' + \alpha^2 \phi), \qquad \breve{\omega}_{z_i}(y) = \mathrm{Im}(-\phi'' + \alpha^2 \phi), \qquad (2.154)$$

and the vorticity becomes

$$\omega_z = -\frac{1}{2}\mathrm{sech}^2 y + \varepsilon e^{\omega_i t}\left[\breve{\omega}_{z_r}\cos(\alpha(x - c_r t)) - \breve{\omega}_{z_i}\sin(\alpha(x - c_r t))\right]. \quad (2.155)$$

A contour plot of the vorticity distribution is shown in Fig. 2.9 for two values of the wavenumber and for $\varepsilon = 0.2$. The value of ε is chosen arbitrarily large to aid in visualizing the dynamics. One value of the wavenumber corresponds to the maximum growth rate ($\alpha_{max} \approx 0.4446$), and the other value corresponds to the neutral mode ($\alpha_N = 1$). Since the flow is periodic in the x-direction, the wavelengths of the disturbances are $\lambda = 2\pi/\alpha_{max} \approx 14.1322$ and $\lambda = 2\pi/\alpha_N = 2\pi$, respectively. Note, for the case corresponding to the maximum growth rate, there are two vortices within a single period. The vorticity distribution shows the initial stages of the roll-up of these two neighboring vortices, characteristic of Kelvin–Helmholtz instabilities. By contrast, the case of the neutral mode there is only a single vortex, which does not roll up.

2.6.2 *Laminar Mixing Layer*

The hyperbolic tangent profile of the previous section is only an approximation to a real mixing layer since it does not satisfy the equations of motion. A more realistic approximation to the mixing layer, which does satisfy the equations of motion under certain assumptions, is called the laminar mixing layer profile.

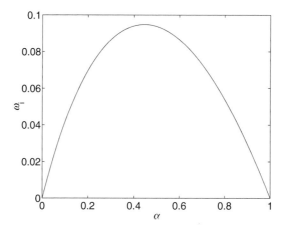

Figure 2.8 Growth rate $\omega_i = \alpha c_i$ as a function of α for the hyperbolic tangent profile.

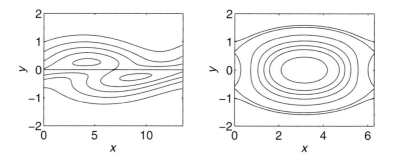

Figure 2.9 Lines of constant vorticity for the wavenumbers of maximum amplification $\alpha_{max} = 0.4446$ (left) and of the neutral disturbance $\alpha_N = 1$ (right) at time $t = 0$ with a disturbance amplitude of $\varepsilon = 0.2$. The contour levels for the left figure are $-0.6, -0.5, -0.4, -0.3, -0.2$, with the closed contours being -0.6 and increasing outward. The contour levels for the right figure are $-0.1, -0.1157, -.2, -.3, -.4, -.5, -.7$, with the innermost closed contour being -0.7 and increasing outward. The mean flow is given by the hyperbolic tangent profile. (After Michalke, 1964, reproduced with permission.)

The derivation is based on the boundary layer approximation, and was first constructed by Lessen (1950), and extended to include different densities and viscosities by Lock (1951).

Begin by considering the nondimensional equations of motion given by (2.6)–(2.8), and set the time derivatives to zero, which is appropriate for steady flows. The boundary layer approximation assumes that variations in the normal direction y are small compared to variations in the streamwise direction (see White, 1974, for example). Therefore let

$$y = \bar{y}/\sqrt{Re} \quad \text{and} \quad v = \bar{v}/\sqrt{Re}. \tag{2.156}$$

The leading-order equations for $Re \gg 1$ are

$$u_x + \bar{v}_{\bar{y}} = 0, \tag{2.157}$$

$$uu_x + \bar{v}u_{\bar{y}} + p_x = u_{\bar{y}\bar{y}}, \tag{2.158}$$

$$p_{\bar{y}} = 0. \tag{2.159}$$

From (2.159), the leading-order pressure term can at most be a function of x and is determined by examining the outer inviscid flow. Outside a boundary layer all variations in y are zero, and the momentum equation (2.158) reduces to

$$\frac{dp}{dx} = -u\frac{du}{dx}. \tag{2.160}$$

Thus, one can either specify a pressure distribution $p(x)$ or a streamwise velocity distribution $u(x)$. In this sense the pressure is considered a known function of x and is imposed on the boundary layer as a source term. Here we assume that the streamwise velocity in the freestream is constant ($u = 1$) and take $p = 1$.

Equations (2.157)–(2.159) are usually referred to as the boundary layer equations and are parabolic in nature. One can numerically solve these equations directly by marching in the streamwise x-direction, given some inlet profile[5] with appropriate boundary conditions in the y-direction. An alternative approach, and one that is commonly employed, is to assume that the solution can be written in terms of a similarity variable. Define

$$\eta = \frac{\bar{y}}{\sqrt{x}} \tag{2.161}$$

where η is the similarity variable.[6] The derivatives transform as

$$\frac{\partial}{\partial y} = \frac{\partial\eta}{\partial y}\frac{\partial}{\partial\eta} = \frac{1}{\sqrt{x}}\frac{\partial}{\partial\eta}, \tag{2.162}$$

$$\frac{\partial}{\partial x} = \frac{\partial}{\partial x} + \frac{\partial\eta}{\partial x}\frac{\partial}{\partial\eta} = \frac{\partial}{\partial x} - \frac{\eta}{2x}\frac{\partial}{\partial\eta}. \tag{2.163}$$

---

[5] One cannot start at $x = 0$, because in this region x and y are of the same order, and so the full viscous problem must be solved. An inlet profile is therefore assumed at some $x_o > 0$.

[6] An alternative definition is $\eta = \bar{y}/\sqrt{2x}$. Either definition is valid, but once a choice is made, care must be taken to be consistent throughout the entire subsequent analysis.

The boundary layer equations (2.157) and (2.158) in the similarity coordinate system are given by

$$u_x - \frac{\eta}{2x} u_\eta + \frac{1}{\sqrt{x}} \bar{v}_\eta = 0, \tag{2.164}$$

and

$$u \left(u_x - \frac{\eta}{2x} u_\eta \right) + \frac{\bar{v}}{\sqrt{x}} u_\eta = \frac{1}{x} u_{\eta\eta}. \tag{2.165}$$

Assume that the streamwise velocity is only a function of the similarity variable, and let

$$u = f'(\eta), \tag{2.166}$$

where the prime denotes differentiation with respect to η. Substituting (2.166) into the continuity equation (2.164) gives

$$\bar{v}_\eta = \frac{\eta}{2\sqrt{x}} f'', \tag{2.167}$$

and, upon integration, reduces to

$$\bar{v} = \frac{1}{2\sqrt{x}} \left(\eta f' - f \right). \tag{2.168}$$

In the integration process we assumed quite arbitrarily that the dividing streamline is centered at zero, and thus $\bar{v}(0) = 0$, implying $f(0) = 0$. From the momentum equation (2.165) we now have

$$2 f''' + f f'' = 0, \tag{2.169}$$

plus appropriate boundary conditions.[7] For the mixing layer the boundary conditions are

$$f'(-\infty) = 0, \quad f(0) = 0, \quad f'(+\infty) = 1. \tag{2.170}$$

Equation (2.169) with boundary conditions (2.170) were originally proposed by Lessen (1950) to describe the mixing layer when the Reynolds number is large. It remains a useful approximation today. Since the equation is third order and nonlinear, no analytical solutions have been found. Instead the solution is obtained numerically. There are several ways to do the integration and the one that we shall give here involves a Runge–Kutta method. Here integration is started at $-\infty$ and integrated toward $+\infty$ using the asymptotic boundary condition (Lessen, 1950; Lock, 1951)

$$f = ag(\xi), \quad \xi = a\eta + b, \tag{2.171}$$

---

[7] If $\eta = \bar{y}/\sqrt{2x}$, then one gets $f''' + f f'' = 0$. The boundary conditions remain unchanged.

where

$$g(\xi) = -1 + e^{\xi/2} - \frac{1}{4}e^{\xi} + \frac{5}{72}e^{3\xi/2} - \frac{17}{864}e^{2\xi} + \cdots. \qquad (2.172)$$

The value of a is determined by requiring $f'(+\infty) = 1$ and the value of b by requiring $f(0) = 0$. An iteration process is performed for each using a Secant method. Table 2.2 shows the numerical results at various values of η. The profile is shown in Fig. 2.10, and for comparison the hyperbolic tangent profile is also shown as a dashed line. From the table and figure we see that the tail of the laminar mixing layer profile is much longer than that of the hyperbolic tangent profile.

Table 2.2 *Values for the laminar mixing layer profile with $a = 1.238494$ and $b = 0.553444$.*

η	f	f'	f''
-20	-1.23849	0.00000	0.00000
-18	-1.23847	0.00001	0.00001
-16	-1.23841	0.00005	0.00003
-14	-1.23821	0.00017	0.00011
-12	-1.23753	0.00060	0.00037
-10	-1.23516	0.00207	0.00128
-8	-1.22700	0.00710	0.00438
-6	-1.19905	0.02423	0.01477
-4	-1.10499	0.08046	0.04716
-2	-0.80602	0.24490	0.12599
0	0.00000	0.58727	0.19971
2	1.53515	0.91210	0.09794
4	3.47304	0.99588	0.00811
6	5.47115	0.99997	0.00009
8	7.47114	1.00000	0.00000

With the laminar mixing layer profile known at least numerically, the stability can be examined, and was first considered by Lessen (1950). Note that the inflection point is located by setting $U'' = 0$, which is equivalent to setting $f''' = 0$. Since f''' is proportional to f, and $f(0) = 0$, we see that the inflection point is at the origin, $y_s = 0$. The neutral phase speed is thus given by $c_N = U(0) = f'(0) = 0.58727$. The corresponding neutral wavenumber and frequency must be found numerically; the result is the neutral mode where

$$c_N \approx 0.587271, \quad \alpha_N \approx 0.395380, \quad \omega_N \approx 0.232195. \qquad (2.173)$$

For values in the unstable region away from the neutral mode, Rayleigh's equation must be solved numerically. Table 2.3 shows the results of the numerical

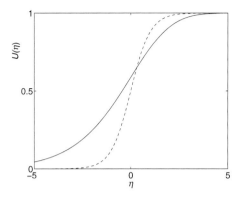

Figure 2.10 Laminar mixing layer profile (solid) and the hyperbolic tangent profile (dash).

solution and Fig. 2.11 plots the growth rate as a function of the wavenumber. Comparing the growth rate curve of Fig. 2.11 for the laminar mixing layer to that of Fig. 2.8 for the hyperbolic tangent profile shows that the two results are in qualitative agreement, but do not agree quantitatively. In particular, the more realistic laminar mixing layer profile has a smaller growth rate, over a shorter range of wavenumbers. For experimentalists who wish to use *linear stability theory* as a guide for flow control, it is imperative to use as realistic a mean profile as possible. However, for theoreticians, qualitative agreement is often sufficient.

Table 2.3 *Phase speeds and frequencies as a function of the wavenumber α for the laminar mixing layer profile.*

α	c_r	c_i	$\omega_r = \alpha c_r$	$\omega_i = \alpha c_i$
0.395	0.587271	0	0.232195	0
0.39	0.585880	0.004364	0.228493	0.001702
0.35	0.575833	0.037869	0.201542	0.013254
0.30	0.563940	0.082773	0.169182	0.024832
0.25	0.552682	0.131790	0.138171	0.032947
0.20	0.541923	0.186028	0.108385	0.037206
0.15	0.531497	0.247054	0.079725	0.037058
0.10	0.521205	0.317135	0.052121	0.031714
0.05	0.510807	0.399673	0.025540	0.019984
0.01	0.502215	0.478137	0.005022	0.004781

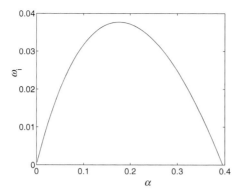

Figure 2.11 Growth rate $\omega_i = \alpha c_i$ as a function of α for the laminar mixing layer profile.

2.7 Exercises

Exercise 2.1   Substitute the instantaneous flow quantities (2.10) into the Navier–Stokes equations (2.6)–(2.8) and show how the linear disturbance equations (2.14)–(2.16) result.

Exercise 2.2   Derive the linearized disturbance equation (2.18) in terms of the $\tilde{v}$ velocity.

Exercise 2.3   Derive the linearized disturbance equation (2.20) in terms of the perturbation streamfunction $\tilde{\psi}$.

Exercise 2.4   Derive the linearized disturbance equation (2.22) in terms of the pressure $\tilde{p}$.

Exercise 2.5   Derive the linearized disturbance equation (2.24) in terms of the vorticity $\tilde{\omega}_z$.

Exercise 2.6   From the linearized disturbance equation (2.18) and the normal mode form (2.25), derive the Orr–Sommerfeld (2.28) and Rayleigh (2.29) equations.

Exercise 2.7   From the nonlinear disturbance equations (2.39)–(2.42) and the normal mode form (2.44), derive the three-dimensional Orr–Sommerfeld equation (2.50).

Exercise 2.8   Consider the two eigenvalue problems:

(1) $y'' - \alpha^2 y = 0$, $y(0) = 0$, $y(h) = 0$;
(2) $y'' - \alpha^2 y = 0$, $y(0) = 0$, $y(\infty)$ is bounded.

Determine the eigenvalues and show that the eigenvalues are different for bounded and unbounded regions. What happens in the limit $h \to \infty$ in the first case? Do the eigenvalues approach those of the second? Explain.

This example illustrates that care must be exercised when approximating unbounded regions by bounded domains for numerical considerations.

Exercise 2.9   Consider the simple pendulum as shown in Fig. 2.12. The differential equation governing the motion of the pendulum is given by

$$mL^2 \frac{d\theta^2}{dt^2} + k\frac{d\theta}{dt} + mgL\sin\theta = 0,$$

where the first term is the acceleration term, the second term is the viscous damping term, and the last term is the restoring moment.

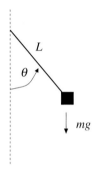

Figure 2.12 Simple Pendulum.

(a) Nondimensionalize the equation and show that only one parameter exists that governs the behavior of the system.
(b) Compute the two equilibrium states.
(c) Determine the stability of each of the two states.

Exercise 2.10   Compute the temporal inviscid stability characteristics of the following piecewise linear profiles: (a) Top hat jet, (b) Triangular jet and (c) Linear shear layer profile.

(a) Top hat (or rectangular) jet.

$$U(y) = \begin{cases} 0 & L < y < \infty \\ U_c & -L < y < L \\ 0 & -\infty < y < -L. \end{cases}$$

Sketch the velocity profile. Note that there are two modes, one which is even about the $y = 0$ axis (i.e., $\hat{v}'(0) = 0$), and one which is odd (i.e., $\hat{v}(0) = 0$). Determine the eigenrelation for each mode.

(b) Triangular jet.

$$U(y) = \begin{cases} 0 & 1 < y < \infty \\ 1 - y & 0 < y < 1 \\ 1 + y & -1 < y < 0 \\ 0 & -\infty < y < -1. \end{cases}$$

Sketch the velocity profile. Again there are two modes, one which is even ($v'(0) = 0$), and one which is odd ($v(0) = 0$). Determine the eigenrelation for each mode, and plot the growth rate curve in the (α, ω_i)-plane.

(c) Linear shear layer profile.

$$U(y) = \begin{cases} 1 & 1 < y < \infty \\ \frac{1}{2}[1 + \beta_U + (1 - \beta_U)y] & -1 < y < 1 \\ \beta_U & -\infty < y < -1. \end{cases}$$

Sketch the velocity profiles. Here, $\beta_U = U_{-\infty}/U_{+\infty}$ is the ratio of freestream velocities. Sketch the velocity profile for several values of the parameter β_U. Determine the eigenrelation, and show that it reduces to that found by Rayleigh when $\beta_U = -1$. Plot the growth rate curve in the (α, ω_i)-plane for $\beta_U = 0.5, 0, -0.5, -1$.

Exercise 2.11 Note that the equations of motion (2.1)–(2.3) for an inviscid flow, with gravity acting in the $-\hat{j}$-direction, can be written as

$$\nabla \cdot \underline{u} = 0,$$

$$\frac{\partial \underline{u}}{\partial t} + (\nabla \times \underline{u}) \times \underline{u} = -\nabla \left(\frac{p}{\rho} + \frac{1}{2}\underline{u} \cdot \underline{u} + gy \right),$$

where $\underline{u} = (u, v)$.

(a) Assume the flow is irrotational. Show for this case that there exists a potential function $\phi(x, y, t)$, defined by $\underline{u} = \nabla \phi$, which satisfies the equations

$$\nabla^2 \phi = 0,$$

$$\frac{\partial \nabla \phi}{\partial t} + \nabla \left(\frac{p}{\rho} + \frac{1}{2}|\nabla \phi|^2 + gy \right) = 0.$$

(b) Integrate the second equation to get

$$\frac{\partial \phi}{\partial t} + \frac{p}{\rho} + \frac{1}{2}|\nabla\phi|^2 + gy = c,$$

where c is a constant of integration. This last equation is Bernoulli's equation for unsteady incompressible flows.

(c) Using the equations in (a,b), we now wish to investigate the temporal stability characteristics for the following (irrotational) mean flow profile

$$U(y) = \begin{cases} U_1 & 0 < y < \infty \\ U_2 & -\infty < y < 0. \end{cases}$$

Using Rayleigh's equation (2.32) and assuming normal mode solutions, show that the eigenrelation is given by

$$\omega = \frac{\alpha(\rho_1 U_1 + \rho_2 U_2)}{\rho_1 + \rho_2} \pm i\sqrt{\frac{\alpha^2 \rho_1 \rho_2 (U_2 - U_1)^2}{(\rho_1 + \rho_2)^2} - \frac{\alpha g(\rho_2 - \rho_1)}{\rho_1 + \rho_2}}.$$

Hint: Write ϕ in terms of normal modes, and then solve the Laplacian equation to determine the perturbation flow. Next, use Bernoulli's equation to get a jump condition that involves gravity.

(d) For internal gravity waves, set $U_1 = U_2 = 0$. Give the eigenrelation, and state the conditions for stability.

(e) For surface gravity waves, set $\rho_1 = 0$ and $U_1 = U_2 = 0$. Give the eigenrelation, and state the conditions for stability.

(f) If surface tension is present then the jump in pressure across the interface is no longer zero, but must be modified to

$$\|p\| = T\frac{\partial^2 f}{\partial x^2},$$

where T is the coefficient for surface tension and $F = y - f(x,t)$ gives the location of the surface. Show that the eigenrelation can be written as

$$\omega = \frac{\alpha(\rho_1 U_1 + \rho_2 U_2)}{\rho_1 + \rho_2}$$

$$\pm i\sqrt{\frac{\alpha^2 \rho_1 \rho_2 (U_2 - U_1)^2}{(\rho_1 + \rho_2)^2} - \frac{\alpha^2}{\rho_1 + \rho_2}\left[\frac{g(\rho_2 - \rho_1)}{\alpha} + \alpha T\right]}.$$

What role does surface tension play in regards to the stability of the flow?

Exercise 2.12   Prove Result 2.7. Use the following to aid in the construction of the proof.

(a) Let $G(y) = \phi(y)/\sqrt{U - c}$, and assume $c_i > 0$.

(b) Show that G satisfies

$$[(U-c)G']' - \left[\frac{1}{2}U'' + \alpha^2(U-c) + \frac{1}{4}\frac{(U')^2}{U-c}\right]G = 0.$$

(c) Show

$$-\int_a^b \left[|G'|^2 + \alpha^2|G|^2\right]dy + \frac{1}{4}\int_a^b \frac{U'^2}{|U-c|^2}|G|^2dy = 0.$$

(d) Note that

$$|U-c|^2 = (U-c_r)^2 + c_i^2 \geq c_i^2$$

and thus

$$|U-c|^{-2} \leq c_i^{-2}.$$

Exercise 2.13 Why might using the Ricatti transformation $G = \phi'/\alpha\phi$ instead of $G = \phi'/\phi$ not be a good idea? See Appendix B for a discussion of the Ricatti transformation for solving Rayleigh's equation.

Exercise 2.14 A numerical method for solving Rayleigh's equation is presented in Appendix B.1. Compute the temporal inviscid stability characteristics of the following continuous profiles. In each case plot the mean profile, and give the location of the inflection point and the corresponding neutral phase speed.

(a) Error function profile.

$$U(y) = \text{erf}(y) \equiv \frac{2}{\sqrt{\pi}}\int_0^y e^{-u^2}du.$$

The error function profile is sometimes used as an approximation to the laminar mixing layer downstream of a splitter plate. Compute the temporal stability and compare graphically the result you obtain with the result of Rayleigh as given by (2.82) and shown in Fig. 2.4.

(b) Hyperbolic tangent profile.

$$U(y) = \frac{1}{2}[1 + \beta_U + (1 - \beta_U)\tanh y]$$

with $\beta_U = 0.5, 0, -0.2$. Here, $\beta_U = U_{-\infty}/U_{+\infty}$ is the ratio of freestream velocities.

(c) Laminar mixing layer.

$$2f''' + ff'' = 0, \quad f'(-\infty) = \beta_U, \quad f(0) = 0, \quad f'(+\infty) = 1$$

with $\beta_U = 0.5, 0$.

(d) Symmetric jet.

$$U(y) = \text{sech}^2(y).$$

Show that two neutral modes exist with

$$\phi_I = \text{sech}^2(y), \quad \alpha_N = 2, \quad c_N = 2/3,$$

$$\phi_{II} = \text{sech}(y)\,\tanh(y), \quad \alpha_N = 1, \quad c_N = 2/3.$$

(e) Symmetric wake.

$$U(y) = 1 - Q\text{sech}^2 y$$

with $Q = 0.3, 0.6, 0.9$. Here, Q is a measure of the wake deficit. Show that two modes exist, and plot the growth rate curves for each.

(f) Gaussian jet.

$$U(y) = e^{-y^2 \ln 2}.$$

Show that two modes exist, and plot the growth rate curves for each. Compare the stability characteristics to that of the symmetric jet in (d).

(g) Combination shear layer and jet.

$$U(y) = \frac{1}{2}[1 + \beta_U + (1 - \beta_U)\tanh y] - Qe^{-y^2 \ln 2}$$

with $\beta_U = 0.5$ and $Q = 0.4, 0.8$. Show that two modes exist if $Q > 0$, and plot the growth rate curves for each.

(h) Asymptotic suction boundary-layer profile.

$$U(y) = 1 - e^{-y}, \quad 0 < y < \infty.$$

(i) Falkner–Skan flow.

$$2f''' + ff'' + \beta(1 - f'^2) = 0, \qquad f(0) = f'(0) = 0, \quad f'(\infty) = 1$$

with the parameter β a measure of the pressure gradient (see, e.g., Acheson, 1990). The value of β can range from $\beta = -0.19884$ (flow separation), to $\beta = 0$ (Blasius), to 1.0 (2D stagnation-point profile). For values of $-0.19884 < \beta < 0$ show that the profile has an inflection point within the flow region, and hence may be unstable to inviscid disturbances. Compute the inviscid temporal stability characteristics, and plot the growth rate curve, for $\beta = -0.1$ and -0.16.

Exercise 2.15 Derive the corresponding Rayleigh equation for an incompressible, density stratified flow

$$\hat{v}'' + (\rho'/\rho)\hat{v}' - \left[\alpha^2 + \frac{U'' + \rho'U'/\rho}{U - c}\right]\hat{v} = 0.$$

Hint: Assume that the mean density is a function of y, but ignore density fluctuations when deriving Rayleigh's equation. The same equation results for compressible flows in the limit of small Mach number, as will be shown later in Chapter 5.

Assume the mean flow profile

$$U(y) = \frac{1}{2}\left[1 + \beta_U + (1 - \beta_U)\tanh y\right],$$

$$\rho(y) = \frac{1}{2}\left[1 + \beta_\rho + (1 - \beta_\rho)\tanh y\right].$$

Solve the temporal stability problem for $\beta_U = 0.5$ and $\beta_\rho = 0.5$, 1.0, 1.5 and graph the eigenrelation in the (α, ω_i)-plane.

3

Temporal Stability of Viscous Incompressible Flows

Nothing in life is to be feared, it is only to be understood.

– Marie Curie (1867–1934)

3.1 Introduction

When a parallel or nearly parallel mean flow does not have an inflection point, viscous effects are important and the Orr–Sommerfeld equation (2.31) must be considered in order to determine the stability of the flow. This is in contrast to solving the much simpler Rayleigh's equation (2.32) where it is generally believed that, for flows with an inflection point, the most unstable mode is inviscid in nature. The latter was noted in the previous chapter.

In this chapter, the temporal viscous stability characteristics will be examined for various well-known profiles. These profiles include bounded flows, such as plane Poiseuille and Couette flows, semi-bounded flows, such as the Blasius boundary layer and the more general Falkner–Skan family, and unbounded flows, such as jets, wakes and mixing layers. Other well-known profiles are given in the exercise section. Attention is restricted to two-dimensional disturbances, since, according to Squire's theorem, if a three-dimensional disturbance is unstable, there corresponds a more unstable two-dimensional disturbance.

For bounded flows, DiPrima & Habetler (1969) proved that the spectrum of the Orr–Sommerfeld equation consists of an infinite number of discrete eigenvalues, and that the spectrum is complete. Contrary to this, if perturbations for this flow are considered inviscidly, then there are no unstable modes. As a result, only a continuous spectrum is possible. Thus, any arbitrary initial disturbance can be decomposed and expressed as a linear combination of eigenfunctions. For unbounded flows, stability calculations for various flows have uncovered only a finite number of eigenvalues.[1] Since any initial disturbance cannot be written in terms of a finite number of modes, a continuum of modes must exist (Grosch & Salwen, 1978, and Salwen & Grosch, 1981). Later, Miklavčič & Williams (1982) and Miklavčič (1983) proved rigorously that, if the mean profile decays exponentially to a constant in the freestream, then only a finite number of eigenvalues exists for any finite Reynolds number while, if the mean profile decays algebraically, then there exists an infinite discrete set of eigenvalues (see also the discussion by Herron, 1987). In the first case, a continuum must exist for a complete set to span the solution space, while in the latter case no continuum exists. Since the continuum is an important concept for receptivity, we present the ideas in the last section of this chapter.

3.2 Channel Flows

Historically speaking, the study of channel flows has been an inspiration and a challenge to generations of applied mathematicians. By definition, a channel flow is confined by two walls, and therefore the boundary conditions are applied at two finite values of y. As a result, the problem is well posed and the mathematical analysis is considerably simplified.

It should be realized that the channel flows we are discussing are those of the fully developed variety in which U is a function of y only. In practice, however, there is a long entrance region beginning with a flow of constant velocity between two thin boundary layers along each wall. The stability of the entrance flow has been studied by Tatsumi (1952), for example.

Eventually the two boundary layers merge and the mean flow becomes truly independent of the variable x (except for the pressure). There are two well-known examples of this system: (a) Poiseuille flow, in which both walls are at rest and the flow is driven by a constant pressure gradient, and (b) Couette flow, in which there is no pressure gradient and the motion of the walls is a parallel flow of one wall with respect to the other.

3.2.1 Plane Poiseuille Flow

The nondimensional plane Poiseuille flow is a parabolic flow due to the presence of a mean streamwise pressure gradient and is defined by

$$U(y) = y(2-y), \quad \frac{dP}{dx} = \text{constant}, \quad \text{for} \quad 0 < y < 2. \tag{3.1}$$

---

[1] See, e.g., Mack (1976) for Blasius boundary layer flow.

This is an exact solution of the system (2.1)–(2.3). We choose U_0, the center-line velocity, and L, the channel half-width, as our units to nondimensional-ize the system. The temporal stability characteristics can be found by solving the Orr–Sommerfeld equation (2.31) subject to the boundary conditions at the walls

$$\phi = \phi' = 0, \qquad \text{at} \qquad y = 0, 2. \tag{3.2}$$

Any function can be represented as the sum of even and odd functions. By solving for the even and odd parts of the solution separately, it is possible to impose even or odd boundary conditions at the centerline of the channel and seek solutions only over half the range. That is, for the even part the odd derivatives are zero at the centerline, and for the odd part the function and even derivatives are zero at the centerline. We shall refer to the even modes as the *Symmetric modes*, and the odd modes as the *Antisymmetric modes*. This classification is with respect to the disturbance velocity $\tilde{u}$; see (2.19) and (2.30). Some authors classify the modes with respect to the disturbance velocity $\tilde{v}$; i.e., Symmetric modes have the function and the even derivatives set to zero, while Antisymmetric modes have the odd derivatives set to zero at the centerline of the channel. In this book we shall generally follow the first classification scheme. This lack of conformity among various authors can cause confusion and the student is advised to exercise much care when reading research articles and books. At the centerline of the channel

$$\text{Symmetric Mode:} \quad \phi' = \phi''' = 0, \qquad \text{at} \qquad y = 1,$$

$$\text{Antisymmetric Mode:} \quad \phi = \phi'' = 0, \qquad \text{at} \qquad y = 1.$$

To solve the Orr–Sommerfeld equation, the Compound Matrix scheme by Ng & Reid (1979) is chosen because of its ease in implementation. The essen-tial idea is to rewrite the Orr–Sommerfeld equation as a system of equivalent first-order equations, integrate from $y = 1$ to $y = 0$ using a standard Runge–Kutta constant-step size integrator, and then determine the eigenvalue c for a fixed value of α and Re that satisfies the wall boundary conditions to within a specified tolerance. The details of the method are outlined in Appendix B. This type of procedure is usually called a *shooting method*, since one guesses the eigenvalue, shoots (i.e., integrates forward) to the other end of the domain to see if it satisfies the proper boundary condition, and then updates the guess using a root-finding technique. The condition that must be satisfied is usually called the discriminate D. Unfortunately, the shooting method only determines one eigenvalue at a time, and the initial guess usually needs to be very close to the exact eigenvalue, which is not known a priori. A more systematic procedure for finding all of the eigenvalues in a given region of the complex c-plane is to compute D at enough points so that contours of $D_r = 0$ and $D_i = 0$ can be drawn (see, for example, Mack, 1976). Intersection points of the curve $D_r = D_i = 0$ provides the eigenvalues. This method can be impracticable if there exists a large number of contours, but does have the advantage that all eigenvalues will be found within the given domain. The eigenvalues determined in this manner are usually only accurate to a few significant digits, but can be used as the initial guess of a root-finding procedure, which then computes each eigenvalue to within a prescribed tolerance (often termed *polishing*).

The above search algorithm described in the preceding paragraph is applied for the Symmetric modes of the plane Poiseuille flow; the Antisymmetric modes are known to be stable.[2] Figure 3.1 shows the contour curves of $D_r = 0$ and $D_i = 0$ in the complex c-plane; the intersection points are identified by circles. These initial guesses are then substituted into the Orr–Sommerfeld solver to gain better accuracy. The eigenvalues for the Symmetric mode are displayed in Fig. 3.2 and listed in Table 3.1. The classification scheme is the one suggested by Mack, that three families of modes exist. The A and P families are finite, and the S family has an infinite number of stable modes. By varying the wavenumber α, the maximum growth rate curve can be traced out in the (ω_i, α)-plane, where $\omega_i = \alpha c_i$ is the temporal growth rate. The growth rate curve for $Re = 10,000$ is shown in Fig. 3.3. This method was also used to identify the neutral stability boundary in the (c_r, Re)-plane, shown in Fig. 3.4. The region inside the boundary corresponds to instability while the region outside the boundary corresponds to stability. Note that there exists a minimum Reynolds number, called the critical Reynolds number Re_{crit}, for which all Reynolds numbers below this value correspond to flows that are stable to all infinitesimal disturbances. The critical Reynolds number has been computed by a number of authors, and the accepted value is $Re_{crit} = 5772.22$ with corresponding values $\alpha_{crit} = 1.02056$ and $c_{crit} = 0.2640$ (Orszag, 1971). Degeneracies, eigenmodes of order two, are discussed by Koch (1986) and Shanthini (1989).

3.2.2 Plane Couette Flow

For plane Couette flow the mean dimensional velocity varies linearly between two plates moving in opposite directions and with equal speed, U_0, with one plate located at $y = 0$ and the other plate located at $y = 2L$. Using the speed U_0 and the half-width of the channel L to nondimensionalize the system, we get

$$U(y) = y - 1, \quad P = \text{constant}, \quad \text{for} \quad 0 < y < 2, \qquad (3.3)$$

[2] In the terminology of Mack (1976) the modes are Antisymmetric and Symmetric, respectively.

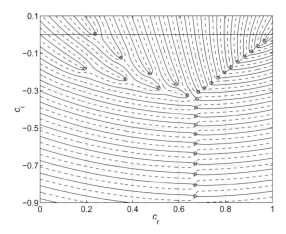

Figure 3.1 Zero contour lines of the discriminate of plane Poiseuille flow at $\alpha =$ 1 and $Re = 10,000$; Symmetric Mode; $D_r = 0$ solid; $D_i = 0$ dash. The circles denote intersection points where $D_r = D_i = 0$. (After Mack, 1976, reproduced with permission.)

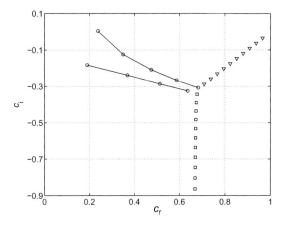

Figure 3.2 Distribution of eigenvalues of plane Poiseuille flow at $\alpha = 1$ and $Re = 10,000$; Symmetric Mode. $\bigcirc$, A family; $\bigtriangledown$, P family; $\odot$, S family. (After Mack, 1976, reproduced with permission.)

and an exact solution of (2.1)–(2.3). Since $U'' = 0$ everywhere, we immediately notice from (2.24) that there is no production of vorticity fluctuations. In fact, the vorticity fluctuations are merely transported and diffused. Therefore, this flow is completely stable for all small disturbances. This is borne out in Fig. 3.5 where we plot the contour curves of $D_r = 0$ and $D_i = 0$, similar to

Table 3.1 *First 30 eigenvalues of plane Poiseuille flow at* $\alpha = 1$ *and* $Re = 10,000$; *Symmetric Mode.*

Mode	c_r	c_i	Mode	c_r	c_i
1	0.23753	0.00374	16	0.51292	−0.28663
2	0.96464	−0.03519	17	0.70887	−0.28766
3	0.93635	−0.06325	18	0.68286	−0.30761
4	0.90806	−0.09131	19	0.63610	−0.32520
5	0.87976	−0.11937	20	0.67764	−0.34373
6	0.34911	−0.12450	21	0.67451	−0.38983
7	0.85145	−0.14743	22	0.67321	−0.43580
8	0.82314	−0.17548	23	0.67232	−0.48326
9	0.19006	−0.18282	24	0.67159	−0.53241
10	0.79482	−0.20353	25	0.67097	−0.58327
11	0.47490	−0.20873	26	0.67043	−0.63588
12	0.76649	−0.23159	27	0.66997	−0.69025
13	0.36850	−0.23882	28	0.66957	−0.74641
14	0.73812	−0.25969	29	0.66923	−0.80439
15	0.58721	−0.26716	30	0.66894	−0.86418

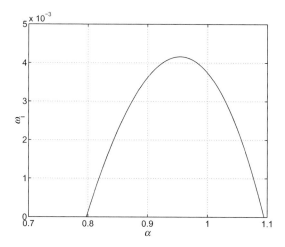

Figure 3.3 Temporal growth rate $\omega_i = \alpha c_i$ as a function of α for plane Poiseuille flow at $Re = 10,000$; Symmetric Mode.

Fig. 3.1 for plane Poiseuille flow. Since $U(y)$ varies between ± 1 so does the phase speed c_r. Note that all eigenvalues lie below the $c_i = 0$ line indicating stable solutions. Indeed, Romanov (1973) proved theoretically that all modes are stable. There is, however, little doubt that this flow will in reality become

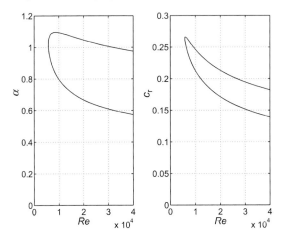

Figure 3.4 Neutral stability curve in the (α, Re)- and (c_r, Re)-planes, respectively, for plane Poiseuille flow.

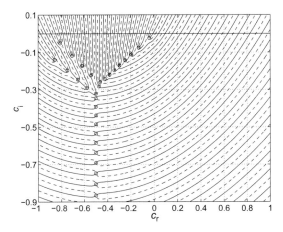

Figure 3.5 Zero contour lines of the discriminate of plane Couette flow at $\alpha = 1$ and $Re = 10,000$; Symmetric Mode; $D_r = 0$ solid; $D_i = 0$ dash. The circles denote intersection points where $D_r = D_i = 0$.

turbulent at large enough Reynolds number. We discuss one possible mechanism for this in Chapter 8.

3.2.3 Generalized Channel Flow

The plane Poiseuille and Couette flows can be combined to yield a new flow, given by

$$U = Ay(2-y) + B(y-1), \quad P = c_1 + c_2 x, \quad \text{for} \quad 0 < y < 2, \quad (3.4)$$

an exact solution of (2.1)–(2.3) (Deardorff, 1963), where c_1 and c_2 are constants. This two parameter family can be reduced to a one-parameter family by requiring the maximum value of the mean velocity U to be 1, leading to the constraint

$$B = \begin{cases} 1 & 0 \le A \le 1/2, \\ 2\sqrt{A(A-1)} & 1/2 \le A \le 1. \end{cases} \quad (3.5)$$

The temporal stability characteristics have been examined by Potter (1966), Reynolds and Potter (1967) and Hains (1967). When $A = 0$, $B = 1$ the profile corresponds to plane Couette flow and is stable. Thus, the critical Reynolds number is simply $Re_{\text{crit}} = \infty$. For $A = 1$, $B = 0$ the profile corresponds to plane Poiseuille flow, and the critical Reynolds number is $Re_{\text{crit}} = 5,772$. Consequently, there exists an intermediate value of B, say B^*, for which the profile first becomes unstable; the region of instability then corresponds to $0 < B < B^*$ and occurs at $B^* = 0.341, A^* = 0.970$, which shows that only a modest component of plane Couette flow is sufficient to completely stabilize plane Poiseuille flow.

3.3 Blasius Boundary Layer

The Blasius boundary layer profile is given by

$$U^\dagger = U_0 f'(\eta), \quad P^\dagger = \text{constant}, \quad \eta = y^\dagger \sqrt{\frac{U_0}{v^\dagger x^\dagger}}, \quad (3.6)$$

where

$$2f''' + ff'' = 0, \quad 0 \le \eta < \infty, \quad (3.7)$$

subject to the boundary conditions

$$f(0) = f'(0) = 0, \quad f'(\infty) = 1. \quad (3.8)$$

Here, $\dagger$ denotes a dimensional quantity, and U_0 is the value of the freestream velocity. Then the Reynolds number is defined as $Re = U_0 L / v^\dagger$. No known analytical solutions exist and therefore the mean profile must be determined numerically. Although there are several methods from which to choose, the

simplest method involves integrating (3.7) as an initial value problem starting
with $f(0) = f'(0) = 0$, $f''(0) = a$. The value of a is determined by requiring
$f'(\infty) = 1$, and is found as part of an iteration procedure. The value of "∞" is
taken to be 13, since then $f'(13)$ and $f''(13)$ deviate less than 10^{-9} from the
asymptotic values of 1 and 0, respectively. Using a Runge–Kutta constant-step
size integrator, it is found that $a = 0.332057336$.

With the Blasius boundary layer profile known at least numerically, the stability can be examined. The stability characteristics of this flow have a long history, beginning with approximations based on asymptotic methods, followed
by a number of publications using machine calculations (for a brief history see,
for example, Drazin & Reid, 1984, or White, 1974). Earlier results were shown
in Fig. 1.6. Today the stability of the Blasius boundary layer is so routine that
we only state the main results.

Mack has shown numerically that only a finite number of eigenvalues exist
at any given Reynolds number and that this number increases as the Reynolds
number increases with only one eigenvalue surviving in the limit of inviscid
flow. The phase speed of all the other eigenvalues save this one approaches
zero (Mack, 1976). Here we reproduce a subset of his results.

The Compound Matrix method is modified to account for the unbounded
domain by integrating the Orr–Sommerfeld equation, beginning in the far-field
with appropriate asymptotic conditions, and then integrating in toward $y = 0$
using a Runge–Kutta constant-step size integrator (Ng & Reid, 1980). The
eigenvalue c is then iterated on until the discriminate D at $y = 0$ is satisfied
to within a prescribed tolerance or, if more than one eigenvalue is required,
we compute D at enough points in a given region of the complex c-plane so
that contours of $D_r = 0$ and $D_i = 0$ can be drawn. Figure 3.6 shows the contour curves for $\alpha = 0.179$ and $Re = 580$. The intersection points $D_r = D_i = 0$
(identified by circles) identify all the discrete eigenvalues in the given domain.
These discrete eigenvalues are listed in Table 3.2 and compare favorably with
those of Mack (1976). In addition to the finite discrete set of eigenvalues, there
also exists a continuum located along the $c_r = 1$ line (see equation 3.26). The
existence of this continuum will be explained in more detail in Section 3.6.
Figure 3.7 plots the growth rate $\omega_i = \alpha c_i$ as a function of the wavenumber α
at $Re = 580$. Figure 3.8 plots the corresponding amplitude of the normalized
eigenfunction $\hat{u} = \phi'$ for $\alpha = 0.179$.

Figure 3.9 is a plot of the neutral stability boundary for the Blasius boundary layer flow. Instead of plotting the neutral stability boundary in the (α, Re)-
plane, say, it is customary to plot the neutral stability boundary in either the
$(\alpha_\delta, Re_\delta)$-plane or the (c_r, Re_δ)-plane, respectively, where $\delta = 1.72078764$

Table 3.2 *Discrete eigenvalues of Blasius boundary layer flow at* $\alpha = 0.179$
and Re = 580.

Mode	c_r	c_i
1	0.36412269	0.00795979
2	0.28971430	−0.27686644
3	0.48392943	−0.19206759
4	0.55719662	−0.36534237
5	0.68626129	−0.33077057
6	0.79365088	−0.43408876
7	0.88736417	−0.41474479

is the displacement thickness.[3] Here, we have the relationships $Re_\delta = \delta Re$
and $\alpha_\delta = \delta\alpha$. The critical Reynolds number and wavenumber are found to
be $Re_{\mathrm{crit}} = 301.641$, $\alpha_{\mathrm{crit}} = 0.1765$; i.e., $Re_{\delta,\mathrm{crit}} = 519.060$, $\alpha_{\delta,\mathrm{crit}} = 0.30377$
(Davey, 1982). For a further discussion of the temporal stability of the Blasius
boundary layer flow at even larger values of the Reynolds number (up to 10^6),
see Davey (1982) and Healey (1995).

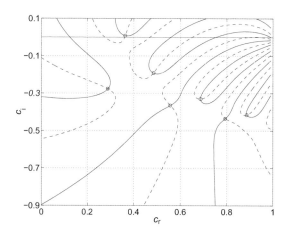

Figure 3.6 Zero contour lines of the discriminate of Blasius boundary layer flow
at $\alpha = 0.179$ and $Re = 580$; $D_r = 0$ solid; $D_i = 0$ dash. The circles denote intersection points where $D_r = D_i = 0$. (After Mack, 1976, reproduced with permission.)

---

[3] The displacement thickness is defined by $\delta = \int_0^\infty (1 - f')d\eta$. If the scaling $\eta = y^\dagger \sqrt{U_0/2\nu^\dagger x^\dagger}$
is employed, then the Blasius boundary layer equation becomes $f''' + ff'' = 0$, and we would
choose $\delta = 1.72078764/\sqrt{2} = 1.2167806$ as the scaling parameter.

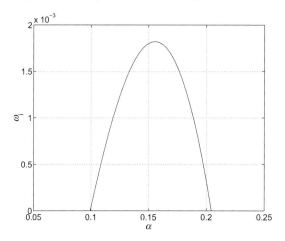

Figure 3.7 Temporal growth rate $\omega_i = \alpha c_i$ as a function of α for Blasius boundary layer flow at $Re = 580$.

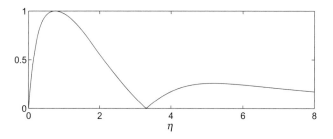

Figure 3.8 Amplitude of the normalized eigenfunction $\hat{u} = \phi'$ as a function of η for Blasius boundary layer flow; Mode 1 of Table 3.2, $Re = 580$, $\alpha = 0.179$.

3.4 Falkner–Skan Flow Family

The more general Falkner–Skan boundary layer flow family is given by

$$2f''' + ff'' + \beta(1 - f'^2) = 0, \tag{3.9}$$

subject to the boundary conditions

$$f(0) = f'(0) = 0, \ f'(\infty) = 1, \tag{3.10}$$

with the parameter β a measure of the pressure gradient (see, e.g., Acheson, 1990 or White, 1974). The value of β can range from $\beta = -0.19884$ (flow separation), to $\beta = 0$ (Blasius), to 1.0 (two-dimensional stagnation-point profile). A plot of the mean profile $U(\eta)$ is shown in Fig. 3.10.

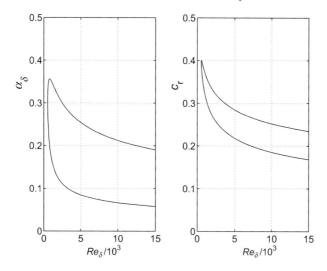

Figure 3.9 Neutral stability curve in the $(\alpha_\delta, Re_\delta)$ and (c_r, Re_δ)-planes, respectively, for Blasius boundary layer flow.

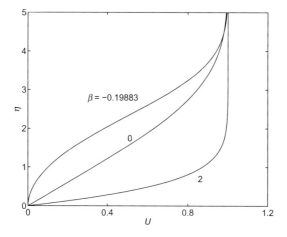

Figure 3.10 Plot of the mean profile as a function of β for the Falkner–Skan flow family.

The temporal stability characteristics of the Falkner–Skan family was considered by Wazzan, Okamura & Smith (1968) and Obremski, Morkovin & Landahl (1969). Rather than generate a myriad number of stability plots as the pressure gradient parameter β is varied, it is more conventional to make a single plot that summarizes the main findings. This is done by computing

Table 3.3 *Critical Reynolds number as a function of β for the Falkner–Skan family (Wazzan, Okamura & Smith, 1968; the value at $\beta = +\infty$ is from Drazin & Reid, 1984).*

β	Re_δ	Re_θ
$+\infty$	21675	10473
1.0	12490	5636
0.8	10920	4874
0.6	8890	3909
0.5	7680	3344
0.4	6230	2679
0.3	4550	1927
0.2	2830	1174
0.1	1380	556
0	520	201
−0.05	318	119
−0.1	199	71
−0.14	138	47
−0.1988	67	17

the critical Reynolds number for each value of β, and then plotting the critical Reynolds number against the shape function $H = \delta/\theta$, where δ is the displacement thickness and θ the momentum thickness.[4] The value of H is unique for each value of β. The critical Reynolds numbers are given in Table 3.3 as a function of β, and graphed in Fig. 3.11. Note that, for values of $-0.19884 < \beta < 0$, the profile has an inflection point within the flow region and may be inviscidly unstable.

3.5 Unbounded Flows

For unbounded flows, such as jets, wakes and mixing layers, where the mean profile has an inflection point, it is known that the largest growth rate is inviscid in nature, and that viscosity is only a dampening effect. For example, Betchov & Szewczyk (1963) have shown for the mixing layer that the effects of viscosity are felt below a Reynolds number of approximately 50, and that no critical value of the Reynolds number exists (see Fig. 3.12). For jets, Tatsumi & Kakutani (1958) and Kaplan (1964) found that viscosity has a stabilizing influence, beginning below a Reynolds number of 100. For this flow a critical Reynolds number does exist and was calculated to be approximately 4 with a wavenum-

---

[4] The momentum thickness is defined by $\theta = \int_0^\infty (1 - f')f'd\eta$.

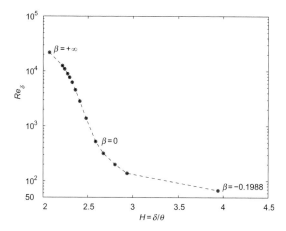

Figure 3.11 Plot of the critical Reynolds number Re_δ against $H = \delta/\theta = Re_\delta/Re_\theta$ for the Falkner–Skan flow family.

ber of 0.2. The same general observations are true for the wake profile. Because the largest growth rate is often of most interest, stability calculations for free shear layers are almost always restricted to solving Rayleigh's equation, and so such cases will not be considered further here because sufficient examples have been solved in Chapter 2.

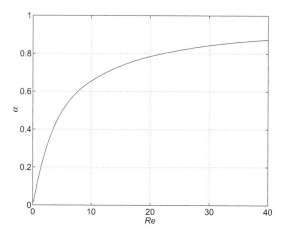

Figure 3.12 Plot of the neutral stability boundary for the mixing layer $U(y) = \tanh(y)$ in the (α, Re) plane. (After Betchov & Szewczyk, 1963, with the permission of AIP Publishing.)

3.6 Discrete and Continuous Spectra

For profiles on a bounded domain, DiPrima & Habetler (1969) have shown that there exists an infinite set of discrete temporal modes of the Orr–Sommerfeld equation, and that this set is complete. Since the normal modes span the solution space, any initial disturbance can be expanded to this set. Thus the complete solution can be described in terms of normal modes. For unbounded domains, general completeness theorems do not exist. However, Miklavčič & Williams (1982) and Miklavčič (1983) did prove rigorously that if the mean profile decays exponentially to a constant in the freestream ($U \to 1 + O(e^{-ay})$, $a > 0$), then only a finite number of eigenvalues exists for any finite Reynolds number, while if the mean profile decays algebraically ($U \to 1 + O(y^{-a})$, $a > 0$), then there exists an infinite discrete set of eigenvalues (see also the discussion by Herron, 1987). In the first case, a continuum must exist for a complete set to span the solution space, while in the latter case no continuum exists.

The theory gives a solid foundation to previous numerical work, where it was shown that for various profiles that decay exponentially in the freestream, such as the boundary layer, the mixing layer and the jet and wake profiles, only a finite number of discrete modes exist. Since a finite set of modes on the unbounded domain are not complete, they cannot be used to describe an arbitrary initial disturbance. Therefore one must consider the presence of a continuum. Grosch & Salwen (1978) and Salwen & Grosch (1981) have shown that the set consisting of the discrete modes and the continuum is complete. Their work also provides the necessary mathematical foundation for the analysis of the receptivity problem, namely how acoustic disturbances or turbulence in the freestream interact with the boundary layer to excite instabilities. Since the work of Grosch & Salwen is of such importance for a proper understanding of the nature of the solution set of the Orr–Sommerfeld equation, we briefly present their analysis here. A review of these papers, and the implications to the receptivity problem, can be found in Hill (1995).

Recall from Section 2.1 that, for a two-dimensional flow, the disturbance velocity components $\tilde{u}$ and $\tilde{v}$ can be expressed in terms of a disturbance streamfunction, $\tilde{\psi}(x,y,t)$, in the usual manner with

$$\tilde{u} = \tilde{\psi}_y, \qquad \tilde{v} = -\tilde{\psi}_x. \tag{3.11}$$

Use of the streamfunction $\tilde{\psi}$ satisfies the continuity equation (2.1) exactly. By substitution of $\tilde{\psi}$ into the momentum equations (2.2) and (2.3) and elimination of the pressure, then the single partial differential equation for $\tilde{\psi}$ can be found to be

$$\left(\frac{\partial}{\partial t} + U\frac{\partial}{\partial x}\right)\nabla^2\tilde{\psi} - U_{yy}\tilde{\psi}_x = Re^{-1}\nabla^4\tilde{\psi}. \tag{3.12}$$

The boundary conditions at the wall are given by

$$\tilde{\psi}_x(x,0,t) = -\tilde{v}(x,0,t) = 0,$$

$$\tilde{\psi}_y(x,0,t) = \tilde{u}(x,0,t) = 0, \tag{3.13}$$

and, for unbounded flows, a finiteness condition at infinity must be imposed, or

$$\int_{-\infty}^{+\infty}\int_0^{+\infty}\left(\tilde{\psi}_x^2 + \tilde{\psi}_y^2\right)dxdy < \infty. \tag{3.14}$$

Physically this means the perturbation energy is finite. Mathematically, this inequality ensures that the Fourier integral expansion of $\tilde{\psi}$,

$$\tilde{\psi}(x,y,t) = \int_{-\infty}^{+\infty}\psi_\alpha(y,t)e^{i\alpha x}d\alpha \tag{3.15}$$

exists, where α is real, according to temporal stability theory.

We now assume that ψ_α is separable and has the form

$$\psi_\alpha(y,t) = \phi_\alpha(y)e^{-i\omega t}. \tag{3.16}$$

This is tantamount to taking a Laplace transform in time, defined as a complex integral in time and then, when the inversion is made, not only poles but branch cuts are made and hence algebraic behavior in time as well as exponential growth or decay due to poles can occur. Here, ϕ_α is the solution to the Orr–Sommerfeld equation

$$(U - c)(\phi'' - \alpha^2\phi) - U''\phi = (i\alpha Re)^{-1}(\phi^{iv} - 2\alpha^2\phi'' + \alpha^4\phi), \tag{3.17}$$

and ω is the complex eigenvalue. The discrete eigenvalues ω_n and the corresponding eigenfunctions ϕ_{α_n} satisfy the Orr–Sommerfeld equation with boundary conditions

$$\phi_{\alpha_n} = \phi'_{\alpha_n} = 0 \quad \text{at} \quad y = 0,$$

$$\phi_{\alpha_n} = \phi'_{\alpha_n} \to 0 \quad \text{as} \quad y \to \infty. \tag{3.18}$$

For a particular mean flow the number of discrete modes, $N(\alpha)$, depends on both the Reynolds number and the wavenumber; $N(\alpha)$ can either be finite or zero.

Since the Orr–Sommerfeld equation is fourth order and linear, there will be four linearly independent solutions $\phi_j(y)$; $j = 1, 2, 3, 4$. The character of each of these solutions can be determined by examining their behavior as $y \to \infty$. By writing

$$\phi_j(y) \approx e^{\lambda_j y}, \qquad \text{as} \qquad y \to \infty, \tag{3.19}$$

and substituting into the Orr–Sommerfeld equation (3.17), we see that

$$\lambda_1 = -Q^{1/2}, \quad \lambda_2 = +Q^{1/2}, \quad \lambda_3 = -\alpha, \quad \lambda_4 = +\alpha, \tag{3.20}$$

where

$$Q = i\alpha Re(U_1 - c) + \alpha^2, \tag{3.21}$$

and it is assumed that $Re(Q) \geq 0$. We have also assumed that $U \to U_1, U', U'' \to 0$ as $y \to \infty$. The constant U_1 is unity for a boundary layer, a mixing layer or a wake, and is zero for a jet. The eigenfunctions ϕ_1 and ϕ_2 are called the viscous solutions since their asymptotic behavior at infinity depends on the Reynolds number while the eigenfunctions ϕ_3 and ϕ_4 are the inviscid solutions. Note that the eigenfunctions ϕ_2 and ϕ_4 are unbounded as y becomes large, and so must be dropped from the solution set. Thus, it is a linear combination of ϕ_1 and ϕ_3, the viscous and inviscid solutions, that must be required to satisfy the two boundary conditions at the wall $y = 0$. In the inviscid limit, the viscous eigenfunction ϕ_1 is not present and the solution to the Rayleigh problem is given only in terms of ϕ_3. For the more general case of an unbounded region $-\infty < y < +\infty$, such as a mixing layer, wake or a jet, we would keep ϕ_3 as the solution that decays as $y \to +\infty$, and keep ϕ_4 as the solution that decays as $y \to -\infty$, and enforce a matching condition at the critical layer.

The continuous part of the solution satisfies the Orr–Sommerfeld equation with boundary conditions

$$\phi_\alpha = \phi_\alpha' = 0 \quad \text{at} \quad y = 0,$$

$$\phi_\alpha, \phi_\alpha' \quad \text{bounded} \quad \text{as} \quad y \to \infty. \tag{3.22}$$

It is important to keep in mind that the difference between the discrete spectra and the continuous spectra is their behavior at infinity; the discrete spectra is required to vanish as $y \to \infty$, while the continuous part is only required to be bounded. We again look for solutions in the far-field by writing

$$\phi_{\alpha_k} \approx e^{\pm iky}, \qquad \text{as} \qquad y \to \infty, \tag{3.23}$$

with $k > 0$ real and positive. Substitution into (3.17) results in the algebraic equation

$$(-k^2 - \alpha^2)\left[-k^2 - \alpha^2 - iRe(\alpha U_1 - \omega) \right] = 0. \tag{3.24}$$

This equation is a linear equation for ω in terms of the real parameter k and hence has one root,

$$\omega_k = \alpha U_1 - iRe^{-1}(k^2 + \alpha^2), \tag{3.25}$$

or, in terms of the phase speed $c = \omega/\alpha$,

$$c = U_1 - i(\alpha Re)^{-1}(k^2 + \alpha^2). \tag{3.26}$$

The temporal continuum branch is shown in Fig. 3.13 for the specific case of Blasius boundary layer flow ($U_1 = 1$) and $\alpha = 0.179$, $Re = 580$. Note that, in addition to the continuum, there are seven discrete eigenvalues, one unstable and six stable. As noted by Mack (1976), as the Reynolds number decreases, one by one the eigenvalues move onto the continuous spectrum until, when at some minimum value of the Reynolds number, there are no discrete eigenvalues.[5] Another way of stating this is, as the Reynolds number increases, new stable modes "pop off" the continuum. At least one of these new modes will move into the upper half of the complex c-plane, giving rise to unstable solutions. This behavior is shown graphically in Fig. 3.14 where we plot the location of the eigenvalues in the complex c-plane as the Reynolds number varies. Note that the number of eigenvalues decreases with the decreasing Reynolds number, and that each eigenvalue disappears as it moves toward the continuum. So, for example, the unstable eigenvalue shown at $Re = 580$ (filled disk) moves into the stable region (defined as $c_i < 0$) between $Re = 350$ and $Re = 250$, then continues to move downward and to the right as Re further decreases ($Re = 150, 50, 25, 5, 4$), until it merges onto the continuum at a Reynolds number slightly less than 4.[6]

The continuum eigenfunction corresponding to ω_k is now given by

$$\psi_{\alpha_k}(y,t) = \phi_{\alpha_k}(y)e^{-i\omega_k t}. \tag{3.27}$$

The general solution in Fourier space is the sum of the discrete spectrum and the continuum, and can be written as

$$\psi_\alpha(y,t) = \sum_{n=1}^{N(\alpha)} \psi_{\alpha_n}(y,t) + \int_0^\infty \psi_{\alpha_k}(y,t)dk$$

$$= \sum_{n=1}^{N(\alpha)} A_{\alpha_n}\phi_{\alpha_n}(y)e^{-i\omega_n t} + \int_0^\infty A_{\alpha_k}\phi_{\alpha_k}(y)e^{-i\omega_k t}dk, \tag{3.28}$$

where the coefficients A_{α_n} and A_{α_k} are found by taking the inner products

---

[5] By contrast, for channel flows where the spectrum is an infinite set of discrete modes, the discrete modes approach zero as the Reynolds number goes to infinity.

[6] At such a low Reynolds number the Blasius boundary layer approximation ceases to be valid, but we ignore this here in order to make our illustration clear.

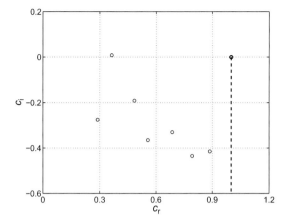

Figure 3.13 Continuum branch (dashed) for Blasius boundary layer flow with $U_1 = 1$, $Re = 580$ and $\alpha = 0.179$. Also shown as circles are the associated discrete eigenvalues.

with respect to the eigenfunctions of the associated adjoint problem. Details and formulas for these coefficients can be found in Grosch & Salwen (1978) and Salwen & Grosch (1981). The complete solution in physical space can be written as

$$\tilde{\psi}(x,y,t) = \int_{-\infty}^{+\infty} \left\{ \sum_{n=1}^{N(\alpha)} A_{\alpha_n} \phi_{\alpha_n}(y) e^{-i\omega_n t} \right.$$

$$\left. + \int_0^\infty A_{\alpha_k} \phi_{\alpha_k}(y) e^{-i\omega_k t} dk \right\} e^{i\alpha x} d\alpha. \qquad (3.29)$$

3.7 Exercises

Exercise 3.1   Develop a numerical code based on the Compound Matrix method (see Appendix B) that solves the generalized fourth-order equation (B.8). Be sure to write the code in double precision (16 significant digits).

(a) Verify the correctness of the new code by reproducing Figs. 3.1 and 3.2 and Table 3.1 for plane Poiseuille flow.
(b) Extend the code to unbounded domains, and verify the code by reproducing Figs. 3.6 and 3.8 and Table 3.2 for Blasius boundary layer flow.
(c) Compute the eigenfunction for plane Poiseuille flow with $Re = 10,000$, $\alpha = 1$ and phase speed c corresponding to the unstable mode.

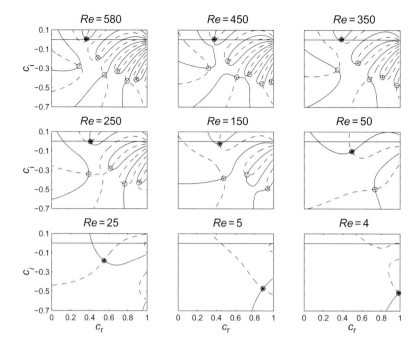

Figure 3.14 Zero contour lines of the discriminate of Blasius boundary layer flow with $U_1 = 1$ and $\alpha = 0.179$ as a function of the Reynolds number; $D_r = 0$ solid; $D_i = 0$ dash. The circles denote intersection points where $D_r = D_i = 0$. The right domain $c_r = 1$, $c_i < 0$ corresponds to the continuum.

(d) Compute the eigenfunction for the Blasius boundary layer with $Re = 580$, $\alpha = 0.179$ and phase speed c corresponding to the unstable mode.

[Note: Verification and Validation are two important aspects of Computational Sciences. Verification asks the question, "Does the code solve the equations correctly?" This is usually answered by comparing the output of the new code to solutions generated by an older code, performing grid-resolution checks, or comparing numerical solutions to simple known analytical solutions. Validation asks the question, "Does the code have the right physics?" or, "Does the model you are trying to solve numerically have the correct equations?" Both aspects, verification and validation, should be kept in mind when dealing with numerical solutions.]

Exercise 3.2 The Asymptotic Suction Profile is an exact solution of the Navier–Stokes equations under the assumptions

$$U(\infty) = U_0, \quad U(0) = 0, \quad V(0) = -V_s,$$

where U_0 is the freestream crossflow velocity, and V_s is the blowing ($V_s < 0$) or suction ($V_s > 0$) parameter.

(a) Assuming the mean profile is a function of y only, deduce the nondimensional solution

$$U = 1 - e^{-y}, \quad V = -Re^{-1}, \quad P = \text{constant},$$

where the reference length is $L = v/V_s$, the reference velocity is U_0, and the Reynolds number is defined as $Re = U_0 L/v = U_0/V_s$.

(b) Using the Orr–Sommerfeld equation (2.31), compute the temporal stability characteristics. In particular, show that the critical Reynolds number is $Re_{crit} = 47{,}047$, with $\alpha_{crit} = 0.1630$ and $c_{crit} = 0.1559$ (see Hughes & Reid, 1965a,b; Drazin & Reid, 1984). Note that e^{-y} has a rather slow decay rate as $y \to \infty$, and so typically one must choose $y_2 = 16$ or larger. Note that $V_{s,crit} = U_0 Re_{crit}^{-1} = 2.13 \times 10^{-5} U_0$, and so only a small fraction of the suction parameter is needed to stabilize the flow when compared to the Blasius boundary layer flow.

(c) Derive, from first principles, the modified Orr–Sommerfeld equation

$$(U - c)(D^2 - \alpha^2)\phi - U''\phi = (i\alpha Re)^{-1} \left[(D^2 - \alpha^2)^2 + (D^2 - \alpha^2)D \right] \phi,$$

for the Asymptotic Suction profile (Hughes & Reid, 1965a,b). Here, $D = d/dy$ and prime denotes differentiation with respect to y.

(d) For the modified Orr–Sommerfeld equation, compute the continuous spectrum and sketch it in the complex c-plane.

(e) For the modified Orr–Sommerfeld equation, compute the temporal stability characteristics. Show graphically that there are a finite number of stable modes, and that the stable modes pop off the continuum as the Reynolds number increases (alternatively, show that the eigenvalues merge toward the continuum as the Reynolds number decreases, starting at a sufficiently large Reynolds number so as to have more than one mode). Finally, show that the critical Reynolds number is $Re_{crit} = 54{,}370$ with $\alpha_{crit} = 0.1555$ and $c_{crit} = 0.150$ (Hocking, 1975).

(f) Comparing the critical Reynolds number obtained in parts (b) and (e), discuss the reason for the differences. What lesson should be learned here?

Exercise 3.3   Consider the Falkner–Skan family (3.9). Using asymptotic expansions, show that in the limit $\beta \to \infty$, the solution of (3.9) is given by

$$U = 3 \tanh^2 \left[\frac{z}{2} + \tanh^{-1} \sqrt{2/3} \right] - 2,$$

where $\eta = z/\sqrt{\beta}$ is the scaled coordinate. Using this profile, compute the temporal stability characteristics using the Orr–Sommerfeld equation (2.31). In particular, show that the critical Reynolds number is $Re_{\delta,\mathrm{crit}} = 21{,}675$, with $\alpha_{\delta,\mathrm{crit}} = 0.1738$ and $c_{\mathrm{crit}} = 0.1841$.

Exercise 3.4 Consider the Falkner–Skan profile (3.9).

(a) Make a plot of the neutral stability boundaries in the (α, Re_δ)-plane for $\beta = 10, 1, 0.5, 0, -0.1, -0.19$.
(b) For the values of β just listed, confirm the critical Reynolds numbers with those listed in Table 3.3.
(c) For the negative values of β, compute the inviscid limit using Rayleigh's equation (2.32).

Exercise 3.5 Consider the constant mean profile

$$U = U_1, \quad V = 0, \quad P = \text{constant}, \quad 0 < y < \infty,$$

which allows slip along a bounding plate at $y = 0$. Although the velocity does not vanish at the plate, assume that the disturbance velocity does.

(a) Show that the general solution to the Orr–Sommerfeld equation (2.31) can be written as

$$\phi = Ae^{-\alpha y} + Be^{+\alpha y} + Ce^{-py} + De^{+py},$$

where $p^2 = \alpha^2 + i\alpha Re(U_1 - c)$.
(b) If $Re(p) > 0$, show that the only solution that satisfies the boundary conditions $\phi(0) = \phi'(0) = 0$ is the trivial solution.
(c) Show that only a continuum exists if $p = ik$, $0 < k < \infty$ is purely imaginary, and deduce the solution

$$\phi_k = A\left[e^{-\alpha y} - \cos(ky) + \alpha k^{-1} \sin(ky)\right],$$

with eigenvalue

$$\omega_k = \alpha U_1 - i(Re)^{-1}(\alpha^2 + k^2).$$

4

Spatial Stability of Incompressible Flows

Equations are more important to me, because politics is for the present,
but an equation is something for eternity.

– Albert Einstein (1879–1955)

4.1 Introduction

So far the discussion has only dealt with situations when the oscillations are periodic in space ($\alpha_i = 0$) and growing, decaying or remaining neutral in time as $e^{\omega_i t}$. In reality, most fluid oscillations have an amplitude that is constant with time but grow, or amplify, in some spatial direction. Examples include boundary layer and free shear flows, such as mixing layers, jets and wakes. Hence, this situation corresponds to equations in which α is complex and the frequency ω is real, with the previous definition $\omega = \alpha c$. It should be noted that if α is complex, then ω cannot be real unless c is also complex. That is, if ω is real, then $\omega = \omega_r$, $\omega_i = 0$, and from (2.27) we have the relations

$$\omega_r = \alpha_r c_r - \alpha_i c_i; \qquad \omega_i = \alpha_r c_i + \alpha_i c_r = 0.$$

From the point of view of mathematical analysis, it is much more convenient to have a complex ω than a complex α because ω appears to first order and α to second or higher order. This becomes even more significant when viscous effects are considered. Furthermore, from a fundamental point of view, we note from the governing equations (2.1) to (2.3), that the time derivative is always first order and has a coefficient of unity. On the other hand, the transition to complex α adds only a few more operations to the program when solved numerically. The searching procedure could be modified in such a way that, given a particular ω, the code will search for the complex α that will satisfy the boundary conditions. Thus it is α that becomes an eigenvalue in place of ω.

It is important to note that, whether the growth occurs in time or in space, the neutral line is the same. Indeed, if $\alpha_i = c_i = 0$, we have $\omega_i = 0$.

In the context of hydrodynamic stability, the first correct treatment of spatial stability theory was done by Gaster (1962, 1965a,b).[1] Prior to 1962, spatial stability calculations were carried out by first computing temporal modes and then relating those to spatial modes by means of the phase velocity; i.e., $\omega_i = -c_r\alpha_i$ (see the review article by Dunn, 1960). Such a simple relation is not valid in general, and very little progress was made as a result. Gaster corrected the relationship by making use of the group velocity instead of the phase velocity. In the first paper, Gaster noted that there exists an asymptotic relation between temporally amplifying and spatially amplifying disturbances. This relation is now referred to as the Gaster transformation in honor of this important contribution to hydrodynamic stability theory, and is presented in the next section. In the second paper, a general discussion was put forth together with a specific example. Gaster's third paper specifically demonstrated that the spatial problem corresponds more correctly to the experiments of Schubauer & Skramstad (1943) for the laminar boundary layer. Since Gaster's contributions, there have been many important contributions to spatial stability theory, and some of these are presented in subsequent sections.

4.2 Gaster's Transformation

The general solution to the Orr–Sommerfeld equation at a fixed Reynolds number yields a dispersion relation between the wavenumber and frequency,[2] given by

$$F(\alpha, \omega) = 0. \tag{4.1}$$

Suppose now that the frequency ω is an analytic function of the wavenumber α, so that $\omega = \omega(\alpha)$. Then, according to the Cauchy–Riemann equations, we have

$$\frac{\partial \omega_r}{\partial \alpha_r} = \frac{\partial \omega_i}{\partial \alpha_i}, \qquad \frac{\partial \omega_r}{\partial \alpha_i} = -\frac{\partial \omega_i}{\partial \alpha_r}. \tag{4.2}$$

---

[1] Historically, the topic of spatial stability was a part of Gaster's thesis but was never published because of little acceptance of the theory at the time. However, the concept of spatial stability was previously noted in the physics literature (see Briggs, 1964Briggs), but apparently unknown to the fluids community until later.

[2] The following discussion is true for any homogeneous, nonconservative system where a dispersion relation is defined.

With these relations integrate over α_i from the S state shown in Fig. 4.1 to the T state, keeping α_r fixed. The S state corresponds to ω real and α complex, while the T state corresponds to ω complex and α real.

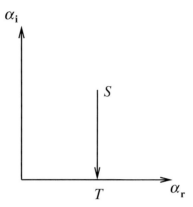

Figure 4.1 Sketch of the integration path in the complex α plane.

Integration of the Cauchy–Riemann equations gives

$$\omega_i \Big|_S^T = \int_S^T \frac{\partial \omega_r}{\partial \alpha_r} d\alpha_i \tag{4.3}$$

and

$$\omega_r \Big|_S^T = -\int_S^T \frac{\partial \omega_i}{\partial \alpha_r} d\alpha_i. \tag{4.4}$$

Equation (4.3) can be simplified to

$$\omega_i(T) = \int_S^0 \frac{\partial \omega_r}{\partial \alpha_r} d\alpha_i, \tag{4.5}$$

because $\omega_i(S) = 0$ and on the right-hand side the T state corresponds to $\alpha_i = 0$. We thus have

$$\omega_i(T) = -\int_0^S \frac{\partial \omega_r}{\partial \alpha_r} d\alpha_i,$$

$$\approx -\frac{\partial \omega_r}{\partial \alpha_r}\Big|_{\alpha_i^*} \alpha_i(S), \tag{4.6}$$

provided the S state is close to zero; i.e., for an S state close to the neutral state where $\alpha_i = 0$ we can expand the integral in a Taylor series about S. Here, α_i^* is any value between zero and S. For dispersive systems we can define a group velocity as

$$c_g = \frac{\partial \omega_r}{\partial \alpha_r},$$

(4.7)

and thus the above equation can be rewritten as

$$\omega_i(T) = -\alpha_i(S)\, c_g.$$

(4.8)

Equation (4.8) is called the Gaster transformation (Gaster, 1962) and relates the growth rate ω_i of temporal calculations to the growth rate $-\alpha_i$ of spatial calculations by means of the group velocity c_g. It is important to keep in mind that this relationship is with respect to the group velocity c_g and not the phase speed c_r.

In a similar fashion, equation (4.4) can be simplified to

$$\omega_r(T) - \omega_r(S) = \int_0^S \frac{\partial \omega_i}{\partial \alpha_r} d\alpha_i.$$

(4.9)

Expanding the integral in a Taylor series, and then taking the maximum of both sides, results in

$$\left| \omega_r(T) - \omega_r(S) \right| \leq \max \left| \frac{\partial \omega_i}{\partial \alpha_r} \right| |\alpha_i(S)| + \cdots.$$

(4.10)

Because the first term of the product on the right-hand side is bounded and the last term of the product is small near a neutral mode, then

$$\omega_r(T) \approx \omega_r(S).$$

(4.11)

Thus, the real part of the frequency for temporal modes is approximately equal to the real part of the frequency for spatial modes, provided one is in a small neighborhood about the neutral line.

One final comment is in order here. Caution must be exercised to ensure that the group velocity is positive when attempting to apply the Gaster transformation for the usual type of flows considered here. Positive group velocities corresponds to waves traveling downstream, in harmony with the assumption that disturbances grow in the downstream direction. Negative group velocities travel upstream, and it would be incorrect to assume in this case that the flow is unstable.

4.3 Incompressible Inviscid Flow

In this section, spatial stability results are presented for various mean profiles of incompressible inviscid flows. These include the mixing layer, jet and wake.

4.3.1 Hyperbolic Tangent Profile

Consider the mixing layer given by the hyperbolic tangent profile

$$U(y) = \frac{1}{2}\left[1 + \tanh y\right]. \tag{4.12}$$

Michalke (1965) first considered the spatial stability for this profile, and much of this section draws upon his work. His primary motivation was that his earlier temporal stability calculations, presented in Subsection 2.6.1, did not agree well with the existing experimental data. In particular, the temporal calculations show that the phase velocity c_r was a constant and hence the wavenumber and frequency were proportional to each other. This was found to be contrary to what was observed experimentally. Thus, it was thought that perhaps spatial calculations would better correlate to the physics. This was indeed the case, as will be shown.

Recall from Subsection 2.6.1 that the neutral mode for the hyperbolic tangent profile is given by

$$\omega_N = 0.5, \quad \alpha_N = 1, \quad c_N = 0.5, \tag{4.13}$$

with the corresponding eigenfunction

$$\hat{v}_N = \operatorname{sech} y. \tag{4.14}$$

For values in the unstable region away from the neutral mode, Rayleigh's equation must be solved numerically. Table 4.1 shows the results of the numerical solution, and Fig. 4.2 plots the spatial growth rate as a function of frequency. We remark here that the phase velocity reported in Table I of Michalke (1965) used the relationship $c_{ph} = \omega/\alpha_r$, a physically meaningful quantity that can be measured in the laboratory. By contrast, the real part of the phase speed is defined as $c_r = (\omega + \alpha_i c_i)/\alpha_r \equiv \omega\alpha_r/(\alpha_r^2 + \alpha_i^2)$, consistent with the definition $\omega = \alpha c$ with ω real and α, c complex. Comparing the two, we see that c_{ph} and c_r are not the same for spatial theory, and caution should be exercised so as not to confuse the two. As mentioned, the phase velocity c_{ph} depends strongly on the frequency, consistent with experimental observations.

Michalke also showed that the eigenfunctions ϕ_r and ϕ_i are neither symmetric nor asymmetric about the origin, as was the case for temporal stability. Thus, the derivatives ϕ_r' and ϕ_i' have zeros away from the critical layer located at $y = 0$, implying that there is flow reversal away from the critical layer, again consistent with experimental observations.

The discussion in this section suggests that spatial stability theory, and not temporal theory, is better able to describe the observations seen in experiments.

As in the temporal case, it is of interest to examine the vorticity distribution

Table 4.1 *Wavenumber and phase speed as a function of the frequency ω for the hyperbolic tangent profile.*

ω	α_r	α_i	c_r	c_i
0.50	1.0	0.0	0.5	0.0
0.40	0.844361	−0.091617	0.468219	0.050804
0.30	0.649548	−0.180225	0.428845	0.118989
0.20	0.382625	−0.227690	0.386014	0.229706
0.10	0.128090	−0.120375	0.414570	0.389600
0	0.0	0.0	0.5	0.5

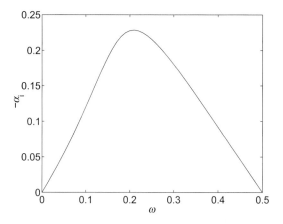

Figure 4.2 Spatial growth rate $-\alpha_i$ as a function of ω for the hyperbolic tangent profile.

since it gives us a means to visualize the dynamics of the mixing layer. Define the spanwise vorticity in the usual fashion as

$$\omega_z = v_x - u_y. \tag{4.15}$$

In terms of the normal mode approach, equation (4.15) becomes

$$\omega_z = -\left[U'(y) + A\tilde{u}_y(x,y,t)\right] + A\tilde{v}_x(x,y,t)$$

$$= -U' + A\left[-\tilde{\psi}_{yy} - \tilde{\psi}_{xx}\right]$$

$$= -U' + A\left[-\phi'' + \alpha^2\phi\right]e^{i\alpha(x-ct)}, \tag{4.16}$$

where A is the initial amplitude of the disturbance and $\tilde{\psi}$ is decomposed according to (2.30). Note, often in the literature ε is used to depict a small amplitude; however, here A is used because later in Chapter 9 it will become finite and the discussion will move to nonlinear disturbances. If we define

$$\tilde{\omega}_{z_r}(y) = \text{Re}(-\phi'' + \alpha^2\phi), \quad \tilde{\omega}_{z_i}(y) = \text{Im}(-\phi'' + \alpha^2\phi), \quad (4.17)$$

then the total vorticity can be written as

$$\omega_z = -\frac{1}{2}\text{sech}^2y + Ae^{-\alpha_i x}\left[\tilde{\omega}_{z_r}\cos(\alpha_r x - \omega t) - \tilde{\omega}_{z_i}\sin(\alpha_r x - \omega t)\right]. \quad (4.18)$$

Note that the vorticity distribution is periodic in time and grows exponentially in the downstream direction. A plot of the normalized eigenfunction ϕ and perturbation vorticity $\tilde{\omega}_z$ are shown in Fig. 4.3, corresponding to the most unstable wave. A contour plot of the vorticity distribution ω_z is shown in Fig. 4.4 at two different times and for $A = 0.0005$. The value of the wavenumber corresponds to the maximum growth rate ($\omega_{\text{max}} = 0.2067$). Since the flow is periodic in time, the period is given by $T = 2\pi/\omega_{\text{max}} = 30.40$. Note that with increasing x, two peaks of vorticity are formed, which will ultimately induce a rotational motion on the base flow, showing the mechanism of spatial instability.

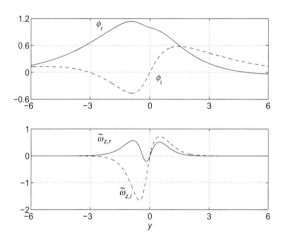

Figure 4.3 Plot of the eigenfunction ϕ (top) and vorticity perturbation $\tilde{\omega}_z$ (bottom) as a function of y for $\omega_{\text{max}} = 0.2067$, $\alpha_r = 0.4031$, $\alpha_i = -0.2284$. The mean flow is given by the hyperbolic tangent profile. (After Michalke, 1965, reproduced with permission.)

4.3.2 Symmetric Jet

The symmetric jet is given by the profile

$$U(y) = \text{sech}^2y, \quad (4.19)$$

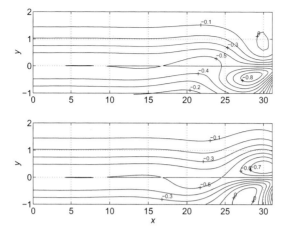

Figure 4.4 Lines of constant vorticity for the case of maximum amplification ω_{max} = 0.2067, $\alpha_r = 0.4031$, $\alpha_i = -0.2284$ at times $t = T$ (top) and $t = 1.5T$ (bottom) with $A = 0.0005$. The mean flow is given by the hyperbolic tangent profile. (After Michalke, 1965, reproduced with permission.)

which can be obtained from a similarity analysis of the boundary layer type equations (cf., White, 1974). Betchov & Criminale (1966) first considered the spatial stability of this profile, and much of this section draws from their work. The inflection point or critical layer is found by setting $U'' = 0$, i.e.,

$$U'' = -2\text{sech}^2 y \left(3\text{sech}^2 y - 2\right) = 0, \tag{4.20}$$

thus,

$$\text{sech}^2 y = \frac{2}{3}, \tag{4.21}$$

with the solution $y_s = \pm 0.6585$. Thus, there are two neutral modes with phase speeds given by $c_N = U(y_s) = 2/3$. The neutral modes with corresponding eigenfunctions are given by

Mode I: $\alpha_N = 2$, $c_N = 2/3$, $\omega_N = 4/3$, $\phi_N = \text{sech}^2 y$,

Mode II: $\alpha_N = 1$, $c_N = 2/3$, $\omega_N = 2/3$, $\phi_N = \sinh y \, \text{sech}^2 y$.

Some authors classify Mode I as the even or symmetric mode since the perturbation streamfunction (and $\hat{v}$) is an even function about the origin, while Mode II is classified as an odd or asymmetric mode. Other authors classify the modes with respect to $\hat{u}$; through continuity we see that this is opposite for that of $\hat{v}$; hence, Mode I would be classified as the odd or asymmetric mode while

Mode II would be classified as the even or symmetric mode. We shall avoid confusion by simply using the notation of Mode I and Mode II.

For values in the unstable region away from the neutral mode, Rayleigh's equation must be solved numerically. The appropriate boundary conditions are

$$\text{Mode I: } \phi'(0) = \phi(\infty) = 0,$$

$$\text{Mode II: } \phi(0) = \phi(\infty) = 0.$$

Note that the boundary condition $\phi = 0$ at $y = -\infty$ is replaced by boundary conditions along the centerline; the boundary condition for Mode I implies that we are searching for an even function about $y = 0$, while the boundary condition for Mode II implies that we are searching for an odd solution. Tables 4.2 and 4.3 shows the results of the numerical solution for each mode, and Fig. 4.5 plots the spatial growth rate as a function of wavenumber, respectively. As in the temporal case, Mode I is the most unstable since it has the largest growth rate.

Table 4.2 *Wavenumber and phase speed as a function of frequency ω for Mode I of the symmetric jet.*

ω	α_r	α_i	c_r	c_i
1.33333	2.0	0.0	0.666666	0.0
1.2	1.871369	−0.029338	0.641084	0.010050
1.0	1.668919	−0.078618	0.597864	0.028164
0.8	1.449709	−0.134110	0.547152	0.050616
0.6	1.203398	−0.194503	0.485895	0.078534
0.4	0.908688	−0.253103	0.408502	0.113783
0.2	0.516928	−0.269983	0.303981	0.158764
0.1	0.270462	−0.206506	0.233571	0.178338
0.0	0.0	0.0	0.0	0.0

4.3.3 Symmetric Wake

The symmetric wake is given by the profile

$$U(y) = 1 - Q \operatorname{sech}^2 y, \tag{4.22}$$

where Q measures the wake deficit. A sketch of the velocity profile is shown in Fig. 4.6. This profile can also be obtained from a similarity analysis of the boundary layer type equations (cf., White, 1974). Note that Q is a parameter of the mean flow and in this sense the stability characteristics will change with changes in the value of Q. Betchov & Criminale (1966) first considered the

Table 4.3 *Wavenumber and phase speed as a function of frequency ω for Mode II of the symmetric jet.*

ω	α_r	α_i	c_r	c_i
0.66667	1.0	0.0	0.666666	0.0
0.6	0.901124	−0.026220	0.665272	0.019357
0.5	0.741506	−0.059703	0.669960	0.053942
0.4	0.569285	−0.077614	0.689814	0.094046
0.3	0.396154	−0.070126	0.734273	0.129978
0.2	0.241420	−0.043023	0.802932	0.143088
0.1	0.110847	−0.014792	0.886360	0.118279
0.0	0.0	0.0	1.0	0.0

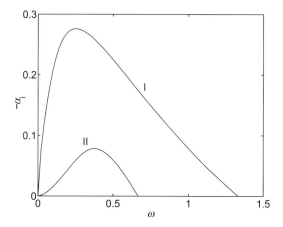

Figure 4.5 Spatial growth rate $-\alpha_i$ as a function of ω for Modes I and II of the symmetric jet.

spatial stability of this profile, and much of this section draws from their work. Before their work is presented, however, note that the $\mathrm{sech}^2 y$, as well as the Gaussian $1 - Q\, e^{-by^2}$, velocity distributions are the far-wake representation of the mean flow. Papageorgiou & Smith (1989) examined the wake stability characteristics of the near-wake region using a corrected mean flow that satisfies the equations of motion for large Reynolds numbers (the wake boundary layer equations of Goldstein (1930)). The nonlinear development was analyzed in Papageorgiou & Smith (1988). The overall picture that emerges for incompressible wakes is that the disturbances grow linearly and two-dimensionally just downstream of the trailing edge of the plate. These two-dimensional disturbances then become nonlinear before three-dimensional effects lead to transition to turbulence. The important point of the work is that the stability of the wake is highly sensitive to the undisturbed flow. Since Gaussian or $\text{sech}^2 y$ profiles are far-wake representations of the mean flow, their range of applicability is limited. It is likely, therefore, that at positions where these profiles can be used rationally the flow is already nonlinear and a substantial history of the evolution is lost.

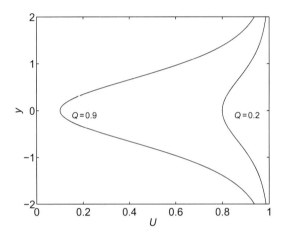

Figure 4.6 Sketch of the symmetric wake profile for $Q = 0.2$ and $Q = 0.9$.

To find the neutral phase speeds first find the value of y_s where $U''(y_s) = 0$. This yields the same locations as in the symmetric jet described earlier this section, and thus $y_s = \pm 0.6585$. Setting $c_N = U(y_s)$ we see that there are two neutral modes with phase speeds given by $c_N = 1 - \frac{2}{3}Q$. The neutral modes with corresponding eigenfunctions are given by

$$\text{Mode I:} \ \alpha_N = 2, \ c_N = 1 - \frac{2}{3}Q, \ \omega_N = \alpha_N c_N, \ \phi_N = \text{sech}^2 y,$$

$$\text{Mode II:} \ \alpha_N = 1, \ c_N = 1 - \frac{2}{3}Q, \ \omega_N = \alpha_N c_N, \ \phi_N = \sinh y \, \text{sech}^2 y.$$

For values in the unstable region away from the neutral mode, Rayleigh's equation must be solved numerically. The appropriate boundary conditions are

$$\text{Mode I:} \ \phi'(0) = \phi(\infty) = 0,$$

$$\text{Mode II:} \ \phi(0) = \phi(\infty) = 0.$$

Tables 4.4 and 4.5 shows the results of the numerical solution for $Q = 0.9$, and Figs. 4.7 and 4.8 plot the spatial growth rate as a function of wavenumber for various values of Q. As in the temporal case, and similar to the symmetric jet profile results, Mode I has the largest growth rate for fixed value of Q.

Table 4.4 *Wavenumber and phase speed as a function of the frequency ω for Mode I of the symmetric wake with $Q = 0.9$.*

ω	α_r	α_i	c_r	c_i
0.8	2.0	0.0	0.4	0.0
0.7	1.669684	−0.398176	0.396682	0.094598
0.6	1.087681	−0.548301	0.439857	0.221732
0.5	0.696734	−0.376460	0.555467	0.300130
0.4	0.487767	−0.235489	0.665050	0.321080
0.3	0.336952	−0.138398	0.761815	0.312903
0.2	0.212425	−0.070076	0.849107	0.280107
0.1	0.102224	−0.023607	0.928719	0.214471
0.0	0.0	0.0	1.0	0.0

Table 4.5 *Wavenumber and phase speed as a function of the frequency ω for Mode II of the symmetric wake with $Q = 0.9$.*

ω	α_r	α_i	c_r	c_i
0.4	1.0	0.0	0.4	0.0
0.3	0.784912	−0.064028	0.379682	0.030972
0.2	0.555653	−0.093738	0.349977	0.059041
0.1	0.312095	−0.083167	0.299171	0.079723
0.0	0.0	0.0	0.15	0.0

Betchov & Criminale (1966) noted that for Mode I, spatial calculations could not be carried out for $0.94 < Q < 1$. To investigate the reason for this they plotted the eigenrelation in the complex (α, ω) plane for $Q = 1$, as shown in Fig. 4.9. From this figure we see the curves of constant α_r (and ω_r) and constant α_i (and ω_i) are orthogonal, implying that α (and ω) is an analytic function of ω (and α), except at some special points where the relationship is singular. A similar analysis for Mode II revealed that no such singularity existed in the eigenrelation. Thus, for Mode I of the symmetric wake, we have:

$0 < Q < 0.94$: singular point does not appear in the spatial calculations;

$Q \approx 0.94$: singular point first appears in the spatial branch:
cusp forms in the eigenrelation;

$0.94 < Q < 1$: the spatial branch can no longer be calculated.

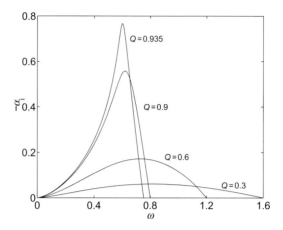

Figure 4.7 Spatial growth rate $-\alpha_i$ as a function of ω for Mode I of the symmetric wake for various values of Q.

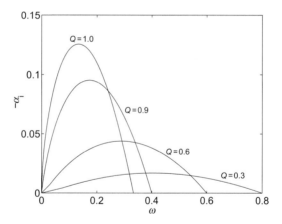

Figure 4.8 Spatial growth rate $-\alpha_i$ as a function of ω for Mode II of the symmetric wake for various values of Q.

We remark here that the singular behavior appears in the upper half plane for $Q < 0.94$, where spatial stability is valid, and then moves down into the lower half plane for $Q > 0.94$.

To investigate the singularity, Betchov & Criminale gave a possible explanation by carrying out the following simple analysis. Assume that ω is an analytic

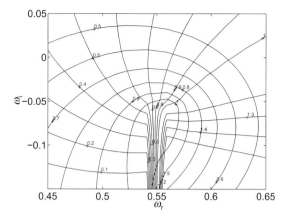

Figure 4.9 Eigenrelation in the (ω) plane for Mode I of the symmetric wake with $Q = 1$.

function of α. Then a Laurent series expansion yields

$$\omega = \omega_o + \left.\frac{d\omega}{d\alpha}\right|_{\alpha_o} (\alpha - \alpha_o) + \frac{1}{2} \left.\frac{d^2\omega}{d\alpha^2}\right|_{\alpha_o} (\alpha - \alpha_o)^2 + \cdots . \tag{4.23}$$

Now assume that the first derivative vanishes at the point α_o

$$\left.\frac{d\omega}{d\alpha}\right|_{\alpha_o} = 0,$$

then the above expansion reduces to

$$\omega = \omega_o + \frac{1}{2} \left.\frac{d^2\omega}{d\alpha^2}\right|_{\alpha_o} (\alpha - \alpha_o)^2 + \cdots . \tag{4.24}$$

Define $\triangle = (\ ) - (\ )_o$, then (4.24) can be rearranged to yield

$$\frac{2\triangle\omega}{d^2\omega/d\alpha^2|_{\alpha_o}} = (\triangle\alpha_r)^2 - (\triangle\alpha_i)^2 + 2i\triangle\alpha_r\triangle\alpha_i. \tag{4.25}$$

Note that both the real and imaginary parts of the right-hand side separately are constant along hyperbolas. Thus there is a saddle point at α_o where the first derivative vanishes. But the first derivative is the definition of the complex group velocity; the saddle point occurs when the complex group velocity vanishes. Defining c_g to be the real part of the group velocity, we have

$$0 < Q < 0.94 \quad c_g > 0 \text{ the group velocity is positive}$$

$$Q \approx 0.94 \quad c_g = 0 \text{ the group velocity first becomes zero}$$

Betchov & Criminale state that *"The occurrence of the singularities was completely unexpected."* They suggested that these singular points had some special significance regarding likely modes of instability, but they were unable to explain in what way the flow was influenced by singularities in the eigenvalue relationships. The appearance of these singularities is the first reported occurrence of this behavior in hydrodynamic stability. Further discussion can be found in Mattingly & Criminale (1972) where experiment as well as calculations were made to confirm this result.

Gaster gave a simple interpretation of the singularities found by Betchov & Criminale in terms of an impulse function at $t = 0$ (Gaster, 1968). This work is of such importance that we outline it in the next section. Today, the concept that singularities or saddle points can develop in the eigenrelation as a physical parameter varies plays a pervasive role in our understanding of global instabilities, feedback mechanisms and, to some extent, flow control.

4.4 Absolute and Convective Instabilities

To explain the significance of the singularities found by Betchov & Criminale (1966) in the dispersion relation of the symmetric wake, Gaster (1968) considered the motion generated by an impulse

$$\tilde{v}(x,0,t) = \delta(x)\delta(t), \qquad (4.26)$$

where δ is the Dirac delta function. Such a disturbance will necessarily excite all modes, and thus any significant irregularities in the dispersion relation will be reflected in the flow. The solution of (2.18), subject to the initial condition (4.26), plus appropriate boundary conditions, will be an integral of traveling wave modes evaluated over all wavenumbers. Starting with this integral representation of the solution, Gaster then makes an asymptotic expansion in the limit as $t \to \infty$ using the method of steepest descent to explain the significance of the singularities. A more rigorous derivation can be found in Huerre & Monkewitz (1985). Below we briefly outline the main points of Gaster and Huerre & Monkewitz.

The solution to the impulse problem is given by

$$\tilde{v}(x,y,t) = \int_{C_\omega} \int_{C_\alpha} \frac{S(\alpha,y,\omega)}{D(\alpha,\omega)} e^{i(\alpha x - \omega t)} d\alpha d\omega, \qquad (4.27)$$

where $D(\alpha,\omega)$ is the complex dispersion relation, $S(\alpha,y,\omega)/D(\alpha,\omega)$ is the solution in Fourier space, and the contours C_α and C_ω are the paths of integration for the inversion integrals in the α and ω planes, respectively (see

Fig. 4.10). In general the dispersion relation D will have a finite number of zeros and branch cuts, giving rise to a discrete spectrum and a continuum (Grosch
& Salwen, 1978; Salwen & Grosch, 1981). For the Fourier inversion to be
valid, we first close the C_ω contour, which must lie above all the zeros and
branch cuts of D, in the upper half plane $\omega_i > 0$, since then $e^{-i\omega t} = e^{-i\omega_r t} e^{\omega_i t}$
decays for $t < 0$; this satisfies causality in that no disturbances can originate
from negative time. For $t > 0$, the contour is closed below C_ω as is shown in
Fig. 4.10(a) and, as already mentioned, the solution can be written as a sum
over all the discrete modes (poles in the complex plane) plus the continuum.
Since for large time the continuum decays, the only contribution that remains
is that corresponding to the pole which has the largest, positive imaginary part,
say $\omega_1(\alpha)$.

For the inversion in space to be valid, the contour must be closed as shown in
Fig. 4.10(b). The contour must be closed below for $x < 0$ since $e^{i\alpha x} = e^{i\alpha_r x} e^{-\alpha_i x}$
must decay, and must be closed above for $x > 0$ for the same reasoning. The
reason why the quarter circle is chosen for the integration path instead of a
semicircle is because $S(\alpha, y, \omega)$ is nonanalytic in α on the imaginary axis
$\alpha_r = 0$. This stems from the fact that the far-field solutions of Rayleigh's
equation are $(D^2 - \alpha^2)\phi \approx 0$, or $\phi \approx e^{-\text{sgn}(\alpha_r)\alpha_r y}$ as $y \to \infty$ and similarly as
$y \to -\infty$. These solutions are not analytic at $\alpha_r = 0$ (see Fig. 4.11). Therefore,
the contour C_α must be restricted to values of α with positive real part.

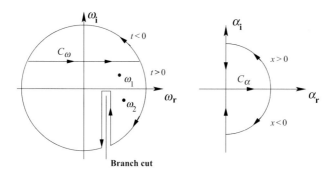

Figure 4.10 Sketch of the integration paths in the complex ω- and α-planes.

Applying the residue theorem to the integral over C_ω gives

$$\int_{C_\omega} \frac{S(\alpha, y, \omega)}{D(\alpha, \omega)} e^{i(\alpha x - \omega t)} d\omega = -2\pi i \frac{S(\alpha, y, \omega_1)}{\partial D/\partial \omega|_{\omega_1}} e^{i(\alpha x - \omega_1 t)}, \qquad (4.28)$$

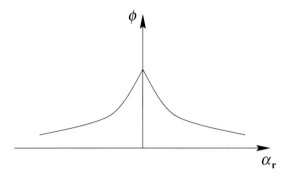

Figure 4.11 Graph of $\phi = e^{-\mathrm{sgn}(\alpha_r)\alpha_r y}$ showing the nonanalytic nature at $\alpha_r = 0$.

and allows the solution to be rewritten as

$$\tilde{v}(x,y,t) = -2\pi i \int_{C_\alpha} \frac{S(\alpha,y,\omega_1)}{\partial D/\partial \omega|_{\omega_1}} \, e^{i(\alpha x - \omega_1 t)} d\alpha. \tag{4.29}$$

The negative sign is because the convention is to close the contour in a counterclockwise fashion.

Examine now the behavior of $\tilde{v}$ as $t \to \infty$. For large values of t an asymptotic expansion of the integral in (4.29) can be obtained by the method of steepest descent, which involves expanding about the saddle point of the exponent where

$$\frac{d}{d\alpha}\left(\alpha x/t - \omega_1 \right)$$

is zero; i.e., the saddle point is given by the relation

$$\frac{d\omega_1}{d\alpha}\bigg|_{\alpha^*} - \frac{x}{t} = 0.$$

Begin by expanding about the saddle point $\alpha = \alpha^*$

$$\omega_1(\alpha) = \omega_1(\alpha^*) + \frac{d\omega_1}{d\alpha}\bigg|_{\alpha^*}(\alpha - \alpha^*) + \frac{1}{2}\frac{d^2\omega_1}{d\alpha^2}\bigg|_{\alpha^*}(\alpha - \alpha^*)^2 + \cdots, \tag{4.30}$$

so that

$$i(\alpha x - \omega_1 t) \approx i\,t\left\{ \alpha^*\frac{x}{t} - \omega_1(\alpha^*) - (\alpha - \alpha^*)\left(\frac{d\omega_1}{d\alpha}\bigg|_{\alpha^*} - \frac{x}{t} \right) \right.$$
$$\left. - \frac{1}{2}\frac{d^2\omega_1}{d\alpha^2}\bigg|_{\alpha^*}(\alpha - \alpha^*)^2 \right\}. \tag{4.31}$$

With the definition of the saddle point, the solution (4.29) becomes

$$\tilde{v} \approx -2\pi i \; \frac{S(\alpha^*, y, \omega_1(\alpha^*))}{\partial D/\partial \omega|_{\omega_1(\alpha^*)}} \; e^{i[\alpha^* x - \omega_1(\alpha^*)t]} \int e^{-\frac{it}{2}\frac{d^2\omega_1}{d\alpha^2}\big|_{\alpha^*}(\alpha-\alpha^*)^2} \, d\alpha$$

$$\approx -2\pi i \; \frac{S(\alpha^*, y, \omega_1(\alpha^*))}{\partial D/\partial \omega|_{\omega_1(\alpha^*)}} \; \frac{\sqrt{\pi} e^{i[\alpha^* x - \omega_1(\alpha^*)t]}}{\sqrt{\dfrac{it}{2}\dfrac{d^2\omega_1}{d\alpha^2}\Big|_{\alpha^*}}}. \tag{4.32}$$

The character of the solution can be described by defining

$$I(x,t) = \sqrt{\frac{2\pi}{t\, d^2\omega_1/d\alpha^2|_{\alpha^*}}} \; e^{\Sigma t}, \tag{4.33}$$

where

$$\Sigma = i\left(\alpha^* \frac{x}{t} - \omega(\alpha^*)\right). \tag{4.34}$$

The values of α^* can be found by satisfying the system of equations

$$\left. \begin{array}{rcl} \dfrac{\partial \omega_r}{\partial \alpha_r}(\alpha^*) &=& \dfrac{x}{t} \\[2mm] \dfrac{\partial \omega_i}{\partial \alpha_i}(\alpha^*) &=& 0, \end{array} \right\} \tag{4.35}$$

which is just the definition of the saddle point restated. Sets of $[\alpha^*, \omega(\alpha^*)]$ can now be found for any x/t. Once these have been determined, the real part of Σ gives the growth rate of the packet; the unstable region corresponds to $\mathrm{Re}(\Sigma) > 0$, the neutral points at $\mathrm{Re}(\Sigma) = 0$ and the stable region to $\mathrm{Re}(\Sigma) < 0$. The values of $c_g = x/t$, which yield $\mathrm{Re}(\Sigma) = 0$ yields neutral rays. The schematic shown in Fig. 4.12 show three possible cases, which are listed as follows.

1. $c_g > 0$ for both the leading and trailing edges of the wave packet. Wave packet moves downstream from the source as time increases. Flow is convectively unstable (see Fig. 4.12(a)).
2. $c_g > 0$ for the leading edge and $c_g = 0$ for the trailing edge of the wave packet. Transition case between a convectively unstable flow to an absolutely unstable flow (see Fig. 4.12(b)).
3. $c_g > 0$ for the leading edge and $c_g < 0$ for the trailing edge of the wave packet. Wave packet moves downstream and upstream from the source as time increases. Flow is absolutely unstable (see Fig. 4.12(c)).

Finally, the real part of I defines the wave packet while $|I|$ defines the envelope. Three-dimensional wave packets can be derived in a similar fashion (Gaster & Davey, 1968).

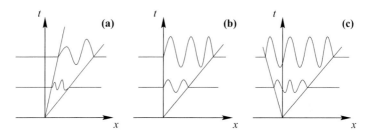

Figure 4.12 Sketch showing (a) convectively, (b) transition and (c) absolutely unstable flow.

4.4.1 Mixing Layer Revisited

We return briefly to the mixing layer of Subsection 4.3.1, but add an additional complexity by means of a parameter β_U. Let

$$U(y) = \frac{1}{2}\left[1 + \beta_U + (1 - \beta_U)\tanh y\right], \qquad (4.36)$$

where the parameter β_U is the velocity ratio defined by the velocity of the freestream at $-\infty$ divided by the velocity of the freestream at $+\infty$. The case $\beta_U > 0$ corresponds to coflow, while if $\beta_U < 0$, the mixing layer has a region of reversed flow. This profile was originally considered by Monkewitz & Huerre (1982), and later by Huerre & Monkewitz (1985).

The neutral mode is given by

$$c_N = \frac{1 + \beta_U}{2}, \quad \alpha_N = 1, \quad \omega_N = \alpha_N c_N. \qquad (4.37)$$

The results of both temporal and spatial calculations are presented in Fig. 4.13 for various values of the velocity ratio β_U. Note that a singularity occurs in the spatial branch as β_U approaches the value -0.135. Thus, the flow changes from being convectively unstable to being absolutely unstable as $\beta_U \to -0.135$.

4.5 Incompressible Viscous Flow

In this section, the spatial stability results are presented for the Blasius boundary layer flow. The asymptotic suction profile and the Falkner–Skan family can be solved in a similar manner and are found at the end of this chapter as exercises.

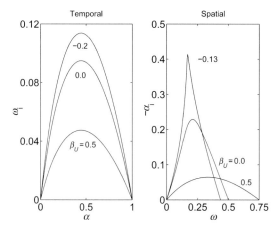

Figure 4.13 Temporal and spatial growth rates for the hyperbolic tangent profile and for various values of β_U.

4.5.1 Spatial Stability

As mentioned in Chapter 3, the Blasius boundary layer profile determined by (3.6) to (3.8) must be determined numerically and, once this is done, the stability characteristics can then be examined. Of course the neutral stability boundary is the same for temporal or spatial theory, so Fig. 3.9 is still relevant. What is needed is to show how Fig. 3.6 and 3.7 are changed when going from temporal to spatial theory.

To compute the spatial stability characteristics for Blasius boundary layer flow, the Compound Matrix method is modified to allow the wavenumber α to be complex while fixing the frequency ω to be real. The eigenvalue that is to be determined is still the phase speed c. Thus, for spatial theory, a (real) value of ω is fixed, and search for values of c, with $\alpha = \omega/c$, which satisfies the Orr–Sommerfeld equation with appropriate boundary conditions.

In Fig. 4.14, contours of $D_r = 0$ and $D_i = 0$ in the complex c-plane are plotted for $Re = 580$ and $\omega = 0.055$.[3] The intersection points are shown as circles, and represent the eigenvalue for which $D_r = D_i = 0$. In particular, three distinct eigenvalues are found (listed in Table 4.6), one unstable and two stable. In addition, a number of eigenvalues are found that seem to lie on a semicircle in the lower half plane. The significance of this is attributed to the presence of a continuum, and will be discussed in more detail in the next section. Finally, in

---

[3] The results presented here are in terms of the values Re and ω; in terms of the displacement thickness, see the discussion in Section 3.3.

Table 4.6 *Wavenumber and phase speed for Re = 580 and ω = 0.055 for the Blasius boundary layer profile.*

α_r	α_i	c_r	c_i
0.15515311	−0.00432824	0.35421288	0.00988133
0.07275631	0.05991724	0.45044992	−0.37096044
0.06729441	0.09298631	0.28092562	−0.38817839

Fig. 4.15 we plot as the continuous curve the spatial growth rate as a function of ω for $Re = 580$. Note that the maximum growth rate occurs near $\omega_{max} \approx 0.05125$.

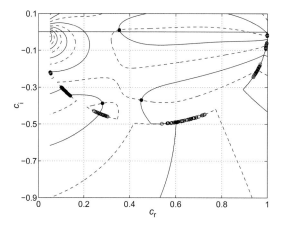

Figure 4.14 Zero contour lines of the discriminate of Blasius boundary layer flow for $\omega = 0.055$ and $Re = 580$; $D_r = 0$ solid; $D_i = 0$ dash. The circles denote intersection points where $D_r = D_i = 0$.

4.5.2 The Gaster Transformation

The Gaster transformation (4.8) can be used to calculate the spatial growth rate from the results of temporal theory. Recall that the spatial growth rate is related to the temporal growth rate via

$$\alpha_i(S) = -\omega_i(T)/c_g, \qquad (4.38)$$

where c_g is the group velocity defined by (4.7). We compute the group velocity numerically as follows. First, fix a value of the Reynolds number, 580 say.

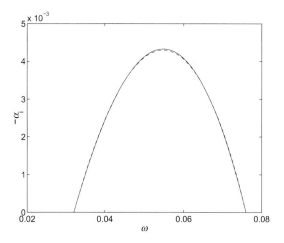

Figure 4.15 Spatial growth rate $-\alpha_i$ for the Blasius boundary layer flow with $Re = 580$; from spatial theory (solid), from Gaster's transformation (dash).

Then, from temporal theory, compute the complex frequency ω given a real value of α that lies on the temporal growth rate curve (see Fig. 3.7). For each value of α on the curve, compute the group velocity as follows:

$$c_g = \frac{\partial \omega}{\partial \alpha} = -\frac{F_\alpha}{F_\omega}, \tag{4.39}$$

where $F(\alpha, \omega) = 0$ is the dispersion relation, and the subscript denotes a partial derivative. The negative sign comes from expanding the dispersion relation about a point (α_0, ω_0),

$$0 = (\alpha - \alpha_0)F_\alpha + (\omega - \omega_0)F_\omega + \cdots, \tag{4.40}$$

and differentiating to get the above result. The derivatives are computed using a standard second-order finite difference scheme,

$$F_\alpha = \frac{F(\alpha + \delta) - F(\alpha - \delta)}{2\delta}, \quad F_\omega = \frac{F(\omega + \delta) - F(\omega - \delta)}{2\delta}, \tag{4.41}$$

and take $\delta = 1.0 \times 10^{-4}$. In general the group velocity, as computed above, will be complex. In applying (4.39), we simply take the real part as the true value for c_g, then substitute into (4.38) to compute $-\alpha_i$. The result of computing the spatial growth rate from temporal theory is shown as the dashed curve in Fig. 4.15. Recall from (4.11) that $\omega_r(T) = \omega_r(S)$. Note the excellent agreement between the two curves, bearing in mind that Gaster's transformation is only a leading-order approximation. For profiles that have an inflection point, the inviscid temporal growth rates are much larger by orders of magnitude than the temporal growth rates of the boundary layer. For this reason, the agreement between the spatial eigenvalues computed using Gaster's transformation and spatial theory is not satisfactory. Gaster's transformation is of historical significance, but today, one simply calculates the spatial growth rate curve directly.

4.5.3 Wave Packets

Insight into the transition process from laminar to turbulent flow can be gained via wave packets. To illustrate this, the wave packet for the Blasius boundary layer is computed using (4.33) to (4.35). In general, both α and ω are complex, except for the single point where the temporal amplification rate is a maximum (e.g., see Fig. 3.7 where $\partial \omega_i / \partial \alpha = 0$ at $\alpha = 0.1554$; the group velocity at this point is $\partial \omega_r / \partial \alpha \approx 0.424$.) This is done numerically as follows. For each (real) value of x/t, find α^* such that $d\omega/d\alpha = x/t$. The value of α^* is found by a root finding procedure, such as Muller's method for complex roots. We guess an initial (complex) value of α, compute the corresponding ω (complex) from the Orr–Sommerfeld equation, use difference formulas to determine $dF/d\alpha$ and $dF/d\omega$, define $c_g = -(dF/d\alpha)/(dF/d\omega)$, and iterate until $c_g - x/t$ is less than some prescribed tolerance. This then defines a set of $\{\alpha^*, \omega(\alpha^*)\}$ pairs which can then be used in (4.33). Figure 4.16 shows the wave packets for four values of x. Note that, as x increases, the amplitude and frequency also increases. Gaster compared the wave packet determined from theory and that from experiments (Gaster & Grant, 1975) and found good qualitative agreement, at least for regions not too far downstream of the initial source disturbance.

4.6 Discrete and Continuous Spectra

For temporal stability, many existence and completeness theorems exist, some of which are mentioned in the previous chapter. We are unaware of any theoretical work about existence or completeness for spatial stability theory, either for a bounded or an unbounded domain (which is still the case since the first edition of this text was published in 2003). For profiles on an unbounded domain, such as the boundary layer, the mixing layer and the jet and wake profiles, all numerical work to date suggests that there are only a finite number of discrete modes; in some cases there may be only one. Since a finite set of modes on the unbounded domain is not complete, they cannot be used to

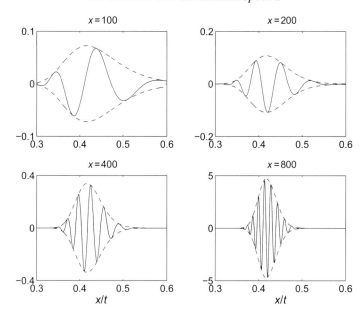

Figure 4.16 Wave packets for $Re_\delta = 1000$.

describe an arbitrary initial disturbance. Therefore one must consider the existence of a continuum. Grosch & Salwen (1978) and Salwen & Grosch (1981) have shown (but not rigorously) that the set consisting of the discrete modes and the continuum is complete. Their work also provides the necessary mathematical foundation for the analysis of the receptivity problem. In short, how do acoustic disturbances or freestream turbulence in the freestream interact with the boundary layer to excite instabilities? Since the work of Grosch & Salwen is of such importance for a proper understanding of the nature of the solution set of the Orr–Sommerfeld equation, we briefly present their analysis next.

Recall from Section 2.1 that, for a two-dimensional flow, the disturbance velocity components $\tilde{u}$ and $\tilde{v}$ can be expressed in terms of a perturbation streamfunction, $\tilde{\psi}(x,y,t)$, in the usual manner with

$$\tilde{u} = \tilde{\psi}_y, \quad \tilde{v} = -\tilde{\psi}_x, \tag{4.42}$$

which satisfies the continuity equation (2.14) exactly. Substitution of $\tilde{\psi}$ into the momentum equations (2.15) and (2.16) and elimination of the pressure results in the single partial differential equation for $\tilde{\psi}$ or

$$\left(\frac{\partial}{\partial t} + U\frac{\partial}{\partial x}\right)\nabla^2\tilde{\psi} - U''\tilde{\psi}_x = Re^{-1}\nabla^4\tilde{\psi}. \tag{4.43}$$

The boundary conditions at the wall are given by

$$\begin{aligned}
\tilde{\psi}_x(x,0,t) &= -\tilde{v}(x,0,t) = 0, \\
\tilde{\psi}_y(x,0,t) &= \tilde{u}(x,0,t) = 0,
\end{aligned} \right\} \tag{4.44}$$

and, at infinity, a finiteness condition must be imposed

$$\int_{-\infty}^{+\infty} \int_{0}^{+\infty} \left(\tilde{\psi}_x^2 + \tilde{\psi}_y^2 \right) dy \, dt < \infty. \tag{4.45}$$

This inequality ensures that the Fourier integral expansion of $\tilde{\psi}$,

$$\tilde{\psi}(x,y,t) = \int_{-\infty}^{+\infty} \psi_\omega(x,y,\omega) e^{-i\omega t} d\omega, \tag{4.46}$$

exists, where ω is real according to spatial stability theory.

Assume that ψ_ω is separable and has the form

$$\psi_\omega(x,y) = \phi_\omega(y) e^{i\alpha x}, \tag{4.47}$$

where α is the complex eigenvalue and the eigenfunction ϕ_ω is the solution to the Orr–Sommerfeld equation

$$(U-c)(\phi_w'' - \alpha^2 \phi_w) - U'' \phi_w = (i\alpha Re)^{-1}(\phi_w'''' - 2\alpha^2 \phi_w'' + \alpha^4 \phi_w). \tag{4.48}$$

The discrete eigenvalues α_n and the corresponding eigenfunctions ϕ_{ω_n} satisfy the Orr–Sommerfeld equation with boundary conditions

$$\begin{aligned}
\phi_{\omega_n} = \phi_{\omega_n}' = 0 \quad &\text{at} \quad y = 0, \\
\phi_{\omega_n} = \phi_{\omega_n}' \to 0 \quad &\text{as} \quad y \to \infty.
\end{aligned} \right\} \tag{4.49}$$

For a particular mean flow, the number of discrete modes, $N(\omega)$, depends on both the Reynolds number and the frequency. As in the temporal problem, $N(\omega)$ can either be finite or zero.

Since the Orr–Sommerfeld equation is fourth order and linear, there will be four linearly independent solutions $\phi_j(y)$; $j = 1, 2, 3, 4$. The character of each of these solutions can be determined by examining their behavior as $y \to \infty$. Writing

$$\phi_j(y) \approx e^{\lambda_j y}, \quad \text{as} \quad y \to \infty, \tag{4.50}$$

and substituting into the Orr–Sommerfeld equation (4.48), we see that

$$\lambda_1 = -Q^{1/2}, \quad \lambda_2 = +Q^{1/2}, \quad \lambda_3 = -\alpha, \quad \lambda_4 = +\alpha, \tag{4.51}$$

where

$$Q = i\alpha Re(U_1 - c) + \alpha^2, \tag{4.52}$$

and it is assumed that $Re(Q) \geq 0$ and $U \to U_1$, U', $U'' \to 0$ as $y \to \infty$. The constant U_1 is unity for a boundary layer, a mixing layer or a wake, and is zero for a jet. The eigenfunctions ϕ_1 and ϕ_2 are called the viscous solutions since their asymptotic behavior at infinity depends on the Reynolds number, while the eigenfunctions ϕ_3 and ϕ_4 are called the inviscid solutions. Note that the eigenfunctions ϕ_2 and ϕ_4 are unbounded as y becomes large, and so must be dropped from the solution set. Thus, it is a linear combination of ϕ_1 and ϕ_3, the viscous and inviscid solutions, that must be required to satisfy the two boundary conditions at the wall $y = 0$. In the inviscid limit, the viscous eigenfunction ϕ_1 is not present and the solution to the Rayleigh problem is given only in terms of ϕ_3. For the more general case of an unbounded region $-\infty < y < +\infty$, such as a mixing layer, wake or a jet, we would keep ϕ_1 and ϕ_3 as the solutions that decay as $y \to +\infty$, and keep ϕ_2 and ϕ_4 as the solutions that decay as $y \to -\infty$, and enforce matching conditions at the critical layer.

The continuum part of the solution satisfies the Orr–Sommerfeld equation with boundary conditions

$$\phi_\omega = \phi_\omega' = 0 \quad \text{at} \quad y = 0, \qquad \phi_\omega, \phi_\omega' \text{ bounded as } y \to \infty. \tag{4.53}$$

It is important to keep in mind that the difference between the discrete spectra and the continuum is the behavior at infinity; the discrete spectra is required to vanish as $y \to \infty$, while the continuum is only required to be bounded. Again, look for solutions in the far-field by writing

$$\phi_{\omega_k} \approx e^{\pm iky}, \qquad \text{as} \qquad y \to \infty, \tag{4.54}$$

with $k > 0$ a real and positive parameter. Substitution of (4.54) into (4.48) results in the algebraic equation

$$(k^2 + \alpha^2)\left[k^2 + \alpha^2 + iRe(\alpha U_1 - \omega)\right] = 0. \tag{4.55}$$

We see that this equation is a quartic equation for α in terms of the real parameter k and hence has four roots $\alpha^{(j)}$; $j = 1, 2, 3, 4$. It should be noted that, whereas the continuum has only one branch for temporal stability theory, the continuum has four branches for spatial stability theory. Two of these four roots, $\alpha^{(1)}$ and $\alpha^{(2)}$, satisfy the quadratic equation

$$k^2 + \alpha^{(j)^2} + iRe\left[\alpha^{(j)} U_1 - \omega\right] = 0, \tag{4.56}$$

or

$$\alpha^{(j)} = \begin{cases} \dfrac{iReU_1}{2}\left\{-1\pm\sqrt{1+4\left(k^2-i\omega Re\right)/\left(Re^2U_1^2\right)}\right\} & U_1 > 0, \\[2em] \pm\sqrt{i\omega Re - k^2} & U_1 = 0. \end{cases}$$
$$\tag{4.57}$$

Order the two roots by defining $\alpha^{(1)}$ with positive real part and $\alpha^{(2)}$ with negative real part. Note that these two roots have branch cuts in the complex α plane. The other two roots to the original quartic are

$$\alpha^{(3)} = ik, \qquad \alpha^{(4)} = -ik. \tag{4.58}$$

The continuum of eigenfunctions corresponding to $\alpha^{(1)}$,

$$\psi_{\omega_k}^{(1)}(x,y)e^{-i\omega t} = \phi_{\omega_k}^{(1)}(y)e^{i(\alpha^{(1)}x - \omega t)}, \tag{4.59}$$

are waves propagating downstream from the source and decay in amplitude as they travel. In the same manner, the continuum eigenfunctions corresponding to $\alpha^{(2)}$ are waves propagating upstream and decaying. The continuum eigenfunctions corresponding to $\alpha^{(3)}$,

$$\psi_{\omega_k}^{(3)}(x,y)e^{-i\omega t} = \phi_{\omega_k}^{(3)}(y)e^{i(\alpha^{(3)}x - \omega t)}$$

$$= \phi_{\omega_k}^{(3)}(y)e^{-kx - i\omega t}, \tag{4.60}$$

are standing waves that decay in amplitude downstream. In the same manner, the continuum of eigenfunctions corresponding to $\alpha^{(4)}$ are standing waves and decay in amplitude upstream.

The four branches for the continuum can be viewed graphically as follows. For the case $U_1 = 1$, we see that as $k \to 0$ with $\omega/Re \ll 1$

$$\alpha^{(1)} \to \omega + ik^2Re^{-1}, \qquad \alpha^{(2)} \to -\omega - iRe - ik^2Re^{-1}, \tag{4.61}$$

and for $k \to \infty$,

$$\alpha^{(1)} \to 0 + ik, \qquad \alpha^{(2)} \to 0 - ik. \tag{4.62}$$

The four continuum branches are sketched in Fig. 4.17. Note that the real axis does not cross any of the branch cuts. The circles denote the limit points as $k \to 0$. We also show the continuum branches, as well as the eigenvalues found earlier in Section 4.5, in Fig. 4.18 for the Blasius boundary layer with $Re = 580$ and $\omega = 0.055$. From this figure we see that the eigenvalues that formed a semicircle in Fig. 4.14 actually approximate the continuum. In this sense they are not discrete and should not be classified as "eigenvalues". Thus, for these particular values of the Reynolds number and frequency, there are only three discrete eigenvalues. This example shows that care must be exercised when computing the eigenvalues numerically. It is important to distinguish between those modes that are discrete and those that approximate the continuum (due to numerics). As in the temporal case, as the Reynolds number increases, more modes will move off the continuum and move into the stable regime.

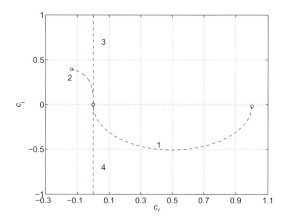

Figure 4.17 Four continuum branches for spatial theory with $U_1 = 1$.

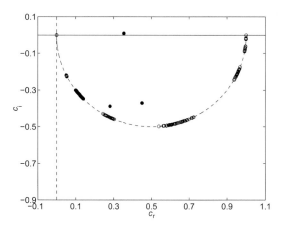

Figure 4.18 Continuum branches for Blasius boundary layer flow with $Re = 580$ and $\omega = 0.055$.

For the case $U_1 = 0$, we see that

$$\alpha^{(1)} \to \frac{1}{\sqrt{2}}(1+i)\sqrt{\omega Re}, \qquad \alpha^{(2)} \to -\frac{1}{\sqrt{2}}(1+i)\sqrt{\omega Re}, \qquad (4.63)$$

as $k \to 0$, and

$$\alpha^{(1)} \to 0 + ik, \qquad \alpha^{(2)} \to 0 - ik, \qquad (4.64)$$

as $k \to \infty$.

The spatial eigenfunctions that form a complete set are formed from the sum of the discrete spectrum and the continuum, and can be written in Fourier space as

$$\psi_\omega(x,y) = \sum_{n=1}^{N(\omega)} \psi_{\omega_n}(x,y) + \sum_{j=1}^{4} \int_0^\infty \psi_{\omega_k}^{(j)}(x,y)dk$$

$$= \sum_{n=1}^{N(\omega)} A_{\omega_n} \phi_{\omega_n}(y)e^{i\alpha_n x} + \sum_{j=1}^{4} \int_0^\infty A_{\omega_k}^{(j)} \phi_{\omega_k}^{(j)}(y)e^{i\alpha_k^{(j)}x}dk, \quad (4.65)$$

where the coefficients A_{ω_n} and $A_{\omega_k}^{(j)}$ are found by taking the inner products with respect to the eigenfunctions of the associated adjoint problem. Details and formulas for these coefficients can be found in both the Grosch & Salwen papers as well as that of Hill (1995). The complete solution in physical space can be written as

$$\tilde\psi(x,y,t) = \int_{-\infty}^{+\infty} \left\{ \sum_{n=1}^{N(\omega)} A_{\omega_n} \phi_{\omega_n}(y)e^{i\alpha_n x}e^{-i\omega t}d\omega \right.$$

$$\left. + \sum_{j=1}^{4} \int_0^\infty A_{\omega_k}^{(j)} \phi_{\omega_k}^{(j)}(y)e^{i\alpha_k^{(j)}x}dk \right\}e^{-i\omega t}d\omega. \qquad (4.66)$$

4.7 Exercises

Exercise 4.1 For inviscid disturbances, modify your numerical code built in Chapter 2 for Rayleigh's equation to allow for spatial stability calculations. Be sure to use double precision. Verify your code by computing the inviscid, spatial stability characteristics for the following flows and obtain the figures in this chapter.

(a) The symmetric jet.
(b) The symmetric wake.

Exercise 4.2 For the Gaussian wake profile

$$U(y) = 1 - Qe^{-y^2 \ln 2},$$

compute the inviscid spatial stability characteristics for various values of the wake deficit parameter Q. Show that the flow becomes absolutely unstable when $Q \geq 0.943$ (see Hultgren & Aggarwal, 1987).

Exercise 4.3 Recall the laminar mixing layer of Subsection 2.6.2, given by

$$2f''' + ff'' = 0,$$

where

$$f'(-\infty) = \beta_U, \quad f(0) = 0, \quad f'(+\infty) = 1.$$

The profile was shown in Fig. 2.10. Compute the spatial growth rate curves for $\beta_U = 0$ and 0.5, and compare to that obtained using the hyperbolic tangent profile (4.36).

Exercise 4.4 For viscous disturbances, modify your numerical code built in Chapter 3 for the Orr–Sommerfeld equation to allow for spatial stability calculations. Be sure to use double precision. Use the code to solve the following problems.

(a) Compute the spatial growth rate for the Falkner–Skan profile of (3.9). Take $\beta = 10, 1, 0.5, -0.1$, and -0.19 with $Re = 1000$. For one of the values of β, compute the temporal growth rate and use Gaster's transformation to compare to the curve obtained using the spatial code.
(b) Compute the continuum and the spatial growth rate for the asymptotic duction profile of Exercise 3.2 using the modified Orr–Sommerfeld equation.

5

Stability of Compressible Flows

I am, and ever will be, a white-socks, pocket-protector, nerdy engineer, born under the second law of thermodynamics, steeped in steam tables, in love with free-body diagrams, transformed by Laplace and propelled by compressible flow.

– Neil Armstrong (1930–2012)

5.1 Introduction

The consideration of flows when the fluid is compressible presents many difficulties. The basic mathematics requires far more detail in order to make a rational investigation. The number of dependent variables is increased because of compressibility. The boundary conditions can be quite involved regardless of the specific mean flow that is under scrutiny. Such observations will become more than obvious as the bases are established for examining the stability of such flows.

By their nature, the physics of compressible flows implies that there are now fluctuations in the density as well as the velocity and pressure, whereby the density can be altered by pressure forces and the temperature. As a result, the laws of thermodynamics must be considered along with the equations for the conservation of mass and momentum. Consequently, we must derive a new set of governing equations. Moreover, this set of equations must be valid for flows that range from slightly supersonic to flows that are hypersonic; i.e., M, the Mach number defined for the flow is of order one or larger. Flows that are characterized by a Mach number that is small compared to one, and that simply have a mean density that is inhomogeneous, are those flows that satisfy what is known as the Boussinesq approximation, and will be examined in Chapter 7.

For now, suffice it to say that it is the force of gravity that plays a key role in such cases.

The stability of compressible flows was initially analytically studied by Lees & Lin (1946), Lees (1947), and Dunn & Lin (1955); a general review of this work has been given by Lin (1955). Later, theoretical work combined with numerical schemes was made by Reshotko (1960) and Lees & Reshotko (1962). A full numerical integration of the compressible stability equations was first achieved by Brown 1961b) and major work using numerical methods in this area is due to Mack (1960, 1965a,b, 1966). Experimentally, evidence for instabilities in compressible boundary layers is due to Laufer & Vrebalovich (1957, 1958, 1960) and Demetriades (1958).

As mentioned, any general study of the stability of a compressible flow presents a most complicated problem because one must consider a significant increase in the number of parameters. For the compressible boundary layer, the mean velocity profile depends upon the conditions at the wall. For example, the wall can be insulated or there can be cooling at the wall, etc. The coefficients of viscosity and thermal conductivity are both functions of temperature. The outer boundary conditions are also intertwined. In this way the basic thermodynamics is now essential to posing any problem in compressible flow. For the most part this means the assumption of a perfect gas for both the mean flow and the perturbations in order to form a closed set of equations for the dependent variables. The basic needs for this approach can be found in the example of Shen (1952), and a more detailed discussion is provided by Mack (1965a). Betchov & Criminale (1967) present the fundamentals.

Due to the complexities inherent with any discussion on the stability of compressible flows, the stability characteristics of the compressible mixing layer and the compressible boundary layer are presented in the following two sections. Other compressible flows not examined here include, but not limited to, the studies of the compressible wake (Papageorgiou, 1990a,b,c; Chen, Cantwell & Mansour, 1989, 1990); the compressible jet (Kennedy & Chen, 1998); and the compressible Couette flow (Glatzel, 1988, 1989; Girard, 1988; Duck, Erlebacher & Hussaini, 1994).

5.2 Compressible Mixing Layer

In this section, we examine the inviscid stability of a compressible mixing layer, which is the interfacial region between two moving gases. The basic formulation of the theory for the stability of compressible shear flows, both free and wall bounded, is due to Lees & Lin (1946). Later, Dunn & Lin (1955) were the first to show the importance of three-dimensional disturbances for the stability of these flows.

Early studies of the stability of compressible mixing layers include those of Lessen, Fox & Zien (1965, 1966) and Gropengiesser (1969). The inviscid temporal stability of the compressible mixing layer to two- and three-dimensional disturbances was studied by Lessen, Fox & Zien for subsonic disturbances (1965) and supersonic disturbances (1966). Lessen *et al.* assumed that the flow was iso-energetic and, as a consequence, the temperature of the stationary gas was always greater than that of the moving gas. In fact, because the ratio of the temperature far from the mixing region varies as the square of the Mach number, the stationary gas is much hotter than the moving gas at even moderately supersonic speeds. Gropengiesser (1969) reexamined this problem without having to use the iso-energetic assumption. Consequentially, he was able to treat the ratio of the temperatures of the stationary and moving gas as a parameter. He carried out inviscid spatial stability calculations for the compressible mixing layer using a generalized hyperbolic tangent profile (see his equation (2.27)) to approximate the Lock profile for temperature ratios of 0.6, 1.0 and 2.0 and for Mach numbers between 0 and 3. Gropengiesser found that, for low and moderate Mach numbers, the flow becomes less unstable as the stationary gas becomes hotter. He also found that the spatial growth rates decrease with increasing Mach number over the range of Mach numbers that he studied. Gropengiesser also found a second unstable mode for two-dimensional waves in a narrow range of Mach numbers, $1.54 < M < 1.73$. Ragab & Wu (1989) recomputed many of the stability results of Gropengiesser, and verified the accuracy of the earlier numerical work.

Blumen (1970), Blumen, Drazin & Billings (1975), Drazin & Davey (1977) and Shivamoggi (1979) all investigated the temporal stability of a compressible mixing layer. A hyperbolic tangent profile was used for the velocity. In the first three studies the temperature was assumed to be a constant throughout the layer, while in the last paper the iso-energetic assumption was made. In the 1975 and 1977 papers, multiple instability modes were found near a Mach number of one. In particular, they showed that the hyperbolic tangent shear layer is unstable at each value of the Mach number, however large. These modes were investigated numerically and theoretically in the long-wave approximation ($\alpha \to 0$). The fourth paper also investigated the long-wave solution.

As a historical note, little else was done until the late 1980s, when a number of theoretical studies on the stability of compressible free shear layers were conducted. The increased interest in such flows was due mainly to the projected use of the scramjet engine for the propulsion of hypersonic aircraft. It is impossible to give a chronological order of events since over 30 archival (and many more non-archival) publications appeared within a five-year time frame. We briefly comment on only a few publications that are directly relevant to the contents of this chapter, namely the inviscid stability of two-dimensional compressible mixing layers to two-dimensional disturbances, and present the rest of the publications as a list.

Earlier studies include that of Jackson & Grosch (1989) who examined the inviscid stability of a compressible two-dimensional mixing layer to two- and three-dimensional disturbances. The mean flow velocity was taken to be a hyperbolic tangent while the temperature was determined using Crocco's relation. The classification of neutral and unstable modes over the Mach number range of zero to ten was determined, thus effectively extending and completing the results of Gropengiesser (1969). In particular, they verified the result that, for subsonic convective Mach numbers, only one subsonic mode existed (except at possible high angles of skewness of the disturbances) and that the growth rate decreased as the Mach number increased. These modes were classified as "subsonic" modes. For supersonic convective Mach numbers, they clarified the second mode found by Gropengiesser (1969), Blumen, Drazin & Billings (1975) and Drazin & Davey (1977). Multiple modes were also found in a temporal stability analysis of the compressible mixing layer without invoking the assumptions of a hyperbolic tangent velocity profile and that of a constant temperature throughout the layer (Macaraeg, Streett & Hussaini, 1988; see also Macaraeg & Streett, 1991). Consider the classification scheme of Jackson & Grosch (1989), which is relevant for supersonic convective Mach numbers when two unstable modes exist. One mode is termed the "Fast" mode, which has a corresponding phase speed greater than

$$c_{\mathrm{N}} = \frac{\beta_U + \beta_T^{1/3}}{1 + \beta_T^{1/3}},$$

where $\beta_U = U_{-\infty}^{\dagger}/U_{+\infty}^{\dagger}$ and $\beta_T = T_{-\infty}^{\dagger}/T_{+\infty}^{\dagger}$ are the velocity and temperature ratios in the freestream at $-\infty$ to that in the freestream at $+\infty$, respectively. The phase speed c_{N} is derived from a vortex sheet analysis (see Subsection 5.2.4), and the $\dagger$ denotes a dimensional quantity. The "Slow" mode has a corresponding phase speed less than c_{N}. The authors also indicated numerically how the two modes come about as the convective Mach number approaches one, which was in general agreement with the theoretical findings of Blumen, Drazin & Billings (1975) and Drazin & Davey (1977). Zhuang, Kubota & Dimotakis (1988) also studied the mixing layer with the hyperbolic tangent profile and found decreasing amplification with increasing Mach number.

Ragab (1988) numerically solved the two-dimensional compressible Navier–
Stokes equations for the wake/mixing layer behind a splitter plate, and then
made a linear stability analysis of the computed mean flow. He found that increasing the Mach number leads to a strong stabilization of the flow and that
the disturbances have large dispersion near the splitter plate and smaller dispersion downstream. Ragab & Wu (1989) examined the viscous and inviscid
stability of a compressible mixing layer using both the hyperbolic tangent and
Sutherland profiles. They found that, if the Reynolds number was greater than
1,000, the disturbances could be calculated very accurately from inviscid theory. In addition, they reported that nonparallel effects are negligible. It seems
that in this study their main interest was in determining the dependence of the
maximum growth rate of the disturbances on the velocity ratio of the mixing
layer. They concluded that the maximum growth rate depends on the velocity
ratio in a complex way, with the maximum growth rate appearing at a particular
nonzero velocity ratio.

Tam & Hu (1989) examined the stability of the compressible mixing layer
in a channel using the hyperbolic tangent profile. They showed that the presence of walls for supersonic convective Mach numbers introduces two new
families of unstable modes. These new modes were classified either as "Class
A" (modes with phase speeds that decrease as the wavenumber and frequency
increase) or "Class B" (modes with phase speeds that increase as the wavenumber and frequency increase).

Additional investigations into the linear stability of compressible mixing
layers include, but are not limited to[1]:

  (i) three-dimensional instabilities (Sandham & Reynolds, 1990);

  (ii) three-dimensional mixing layers (Grosch & Jackson, 1991; Macaraeg, 1991; Lu & Lele, 1993);

  (iii) effect of thermodynamics (Jackson & Grosch, 1991);

  (iv) effect of chemical reactions (Jackson & Grosch, 1990a; Shin & Ferziger, 1991; Planche & Reynolds, 1991; Shin & Ferziger, 1993; Jackson & Grosch, 1994; Day, Reynolds & Mansour, 1998a,b; Papas, Monkewitz & Tomboulides, 1999);

  (v) correlations of the growth rates with the convective Mach number (Ragab & Wu, 1989; Zhuang, Kubota & Dimotakis, 1990a; Jackson & Grosch, 1990b; Lu & Lele, 1994);

  (vi) absolute-convective instabilities (Pavithran & Redekopp, 1989; Jackson & Grosch, 1990b; Hu et al. 1993; Peroomian & Kelly, 1994);

[1] Note here that many of the papers listed can well fit into more than one category; we apologize to the authors for the over-simplification of their work.

(vii) effect of walls (Tam & Hu, 1989; Greenough *et al.* 1989; Macaraeg & Streett, 1989; Zhuang, Dimotakis & Kubota, 1990b; Macaraeg, 1990; Jackson & Grosch, 1990c; Morris & Giridharan, 1991);

(viii) stability of binary gases (Kozusko *et al.* 1996);

(ix) nonhomentropic flows (Djordjevic & Redekopp, 1988);

(x) high Mach number studies (Balsa & Goldstein, 1990; Cowley & Hall, 1990; Goldstein & Wundrow, 1990; Smith & Brown, 1990; Blackaby, Cowley & Hall, 1993);

(xi) the combined effect of a wake with a mixing layer (Koochesfahani & Frieler, 1989).

5.2.1 Mean Flow

Consider the two-dimensional compressible mixing layer with zero pressure gradient, which separates two streams of different speeds and temperatures, and assume that the mean flow is governed by the compressible boundary-layer equations. Let (U,V) be the nondimensional velocity components in the (x,y) directions, respectively, ρ the density and T the temperature. All of the variables are made dimensionless using the magnitudes of the freestream values at $y = +\infty$; i.e., $U_{+\infty}^\dagger$, $\rho_{+\infty}^\dagger$, and $T_{+\infty}^\dagger$. The mean flow equations are first transformed into the incompressible form by means of the Howarth–Dorodnitzyn transformation, namely

$$Y = \int_0^y \rho \, dy, \quad \text{and} \quad \hat{V} = \rho V + U \int_0^y \rho_x \, dy, \tag{5.1}$$

yielding

$$\rho T = 1, \tag{5.2}$$

$$U_x + \hat{V}_Y = 0, \tag{5.3}$$

$$U U_x + \hat{V} U_Y = (\rho \mu U_Y)_Y, \tag{5.4}$$

$$U T_x + \hat{V} T_Y = \left(\frac{\rho \mu}{Pr} T_Y\right)_Y + (\gamma - 1) M^2 \rho \mu U_Y^2. \tag{5.5}$$

Here, the nondimensional viscosity μ is assumed to be a function of temperature. The nondimensional parameters appearing in equations (5.1)–(5.5) are the Prandtl number $Pr = c_p^\dagger \mu^\dagger / \kappa^\dagger$ where $c_p^\dagger$ is the specific heat at constant pressure and $\kappa^\dagger$ is the thermal conductivity, the Mach number $M = U_{+\infty}^\dagger / a_{+\infty}^\dagger$ and the ratio of specific heats γ. Note that the last term in the energy equation is due to viscous heating and has an important effect when the Mach number is large.

Solutions to this system can be found by numerically marching in the x

direction subject to appropriate initial and boundary conditions. Alternatively, one can assume that a self-similar solution exists. Here, the second approach is taken and solutions are sought in terms of the variable

$$\eta - \eta_0 = \frac{Y}{2\sqrt{x}}, \tag{5.6}$$

which is the similarity variable for the chemically frozen heat conduction problem, and η_0 corresponds to a shift in the origin. For the case of both streams being supersonic, η_0 is determined uniquely from a compatibility condition found by matching the pressure across the mixing layer, while if both streams are subsonic, the compatibility condition is trivially satisfied and thus η_0 would remain indeterminate (Ting, 1959; Klemp & Acrivos, 1972).

Using the transformation (5.6), with

$$U = f'(\eta), \quad \hat{V} = (\eta f' - f)/\sqrt{x}, \quad \text{and} \quad T = T(\eta), \tag{5.7}$$

equations (5.1)–(5.5) become

$$\left(\frac{\mu}{T}f''\right)' + 2ff'' = 0 \tag{5.8}$$

and

$$\left(\frac{\mu}{PrT}T'\right)' + 2fT' + (\gamma - 1)M^2\frac{\mu}{T}(f'')^2 = 0, \tag{5.9}$$

subject to the boundary conditions

$$U(\infty) = f'(\infty) = 1, \quad U(-\infty) = f'(-\infty) = \beta_U, \tag{5.10}$$

$$T(\infty) = 1, \quad T(-\infty) = \beta_T, \tag{5.11}$$

where $\beta_U = U_{-\infty}^{\dagger}/U_{\infty}^{\dagger} \leq 1$ and $\beta_T = T_{-\infty}^{\dagger}/T_{\infty}^{\dagger}$ are the velocity and temperature ratios. It should be noted that the system (5.8) and (5.9) constitute a fifth-order boundary-value problem, but there are only four boundary conditions. As stated earlier, a fifth boundary condition can be given if both streams are supersonic. Here, we assume that the shift η_0 is such that the dividing streamline lies at the origin, that is

$$f(0) = 0. \tag{5.12}$$

For a general viscosity law, solutions to the above equations can now be obtained numerically for any given value of the Mach number.

For the special case of the linear viscosity law $\mu = T$ and unit Prandtl number, the energy equation can be solved in closed form

$$T = 1 - (1 - \beta_T)(1 - \psi) + \frac{\gamma - 1}{2}M^2(1 - \beta_U)^2\psi(1 - \psi), \tag{5.13}$$

where

$$
\psi = \begin{cases} (f' - \beta_U)/(1 - \beta_U), & 0 \le \beta_U < 1, \\ (1 + \mathrm{erf}(\eta))/2, & \beta_U = 1. \end{cases}
\tag{5.14}
$$

We further make the simplifying assumption that the mean flow can be modeled by a hyperbolic tangent profile

$$
U = \frac{1}{2}\left[1 + \beta_U + (1 - \beta_U)\tanh(\eta)\right].
\tag{5.15}
$$

In this way, the entire mean flow is known analytically.

5.2.2 Inviscid Fluctuations

In the absence of heat conductivity and viscous effects, the general equations for the velocity, pressure, density and energy are

$$
\frac{\partial \rho}{\partial t} + \frac{\partial(\rho u)}{\partial x} + \frac{\partial(\rho v)}{\partial y} = 0,
\tag{5.16}
$$

$$
\rho\left(\frac{\partial u}{\partial t} + u\frac{\partial u}{\partial x} + v\frac{\partial u}{\partial y}\right) + \frac{1}{\gamma M^2}\frac{\partial p}{\partial x} = 0,
\tag{5.17}
$$

$$
\rho\left(\frac{\partial v}{\partial t} + u\frac{\partial v}{\partial x} + v\frac{\partial v}{\partial y}\right) + \frac{1}{\gamma M^2}\frac{\partial p}{\partial y} = 0
\tag{5.18}
$$

and

$$
\rho\left(\frac{\partial T}{\partial t} + u\frac{\partial T}{\partial x} + v\frac{\partial T}{\partial y}\right) - \frac{\gamma - 1}{\gamma}\left(\frac{\partial p}{\partial t} + u\frac{\partial p}{\partial x} + v\frac{\partial p}{\partial y}\right) = 0.
\tag{5.19}
$$

In addition to these equations, another equation for a perfect gas is

$$
p = \rho T.
\tag{5.20}
$$

In the usual fashion, these equations have been made dimensionless with respect to the freestream values $U^{\dagger}_{+\infty}$, $\rho^{\dagger}_{+\infty}$, $T^{\dagger}_{+\infty}$, $P^{\dagger}_{+\infty} = \rho^{\dagger}_{+\infty}RT^{\dagger}_{+\infty}$ for the velocities, density, temperature and pressure, respectively. The length scale is referenced to L, and the time scale is $L/U^{\dagger}_{+\infty}$. The Mach number is $M = U^{\dagger}_{+\infty}/a^{\dagger}_{+\infty}$, where $a^{\dagger} = \sqrt{\gamma p^{\dagger}/\rho^{\dagger}}$ is the speed of sound and γ is the ratio of specific heats. Our system is now closed, and we have five equations for the five unknowns ρ, u, v, p and T.

Compressible Rayleigh Equation

The flow field is perturbed by introducing two-dimensional wave disturbances in the velocity, pressure, temperature and density with amplitudes that are functions of y. Following the notation used by Lees & Lin (1946), we write using the normal mode approach

$$(u,v,p,T,\rho) = (U,0,1,T,\rho)(y) + (f,\alpha\phi,\Pi,\theta,r)(y)e^{i\alpha(x-ct)}. \qquad (5.21)$$

Substituting (5.21) into (5.16)–(5.20), the linearized equations for the fluctuations are

$$i(U-c)r + i\rho f + (\rho\phi)' = 0, \qquad (5.22)$$

$$\rho\left[i(U-c)f + U'\phi\right] + \frac{i\Pi}{\gamma M^2} = 0, \qquad (5.23)$$

$$i\alpha^2\rho(U-c)\phi + \frac{\Pi'}{\gamma M^2} = 0, \qquad (5.24)$$

$$\rho\left[i(U-c)\theta + T'\phi\right] - i\frac{\gamma-1}{\gamma}(U-c)\Pi = 0, \qquad (5.25)$$

and

$$\Pi = \rho\theta + rT. \qquad (5.26)$$

The usual manipulation of the above set of equations (see the exercise section) results in the following relation for the pressure

$$\Pi'' + \left[\frac{T'}{T} - \frac{2U'}{U-c}\right]\Pi' - \frac{\alpha^2}{T}\left[T - M^2(U-c)^2\right]\Pi = 0, \qquad (5.27)$$

or, in terms of ϕ,

$$\left[\frac{\phi'}{\xi}\right]' - \left[\frac{\alpha^2}{T} + \frac{1}{U-c}\left(\frac{U'}{\xi}\right)'\right]\phi = 0, \qquad (5.28)$$

where

$$\xi = T - M^2(U-c)^2. \qquad (5.29)$$

These two equations are the equivalent counterpart of Rayleigh's equation extended to compressible flows. Although first derived by Lees & Lin (1946), we shall refer to either of the above equations as the "compressible Rayleigh equation" even though Rayleigh did not investigate compressible stability per se. For a discussion in terms of vorticity or pressure, see Betchov & Criminale (1967).

Temporal Theory

Several important theorems concerning the compressible Rayleigh equation are presented in this section. The first result is due to Lees & Lin (1946) and extends Rayleigh's inflection point theorem (see Result 2.1) to compressible flows. The next three results are due to Chimonas (1970) and Shivamoggi (1977), and extend Howard's theorems, Results 2.4, 2.5 and 2.7, to compressible flows. (Result 5.3 was also proved by Blumen (1970).) The bounds presented in the following results may not be the best possible, because they are independent of both the Mach number and the temperature distribution.

Result 5.1 (Generalized Inflection Point Theorem) *A critical layer exists at $U(y_c) - c = 0$, and, provided the disturbances decay exponentially in the freestreams, the solution is regular if the following is true*

$$\frac{d}{dy}\left(\frac{1}{T}\frac{dU}{dy}\right)\bigg|_{y=y_c} = 0, \quad at \ U(y_c) - c = 0,$$

otherwise the solution is singular. Thus, a neutral mode exists and is given by $c_N = U(y_c)$.

Result 5.2 *The phase velocity c_r of an amplified disturbance must lie between the minimum and the maximum values of the mean velocity profile $U(y)$.*

Result 5.3 (Semicircle Theorem) *If $c_i > 0$, then the real and imaginary parts of the phase speed c must lie inside the semicircle*

$$\left(c_r - \frac{U_{\min}+U_{\max}}{2}\right)^2 + c_i^2 \leq \left(\frac{U_{\max}-U_{\min}}{2}\right)^2.$$

Result 5.4 (Upper Bound on the Growth Rate) *If there exists a solution with $c_i > 0$, then an upper bound exists and is given by*

$$\alpha c_i \leq \frac{1}{2}\max\left|U'(y)\right|,$$

where the maximum is taken over the open interval (a,b).

5.2.3 Linear Stability

Formulation

Since the mean flow is given in terms of the similarity variable η, it is convenient to transform the compressible Rayleigh equation (5.27) so that all equations are solved in a single coordinate framework. The compressible Rayleigh equation for the pressure disturbance in the transformed space is given by

$$\Pi'' - \frac{2U'}{U-c}\Pi' - \alpha^2 T\left[T - M^2(U-c)^2\right]\Pi = 0, \tag{5.30}$$

where now $()'$ denote derivatives with respect to η. The boundary conditions for Π are obtained by considering the limiting form as $\eta \to \pm\infty$. The solutions are of the form

$$\Pi \to \exp(\pm\Omega_\pm \eta), \tag{5.31}$$

where

$$\Omega_+^2 = \alpha^2\left[1 - M^2(1-c)^2\right], \qquad \Omega_-^2 = \alpha^2\beta_T\left[\beta_T - M^2(\beta_U - c)^2\right], \tag{5.32}$$

using boundary conditions (5.10) and (5.11). If $\Omega_\pm^2$ are positive, then the disturbances decay exponentially in the freestreams. If Ω_+^2 or Ω_-^2 are negative, then taking a square root results in a complex value, and thus the solution oscillates (i.e., bounded) in the respective freestream. In the former case, the equation in the freestream is elliptic, while in the latter case, the equation is hyperbolic. Thus, for the hyperbolic case the freestream solutions are acoustic waves, and hence supersonic in nature.

Let us define $c_\pm$ to be the phase speed for which $\Omega_\pm^2$ vanishes. Thus,

$$c_+ = 1 - \frac{1}{M}, \qquad c_- = \beta_U + \frac{\sqrt{\beta_T}}{M}. \tag{5.33}$$

Note that c_+ is the phase speed of a sonic disturbance in the fast stream and c_- is the phase speed of a sonic disturbance in the slow stream. At

$$M = M_* = \frac{1 + \sqrt{\beta_T}}{1 - \beta_U}, \tag{5.34}$$

$c_\pm$ are equal.

The nature of the disturbances and the appropriate boundary conditions can now be illustrated by reference to Fig. 5.1, where we plot $c_\pm$ as functions of M. In what follows we assume that $\alpha_r^2 > \alpha_i^2$. These curves divide the (c_r, M)-plane into four regions, where c_r is the real part of c. If a disturbance exists with a M and c_r in region 1, then Ω_+^2 and Ω_-^2 are both positive. The disturbance is subsonic at both boundaries and is classified as a subsonic mode. In region 3, both Ω_+^2 and Ω_-^2 are negative and hence the disturbance is supersonic at both boundaries and is classified as a supersonic–supersonic mode. In region 2, Ω_+^2 is positive and Ω_-^2 is negative and the disturbance is subsonic at $+\infty$ and supersonic at $-\infty$. In this region, the mode is classified as a fast mode. Finally, in region 4, Ω_+^2 is negative and Ω_-^2 is positive so the disturbance is supersonic at

$+\infty$ and subsonic at $-\infty$. In this region, the mode is classified as a slow mode. Note that the terminology "fast" and "slow" is in reference to the magnitude of the phase speed c_r. These modes are also referred to as "outer" modes (see Day, Reynolds & Mansour, 1998a,b).

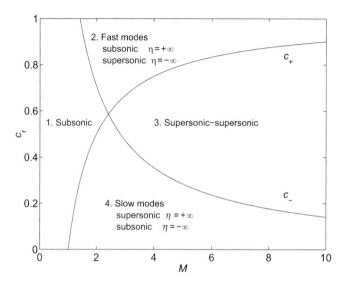

Figure 5.1 Plot of the sonic speeds $c_{\pm}$ versus Mach number M for $\beta_U = 0$ and $\beta_T = 2$.

If the disturbance wave is subsonic at both $\pm\infty$ (region 1), one can choose the appropriate sign for $\Omega_{\pm}$ and have decaying solutions, which leads to an eigenvalue problem. If the disturbance is supersonic at either, or both, boundaries, then the asymptotic solutions are purely oscillatory. These solutions are of two types: incoming and outgoing waves. If one assumes that only outgoing waves are permitted, the problem of finding solutions in regions 2, 3 or 4 is again an eigenvalue problem, whereby one chooses boundary conditions for the compressible Rayleigh equation that lead to solutions that are only outgoing waves in the far field.

However, if one permits both incoming and outgoing waves in the far field, it is obvious that there are always solutions for any c in regions 2, 3 and 4. For a given ω, one can always find a continuum of α such that there is a solution to the compressible Rayleigh equation with constant amplitude oscillations at either or both boundaries. Lees & Lin gave a physical interpretation of this pair of incoming and outgoing waves as an incoming wave and its reflection from the shear layer. Mack (1975) also used this idea in developing a theory for the forced response of the compressible boundary layer. At present, the continuum modes are ignored for the remainder of this section.

One can now see that the appropriate boundary condition for either the damped or outgoing waves in the freestreams $\eta = +\infty$ and $\eta = -\infty$ are, respectively,

$$\Pi \to \exp(-\Omega_+ \eta), \ \text{if } c_r > c_+, \quad \Pi \to \exp(-i\eta \sqrt{-\Omega_+^2}), \ \text{if } c_r < c_+, \quad (5.35)$$

and

$$\Pi \to \exp(\Omega_- \eta), \ \text{if } c_r < c_-, \quad \Pi \to \exp(-i\eta \sqrt{-\Omega_-^2}), \ \text{if } c_r > c_-. \quad (5.36)$$

Generalized Regularity Condition

Recall Result 5.1 where, if a neutral mode is to exist in region 1, the phase speed will be given by $c_N = U(\eta_c)$, where η_c is found from the regularity condition

$$S(\eta) = \frac{d}{d\eta} \left(T^{-2} \frac{dU}{d\eta} \right) = 0. \quad (5.37)$$

The corresponding wavenumber α must be determined numerically. Note that this form differs from that given in Result 5.1 by a factor of T^{-1} because (5.37) is shown in terms of the similarity variable η.

Figure 5.2 is a plot of S over a range of values of η for various values of M and fixed $\beta_U = 0$ and $\beta_T = 2$. From this plot one can see that, for low values of the Mach number, only one real root of S exists. But, as the Mach number increases, three real roots exist. For example, at $M = 0$ one zero exists at $y = 0.347$, at $M = 4$, $y = 0.601$, at $M = 7$, $y = 1.104$ and for $M = 10$ three zeros exist at $y = -0.981$, -0.149 and 1.478. For a root to correspond to a neutral mode, it must lie in region 1 of Fig. 5.1. Because the Mach number at which three real roots first appears is greater than M_* (for $\beta_U = 0$ and $\beta_T = 2$, we have $M_* = 2.414$; see equation (5.34)), these roots cannot correspond to neutral modes.

However, Jackson and Grosch (1989) have shown, for three-dimensional disturbances, that the sonic speeds $c_\pm$ are functions of the angle of propagation of the disturbance. As the angle increases, the sonic curves shift toward higher Mach number. Thus for any value of β_T, there will always be some angle of propagation for which all three zeros of S lie in region 1, and by Result 5.1, there are now three neutral modes with phase speeds equal to the value of U at the corresponding values of η_c. Thus, the significance of the three real zeros of S only becomes apparent at very large angles of propagation.

There can also be supersonic neutral modes. Such modes do not satisfy (5.37) but are solutions of (5.30) with only outgoing or damped waves at $\pm\infty$.

It is obvious that these are singular eigenfunctions. The singularity will be removed by the action of nonzero viscosity. Hence we can regard these singular modes as the limit of some viscous stability modes, as the Reynolds number approaches infinity.

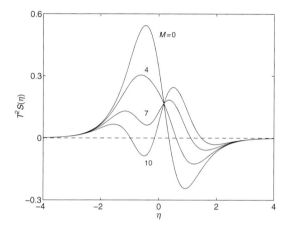

Figure 5.2 Plot of $T^2 S(\eta)$ for $\beta_U = 0$, $\beta_T = 2$ and for various M.

Results

To solve the disturbance equation (5.30), we first transform it to a Riccati equation by setting

$$G = \frac{\Pi'}{\alpha T \Pi}. \tag{5.38}$$

The spatial stability problem is thus to solve the Riccati equation, subject to appropriate boundary conditions, for a given real frequency ω and Mach number M, with U and T defined by (5.15) and (5.13), respectively. The eigenvalue is the complex wavenumber α. Because the equation has a singularity at $U = c_N$, it is convenient to perform the integration in the complex plane, choosing for example the contour $(-L, -1)$ to $(0, -1)$ and $(L, -1)$ to $(0, -1)$, with $L \geq 6$, using a Runge–Kutta scheme with variable step size. Then iterate on α until the boundary conditions are satisfied and the jump in G at $(0, -1)$ is less than a specified small value (e.g., 10^{-6}). All calculations are done in 64-bit precision.

The neutral phase speeds as a function of Mach number are shown in Fig. 5.3. Three different values of the temperature ratio β_T are taken. The top image corresponds to $\beta_T = 0.5$, the middle to 1 and the bottom to 2. Here, $\beta_U = 0$ for all cases. For each value of β_T, there exists one or more unstable regions, bounded between two neutral curves. The classification for the neutral modes are: (1) subsonic, $\alpha_N \neq 0$; (2) subsonic, $\alpha_N = 0$; (3) fast, $\alpha_N \neq 0$; (4) slow, $\alpha_N \neq 0$; (5) constant speed supersonic–supersonic, $\alpha_N = 0$; (6) fast supersonic–supersonic, $\alpha_N = 0$; (7) slow supersonic–supersonic, $\alpha_N = 0$. The sonic curves are shown as dashed.

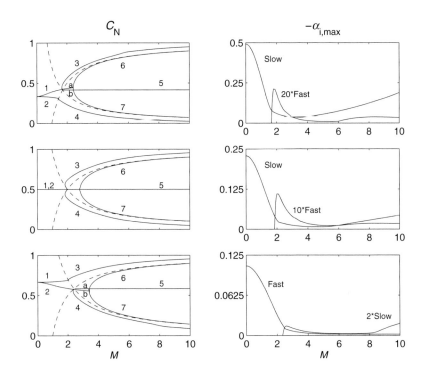

Figure 5.3 Plots of two-dimensional neutral curves (left) and maximum spatial growth rates (right) for $\beta_T = 0.5$ (top), $\beta_T = 1$ (middle) and $\beta_T = 2$ (bottom) as a function of the Mach number M and for $\beta_U = 0$. The neutral mode classification is: (1) subsonic, $\alpha_N \neq 0$; (2) subsonic, $\alpha_N = 0$; (3) fast, $\alpha_N \neq 0$; (4) slow, $\alpha_N \neq 0$; (5) constant speed supersonic–supersonic, $\alpha_N = 0$; (6) fast supersonic–supersonic, $\alpha_N = 0$; (7) slow supersonic–supersonic, $\alpha_N = 0$. The sonic curves are shown as dashed. The spatial growth rates for the Fast modes for $\beta_T = 0.5$ and 1, and for the Slow modes for $\beta_T = 2$, have been scaled to better visualize the curves.

For $M < M_*$, there exists basically only one unstable region. The neutral curve labeled 1 corresponds to the value of c_N determined from the Lees & Lin condition, while the neutral curve labeled 2 has a neutral wavenumber of zero. The unstable region thus lies between curves 1 and 2. Because of the symmetry of the hyperbolic tangent profile with $\beta_T = 1$, curves 1 and 2 coincide; for more general profiles this is not true. For $M > M_*$, there exists more than one unstable region. For $M_* < M < M_{CR}$ and $\beta_T = 1$ (middle figure), the Fast modes lie between the neutral curves 3 ($\alpha_N \neq 0$, determined numerically) and 5 ($\alpha_N = 0$), while the Slow modes lie between the neutral curves 4 ($\alpha_N \neq 0$, determined numerically) and 5 ($\alpha_N = 0$). Mode splitting from subsonic to supersonic speeds at $M = M_*$ have been described by Drazin & Davey (1977). The upper limit M_{CR} of the Mach number is determined from an $\alpha \to 0$ asymptotic analysis, along the lines previously used by Miles (1958), Drazin & Howard (1962) and Blumen, Drazin & Billings (1975) in related studies; the analysis is presented later in the context of a vortex sheet. For $M_* < M < M_{CR}$ and $\beta_T \neq 1$ (top and bottom figures), the Fast modes lie between the neutral curves 3 and 5a, while the Slow modes lie between the neutral curves 4 and 5b. Note that curves 5b for $\beta_T = 0.5$ and 5a for $\beta_T = 2$ are the continuation of the constant speed supersonic–supersonic neutral mode for Mach numbers below M_{CR}, while the curves 5a for $\beta_T = 0.5$ and 5b for $\beta_T = 2$ are the continuation of the subsonic neutral mode through the value of the phase speed where $c_\pm$ are equal, continuing to the minimum value where curves 6 and 7 join at M_{CR}. It is this mode that splits (or bifurcates), creating modes 6 and 7. Thus there is a small stable region between curves 5a and 5b, due to the nonsymmetrical nature of the mean profile. For $M > M_{CR}$, the Fast modes lie between the neutral curves 3 and 6 ($\alpha_N = 0$), while the Slow modes lie between the neutral curves 4 and 7 ($\alpha_N = 0$). The unstable modes associated with the neutral phase speed labeled 5 for $M > M_{CR}$ have never been fully explored, nor has the possibility of neutral modes associated with the freestream speeds $c_N = \beta_U$ or $c_N = 1$ (these would be analogous to the noninflectional modes of the compressible boundary layer found by Mack (1984)).

The corresponding maximum spatial growth rates as a function of Mach number are also shown in Fig. 5.3. For each value of the temperature ratio β_T, the maximum spatial growth rate of the subsonic mode decreases as the Mach number increases from zero to M_*. For Mach numbers greater than M_*, two unstable modes appear and remain relatively small as the Mach number is further increased. The curves also show that by increasing β_T, a decrease in the maximum growth rates results.

Convective Mach Number

A number of experiments (e.g., Brown & Roshko, 1974; Chinzei et al., 1986; and Papamoschou & Roshko, 1986, 1988) and numerical simulations (e.g., Guirguis, 1988; Lele, 1989; Sandham & Reynolds, 1990; and Mukunda et al., 1989) have shown that the compressible mixing layer becomes less unstable with increasing Mach number. In order to correlate the experimental results, a number of experimentalists have used a heuristically defined "convective Mach number." This idea, first introduced by Bogdanoff (1983) for compressible flows, has permeated much of the work on the stability of compressible free shear layers since then. The basic idea is to define a Mach number in a moving frame of reference fixed to the large scale structures of the mixing layer (Bogdanoff, 1983; Papamoschou & Roshko, 1986, 1988). Thus, for streams of equal gases, the convective Mach number is defined as

$$M_c = \frac{U^\dagger_{+\infty} - U^\dagger_{-\infty}}{a^\dagger_{+\infty} + a^\dagger_{-\infty}} = \frac{M(1-\beta_U)}{1+\sqrt{\beta_T}}. \qquad (5.39)$$

More general definitions include the effect of different gases (Bogdanoff, 1983 and Papamoschou & Roshko, 1986, 1988) or to defining the convective Mach number in terms of the most unstable wave (Zhuang, Kubota & Dimotakis, 1988). These definitions were for unbounded flows. Tam & Hu (1989) have applied these concepts to a mixing layer in a channel. When a mixing layer is formed between two different gases, some definitions will yield different values for the convective Mach number in the two gases. It is not obvious which value is the proper one to use in the correlation, or whether some average of the two is appropriate.

It is interesting to note that another interpretation of the convective Mach number exists. Extensive spatial stability calculations for the compressible mixing layer (Jackson & Grosch, 1989, 1991) suggested a way to rigorously derive from linear stability theory a single convective Mach number for a compressible mixing layer for both a single species gas and a multi-species gas (Jackson & Grosch, 1990b). In particular, the definition is based on the freestream Mach number in the laboratory frame and is independent of the speed of the large-scale structures and the speed of the most unstable wave. From Fig. 5.1, we see that subsonic modes exist for $M < M_*$ and supersonic modes that radiate into one or the other stream exist for $M > M_*$, and thus one can define a convective Mach number as follows

$$M_c = \frac{M}{M_*} = \frac{M(1-\beta_U)}{1+\sqrt{\beta_T}}. \qquad (5.40)$$

With this scaling, subsonic modes exist for $M_c < 1$ and supersonic modes exist for $M_c > 1$. This definition is identical to (5.39) given by Bogdanoff (1983).

In order to present the variation of the maximum growth rates with M_c, the growth rates are normalized by defining the normalized growth rate for spatial stability as

$$R = \frac{(-\alpha_i)_{\text{MAX}}\ (\beta_T, \beta_U, M)}{(-\alpha_i)_{\text{MAX}}\ (\beta_T, \beta_U, 0)}. \qquad (5.41)$$

Figure 5.4 is a plot of R versus M_c for the three thermodynamic models and β_T of 0.5, 1 and 2, which is taken from Jackson and Grosch (1991). It can be seen that with these scalings the data collapses onto essentially a single curve for $M_c < 1$, and a narrow band for $M_c > 1$. One should also note that the second unstable supersonic modes appear around $M_c = 1$. This curve is similar to that obtained by Ragab and Wu (1989), who use Bogdanoff's heuristic definition of the convective Mach number. However, in their graph the second supersonic mode is absent.

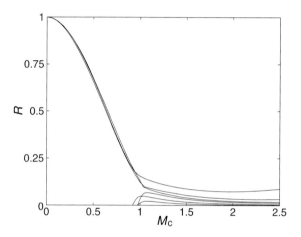

Figure 5.4 Plot of normalized growth rate as a function of the convective Mach number for $\beta_T = 0.5$, 1, and 2, and for $\beta_U = 0$.

5.2.4 Compressible Vortex Sheet

Miles (1958) analyzed the compressible Rayleigh equation in the limit $\alpha \to 0$ for a vortex sheet. As will be shown in the next paragraph, this analysis also corresponds to determining the neutral modes of a compressible mixing layer.

The general solution to the compressible Rayleigh equation (5.30) is given by[2]

$$\Pi = \begin{cases} A_+ F(\eta) \exp(-\alpha \Omega_+ \eta), & \eta > 0, \\ A_- G(\eta) \exp(+\alpha \Omega_- \eta), & \eta < 0, \end{cases} \qquad (5.42)$$

[2] The analysis presented here follows that of R. Klein and A. Majda, and appears in the unpublished report by Grosch, Jackson, Klein, Majda & Papageorgiou, 1991.

where the equations for F and G are given by

$$F'' - \left(\frac{2U'}{U-c} + 2\alpha\Omega_+\right)F' + \left(\alpha^2\Omega_+^2 + \frac{2U'}{U-c}\alpha\Omega_+ - \alpha^2 g\right)F = 0 \quad (5.43)$$

for $\eta > 0$, and

$$G'' - \left(\frac{2U'}{U-c} - 2\alpha\Omega_-\right)G' + \left(\alpha^2\Omega_-^2 - \frac{2U'}{U-c}\alpha\Omega_- - \alpha^2 g\right)G = 0 \quad (5.44)$$

for $\eta < 0$, where

$$g(\eta) = T\left[T - M^2(U-c)^2\right]. \quad (5.45)$$

The appropriate boundary conditions are $F(+\infty) = 1$ and $G(-\infty) = 1$. In order for the above to be a uniformly valid solution over $(-\infty, +\infty)$, the Wronskian must vanish at $\eta = 0$, yielding the required eigenrelation

$$F\left[G' + \alpha\Omega_- G\right] = G\left[F' - \alpha\Omega_+ F\right]. \quad (5.46)$$

Since we are interested in long-wave perturbations ($\alpha \to 0$), the appropriate expansions are found to be

$$\left. \begin{array}{rcl} F & = & F_0 + \alpha F_1 + \alpha^2 F_2 + \cdots, \\ G & = & G_0 + \alpha G_1 + \alpha^2 G_2 + \cdots. \end{array} \right\} \quad (5.47)$$

After substituting (5.47) into (5.43) and (5.44), the solutions to the first two orders are

$$F_0 = 1, \quad F_1 = -\Omega_+ \int_\eta^{+\infty}\left[1 - \left(\frac{U-c}{1-c}\right)^2\right]d\eta, \quad (5.48)$$

and

$$G_0 = 1, \quad G_1 = -\Omega_- \int_{-\infty}^\eta\left[1 - \left(\frac{U-c}{\beta_U - c}\right)^2\right]d\eta. \quad (5.49)$$

The eigenrelation (5.46) is also expanded in powers of α, resulting in the following equations

$$F_0 G_0' = G_0 F_0' \quad (5.50)$$

to zeroth order and

$$F_0(G_1' + \Omega_- G_0) + F_1 G_0' = G_0(F_1' - \Omega_+ F_0) + G_1 F_0' \quad (5.51)$$

to first order. The zeroth-order equation is satisfied trivially, and the first-order equation yields

$$\frac{\Omega_-}{(\beta_U - c)^2} = -\frac{\Omega_+}{(1-c)^2}.$$
(5.52)

Note that this relation is independent of the detailed form of U and T, and only depends on the basic flow characteristics at infinity. This is to be expected from physical arguments since the length scale of the instability is much larger than the length scale over which the undisturbed flow is nonuniform. It is precisely at this order that the vortex sheet results of Miles (1958) is recovered. Corrections can be obtained at the next order (Grosch, Jackson, Klein, Majda & Papageorgiou, 1991). Note that (5.52) requires that both Ω_- and Ω_+ be complex, and hence the analysis is strictly limited to region 3 of Fig. 5.1. With the definitions for $\Omega_{\pm}$, then

$$\beta_T \left[M^2 (\beta_U - c_N)^2 - \beta_T \right] (1 - c_N)^4 = \left[M^2 (1 - c_n)^2 - 1 \right] (\beta_U - c_N)^4, \quad (5.53)$$

which determines the phase speed c_N for a neutral mode. This equation is identical to (5.3a) of Miles (1958) if his result is expressed in our notation.

The following comments now apply:

1. A single real root of (5.53) exists for

$$M \geq M_* \equiv \frac{1 + \sqrt{\beta_T}}{1 - \beta_U},$$
(5.54)

    with phase speed

$$c_N = \frac{\beta_U + \sqrt{\beta_T}}{1 + \sqrt{\beta_T}},$$
(5.55)

    which is classified as a constant speed supersonic–supersonic neutral mode. This solution is independent of the Mach number, and corresponds to the phase speed at which the two sonic speeds $c_{\pm}$ are equal. In this regime there is also a pair of complex conjugate eigenvalues corresponding to one unstable and one stable eigenvalue. The associated instability is analogous to the classical Kelvin–Helmholtz instability for subsonic vortex sheets (Artola & Majda, 1987). This instability vanishes as the Mach number increases.

2. A double root first appears at

$$M_{CR} = \frac{(1 + \beta_T^{1/3})^{3/2}}{1 - \beta_U},$$
(5.56)

with phase speed

$$c_{\rm N} = \frac{\beta_U + \beta_T^{1/3}}{1 + \beta_T^{1/3}}. \tag{5.57}$$

3. There are three distinct real roots for $M > M_{CR}$. One of the roots is (5.55), while the other two roots must be found numerically. For the special case of $\beta_T = 1$, these roots are given by

$$c_{\rm N} = \frac{1 + \beta_U}{2} \pm \frac{1}{2M} \left[M^2(1 - \beta_U)^2 + 4 - 4\sqrt{M^2(1 - \beta_U)^2 + 1} \right]^{1/2}. \tag{5.58}$$

The root that corresponds to the $(+)$ sign is classified as a fast supersonic–supersonic neutral mode, while that which corresponds to the $(-)$ sign is classified as a slow supersonic–supersonic neutral mode.

The phase speeds are plotted in Fig. 5.3 for $\beta_U = 0$ and various values of β_T. The classification scheme is given in the figure caption.

The neutral phase speeds given in equations (5.55), (5.57) and (5.58) are exact for $\alpha = 0$. In order to obtain the higher-order corrections for $\alpha \neq 0$, the value of c must also be expanded in powers of α. When this was done it was found that the overall growth rate was $O(\alpha^2)$; Balsa and Goldstein (1990) also found, numerically, the $O(\alpha^2)$ growth rate for these modes. It was also found that the growth rate at $O(\alpha^2)$ becomes singular at M_{CR}. This singular behavior was studied by expansions about the singular value of M. A connection between the regimes $M_* < M < M_{CR}$ and $M > M_{CR}$ was found and yielded the transition from a stable/unstable pair of eigenmodes plus a supersonic neutral mode for $M < M_{CR}$ to three supersonic neutral modes for $M > M_{CR}$.

5.2.5 *Bounded Compressible Mixing Layer*

The stability characteristics of a bounded mixing layer, that is a mixing layer in a rectangular duct, have been considered by Tam & Hu (1989), Greenough *et al.* (1989), Macaraeg & Streett (1989), Zhuang, Kubota & Dimotakis (1990b), Macaraeg (1990), Jackson & Grosch (1990c), and Morris & Giridharan (1991). The effect of walls becomes apparent at supersonic convective Mach numbers because the coupling between the motion of the mixing layer and the acoustic modes of the channel produce two new instability modes. That is, with boundedness, there will always be incoming and outgoing waves and these must be properly accounted for, whereas, for the unbounded mixing layer, only outgoing waves are considered. This effect was clearly demonstrated by Tam & Hu (1989), who classified the new modes as Class A (modes with phase speeds that decrease as the wavenumber and frequency increase) or Class B (modes with phase speeds that increase as the wavenumber and frequency increase). In the following discussion, a subset of the results of Tam & Hu (1989) are presented within the similarity-variable framework that was just discussed for the unbounded mixing layer.

We begin by considering the inviscid temporal stability of a mixing layer in a channel of width $2H$. Assume that the thickness of the mixing layer is 2δ, with $\delta < H$. The stability problem can be formulated independently of the detailed form of the velocity and temperature profiles. A standard normal mode analysis again leads to the compressible Rayleigh equation (5.30). The appropriate boundary conditions are

$$\Pi'(\pm H) = 0. \tag{5.59}$$

The outer solution, valid in $\delta < \eta \leq H$, is given by

$$\Pi = \cosh\left[\alpha\Omega_+(\eta - H)\right], \tag{5.60}$$

and the outer solution, valid in $-H \leq \eta < -\delta$, is given by

$$\Pi = \cosh\left[\alpha\Omega_-(\eta + H)\right], \tag{5.61}$$

where $\Omega_\pm$ are given by (5.32). Note that, when $\Omega_\pm$ are complex (i.e., the outer flow is supersonic), the outer solutions become cosine functions and thus represent acoustic modes.

The mean flow is given by (5.13) and (5.15). In the temporal stability calculations presented in this section we have taken $\beta_T = 1$, $\beta_U = 0$ and $M = 3.5$. In addition, let $H = 12$, which is twice that of the shear layer thickness δ.

Figures 5.5 and 5.6 plot the phase speeds and temporal growth rates as a function of the wavenumber. In Fig. 5.5, we see that the phase speeds depicted in the top plot decreases from one as the wavenumber increases, and in the terminology of Tam & Hu (1989), are classified as Class A modes, whereas the phase speeds depicted in the bottom plot increases from zero as the wavenumber increases, and are classified as Class B modes. The phase speeds of the two classes are a reflection of each other about the centerline $c_r = 1/2$ due to the mean profiles being symmetrical. The sonic speeds $c_\pm$ are also shown (dashed) and can be used as a reference when classifying the behavior of the disturbances at the walls with regard to Fig. 5.1 (for the bounded case, change the text "∞" to "H" in regions 2 and 3, and the figure then applies when walls are present). For example, Class A modes, which initially have a phase speed greater than c_+, are subsonic at $\eta = +H$ and supersonic at $\eta = -H$. As the

phase speed decreases and eventually crosses the $c = c_+$ sonic line, the disturbances become supersonic at both walls. Similarly, Class B modes, which initially have a phase speed less than c_-, are supersonic at $\eta = +H$ and subsonic
at $\eta = -H$. As the phase speed increases and crosses the $c = c_-$ sonic line,
the disturbances become supersonic at both walls. Thus there is a continuous
change in the character of the modes as the wavenumber is increased in that a
mode which is subsonic at one boundary is transformed into a supersonic mode
at that boundary. The growth rates for the Class A modes are shown in Fig. 5.6.
Note that after an initial increase in the maximum growth rate, the maximum
growth rates generally decrease as the wavenumber increases. It is clear from
this figure that the spectrum is quite complicated, with multiple modes existing
at higher values of α. Because the mean profiles are symmetrical, the growth
rates of the Class A and B modes are identical.

In addition to the Class A and B modes, there exists another set of modes, all
with phase speeds of exactly $1/2$ and independent of α. These are called Class
C modes. The growth rates of these modes are shown in Fig. 5.7. These growth
rates are about 50% greater than those of Class A and B modes. Macaraeg &
Streett (1989) and Macaraeg (1990) also found similar modes in their calculations. However, Tam & Hu (1989) did not report any such modes. Perhaps this
is due to the fact that their mean velocity profile was not symmetric; hence,
these modes are likely of academic interest only.

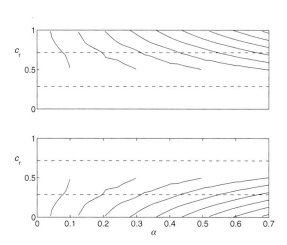

Figure 5.5 Plot of the phase speeds c_r as a function of the wavenumber α for the
Class A modes (top) and the Class B modes (bottom) at $M = 3.5$ and $\beta_T = 1$. The
sonic speeds $c_+ = 0.714$ and $c_- = 0.286$ are shown as the dashed lines.

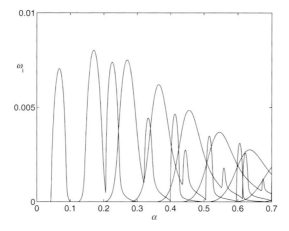

Figure 5.6 Plot of the temporal growth rates ω_i as a function of the wavenumber α for both the Class A and Class B modes at $M = 3.5$ and $\beta_T = 1$.

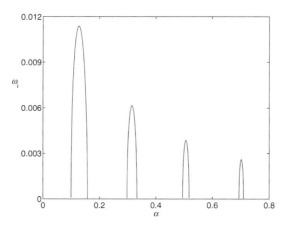

Figure 5.7 Plot of the temporal growth rates ω_i as a function of the wavenumber α for the Class C modes at $M = 3.5$ and $\beta_T = 1$. The corresponding phase speeds are $c_r = 1/2$.

5.3 Compressible Boundary Layer

The stability of compressible boundary layers has received considerable attention since the mid 1940s. In the early years, progress was slow because calculations were limited to asymptotic methods, and the results were often incomplete and sometimes contradictory. Substantial progress began in earnest in the late 1950s and early 1960s as numerical calculations became routine.

However, it still took more than twenty years to unravel most of the subtleties of the stability characteristics of the compressible boundary layer. For a historical review, and results obtained up to the early 1980s, see Mack (1984, 1987). Our aim is to present the elementary basics while illuminating the richness of the subject; a complete review or summary of the subject is well beyond the intent of this book. Thus, in this section we consider only the two-dimensional compressible boundary layer on a flat plate subject to two-dimensional disturbances. More general configurations are presented in the review article by Mack (1984); see also the references contained therein. These extensions include both temporal and spatial theories, three-dimensional disturbances, and three-dimensional boundary layers (e.g., rotating disk, Falkner–Skan–Cooke boundary layers, and swept wing). Finally, we note that the continuous spectra of the compressible boundary layer was discussed by Ashpis & Erlebacher (1990) and Balakumar & Malik (1992).

5.3.1 Mean Flow

Consider the two-dimensional compressible boundary layer flow over a flat plate with zero pressure gradient. A description of the mean flow can be found in Mack (1965a), which is briefly presented here for completeness. Let $(u^\dagger, v^\dagger)$ be the dimensional velocity components in the $(x^\dagger, y^\dagger)$ directions, respectively, $\rho^\dagger$ the density, $T^\dagger$ the temperature and $p^\dagger$ the pressure, a known constant. The dimensional equations governing the flow of a perfect gas are

$$(\rho^\dagger u^\dagger)_{x^\dagger} + (\rho^\dagger v^\dagger)_{y^\dagger} = 0, \tag{5.62}$$

$$\rho^\dagger \left[u^\dagger u^\dagger_{x^\dagger} + v^\dagger u^\dagger_{y^\dagger} \right] = \left(\mu^\dagger u^\dagger_{y^\dagger} \right)_{y^\dagger}, \tag{5.63}$$

$$\rho^\dagger \left[u^\dagger e^\dagger_{x^\dagger} + v^\dagger e^\dagger_{y^\dagger} \right] = \left(\kappa^\dagger T^\dagger_{y^\dagger} \right)_{y^\dagger} + \mu^\dagger (u^\dagger_{y^\dagger})^2, \tag{5.64}$$

and

$$p^\dagger = \rho^\dagger R T^\dagger. \tag{5.65}$$

Here, $\mu^\dagger$ and $\kappa^\dagger$ are the viscosity and thermal conductivity coefficients, assumed to be functions of temperature; R is the gas constant for air. The viscosity coefficient is computed from the Sutherland formula, namely

$$\mu^\dagger = \begin{cases} \dfrac{1.458\,T^{\dagger 3/2}}{T^\dagger + 110.4} \times 10^{-5} \ \text{gm/cm-sec}, & T^\dagger > 110.4\text{K}, \\[2ex] \left(0.693873 \times 10^{-6}\right) T^\dagger \ \text{gm/cm-sec}, & T^\dagger < 110.4\text{K}, \end{cases} \tag{5.66}$$

while the thermal conductivity coefficient can be computed from the formula

$$
\kappa^\dagger = \begin{cases} \dfrac{0.6325\sqrt{T^\dagger}}{1+(245.4/T^\dagger)10^{-12/T^\dagger}} \ \text{cal/cm-sec-C}, & T^\dagger > 80\text{K}, \\[2ex] \left(0.222964 \times 10^{-6}\right) T^\dagger \ \text{cal/cm-sec-C}, & T^\dagger < 80\text{K}. \end{cases}
\tag{5.67}
$$

The viscosity and thermal conductivity coefficients are related via the Prandtl number or

$$
Pr = \mu^\dagger c_P^\dagger / \kappa^\dagger,
\tag{5.68}
$$

which is a function of temperature. The enthalpy, $e^\dagger$, is a function of temperature and is given in terms of $c_P^\dagger$, the specific heat at constant pressure, by

$$
e^\dagger = \int_0^{T^\dagger} c_P^\dagger dT^\dagger.
\tag{5.69}
$$

The equations assume that the temperature everywhere in the flow is below the temperature of dissociation. At high temperature, which can occur within the compressible boundary layer if the Mach number is large enough, the flow dissociates and, as a result, real gas effects must be considered.

In general, tables for a perfect gas are needed for the enthalpy–temperature relation and the Prandtl number. The specific heat, $c_P^\dagger$, is then computed from (5.68). This is the approach taken by Mack, and subsequently used in all of his stability calculations. For purposes of clarity, we assume constant values for $c_P^\dagger$ and Pr; the thermal conductivity can then be omitted from active consideration, as well as the enthalpy in favor of the temperature via $e^\dagger = c_P^\dagger T^\dagger$. The stability calculations presented here will therefore differ only slightly from those of Mack.

To solve the mean flow equations, first introduce the similarity variable

$$
\eta = \frac{y^\dagger}{x^\dagger}\sqrt{Re_x},
\tag{5.70}
$$

where Re_x is the x-Reynolds number

$$
Re_x = \frac{u_\infty^\dagger x^\dagger}{v_\infty^\dagger}, \qquad v_\infty^\dagger = \mu_\infty^\dagger / \rho_\infty^\dagger,
\tag{5.71}
$$

$u_\infty^\dagger$ is the freestream velocity and $v_\infty^\dagger$ is the kinematic viscosity. The partial derivatives thus transform according to

$$
\frac{\partial}{\partial x^\dagger} = \frac{\partial}{\partial x^\dagger} - \frac{\eta}{2x^\dagger}\frac{\partial}{\partial \eta}, \qquad \frac{\partial}{\partial y^\dagger} = \frac{\sqrt{Re_x}}{x^\dagger}\frac{\partial}{\partial \eta}.
\tag{5.72}
$$

The following dimensionless quantities are introduced

$$
U = u^\dagger / u_\infty^\dagger, \quad V = v^\dagger / u_\infty^\dagger, \quad \rho = \rho^\dagger / \rho_\infty^\dagger,
$$

$$\mu = \mu^\dagger/\mu_\infty^\dagger, \quad T = T^\dagger/T_\infty^\dagger, \quad \theta = \frac{T^\dagger - T_\infty^\dagger}{T_0^\dagger - T_\infty^\dagger}, \tag{5.73}$$

where the freestream stagnation temperature, $T_0^\dagger$, is defined by

$$T_0^\dagger = T_\infty^\dagger + \frac{(u_\infty^\dagger)^2}{2c_P^\dagger}. \tag{5.74}$$

From the definition of the scaled temperature θ, we see that

$$T = 1 + \frac{\gamma - 1}{2}M^2\theta, \tag{5.75}$$

where γ is the ratio of specific heats and M the freestream Mach number. Here we made use of the relations (see, e.g., White, 1974)

$$c_P^\dagger = \frac{\gamma R}{\gamma - 1}, \quad M = u_\infty^\dagger/a_\infty^\dagger \quad a^\dagger = \sqrt{\gamma R T^\dagger}. \tag{5.76}$$

Now assume that the horizontal velocity component and the temperature are functions of η only, and write

$$\rho U = 2g'(\eta), \tag{5.77}$$

where the prime denotes differentiation with respect to the similarity variable η. In these terms, the continuity equation (5.62) becomes

$$\rho V = \frac{1}{\sqrt{R_x}}(\eta g' - g) \tag{5.78}$$

for the normal velocity component. The momentum (5.63) and the energy (5.64) equations then reduce to

$$\frac{d}{d\eta}\left(\mu \frac{dU}{d\eta}\right) + g\frac{dU}{d\eta} = 0, \tag{5.79}$$

and

$$\frac{d}{d\eta}\left(\frac{\mu}{Pr}\frac{d\theta}{d\eta}\right) + g\frac{d\theta}{d\eta} = -2\mu\left(\frac{dU}{d\eta}\right)^2. \tag{5.80}$$

The appropriate boundary conditions are given by

$$\text{at } \eta = 0: \quad U = 0, \quad \begin{array}{ll} \text{Case(i)} & \theta'(0) = 0, \\ \text{Case(ii)} & \theta(0) = \text{given,} \end{array} \tag{5.81}$$

$$\text{at } \eta \to \infty: \quad U = 1, \quad \theta = 0.$$

Case (i) corresponds to an insulated wall while case (ii) corresponds to a wall at a specified but fixed temperature. In the second case, the wall can either be cooled or heated, depending on the sign of $T_w^\dagger/T_{ad}^\dagger - 1$, where $T_w^\dagger$ is the wall temperature and $T_{ad}^\dagger$ is the adiabatic wall temperature, the value of the temperature of an insulated wall; the adiabatic wall temperature is also sometimes referred to as the recovery temperature. Here the discussion will focus exclusively on case (i); for a discussion of case (ii) and its stability characteristics, see Mack (1984) and the references contained therein.

In the incompressible limit $M \to 0$ equation (5.75) yields $T = 1$, and hence $T^\dagger = T_\infty^\dagger$ from (5.73), $\mu = 1$ from (5.73), and using (5.77), (5.79) reduces to

$$g''' + gg'' = 0,$$

the well-known Blasius profile.

The solution procedure is as follows. We begin by integrating the system (5.79) and (5.80) from $\eta = 0$ to $\eta = \eta_\delta$, where η_δ is the edge of the boundary layer. (Mack, 1965a, rewrites the system as four first-order equations so that a fourth-order Runge–Kutta integrator can be used, but this is merely a matter of preference.) Since the system is fourth order, the integrator requires four boundary conditions to be specified. For the case of an insulated wall (the other case follows similarly), the boundary conditions are

$$U(0) = 0, \quad U'(0) = U_0, \quad \theta(0) = \theta_0, \quad \theta'(0) = 0, \tag{5.82}$$

where U_0 and θ_0 are determined by an iteration procedure so that the boundary conditions

$$U(\eta_\delta) = 1, \quad \theta(\eta_\delta) = 0, \tag{5.83}$$

at $\eta = \eta_\delta$ are satisfied. The unknown function g is found by simultaneously solving

$$g' = \frac{U}{2T}, \quad g(0) = 0, \tag{5.84}$$

with T defined by (5.75).

To complete the description of the mean flow, $T_\infty^\dagger$ must be specified in order to compute the dimensionless viscosity μ (see 5.73). One can either specify $T_\infty^\dagger$ directly, or specify the stagnation temperature $T_0^\dagger$ (see 5.74). It is customary to specify the stagnation temperature and, with the freestream Mach number given, compute $T_\infty^\dagger$ from the relation

$$\frac{T_0^\dagger}{T_\infty^\dagger} = 1 + \frac{\gamma - 1}{2} M^2. \tag{5.85}$$

In all cases, assume "wind tunnel" conditions (in the nomenclature of Mack) by setting $T_0^\dagger = 311\text{K}$ until, with increasing Mach number, $T_\infty^\dagger$ drops to 50K. For higher Mach numbers, $T_\infty^\dagger$ is held constant at 50K. In all calculations, we use $\gamma = 1.4$ and set $Pr = 0.72$.

Figure 5.8 plots the mean flow quantities U and T as a function of the similarity coordinate η at various values of the Mach number. Note that as the Mach number increases, both the velocity and the temperature are defined over a broader and broader region, while at the plate the temperature increases due to viscous heating.

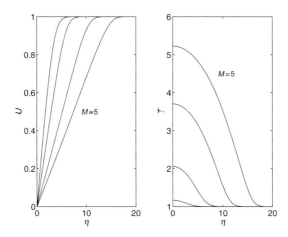

Figure 5.8 Plot of the mean velocity component U and temperature T as a function of η and for $M = 1$, 2.5, 4, 5; insulated wall and wind tunnel conditions, $Pr = 0.72$, $T_0^* = 311$K.

One final comment is in order here. In the original derivation of Mack, the density-weighted transformation was not used, as was the case for the compressible mixing layer presented earlier. There is no difficulty in employing such a transformation, and indeed some subsequent researchers found it convenient to do so. The reasons why Mack did not use the density-weighted transformation is because, first, if one stays in the physical coordinate system, physical insight and interpretation is easier than when working in the transformed space since the inverse transformation must be used to back out the solutions to the physical space. Secondly, if the density-weighted transformation is used, the viscous stability equations become slightly more complicated.

5.3.2 Inviscid Fluctuations

For nonzero Mach numbers, the mean profile has an inflection point, defined by Lees & Lin (1946) when $(U'/T)' = 0$; see Result 5.1. Thus, the compressible boundary layer may be unstable to inviscid disturbances. We therefore begin the stability analysis of the compressible boundary layer by examining the inviscid case.

The compressible Rayleigh equation (5.30) governs the inviscid stability of the compressible boundary layer. The boundary conditions in this limit are

$$\Pi'(0) = 0, \quad \phi(0) = 0 \tag{5.86}$$

at the wall, and

$$\Pi, \phi \approx \exp(-\Omega_+ \eta) \tag{5.87}$$

in the freestream, where

$$\Omega_+^2 = \alpha^2 \left[1 - M^2(1 - c)^2\right]. \tag{5.88}$$

Here we use the same general notation as that used for the compressible mixing layer, and so the subscript $+$ denote values in the freestream $\eta \to +\infty$. Let us define c_+ to be the phase speed for which Ω_+^2 vanishes. Thus

$$c_+ = 1 - \frac{1}{M}. \tag{5.89}$$

As in the case of the compressible mixing layer, the disturbances can either be subsonic $(c > c_+)$ or supersonic $(c < c_+)$ in the freestream. The sonic curve (solid) is shown in Fig. 5.9.

In addition to the sonic condition in the freestream, a sonic condition also occurs at the wall. This comes about by noticing that the compressible Rayleigh equation (5.28) is singular whenever $\xi = 0$. Evaluated at $\eta = \infty$ leads to the sonic speed c_+ defined earlier. However, ξ can also vanish within the compressible boundary layer. This will occur first at the wall and leads to the "wall" sonic speed $c_w = \sqrt{T_w}/M$, where $T_w = T(0)$ is the nondimensional temperature at the wall (note that $U(0) = 0$). For an insulated wall, the wall sonic speed c_w is shown in Fig. 5.9 as the dashed curve. As the Mach number increases, a supersonic region exists extending from the wall to the sonic line, located well within the boundary layer. Thus, disturbances can either be subsonic $(c < c_w)$ or supersonic $(c > c_w)$ in a region adjacent to the wall.

As in the case of the compressible mixing layer, the nature of the disturbances and the appropriate boundary conditions can be illustrated by reference to Fig. 5.9, where we plot the sonic speeds c_+ and c_w versus M. As before, there exists four regions in the (c_r, M)-plane. In region 1, the flow is subsonic everywhere within the boundary layer and in the freestream. The phase speed of a neutral mode is thus given by the Lees & Lin condition $c_s = U(\eta_s)$, where η_s is the generalized inflection point found by solving

$$S(\eta) = (U'/T^2)' = 0.$$

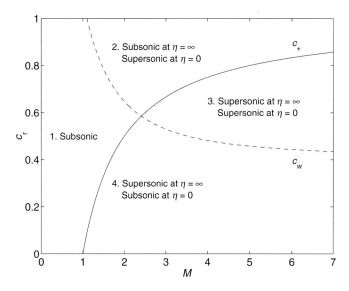

Figure 5.9 Plot of the sonic speed $c_+ = 1 - 1/M$ (solid) and the wall sonic speed $c_w = \sqrt{T_w}/M$ (dashed) versus Mach number. The wall sonic speed is calculated for the case of an insulated wall and wind tunnel conditions.

Fig. 5.10 is a plot of S as a function of η for various values of the Mach number. Note that as the Mach number increases, the point η_s also increases; i.e., the critical layer moves toward the freestream as the Mach number increases. The corresponding unique neutral wavenumber can be found numerically. In region 2, the flow is subsonic in the freestream, but has a supersonic region adjacent to the wall. The Lees & Lin condition still holds in determining the neutral phase speed, but now the wavenumber is no longer unique. In fact, there is an infinite set of wavenumbers, called higher modes. These are primarily acoustic in nature and represent sound waves reflecting from the wall and the relative sonic line. Following the nomenclature of Mack, we label the neutral wavenumbers in regions 1 and 2 as $\alpha_{s,n}$, where the subscript s indicates that the neutral phase speed c_s is determined from the Lees & Lin condition, and n is the mode number. Mack calls these "inflectional" modes. Figure 5.11 is a plot of the neutral phase speed c_s. Note that at a Mach number of about 2.2, the neutral phase speed crosses from region 1 into region 2, where the neutral disturbances first become supersonic at the wall. The corresponding neutral wavenumbers are shown in Fig. 5.12. The associated eigenfunction can be computed for each neutral wavenumber, and it is found that the number of zeros of the eigenfunctions is one less than the mode number.

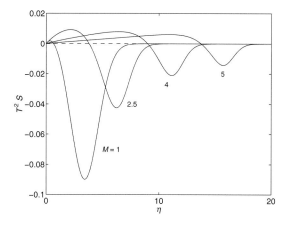

Figure 5.10 Plot of $T^2 S(\eta)$ for various values of the Mach number; insulated wall and wind tunnel conditions.

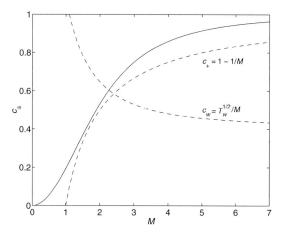

Figure 5.11 Plot of the neutral phase speed c_s as a function of Mach number; insulated wall and wind tunnel conditions. The dashed curves correspond to the sonic speed c_+ and the wall sonic speed c_w.

The maximum temporal growth rates for these modes have also been computed by Mack, and it was found that, below a Mach number of $M = 2.2$, the maximum growth rates are so small that the compressible boundary layer is virtually stable to inviscid disturbances. Above $M = 2.2$, the second mode has the largest growth rate, and it is this mode that will first trigger nonlinearities. Above a Mach number of about $M = 3$, the maximum growth rate of mode 2

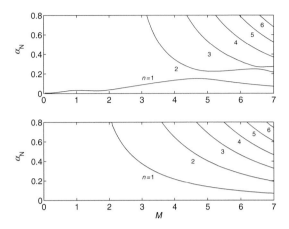

Figure 5.12 Plot of the neutral wavenumbers as a function of Mach number; insulated wall and wind tunnel conditions. Top: inflectional modes $\alpha_{s,n}$ with phase speed c_s. Bottom: noninflectional modes $\alpha_{1,n}$ with phase speed $c_1 = 1$.

becomes larger than that computed from viscous theory. Thus, inviscid instability becomes dominant and viscosity is stabilizing at all Reynolds number, just as in free shear flows. It is this fact that allows the inviscid theory to be used to investigate most of the stability characteristics of the compressible boundary layer.

In addition to the inflectional modes described in the above paragraph for regions 1 and 2, there is also a set of "noninflectional" modes having a neutral phase speed of $c_1 = 1$ of region 2. These modes are thus characterized because the requirement for an inflectional mode, that $(U'/T^2)'$ vanish somewhere within the boundary layer, occurs only in the freestream. The corresponding neutral wavenumbers $\alpha_{1,n}$, where the subscripts 1 indicates that the phase speed is unity and n the mode number, are shown in Fig. 5.12. The importance of the neutral wave with $c_1 = 1$ is that the neighboring waves with $c < 1$ are always unstable. As in the case of inflectional modes, the number of zeros of the associated eigenfunctions for the noninflectional modes is one less than the mode number.

In region 3, the flow disturbance is supersonic both in the freestream and within a region adjacent to the wall, while in region 4 the flow disturbance is subsonic everywhere within the boundary layer but supersonic in the freestream. The appropriate boundary condition is that only outgoing waves are allowed, as was the case for the compressible mixing layer. Although Mack did find solutions in these two regions, their associated maximum growth rates were much smaller than those of regions 1 and 2, and so little attention has been paid to them.

As a final note, we comment here about the various solution techniques employed to compute solutions to the inviscid eigenvalue problem. One solution technique is to solve the compressible Rayleigh equation (5.30) directly with appropriate boundary conditions. Alternatively, one can employ the Riccati transformation

$$G = \frac{\Pi'}{\alpha T \Pi},$$

or some other equivalent form, to get a nonlinear first-order differential equation, as was done previously for the compressible mixing layer. A final technique is to solve the two first-order equations

$$\Pi' = \frac{-i\alpha^2 (U - c)\phi}{T},$$

$$\phi' = \frac{U'\phi}{U - c} + \frac{i\Pi}{U - c}\left[T - M^2(U - c)^2\right],$$

where the prime denotes a derivative with respect to y, the first equation is (5.24) and the second equation is easily derived from the system (5.22)–(5.26); the factor γM^2 has been removed by the scaling of Π. Note that, in this formulation, only the mean flow temperature T and the velocity U and its first derivative are needed. In all cases the contour of numerical integration must be indented below the critical layer into the complex plane for neutral disturbances. The point is this: if one thinks of the solution as a topological surface, and the eigenvalues the zeros of that surface, then it is clear that different forms of the equations will lead to different topologies, and finding the zeros on one topological surface might be easier computationally than on another surface. For this reason, it is generally advised to consider all the various forms of the equations; if difficulties should arise with one set of equations, then try another.

5.3.3 *Viscous Fluctuations*

The dimensional equations governing the flow of a viscous compressible ideal gas in two dimensions are

$$\rho^\dagger_{t^\dagger} + (\rho^\dagger u^\dagger)_{x^\dagger} + (\rho^\dagger v^\dagger)_{y^\dagger} = 0, \qquad (5.90)$$

$$\rho^\dagger \left[u^\dagger_{t^\dagger} + u^\dagger u^\dagger_{x^\dagger} + v^\dagger u^\dagger_{y^\dagger}\right] + p^\dagger_{x^\dagger} = \frac{\partial}{\partial x^\dagger}\left[2\mu^\dagger u^\dagger_{x^\dagger} - \frac{2}{3}\mu^\dagger (u^\dagger_{x^\dagger} + v^\dagger_{y^\dagger})\right]$$

$$+ \frac{\partial}{\partial y^{\dagger}} \left[\mu^{\dagger} (u^{\dagger}_{y^{\dagger}} + v^{\dagger}_{x^{\dagger}}) \right], \tag{5.91}$$

$$\rho^{\dagger} \left[v^{\dagger}_{t^{\dagger}} + u^{\dagger} v^{\dagger}_{x^{\dagger}} + v^{\dagger} v^{\dagger}_{y^{\dagger}} \right] + p^{\dagger}_{y^{\dagger}} = \frac{\partial}{\partial x^{\dagger}} \left[\mu^{\dagger} (u^{\dagger}_{y^{\dagger}} + v^{\dagger}_{x^{\dagger}}) \right]$$

$$+ \frac{\partial}{\partial y^{\dagger}} \left[2\mu^{\dagger} v^{\dagger}_{y^{\dagger}} - \frac{2}{3} \mu^{\dagger} (u^{\dagger}_{x^{\dagger}} + v^{\dagger}_{y^{\dagger}}) \right], \tag{5.92}$$

$$\rho^{\dagger} c_P^{\dagger} \left[T^{\dagger}_{t^{\dagger}} + u^{\dagger} T^{\dagger}_{x^{\dagger}} + v^{\dagger} T^{\dagger}_{y^{\dagger}} \right] - \left[p^{\dagger}_{t^{\dagger}} + u^{\dagger} p^{\dagger}_{x^{\dagger}} + v^{\dagger} p^{\dagger}_{y^{\dagger}} \right]$$

$$= (\kappa^{\dagger} T^{\dagger}_{x^{\dagger}})_{x^{\dagger}} + (\kappa^{\dagger} T^{\dagger}_{y^{\dagger}})_{y^{\dagger}}$$

$$+ \mu^{\dagger} \left[2u^{\dagger}_{x^{\dagger}}{}^2 + 2v^{\dagger}_{y^{\dagger}}{}^2 + 2u^{\dagger}_{y^{\dagger}} v^{\dagger}_{x^{\dagger}} + u^{\dagger}_{y^{\dagger}}{}^2 + v^{\dagger}_{x^{\dagger}}{}^2 \right]$$

$$- \frac{2}{3} \mu^{\dagger} \left[u^{\dagger}_{x^{\dagger}}{}^2 + v^{\dagger}_{y^{\dagger}}{}^2 + 2u^{\dagger}_{x^{\dagger}} v^{\dagger}_{y^{\dagger}} \right], \tag{5.93}$$

and

$$p^{\dagger} = \rho^{\dagger} R T^{\dagger}, \tag{5.94}$$

where $(u^{\dagger}, v^{\dagger})$ are the dimensional velocity components in the $(x^{\dagger}, y^{\dagger})$ directions, respectively, $\rho^{\dagger}$ the density, $T^{\dagger}$ the temperature and $p^{\dagger}$ the pressure. Here, $\mu^{\dagger}$ is the viscosity, $\kappa^{\dagger}$ the thermal conductivity, $c_P^{\dagger}$ the specific heat at constant pressure, assumed constant, and R the gas constant for air. We assume the Prandtl number $Pr = c_P^{\dagger} \mu^{\dagger} / \kappa^{\dagger}$ is constant, and so we eliminate the thermal conductivity from active consideration. For simplicity, the Stokes approximation of zero bulk viscosity has been assumed. As before, these equations are now rendered dimensionless with respect to the freestream values.

The flow field is perturbed by introducing two-dimensional wave disturbances in the velocity, pressure, temperature and density with amplitudes that are functions of η. Following the notation used by Lees & Lin (1946), the fluctuations are given by

$$(u, v, p, T, \rho) = (U, 0, 1, T, \rho)(\eta) + (f, \alpha\phi, \Pi, \theta, r)(\eta)e^{i\alpha(x - ct)}, \tag{5.95}$$

while for the viscosity

$$\mu = \mu(\eta) + s(\eta)e^{i\alpha(x - ct)}. \tag{5.96}$$

The linearized equations for the fluctuations now read as

$$i(U - c)r + i\rho f + (\rho\phi)' = 0, \tag{5.97}$$

$$\rho\left[i(U-c)f+U'\phi\right]+\frac{i\Pi}{\gamma M^2}=\frac{\mu}{\alpha Re}\left[f''+\alpha^2(i\phi'-2f)\right]$$

$$-\frac{2}{3}\frac{\mu\alpha^2}{\alpha Re}(i\phi'-f)+\frac{1}{\alpha Re}\left[sU''+s'U'+\mu'(f'+i\alpha^2\phi)\right],\quad(5.98)$$

$$i\rho(U-c)\phi+\frac{\Pi'}{\alpha^2\gamma M^2}=\frac{\mu}{\alpha Re}\left[2\phi''+if'-\alpha^2\phi\right]$$

$$-\frac{2}{3}\frac{\mu}{\alpha Re}(\phi''+if')+\frac{1}{\alpha Re}\left[isU'+2\mu'\phi'-\frac{2}{3}\mu'(\phi'+if)\right],\quad(5.99)$$

$$\rho\left[i(U-c)\theta+T'\phi\right]+(\gamma-1)(\phi'+if)$$

$$=\frac{\gamma}{\alpha RePr}\left[\mu(\theta''-\alpha^2\theta)+(sT')'+\mu'\theta'\right]$$

$$+\frac{\gamma(\gamma-1)M^2}{\alpha Re}\left[sU'^2+2\mu U'(f'+i\alpha^2\phi)\right],\quad(5.100)$$

and

$$\Pi=\rho\theta+rT.\quad(5.101)$$

As before, primes denote differentiation with respect to the similarity variable η. In addition the viscosity perturbation s can be related to the temperature perturbation via

$$s=\theta\frac{d\mu}{dT}.\quad(5.102)$$

The stability equations can be written in matrix form[3] as

$$\left(\mathbf{A}D^2+\mathbf{B}D+\mathbf{C}\right)\Phi=0,\quad(5.103)$$

where $\Phi=(f,\phi,\Pi,\theta)^\mathrm{T}$, and $\mathbf{A}$ is the 4×4 matrix

$$\mathbf{A}=\begin{bmatrix}1&0&0&0\\0&1&0&0\\0&0&0&0\\0&0&0&1\end{bmatrix}.\quad(5.104)$$

The appropriate boundary conditions are

$$\Phi_1(0)=\Phi_2(0)=\Phi_4(0)=0,\quad\text{at}\quad\eta=0,\quad(5.105)$$

---

[3] The student is asked to compute the elements of the matrices $\mathbf{B}$ and $\mathbf{C}$ in the exercise section.

and

$$\Phi_1, \ \Phi_2, \ \Phi_4 \to 0, \quad \text{as} \quad \eta \to \infty. \tag{5.106}$$

Note that, although the mean profile assumes an insulated wall (i.e., $T'(0) = 0$), it is the temperature perturbation, and not its derivative, that is assumed to vanish (i.e., $\theta(0) = 0$). As pointed out by Malik (1990), this is equivalent to the assumption that the wall will appear insulated on the time scale of the mean flow but not on the short time scales of the disturbances. However, for stationary disturbances (such as crossflow and Görtler), one may need to replace $\theta(0) = 0$ in favor of its derivative or a combination thereof, depending on the physical properties of the solid and the gas.

The stability equations can also be rewritten as a system of first-order equations,[4] given by

$$Z_i' = \sum_{j=1}^{6} a_{i,j} Z_j, \qquad i = 1, \dots, 6, \tag{5.107}$$

where

$$Z_1 = f, \quad Z_2 = f', \quad Z_3 = \phi, \quad Z_4 = \frac{\Pi}{\gamma M^2}, \quad Z_5 = \theta, \quad Z_6 = \theta'.$$

The appropriate boundary conditions here are

$$Z_1(0) = Z_3(0) = Z_5(0) = 0, \quad \text{at} \quad \eta = 0 \tag{5.108}$$

and

$$Z_1, \ Z_3, \ Z_5 \to 0, \quad \text{as} \quad \eta \to \infty. \tag{5.109}$$

Various numerical methods have been presented to solve the compressible stability equations, either the system (5.103) or (5.107). These methods can be classified into initial value methods (IVM) or boundary value methods (BVM). In the IVM, one constructs the solution by integrating the system (5.107) from either $\eta = 0$ to $\eta = \infty$, or in the other direction from $\eta = \infty$ to $\eta = 0$. Initial conditions are needed, and the solution is advanced using a high-order integrator, such as a fourth-order Runge–Kutta method. The eigenvalue is determined by satisfying the appropriate final boundary conditions. Provided a sufficiently good initial guess is known, an iterative procedure can be used to determine the eigenvalue. Eigenvalues are thus determined one at a time. (This method has been extensively discussed in the context of incompressible flows in previous chapters.) An example of the IVM is described by Mack (1965a).

Alternatively, the BVM reduces the differential equations to algebraic equations using either a finite difference discretization or a spectral representation.

---

[4] The student is asked to compute the elements $a_{i,j}$ in the exercise section.

The boundary conditions at both ends are included in a natural way as part of the method. The end result of the BVM leads to the eigenvalue problem of the general form

$$\overline{\mathbf{A}}\Psi = \omega\overline{\mathbf{B}}\Psi, \tag{5.110}$$

or, equivalently,

$$\left|\overline{\mathbf{A}} - \omega\overline{\mathbf{B}}\right| = 0, \tag{5.111}$$

which can be solved by standard matrix eigenvalue packages. Since the number of eigenvalues correspond to the number of grid points, care must be exercised to distinguish true eigenvalues from spurious ones. Various methods are reviewed and compared by Malik (1990).

The effect of the freestream Mach number on the neutral stability curve is shown in Fig. 5.13. For each Mach number, the corresponding inviscid neutral wavenumber is given by the first mode $\alpha_{s,1}$ shown in Fig. 5.12. Here, the Prandtl number is taken to be $Pr = 0.72$. Note that as the Mach number increases, the critical Reynolds number also increases. Note also that the neutral stability curve at $M = 1.6$ looks quite similar to the incompressible case, but at higher values of the Mach number the upper branch turns upward toward the inviscid limit; at a Mach number of 3.8 the viscous upper branch almost entirely vanishes and so the stability characteristics of the flow are governed by inviscid disturbances. As pointed out by Mack in his calculations, inviscid disturbances begin to dominate at $M = 3$ and the stability characteristics are more like those of a free shear layer than of a low-speed zero-pressure gradient boundary layer.

As a final comment, with the neutral curves discussed in this section are determined as a function of Mach number, the instability regions can be examined via either temporal or spatial theory. Early results relied on temporal calculations and used the Gaster transformation to relate the most unstable temporal growth rate to generate the most unstable spatial growth rates, but today one calculates the spatial growth rates directly. Since the boundary layer develops downstream, spatial calculations are more appropriate than temporal calculations.

5.4 Exercises

Exercise 5.1   Consider the linear stability system (5.22)–(5.26).

(a) Show that the system can be reduced to the two first-order equations

$$\alpha^2\rho(U - c)\phi = \frac{i\Pi'}{\gamma M^2}$$

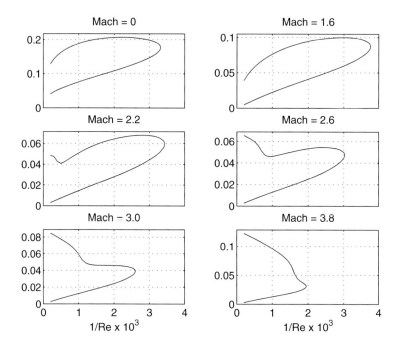

Figure 5.13 Neutral stability curve in the (wavenumber, Reynolds number)-plane for Mach numbers of $M = 0$, 1.6, 2.2, 2.6, 3.0, 3.8; insulated wall and wind tunnel conditions.

and

$$(U - c)\phi' - U'\phi = \frac{i\Pi}{\gamma M^2}\xi,$$

where ξ is defined in (5.29).

(b) Combine these two equations to get the second-order equations (5.27) and (5.28).

(c) Following Gropengiesser (1969), introduce the transformation

$$\chi = \frac{i\Pi}{\gamma M^2 \phi}$$

and derive the following first-order nonlinear differential equation

$$\chi' = \rho\alpha^2(U - c) - \chi\left[\frac{\chi\xi + U'}{U - c}\right],$$

where ξ is defined in (5.29). State the proper boundary conditions in the freestream as $y = \pm\infty$.

Exercise 5.2 Write a numerical code that solves the compressible Rayleigh equation (5.30). Check the code by reproducing Fig. 5.3 using the hyperbolic tangent profile (5.15) with $\beta_T = 1$, $\beta_U = 0$, and $0 < M < 5$. Compare the results with those of the Lock profile by solving (5.8) with $\mu = T$ for the mean velocity and (5.13) for the mean temperature.

Exercise 5.3 Use the following steps to prove Results (5.2)–(5.4).

(a) Show that the compressible Rayleigh equation (5.28) can be written as

$$\left(\frac{\Omega^2 \psi'}{\xi}\right)' - \frac{\alpha^2 \Omega^2}{T}\psi = 0,$$

where

$$\Omega = U - c, \quad \xi = T - M^2 \Omega^2, \quad \psi = \phi/(U - c).$$

(b) To prove Result 5.2 and 5.3, multiply the above equation by ψ^*, the complex conjugate of ψ, integrate over the region $[a,b]$, and set the real and imaginary parts to zero to get the desired results. It is helpful when proving Result 5.3 to use the relation

$$(U - U_{\min})(U - U_{\max}) \leq 0,$$

valid for monotone profiles.

(c) To prove Result 5.4 show that the above equation can be written as

$$\left[\frac{\Omega \chi'}{\xi}\right]' - \chi\left[\frac{1}{2}\left(\frac{U'}{\xi}\right)' + \frac{(U')^2}{4\xi\Omega} + \alpha^2\Omega\right] = 0,$$

where

$$\chi = \Omega^{1/2}\psi.$$

Multiply by χ^*, integrate over the region $[a,b]$, and examine the imaginary part.

Exercise 5.4 Rewrite the compressible Rayleigh equation (5.28) in terms of the similarity variable η.

Exercise 5.5 Compute the elements of the matrices **B** and **C** of equation (5.103).

Exercise 5.6 Compute the elements $a_{i,j}$ of equation (5.107).

Exercise 5.7 Write a numerical solver using the method described by Malik (1990) for the compressible boundary layer with $Pr = 0.72$. Reproduce at least two panels of Fig. 5.13.

6

Centrifugal Stability

Examples ... show how difficult it often is for an experimenter to interpret his results without the aid of mathematics.

– Sir John William Strutt (Lord Rayleigh) (1842–1919)

6.1 Coordinate Systems

Unlike previous chapters, there are many flows that require the formulation of the stability problem to be cast in Cartesian, polar, spherical or cylindrical coordinate systems. For example, pipe flow is perhaps the most notable case where polar coordinates are quite natural, and cylindrical coordinates are the more obvious subset of polar coordinates. There is the case that is now referred to as "stability of Couette flow" and has been examined both theoretically and experimentally by Taylor (1921, 1923). In this case, concentric cylinders rotate relative to each other to produce the flow. Free flows, such as the jet and wake, can be thought of as round rather than planar and, again, cylindrical coordinates are quite natural. If the boundary layer occurs on a curved wall, then Görtler vortices may occur. All of these examples can be described in terms of polar coordinates. And, at the outset, it should be recognized that, not only will the governing mathematics be different from that which has been used up to this point but the resulting physics may have novel characteristics as well.

The prevailing basis for the flows that have been examined up to this point has been that the flows were parallel, or almost parallel, and the governing equations were cast in Cartesian coordinates. Then, the solutions for the disturbances were of the form of plane waves that propagate in the direction of the mean flow or, more generally, obliquely to the mean flow. And, as we have learned, if solid boundaries are present in the flow, viscosity is a cause for instability. Now there are flows that will have curved streamlines and this leads to the possibility of a centrifugal force, and its influence must be considered. A centrifugal force is the apparent force that draws a rotating body away from the center of rotation. For boundary layers on curved walls, this influence is also possible but the flow also has the potential for a secondary instability, which will be discussed in Chapter 9. Such a secondary instability is promoted by the fact that there is now curvature of the stream lines that are due to the Tollmien–Schlichting waves, and hence the dual consequences of a primary mode and a secondary mode.

As an illustration let us consider what may be the stability criterion when there is an effect due to a centrifugal force. For this purpose, consider the flow of a constant density fluid in the absence of viscous effects that is described by curved streamlines. Assume also that a suitable curvilinear coordinate reference system has been established for examining the flow. Then, consider an element of fluid that has a velocity V and with the radius of curvature r for the streamline at that point. If ρ is the density of this element of fluid, a centrifugal force acting on it has the magnitude $\rho V^2/r$ and its angular momentum is rV. The question of stability here centers around the question: If the element of fluid is displaced a small distance in the radial direction to the point $r + \delta r$, where it will have a new velocity $V + \delta V$, will the fluid element return to its original position or not?

Under the assumed conditions for this flow, the angular momentum is conserved and hence

$$r\delta V + V\delta r = 0$$

or

$$\delta V = -(V/r)\delta r. \tag{6.1}$$

And, since $V = V + \delta V$ at the new position $r + \delta r$, use of (6.1) leads to

$$V = V - (V/r)\delta r. \tag{6.2}$$

With these relations, the centrifugal force, F, acting on an element is

$$F = \frac{\rho(V - V/r\,\delta r)^2}{(r + \delta r)}. \tag{6.3}$$

The force on this element due to the pressure gradient is

$$\Delta p = \frac{\rho(V + \delta V)^2}{(r + \delta r)}. \tag{6.4}$$

The question of stability can now be analyzed by examining the balance between these two forces.

Subtracting (6.4) from (6.3) gives

$$F - \Delta p = \frac{\rho(V - V/r\,\delta r)^2}{(r + \delta r)} - \frac{\rho(V + \delta V)^2}{(r + \delta r)}. \tag{6.5}$$

Since $\delta r / r \ll 1$, equation (6.5) can be expanded and, to lowest order, can be written as

$$F - \Delta p = -\frac{2\rho V^2}{r}\left(\frac{\delta r}{r}\right)\left[1 + \frac{r}{V}\frac{\delta V}{\delta r}\right]. \tag{6.6}$$

Stability here implies that the pressure gradient resists the centrifugal force. If this is true, once the fluid parcel is displaced, it will return to its original location. Since there is a negative sign in (6.6), the determination rests with

$$\left(1 + \frac{r}{V}\frac{\delta V}{\delta r}\right) \leq 0. \tag{6.7}$$

In short, if (6.7) is positive, the flow is stable. If (6.7) is zero or negative, the flow is either neutrally stable or unstable.

6.2 Taylor Problem

The problem of determining the stability of the flow that exists between two concentric rotating cylinders is familiarly referred to as the Couette, Taylor or Taylor–Couette problem. Figure 6.1 is an illustration of exactly what Taylor was considering. Note that r is the radial direction, θ is the azimuthal direction and z is the axial direction.

At the outset, the description for this flow should be cast in cylindrical coordinates. For this purpose, let the (x,y)-plane be defined by r and θ where $x = r\cos\theta$ and $y = r\sin\theta$; z will be the common axis along the length of the cylinders. Define u as the velocity in the radial direction r, v as the velocity in the circumferential direction θ and w as the velocity in the axial direction z. With this description, the nondimensional governing equations are

$$\frac{1}{r}\frac{\partial(ru)}{\partial r} + \frac{1}{r}\frac{\partial v}{\partial\theta} + \frac{\partial w}{\partial z} = 0, \tag{6.8}$$

$$\frac{\partial u}{\partial t} + u\frac{\partial u}{\partial r} + \frac{v}{r}\frac{\partial u}{\partial\theta} + w\frac{\partial u}{\partial z} - \frac{v^2}{r} = -\frac{\partial p}{\partial r} + Re^{-1}\left(\nabla^2 u - \frac{u}{r^2} - \frac{2}{r^2}\frac{\partial v}{\partial\theta}\right), \tag{6.9}$$

$$\frac{\partial v}{\partial t} + u\frac{\partial v}{\partial r} + \frac{v}{r}\frac{\partial v}{\partial\theta} + w\frac{\partial v}{\partial z} + \frac{uv}{r} = -\frac{1}{r}\frac{\partial p}{\partial\theta} + Re^{-1}\left(\nabla^2 v - \frac{v}{r^2} + \frac{2}{r^2}\frac{\partial u}{\partial\theta}\right), \tag{6.10}$$

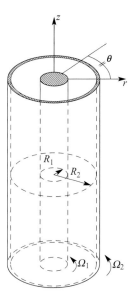

Figure 6.1 Concentric rotating cylinders.

$$\frac{\partial w}{\partial t} + u\frac{\partial w}{\partial r} + \frac{v}{r}\frac{\partial w}{\partial \theta} + w\frac{\partial w}{\partial z} = -\frac{\partial p}{\partial z} + Re^{-1}\nabla^2 w, \qquad (6.11)$$

where

$$\nabla^2 = \frac{\partial^2}{\partial r^2} + \frac{1}{r}\frac{\partial}{\partial r} + \frac{1}{r^2}\frac{\partial^2}{\partial \theta^2} + \frac{\partial^2}{\partial z^2}. \qquad (6.12)$$

All variables have been nondimensionalized with respect to the gap width, $\ell = R_2 - R_1$ (where R_1 and R_2 are the radii of the inner and outer cylinders) and a characteristic velocity U_0; so that the time scale ℓ/U_0 is the inertia time scale. The Reynolds number, $Re = U_0\ell/\nu$, and the density, ρ, are constants.

The solution for the mean flow is one that is in the θ-direction and a function of r only or $V = V(r)$. Likewise, the mean pressure, P, is taken as only a function of r. Thus, $V = Ar + B/r$ is the dimensional solution that meets these requirements and the pressure, P, can be obtained from the relation $dP/dr = \rho V^2/r$. The coefficients A and B are fixed by requiring the value of V to be that of the cylinders at $r = R_1$ and R_2 and are found to be $A = \left[\Omega_2 R_2^2 - \Omega_1 R_1^2\right]/\left[R_2^2 - R_1^2\right]$; $B = \left[\Omega_1 - \Omega_2\right]/\left[R_1^{-2} - R_2^{-2}\right]$, when the cylinders rotate in the same direction. If the rotation is in opposite directions, simply replace Ω_1 by $-\Omega_1$.

Similar to the case for parallel flows, assume that the flow can be decomposed into a laminar basic flow and a fluctuating component that oscillates about the basic flow such that

$$u = \tilde{u}(r,\theta,z,t),$$

$$v = V(r) + \tilde{v}(r,\theta,z,t),$$

$$w = \tilde{w}(r,\theta,z,t),$$

$$p = P(r) + \tilde{p}(r,\theta,z,t),$$

which is substituted into (6.8) to (6.11) and linearized to give

$$\frac{1}{r}\frac{\partial(r\tilde{u})}{\partial r} + \frac{1}{r}\frac{\partial\tilde{v}}{\partial\theta} + \frac{\partial\tilde{w}}{\partial z} = 0, \tag{6.13}$$

$$\frac{\partial\tilde{u}}{\partial t} + \frac{V}{r}\frac{\partial\tilde{u}}{\partial\theta} - 2\frac{V}{r}\tilde{v} = -\frac{\partial\tilde{p}}{\partial r} + Re^{-1}\left(\nabla^2\tilde{u} - \frac{\tilde{u}}{r^2} - \frac{2}{r^2}\frac{\partial\tilde{v}}{\partial\theta}\right), \tag{6.14}$$

$$\frac{\partial\tilde{v}}{\partial t} + \frac{V}{r}\frac{\partial\tilde{v}}{\partial\theta} + \left(\frac{dV}{dr} + \frac{V}{r}\right)\tilde{u} = -\frac{1}{r}\frac{\partial\tilde{p}}{\partial\theta} + Re^{-1}\left(\nabla^2\tilde{v} - \frac{\tilde{v}}{r^2} + \frac{2}{r^2}\frac{\partial\tilde{u}}{\partial\theta}\right), \tag{6.15}$$

$$\frac{\partial\tilde{w}}{\partial t} + \frac{V}{r}\frac{\partial\tilde{w}}{\partial\theta} = -\frac{\partial\tilde{p}}{\partial z} + Re^{-1}\nabla^2\tilde{w}. \tag{6.16}$$

Note that $dV/dr + V/r = 2A$ for Taylor–Couette flow. The solutions of these equations must satisfy the boundary conditions that arise from requiring the velocity fluctuations $(\tilde{u}, \tilde{v}, \tilde{w})$ to vanish on both the inner and the outer cylinder walls.

A general solution to this set of linear equations was not attempted in the original theoretical work of Taylor (1923) because he had significant results from his experiments (Taylor, 1921) to justify simplifications. First, he assumed only axisymmetric disturbances and hence he omitted θ-dependence. Second, he assumed that the cylinders were fixed in such a way that the gap width, ℓ, is small or, in the sense of the nondimensional variables, $\ell/r \ll 1$.

When axisymmetry $(\partial/\partial\theta = 0)$ is incorporated into (6.13)–(6.16) and the pressure is eliminated, then the pair of coupled equations,

$$\left[\frac{\partial}{\partial t} - Re^{-1}\left(\nabla^2 - \frac{1}{r^2}\right)\right]\left(\nabla^2\tilde{u} - \frac{\tilde{u}}{r^2}\right) = 2\left(\frac{V}{r}\right)\frac{\partial^2\tilde{v}}{\partial z^2} \tag{6.17}$$

and

$$\left[\frac{\partial}{\partial t} - Re^{-1}\left(\nabla^2 - \frac{1}{r^2}\right)\right]\tilde{v} = -\left(\frac{dV}{dr} + \frac{V}{r}\right)\tilde{u} \tag{6.18}$$

result. Equations (6.17) and (6.18) bear a strong resemblance to those for perturbations in a parallel mean flow, namely Orr–Sommerfeld and Squire equations. In this case, however, the equations do not uncouple and the full problem remains sixth order. Solutions for $\tilde{u}$ and $\tilde{v}$ can be obtained in terms of normal modes or

$$\{\tilde{u},\tilde{v}\}(r,z,t) = \{\hat{u},\hat{v}\}(r)e^{\sigma t+i\lambda z}. \tag{6.19}$$

As a result of this form for the solutions, (6.17) and (6.18) now become coupled ordinary differential equations.

Even when the assumption of axisymmetry is not used, the general set of equations (6.13)–(6.16), can be reduced to ordinary differential equations and the solutions can be expressed in terms of known eigenfunctions, namely Bessel functions. The resulting eigenvalue problem that determines the stability will be given as a function of the parameters $F(\sigma,\lambda,R_2/R_1,\Omega_2/\Omega_1,Re) = 0$.

The second approximation made by Taylor was to limit the analysis to the case where the gap width is small. In these terms, the operator, ∇^2 is of order $1/\ell^2$ and therefore the two terms, $\hat{u}/r$ and $\hat{v}/r$ as well as $1/r^2$, can be neglected. The net result of axisymmetry, the small gap approximation, and the normal mode solutions reduces (6.17) and (6.18) to

$$\left[Re^{-1}\left(\frac{d^2}{dr^2}-\lambda^2\right)-\sigma\right]\left(\frac{d^2\hat{u}}{dr^2}-\lambda^2\hat{u}\right) = 2\lambda^2\left(\frac{V}{r}\right)\hat{v} \tag{6.20}$$

and

$$\left[Re^{-1}\left(\frac{d^2}{dr^2}-\lambda^2\right)-\sigma\right]\hat{v} = \left(\frac{dV}{dr}+\frac{V}{r}\right)\hat{u}, \tag{6.21}$$

with the boundary conditions $\hat{u} = d\hat{u}/dr = \hat{v} = 0$ at the cylinder walls.

The small gap approximation also affects the variation of the mean velocity profile. Specifically, the part of mean flow that varies as $1/r$ can be neglected just as it was for the operators in the perturbation equations. Thus, $V(r) = Ar$, a linear variation, and is the reason that the flow has been referred to as Couette flow.

Equations (6.20) and (6.21) have been formulated in other ways by other authors (cf. Drazin & Reid, 1984) but the result is the same save for the designated nondimensional parameter. And, instead of the Reynolds number, Re, for example, it is more common to use the Taylor number, Ta. Moreover, no statement was made as to whether or not the cylinders are rotating in the same or in opposite directions. Then, as a minor point, in lieu of the velocity component u, a stream function is sometimes introduced and the two governing equations are in terms of the stream function and the velocity component in the θ-direction. These points will be clarified in more detail but, at this stage, this set of equations can be used to demonstrate what has become known as the principle of exchange of stabilities. Unlike the perturbations in parallel or almost parallel mean flows, it can be shown here that, if the disturbances are periodic in time, then they must decay in time. As a consequence, there are no traveling unstable waves and neutral stability corresponds to time independent or stationary behavior. It should be noted that Taylor's experiments verified these conclusions very well.

First, define $\varepsilon = Re^{-1}$. Then, expand (6.20) and (6.21) so that all terms involving the second and fourth derivatives as well as the dependent variables can be grouped. These reordered equations are then multiplied by the respective complex conjugates of $\hat{u}$ and $\hat{v}$. Once this has been done, the product equation relations are integrated over r from R_1 to R_2; note that since we are dealing with nondimensional equations, R_1 and R_2 are scaled by ℓ and are therefore the nondimensional radii. By integrating by parts and invoking the boundary conditions, it is found that

$$\varepsilon \int_{R_1}^{R_2} |\hat{u}''|^2 dr + (\sigma + 2\varepsilon\lambda^2) \int_{R_1}^{R_2} |\hat{u}'|^2 dr + \lambda^2(\sigma + \varepsilon\lambda^2) \int_{R_1}^{R_2} |\hat{u}|^2 dr$$

$$= 2\lambda^2 \int_{R_1}^{R_2} \frac{V}{r} \hat{u}^* \hat{v} dr \qquad (6.22)$$

and

$$\varepsilon \int_{R_1}^{R_2} |\hat{v}'|^2 dr + (\sigma + \varepsilon\lambda^2) \int_{R_1}^{R_2} |\hat{v}|^2 dr = -\left(\frac{dV}{dr} + \frac{V}{r}\right) \int_{R_1}^{R_2} \hat{u}\hat{v}^* dr. \qquad (6.23)$$

By defining the integrals of $|\hat{u}''|^2$, $|\hat{u}'|$ and $|\hat{u}|^2$ as I_2^2, I_1^2 and I_0^2, all > 0, (6.22) can be written compactly as

$$\varepsilon I_2^2 + (\sigma + 2\varepsilon\lambda^2)I_1^2 + \lambda^2(\sigma + \varepsilon\lambda^2)I_0^2 = 2\lambda^2 \frac{V}{r} \int_{R_1}^{R_2} \hat{u}^* \hat{v} dr. \qquad (6.24)$$

Analogously, for the integrals of $|\hat{v}'|^2$ and $|\hat{v}|^2$ as J_1^2 and J_0^2, all > 0, (6.23) becomes

$$\varepsilon J_1^2 + (\sigma + \varepsilon\lambda^2)J_0^2 = -\left(\frac{dV}{dr} + \frac{V}{r}\right) \int_{R_1}^{R_2} \hat{u}\hat{v}^* dr. \qquad (6.25)$$

If $\sigma = \sigma_r + i\sigma_i$ and the respective right-hand sides of (6.24) and (6.25) are similarly denoted as real and imaginary parts, then the following relations are obtained:

$$\sigma_r(I_1^2 + \lambda^2 I_0^2) + \varepsilon(I_2^2 + 2\lambda^2 I_1^2 + \lambda^4 I_0^2) = 2\lambda^2 \frac{V}{r} \int_{R_1}^{R_2} (\hat{u}^* \hat{v})_r dr \qquad (6.26)$$

and

$$\sigma_i(I_1^2 + \lambda^2 I_0^2) = 2\lambda^2 \frac{V}{r} \int_{R_1}^{R_2} (\hat{u}^* \hat{v})_i dr \qquad (6.27)$$

for (6.24), and

$$\sigma_r J_0^2 + \varepsilon(J_1^2 + \lambda^2 J_0^2) = -\left(\frac{dV}{dr} + \frac{V}{r}\right)\int_{R_1}^{R_2}(\hat{u}\hat{v}^*)_r dr \qquad (6.28)$$

$$\sigma_i J_0^2 = -\left(\frac{dV}{dr} + \frac{V}{r}\right)\int_{R_1}^{R_2}(\hat{u}\hat{v}^*)_i dr \qquad (6.29)$$

for (6.25).

The two imaginary relations for σ_i can be combined by recognizing the relations of complex conjugate pairs. This means the integrals on the right-hand sides of (6.27) and (6.29) differ only by a minus sign and

$$\sigma_i\left[\frac{2\lambda^2(V/r)}{\frac{dV}{dr}+\frac{V}{r}} - \frac{(I_1^2+\lambda^2 I_0^2)}{J_0^2}\right] = 0 \qquad (6.30)$$

results. For this to be true, either $\sigma_i = 0$ or the bracketed terms must balance. If $\sigma_i \neq 0$, then the disturbances are periodic in time. In turn, for this to occur, the ratio $2\lambda^2(V/r)/(dV/dr + V/r) > 0$, implying that the mean profile must increase outward. Certainly, if $V > 0$, $V' > 0$. Let us call the necessary relation as $2\lambda^2(V/r)/(dV/dr + V/r) = P > 0$.

Now, the expression for the real parts of σ_r can be used. In fact, it is found that

$$\sigma_r = -\frac{(E+B/P)}{(D+A/P)}, \qquad (6.31)$$

where $A = (I_1^2 + \lambda^2 I_0^2)$, $B = \varepsilon(I_2^2 + 2\lambda^2 I_1^2 + \lambda^4 I_0^2)$, $D = J_0^2$, $E = \varepsilon(J_1^2 + \lambda^2 J_0^2)$. Thus, $\sigma_r < 0$ and, consequently, a disturbance that is periodic in time must decay. The combination indicates that neutral stability corresponds to $\sigma = \sigma_r + i\sigma_i = 0$. This has become known as the principle of exchange of stabilities.

By using the principle just established, the neutral solution for the Taylor problem can now be determined. After setting $\sigma = 0$ in (6.20) and (6.21) we have

$$(D^2 - \lambda^2)^2\hat{u} = 2\lambda^2\left(\frac{V}{r}\right)Re\,\hat{v}, \qquad (6.32)$$

$$(D^2 - \lambda^2)\hat{v} = \left(\frac{dV}{dr} + \frac{V}{r}\right)Re\,\hat{u}, \qquad (6.33)$$

where the Reynolds number has been moved to the right-hand side for these relations. And, strictly speaking, for the narrow gap approximation, $V/r = A$, with A an angular velocity. These two equations can be combined to form one for $\hat{u}$, namely

$$(D^2 - \lambda^2)^3 \hat{u} = 2\lambda^2 Re^2 \left(\frac{V}{r}\right)\left(\frac{dV}{dr} + \frac{V}{r}\right)\hat{u}. \tag{6.34}$$

Now, the Taylor number emerges and is defined as $Ta = -2Re^2(V/r)(dV/dr + V/r)$. For the small gap approximation, (6.34) can be solved exactly. When the cylinders rotate in opposite directions, then $V = 0$ for some value of r between R_1 and R_2, and the solution must account for this fact and this has been done. Moreover, the flow in concentric cylinders has been investigated with arbitrary gap (Kirchgässner, 1961; Meister, 1962), a pressure gradient acting around the cylinders (DiPrima, 1959), and for non-axisymmetric disturbances (DiPrima, 1961).

The role of viscosity in this system is simpler than that for parallel or almost parallel flows and is only damping and thereby stabilizing. If the Rayleigh (1880) criterion, namely that a necessary condition for instability for centrifugal flow is that the gradient of the mean vorticity must change sign within the bounds of the flow is invoked, this can be seen directly. This need is essentially that of the inflection point criterion for parallel flow. For inviscid disturbances, the criterion for instability for the rotating cylinders was explicitly shown by Synge (1938) and requires

$$\frac{d}{dr}(rV)^2 \equiv \frac{d}{dr}\left(r^2\Omega\right)^2 < 0, \tag{6.35}$$

where $\Omega = V/r$, the angular velocity. When this condition is satisfied, the flow may be unstable. Conversely, if

$$\frac{d}{dr}\left(r^2\Omega\right)^2 > 0 \tag{6.36}$$

everywhere in the flow, then the flow is stable to all axisymmetric perturbations. This result is equivalent to

$$(\Omega_2 R_2^2 - \Omega_1 R_1^2)V(r) > 0. \tag{6.37}$$

That is, $\Omega_2 R_2^2 > \Omega_1 R_1^2$ and, in the inviscid limit, $\Omega_1/\Omega_2 = R_2^2/R_1^2$. This is the celebrated Rayleigh's criteria for instability for rotating flows. For the values of Taylor's experiment, Fig. 6.2 vividly shows how well (6.35) is valid. The additional curve in this plot essentially shows the effects of viscosity, and the plot is valid for cylinders rotating in the same and opposite directions.

With viscosity, and in terms of the Taylor number, there is also a critical value for this parameter. The value obtained is $Ta = 1708$ for $\lambda = 3.13$. When the critical value for Ta is exceeded, then the flow has an instability and Taylor noted that the flow takes on a system of steady state vortices, whose axes

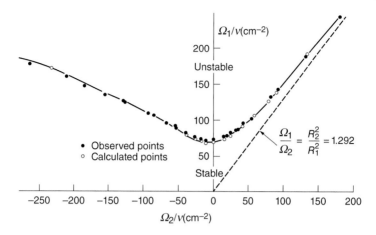

Figure 6.2 Results for narrow gap calculations and experiment (after Taylor, 1923).

are located along the circumference with alternately opposite directions. Figure 6.3 illustrates this result, and it is for this reason that such vortices are now known as Taylor vortices. It should be noted here that there are several other problems that can be investigated within this same framework. One problem involves density perturbations arising from gravitational and mean temperature fields but no mean flow. A second problem is that of the flow in an electrically conducting fluid with a magnetic field. Neither of these cases will be considered here but the flow over a wavy wall be analyzed, a problem that has become known as one that leads to Görtler vortices (Görtler, 1940a,b).

6.3 Görtler Vortices

Görtler (1940a,b) made a significant contribution with his investigation of the stability of a boundary layer on a concave wall. This is a problem that involves centrifugal instability as well as the kind of disturbances that have already been shown to exist in boundary layers on flat walls. Thus, unlike the Blasius boundary layer, both the mathematics for the linear system as well as the consequential physics are different and the differences are directly due to role of the centrifugal force in this flow. The problem has subsequently been refined with the work of Smith (1955), Meksyn (1950), Hämmerlin (1955) and Witting (1958), but the fundamentals remain the same. Experiments in this kind

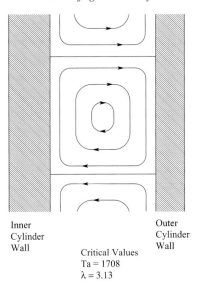

Inner
Cylinder
Wall

Outer
Cylinder
Wall

Critical Values
Ta = 1708
λ = 3.13

Figure 6.3 Neutral stability for rotating cylinders.

of flow were made even earlier than the theory with the work of Clauser &
Clauser (1937), where the effect of curvature on the transition of the boundary layer was considered. Then, the Görtler bases have since provided a means
whereby secondary instability has been examined. In short, secondary instability can result in a flat-plate boundary layer once Tollmien–Schlichting waves
have been excited. The initial contributions that used this concept were due to
Görtler & Witting (1958) and Witting (1958).

Figure 6.4 illustrates the basic flow and the coordinate system that is used.
The linearized perturbation equations for this flow that were developed by
Görtler (1940a,b) invoke approximations: (1) The centrifugal force is neglected
in determining the mean flow. As a result, the mean velocity, U, is taken as almost parallel, a common approximation in boundary layer stability analysis,
and therefore $U = U(y)$ and there is no mean component in the y-direction. (2)
The boundary layer thickness, δ, is taken to be much smaller than the radius of
curvature, R. In fact, the equations are effectively developed with δ/R as the
small parameter. The net result, save for the choice of independent coordinates,
makes this problem effectively the same as the narrow gap concentric cylinder
problem given by Taylor (1923) and discussed in Section 6.2.

It is assumed that the perturbations are independent of x, making the analysis
one that is local. The net result is a linear system that is a coupled sixth-order

problem and can be expressed as the coupling of one fourth-order together with one second-order equation. With solutions given by normal modes where

$$\{\tilde{u}, \tilde{v}, \tilde{w}, \tilde{p}\}(r, z, t) = \{\hat{u}, \hat{v}, \hat{w}, \hat{p}\}(r)e^{i\gamma z + \omega t}.$$

With δ and U_0 the basic scales, these equations, expressed nondimensionally, are

$$\left(\frac{d^2}{d\eta^2} - \alpha^2\right)\left(\frac{d^2}{d\eta^2} - \alpha^2 - \sigma\right)\hat{v} = -2\alpha^2\overline{Re}U\hat{u}, \qquad (6.38)$$

and

$$\left(\frac{d^2}{d\eta^2} - \alpha^2 - \sigma\right)\hat{u} = U'\hat{v}, \qquad (6.39)$$

with $\alpha = \gamma\delta$, $\eta = y/\delta$, $\sigma = \omega\delta^2/\nu$ and $\overline{Re} = Re^2(\delta/R)$; $Re = U_0\delta/\nu$. The boundary conditions require $\hat{v} = \hat{v}' = \hat{u} = 0$ at $\eta = 0$ and approaches zero as $\eta \to \infty$.

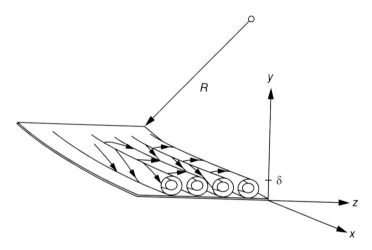

Figure 6.4 Basic flow and coordinate system for Görtler vortices (after Betchov & Criminale, 1967).

The analogy of this perturbation system to that of the narrow gap concentric cylinders is clear except, that is, for the boundary conditions. The condition of the principle of exchange of stabilities has been used although it is not, strictly speaking, a proven tool for this flow. Specific results for various values of the amplification factor have been obtained numerically for a range of specific mean velocity profiles, including Blasius, asymptotic suction, and a piecewise-linear profile. Although results can be expressed graphically in terms of a Görtler number as a function of the wavenumber, they can just as well be displayed as has been done for the Blasius boundary layer but where the Reynolds number is modified to include the δ/Re ratio. Smith (1955) calculated the stability loci for the Blasius profile. Hämmerlin (1955) made calculations to compare the three mean profiles. Regardless, the important conclusion is the fact that even a moderately curved wall will lead to an earlier breakdown of the boundary layer than that on a flat plate.

The problem where there is flow in a curved channel has been termed Taylor–Dean flow. The linear system remains one that is the coupled fourth and second order and the boundary conditions are to applied at the two boundaries. A general reference for results can be found in Drazin & Reid (1984).

6.4 Pipe Flow

The stability of the flow in a pipe of circular cross section has been and continues to be an enigma in the field of hydrodynamic stability for, up to this time, it is a flow that appears to be immune to linear disturbances. Yet this flow prototype is the one that was used by Reynolds to demonstrate the transition from laminar to turbulent flow as illustrated in Fig. 1.3. And it is a flow that is analogous to that of plane Poiseuille flow, a flow that is unstable. It is now believed that the cause of breakdown and transition in the pipe is due to nonlinearity. At the same time, it should be noted, the full three-dimensional problem is rather involved and the linear system does not have the benefit of the Squire transformation, even though this is a parallel flow, strictly speaking.

Consider pipe flow as shown in Fig. 6.5, where the mean flow is given as $\underline{U} = (U(r),0,0)$, with a pressure, P, that is a function of x only. The perturbation variables will be $\underline{u} = (\tilde{u},\tilde{v},\tilde{w})$ in the x, r and θ directions, respectively. Also, $r^2 = y^2 + z^2$ and $\tan\theta = z/y$ in this coordinate system. The mean flow variation is $U(r) = 1/4P(a^2 - r^2)$, where P is the pressure gradient in the x-direction and a is the radius of the pipe.

With this notation, the linearized perturbation equations in nondimensional form become

$$\frac{1}{r}\frac{\partial(r\tilde{v})}{\partial r} + \frac{1}{r}\frac{\partial\tilde{w}}{\partial\theta} + \frac{\partial\tilde{u}}{\partial x} = 0, \tag{6.40}$$

$$\frac{\partial\tilde{v}}{\partial t} + U\frac{\partial\tilde{v}}{\partial x} = -\frac{\partial\tilde{p}}{\partial r} + Re^{-1}\left(\nabla^2\tilde{v} - \frac{\tilde{v}}{r^2} - \frac{2}{r^2}\frac{\partial\tilde{w}}{\partial\theta}\right), \tag{6.41}$$

$$\frac{\partial\tilde{w}}{\partial t} + U\frac{\partial\tilde{w}}{\partial x} = -\frac{1}{r}\frac{\partial\tilde{p}}{\partial\theta} + Re^{-1}\left(\nabla^2\tilde{w} - \frac{\tilde{w}}{r^2} + \frac{2}{r^2}\frac{\partial\tilde{v}}{\partial\theta}\right), \tag{6.42}$$

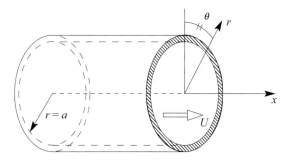

Figure 6.5 Poiseuille pipe flow.

$$\frac{\partial \tilde{u}}{\partial t} + U\frac{\partial \tilde{u}}{\partial x} + \tilde{v}\frac{dU}{dr} = -\frac{\partial \tilde{p}}{\partial x} + Re^{-1}\nabla^2\tilde{u}, \tag{6.43}$$

with ∇^2 defined in polar coordinates, just as was done for the Taylor problem.

As can be seen, this set of equations bears a strong resemblance to those needed for the concentric cylinders. Thus, it is clear that they are fully coupled. However, if the assumption of axisymmetry is made together with normal modes for solutions, then some simplification does result. With the notation,

$$\{\tilde{u},\tilde{v},\tilde{w},\tilde{p}\}(r,x,t) = \{\hat{u},\hat{v},\hat{w},\hat{p}\}(r)e^{i(\alpha x - \omega t)},$$

the above equations are

$$\frac{1}{r}\frac{d(r\hat{v})}{dr} + i\alpha\hat{u} = 0, \tag{6.44}$$

$$i(\alpha U - \omega)\hat{v} = -\hat{p}' + Re^{-1}\left(\Delta\hat{v} - \frac{\hat{v}}{r^2}\right), \tag{6.45}$$

$$i(\alpha U - \omega)\hat{w} = Re^{-1}\left(\Delta\hat{w} - \frac{\hat{w}}{r^2}\right), \tag{6.46}$$

$$i(\alpha U - \omega)\hat{u} + U'\hat{v} = -i\alpha\hat{p} + Re^{-1}\Delta\hat{u} \tag{6.47}$$

with $\Delta = d^2/dr^2 + (1/r)d/dr - \alpha^2$. These equations must be solved subject to the boundary conditions that $\hat{u} = \hat{v} = \hat{w} = 0$ at the boundary of the pipe ($r = a$, in dimensional form) and be finite at $r = 0$.

Observations: (i) the governing equation for the θ component of the velocity, $\hat{w}$, does uncouple from the others, making the problem one that is pseudo two-dimensional; (ii) the equations for $\hat{u}$ and $\hat{v}$ can now be combined to generate one fourth-order ordinary equation; (iii) this equation would normally be in terms of the velocity component $\hat{v}$ as the dependent variable but, with the uncoupling, $\hat{u}$ and $\hat{v}$ can instead be replaced by a perturbation streamfunction, $\tilde{\psi}$, defined as $r\tilde{v} = \partial\tilde{\psi}/\partial x$, $r\tilde{u} = -\partial\tilde{\psi}/\partial r$. And, with the normal mode solution as $\tilde{\psi}(x,r,t) = \hat{\psi}(r)e^{i\alpha x}e^{-i\omega t}$, the governing equation is

$$(\alpha U - \omega)\left(D^2 - \alpha^2\right)\hat{\psi} - \alpha r\left(\frac{U'}{r}\right)'\hat{\psi} = \frac{1}{iRe}\left(D^2 - \alpha^2\right)^2\hat{\psi}, \qquad (6.48)$$

with $D^2 = d^2/dr^2 - (1/r)d/dr$. The boundary conditions are now $\hat{\psi} = \hat{\psi}' = 0$ at the wall of the pipe but now $\hat{\psi}/r$ and $\hat{\psi}'/r$ must be finite at $r = 0$.

Equation (6.48) bears a strong resemblance to the Orr–Sommerfeld equation for plane Poiseuille flow. And, in like manner, this equation is one for vorticity, strictly speaking. Consequently, the flow could be expected to be unstable. But, for pipe flow, the analogy is somewhat deceptive in that a crucial term, namely the term that represents the interaction of the mean vorticity with the perturbation field, $(U'/r)'$, is identically zero for the mean flow profile, $U(r)$, that varies as r^2. This being the case, then (6.48) is more like that for plane Couette flow, a flow that is well known to be stable. All eigenvalues that have been obtained for pipe flow are likewise all damped, as shown by Davey & Drazin (1969) or Drazin & Reid (1984), for example.

6.5 Rotating Disk

Since it involves a wealth of physics, the determination of the stability of the flow on a rotating disk is a problem that has interesting features. First, for the mean steady flow, it combines the flow of a boundary layer on a finite circular flat plate with the additional centrifugal force due to the rotation. As a result, the boundary layer is constrained and the solution for this flow is one that has an exact similarity solution of the Navier–Stokes equations (von Kármàn, 1921). In these terms, the mean profile is three-dimensional and the spatial dependence of the components is independent of the radius in these terms. This is also true for the boundary layer thickness. Then, just as we have seen for other problems where the centrifugal force is important, the perturbations must be analyzed with full three-dimensionality in order to make meaningful conclusions. This flow is also unstable in the inviscid limit due to the fact that the mean velocity profile possesses at least one inflection point and thereby satisfies the criterion dictated by the Rayleigh theorem (Section 2.4) for instability.

The original work in this area was due to Gregory, Stuart & Walker (1955) and included other related flows within this framework, such as that over swept wings or rotating boundary layers on curved boundaries as well as the more idealized flat disk. This collaborative work was both experimental as well as analytical. Since this time, there have been numerous additional contributions, even for the case of a flexible rather than a solid boundary and exploration of spatial as well as temporal stability. Among others of note are those due to Malik, Wilkinson & Orszag (1981), Wilkinson & Malik (1985), Malik (1986). Spatially, see Lingwood (1995, 1996); for the flexible wall disk, see Cooper & Carpenter (1997).

Consider the rotating disk and the coordinate system for this flow as defined in Fig. 6.6. From the list of references for this problem, the most illustrative is that due to Lingwood (1995) and will be followed here. The mean flow determined by von Kármàn (1921) is $\underline{U} = (U,V,W)$ and each component is a function of z, the coordinate normal to the plate due the similarity form of the solutions and defined relative to the disk. These three components, in dimensionless form, are defined as

$$\overline{U}(z) = U/(r\Omega), \quad \overline{V}(z) = V/(r\Omega) \quad \text{and} \quad \overline{W}(z) = W/\sqrt{v\Omega}. \qquad (6.49)$$

Here, Ω is the constant angular frequency of the disk about the axis perpendicular to the disk and v is the kinematic viscosity. A mean length scale, L, is defined in terms of the kinematic viscosity and the angular frequency or $L = (v/\Omega)^{1/2}$. Likewise, z is nondimensional with respect to L but r is the dimensional polar coordinate in the plane (along with the angle θ) of the disk. As determined by Lingwood, these three components of the mean velocity are shown in Fig. 6.7. A more revealing velocity profile is one that is defined in terms of ε, the angle between the radial direction of the flow and rotation. Again, referring to Lingwood, this is defined as $Q(z)$, is termed the resolved velocity and is given by

$$Q(z) = \overline{U}(z)\cos\varepsilon + \overline{V}(z)\sin\varepsilon. \qquad (6.50)$$

This is effectively the radial profile and is displayed in Fig. 6.8 for the full range of values for ε. It is clear that this profile is both inflectional and has reverse flow.

The stability analysis is performed by what is known as local. This means that perturbations are introduced at a specific value of the radius, $r = r_a$. Then, with a Reynolds number defined as $Re = r_a\Omega L/v = r_a/L$, the instantaneous velocity field and pressure are given by

$$\begin{aligned}
u(r,\theta,z,t) &= (rRe^{-1})\overline{U}(z) + \tilde{u}(r,\theta,z,t), \\
v(r,\theta,z,t) &= (rRe^{-1})\overline{V}(z) + \tilde{v}(r,\theta,z,t), \\
w(r,\theta,z,t) &= Re^{-1}\overline{W}(z) + \tilde{w}(r,\theta,z,t), \\
p(r,\theta,z,t) &= Re^{-2}\overline{P}(z) + \tilde{p}(r,\theta,z,t).
\end{aligned} \right\} \qquad (6.51)$$

These variables are substituted into the incompressible Navier–Stokes equations, linearized and the mean flow subtracted so that a coupled set of perturbation equations are found. In order to proceed further, however, further

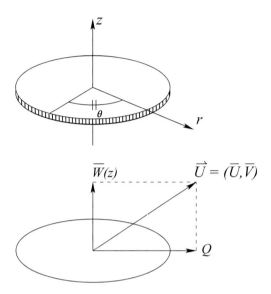

Figure 6.6 Coordinate system for the rotating disk.

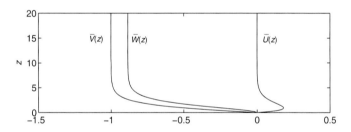

Figure 6.7 Mean velocity components for the rotating disk: $\overline{U}(z)$, $\overline{V}(z)$, $\overline{W}(z)$; $\overline{U}(z) \to 0$, $\overline{V}(z) \to -1$ and $\overline{W}(z) \to -0.8838$ as $z \to \infty$ (after Lingwood, 1995, reproduced with permission).

approximations are made, namely dependence of the Reynolds number due to the radius is ignored. This must be done even though the thickness of the boundary on the rotating disk is constant. With this step, a separable normal mode form for the solutions of the perturbations can be assumed and becomes

$$\{\tilde{u}, \tilde{v}, \tilde{w}, \tilde{p}\}(r, \theta, t, z) = \{\hat{u}, \hat{v}, \hat{w}, \hat{p}\}(z)e^{i\alpha r + i\beta\theta - i\omega t}.$$

Then, in addition to the linearization, all terms of $O(Re^{-2})$ are neglected and a set of coupled sixth-order ordinary differential equations must be solved subject to initial values and boundary conditions. Lingwood examined this set of equations in a thorough manner. First, all terms were collected, fixed in groups

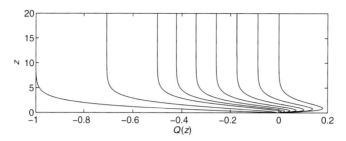

Figure 6.8 From left to right, $Q(z)$ for $\varepsilon = 90°$, $45°$, $30°$, $25°$, $20°$, $15°$, $10°$, $5°$, $0°$ (after Lingwood, 1995, reproduced with permission).

and noted as to the effects due to (a) viscosity, (b) rotation (Coriolis), and (c) streamline curvature, respectively.

If the rotation and streamline curvature effects are ignored, the set of equations uncouple and the more familiar Orr–Sommerfeld and Squire equations are found. If viscous effects are further neglected, then the Rayleigh equation emerges. The only note is that the mean velocity in each of these cases is in terms of the sum of both the $\overline{U}$ and the $\overline{V}$ components, as suggested by the $Q(z)$ definition of (6.50) given earlier. Lingwood went on to numerically integrate the full set of linear equations subject to the initial input of an impulsive line forcing given as $\delta(r - r_s)\delta(t)e^{i\beta\theta}$. The analysis and the computations were performed as one that is a combination of the spatio-temporal stability basis.

Of special interest to Lingwood was the question of absolute instability (cf. Chapter 4). When the system has absolute instability, then, at a fixed point, amplification continues without bounds. As a result, even a weak disturbance at a fixed point will amplify to quite large amplitude. For one that is amplified in the convective sense, it will be swept away and the boundary layer will recover. Lingwood did indeed find such an instability once a certain value of the Reynolds number, $Re > 510.625$, and if the parameter, $\beta/Re \approx 0.126$, are reached. The immediate effect of this is that the perturbations become nonlinear and transition is to be expected. Below this critical Reynolds number, the flow is unstable but not absolutely.

6.6 Trailing Vortex

Another flow of significance that has received attention within the framework of centrifugal stability is that of the trailing vortex. Such a flow is most prominent at the tips of aircraft wings and the ensuing trail behind the aircraft. The trailing line vortex. It is an important example of the class of flows denoted as swirling flows. And, just as other flows with such effects, such as the rotating disk, instability can be found inviscidly as well as in the viscous mode. This result represents a distinction from other free shear flows since viscosity normally has a damping effect (as in the jet, wake or the mixing layer, for example) and the instability that can be determined inviscidly is simply reduced in value when viscous effects are included. Viscous-induced instability has required a solid boundary in the discussions heretofore.

Investigation of swirling flows stems from the work of Howard & Gupta (1962) and was most general in that this effort dealt with both stratified as well non stratified flows and was fully three-dimensional. Sufficient conditions for stability were established together with a semi circle theorem for perturbations that are axisymmetric. This later result was extended to non-axisymmetric perturbations by Barston (1980). Numerical solutions for the inviscid stability problem were determined by Lessen, Singh & Paillet (1974); see also Duck & Foster (1980). Then, Lessen & Paillet (1974), Khorrami (1991) and Duck & Khorrami (1992) provided the work to show that this problem has instabilities that are caused by viscosity as well as those that were found in the inviscid limit. Lastly, the presentation by Mayer & Powell (1992) is one makes an extensive survey and determines the stability of all possible modes numerically.

The basic mean steady flow that has been incorporated in the studies was determined by Batchelor (1964) by means of a similarity solution for the line vortex far downstream of its origin of generation. With the z-axis as the coordinate that coincides with the axis of the vortex and (r, θ) as the polar coordinates in the vortex cross section, the similarity solution is two-dimensional and is given by $\underline{U} = (0, V, W)$, with V the azimuthal component and W is in the axial direction. When expressed in terms of the nondimensional similarity variable, η, with $\eta \sim r/z^{1/2}$, these functions are

$$V(\eta) = q/\eta \, (1 - e^{-\eta^2}), \qquad W(\eta) = W_\infty + e^{-\eta^2}, \qquad (6.52)$$

with q denoting the swirl intensity of the vortex and W_∞ a constant.

As mentioned, the stability analysis due to Mayer & Powell (1992) is one that has explored the full range of possibilities for this problem and will be the reference used here. When the fluid is incompressible, the set of linear equations for the perturbations can be found and solutions for this set are taken in normal mode form as

$$\{\tilde{u}, \tilde{v}, \tilde{w}, \tilde{p}\}(\eta, \theta, z, t) = \{\hat{u}, \hat{v}, \hat{w}, \hat{p}\}(\eta) \, e^{i\alpha z + in\theta - i\omega t}. \qquad (6.53)$$

Boundary conditions for these variables require finiteness at $\eta = 0$ and this will vary according to whether the value of n is zero, $|n| = 1$, or $|n| > 1$. As $\eta \to \infty$, all variables are required to vanish. Inviscidly, the result is a coupled third-order system of ordinary differential equations. When viscous, the system is sixth order and there are no changes in the boundary conditions.

When inviscid, Mayer & Powell determined the most unstable eigenvalue numerically. It was found that $\omega = 0.049718 + (0.202628)i$ and is for the case when the parameters n, q and α have the values $n = 1$, $q = -0.5$ and $\alpha = 0.5$, respectively. These authors have made extensive searches with extreme numerical precision and the most revealing graphics of this work have been provided by contour plots of the growth rates in the (α, q)-plane for fixed mode number n.

As already noted, for the full viscous problem, Khorrami (1991) reported that there are instabilities that are due to viscosity, and these were found when $n = 0$ and $n = 1$. For the $n = 1$ case, the lowest critical Reynolds number was 13.905, far in contrast to one of infinite value. Furthermore, the result for the inviscid instabilities when $n = 1$ were stabilized with increasing viscosity. More importantly, these new viscous instabilities occurred for values of the parameters where no inviscid instability had been found. These results are especially significant for both of the growth rates, and the physical characteristics are in good qualitative agreement with observations in experimental studies of aircraft contrails at high altitudes.

6.7 Round Jet

In many ways, save for the results, pipe flow can be thought of as the finite analogy to plane Poiseuille channel flow. Likewise, just as there are planar examples, such as the plane jet and wake, where the stability can be treated inviscidly, there are examples for flows that are labeled as round that also permit the stability investigation to be determined without viscous effects. It is only when there are solid boundaries present in the flow that viscosity can be a cause of instability; otherwise, viscosity leads to damping. In this sense, the round jet is an example of particular note. And, in fact, this problem even allows for the extension of results that follow theorems determined by Rayleigh for plane inviscid flows. More specifically, the extension deals with the mean velocity profile where, in Cartesian coordinates, an inflection point was needed somewhere within the region of the flow in order to have instability; this will be shown as the equations for the perturbations are developed. In fact, it was (Rayleigh 1880, 1892a,b, 1916a,b) himself who noted the extension. The thorough stability analysis of the round jet was not, however, done until the work by Batchelor & Gill (1962).

Consider the coordinates and perturbations as those defined for pipe flow and illustrated in Fig. 6.5. When considered inviscidly and solutions for the perturbations are assumed to be of the normal mode form, where

$$\{\tilde{u}, \tilde{v}, \tilde{w}, \tilde{p}\}(r, \theta, x, t) = \{\hat{u}, \hat{v}, \hat{w}, \hat{p}\}(r) e^{i\alpha x + in\theta - i\omega t},$$

with $(\hat{u}, \hat{v}, \hat{w}, \hat{p})$ only functions of r, the system becomes

$$i\alpha\hat{u} + \frac{1}{r}\frac{d(r\hat{v})}{dr} + \frac{in}{r}\hat{w} = 0, \tag{6.54}$$

$$i(\alpha U - \omega)\hat{u} + U'\hat{v} = -i\alpha\hat{p}, \tag{6.55}$$

$$i(\alpha U - \omega)\hat{v} = -\frac{d\hat{p}}{dr} \tag{6.56}$$

and

$$i(\alpha U - \omega)\hat{w} = -\frac{in}{r}\hat{p}. \tag{6.57}$$

By eliminating $\hat{u}, \hat{w}$ and $\hat{p}$, just as was done for the system in Cartesian coordinates, one equation for $\hat{v}$ can be found, and this is

$$(\alpha U - \omega)\frac{d}{dr}\left[\frac{r(r\hat{v})'}{\alpha^2 r^2 + n^2}\right] - (\alpha U - \omega)\hat{v} - \frac{d}{dr}\left[\frac{\alpha r U'}{\alpha^2 r^2 + n^2}\right]r\hat{v} = 0. \tag{6.58}$$

It is from (6.58), with $n = 0$ (axisymmetric disturbances), that Rayleigh noted that a necessary but not sufficient condition for instability is the requirement that

$$\frac{d}{dr}\left(\frac{U'}{r}\right) = 0.$$

If it is recalled that $\omega = \alpha c$, this is fully equivalent to the need for the reflection point in the mean velocity profile when $(U - c) = 0$. When $n \neq 0$, then the generalization of this requirement is that

$$\frac{d}{dr}\left[\frac{rU'}{\alpha^2 r^2 + n^2}\right] = 0$$

at some point in the flow. An additional result was found by Batchelor & Gill (1962) for this condition by noting that, when the mode number, n, becomes quite large, then the flow is always stable because this term tends to zero everywhere. This result is precisely that which has already been observed for pipe flow in that it is no longer possible for there to be any generation of perturbation vorticity when the mean profile has such behavior.

The most general conclusion derived for rotating flows is due to Rayleigh and is known as the circulation criterion. Again, for axisymmetric disturbances, if the square of the circulation, defined as

$$\frac{1}{r^3}\frac{d}{dr}\left(r^2\Omega\right)^2,$$

where $U = r\Omega$, does not decrease at any location, then the flow is stable. Just as was outlined in the introductory Section 6.1, this argument is based on the physics of the flows that have centrifugal influence.

Batchelor and Gill used the top hat velocity profile to describe the round jet as well as for the case $U(r) = U_0/[1+(r/r_0)^2]$, where r_0 is the scale of the extent for the jet. The solutions for $\hat{v}$ must vanish as the radius is increased and be finite for $r = 0$. Regardless of the choice for the mean profile, it was also found that only the mode $n = 1$ is unstable. Otherwise, all other modal solutions were found to be stable for these choices of the mean velocity variation.

Batchelor & Gill also showed that, by a unique transformation, that the equivalence of the Squire transformation can be formulated for this problem. In effect, the jet is cast in terms of a helix rather than the standard form of expression in polar coordinates. The helix is defined as $r =$ constant and, in terms of the original coordinates, $\alpha x + n\theta =$ constant. Once this is done, the perturbation variables are defined as

$$\left.\begin{aligned}
\overline{\alpha u} &= \alpha\hat{u} + (n\hat{w}/r), \\
\overline{v} &= \hat{v}, \\
\overline{\alpha w} &= \alpha\hat{w} - (n\hat{u}/r), \\
\overline{p}/\overline{\alpha} &= \hat{p}/\alpha,
\end{aligned}\right\}
\tag{6.59}$$

where, it should be noted, $\overline{u}$ is perpendicular to the radius and helix, $\overline{v}$, is parallel to the radius, and $\overline{w}$ is parallel to the tangent. The wavenumber, $\overline{\alpha}$, is now given by

$$\overline{\alpha} = \left(\alpha^2 + n^2/r^2\right)^{1/2}.$$

In these terms, the governing equations can be written as

$$i\overline{\alpha u} + \frac{1}{r}\frac{d}{dr}(r\overline{v}) = 0,
\tag{6.60}$$

$$i\overline{\alpha}(U - c)\overline{u} + U'\overline{v} = -i\overline{\alpha}\overline{p},
\tag{6.61}$$

$$i\overline{\alpha}(U - c)\overline{v} = -\overline{p}'
\tag{6.62}$$

and

$$i\overline{\alpha}(U - c)\overline{w} - \frac{n}{\alpha r}U'\overline{v} = 0.
\tag{6.63}$$

Since the $\overline{w}$ component does not appear in those for $\overline{u}$ and $\overline{v}$, the system (6.60) to (6.63) is similar to that for parallel or axisymmetric flow. In effect, a two-dimensional perturbation problem as the Squire transformation provides for three-dimensional perturbations in a parallel shear flow. And, although the actual problem considered was that of a round jet, such a mean profile can be adjusted so that it represents that of a round wake with the same conclusions.

Gold (1963) and Lees & Gold (1964) extended the Batchelor & Gill (1962) formulation for the jet and wake when the fluid is compressible. In this case, it was again found that the $n = 1$ mode is the most unstable but, unlike the incompressible problem, other modes are unstable, and there is a strong influence from the temperature of the core in that an increase in the temperature of in this part of the flow is destabilizing.

6.8 Exercises

Exercise 6.1 Derive Rayleigh's stability condition (6.36) in the inviscid limit. You may assume axisymmetric disturbances, but do not invoke the small gap approximation.

Exercise 6.2 Using asymptotic analysis appropriate for the small gap approximation, derive equations (6.20)–(6.21) from (6.17)–(6.18).

Exercise 6.3 Compute the inviscid solutions for the Taylor problem in the small gap approximations and for $\Omega_1 = \Omega_2$. Show that the solutions are stable.

Exercise 6.4 Begin with the nondimensional Navier–Stokes equations and derive the disturbance equations (6.38) and (6.39) for the Göerler problem.

Exercise 6.5 Determine the governing equations for the perturbations for the rotating disk and indicate the terms that are due to viscosity, rotation and streamline curvature.

Exercise 6.6 Determine the linear equations for the trailing vortex. Then, examine the energy equation and that for vorticity.

Exercise 6.7 Determine the solution for the perturbations in pipe flow in the inviscid limit.

Exercise 6.8 For the round jet, answer the following.

(a) Starting with equation (6.58), derive Rayleigh's inflection point theorem.
(b) Starting with equation (6.58), and introducing the transformation $g(r) = r\hat{v}/(U - c)$, derive Howard's semi circle theorem.
(c) Derive the linearized equations (6.60)–(6.63).

7

Geophysical Flow

Truth is ever to be found in simplicity, and not in the multiplicity and confusion of things.

– Sir Isaac Newton (1642–1727)

7.1 General Properties

From the class of flows that are termed geophysical, there are three that are distinct and illustrate the salient properties that such flows possess when viewed from the basis of perturbations.

First, there are stratified flows where there is a mean density variation, which plays a dominant role in the physics because there is now a body force due to gravity. At the same time, the fluid velocity, to a large degree of approximation, remains solenoidal and therefore the motion is incompressible. The net result leads to the production of anisotropic waves, known as internal gravity waves, and which exist in both the atmosphere and the ocean.

Second, because of the spatial scales involved, motion at many locations of the Earth, such as the northern or southern latitudes, is present in an environment where the effects of the Earth's rotation cannot be taken as constant. On the contrary, rotation plays a dominant role, and this combination of circumstances leads to the generation of waves.

Viscous effects can be neglected in the analysis for both the stratified flow and the problem with rotation, but the presence of a mean shear in either flow, as we have already seen so often, does lead to important consequences for the dynamics when determining the stability of the system.

Third, there is the modeled geophysical boundary layer where the rotation is present but taken as constant, the surface is assumed flat. In this case, viscous shear is important and the flow is known as the Ekman layer. The resulting mean flow solution also happens to be an exact solution of the full Navier–Stokes equations even if unsteady. The perturbation problem here leads to quite interesting results for, unlike the Blasius boundary layer, the flow can be unstable inviscidly as well as having the viscous instability that is common for the boundary layer.

The concept of waves has been primarily investigated in terms of the effects of fluid compressibility, as was shown in Chapter 5. For linearized perturbations in compressible flow, this is tantamount to acoustics when referring to the physics. In terms of the mathematics, it is clear that there must be a hyperbolic partial differential equation that governs the dependent variable of note in order to have waves. None of the stability problems discussed as yet, except those dealing with compressibility, could have wave motion in this sense. The unique features of the geophysical flows have different bases for waves and, except for the physics, the concept is no different from the standpoint of the mathematics. The governing equation is hyperbolic, but the waves that are produced are decidedly anisotropic. In addition, the flows can be unstable if there is a mean shear.

Many of the properties or results that have been found in the investigation of the more traditional shear flow instabilities, such as the Squire theorem, cannot necessarily be extended to these problems, and it is the purpose here to probe such flows in more detail, and to not only indicate the novel effects, but to ascertain exactly what established results may or may not be used for these cases.

7.2 Stratified Flow

Fluid motions that transpire in an environment where the density can vary are numerous and occur in nature to a large degree of approximation. Both the oceans and the atmosphere are harbingers of such action. For the ocean such a variation is due primarily to dissolved salts and temperature while, in the atmosphere, this is due more to temperature variation. Since gravity acts in only the vertical direction, then a component of body force in this direction must be incorporated into the dynamics. This can be done and still have the motion remain incompressible. Such an assumption corresponds to motion of a fluid where it is difficult to change the density of a fluid parcel by pressure forces. The resulting governing equations within this framework are the result of what is known as the Boussinesq approximation. A full account of the details needed for this approximation can be found in the book on fluid dynamics by Batchelor (1967), but it will be useful to review the major points here.

Before proceeding, let the reader understand that the coordinate system for this section is different than that used in previous chapters. Here, instead of using x–y as the primary flow direction, the crossflow coordinate, x–z, will be used in this chapter because this is the conventional coordinate system that has long been established for such geophysical flows and is prevalent in the literature.

Basically, the Boussinesq approximation is tantamount to stating that the fluid velocity is solenoidal, or $\nabla \cdot \underline{u} = 0$, which is the major basis for incompressibility. In order for this to be approximately true, then certain conditions should be met. Consider the spatial distribution of the velocity, $\underline{u}$, (and other flow variables as well) to be characterized by a length scale, L, and all variation will involve scales small compared to L. Similarly, a magnitude for the velocity is taken to be U_0. Thus, in these terms, it can be stated that $|\nabla \cdot \underline{u}| = O(U_0/L)$, and then $\underline{u}$ will be considered solenoidal if $|\nabla \cdot \underline{u}| \gg U_0/L$ or, because the divergence of $\underline{u}$ is coupled to variations in the density in the equation for the conservation of mass, this means $\left|\frac{1}{\rho}\frac{D\rho}{Dt}\right| \gg U_0/L$ must also be true. As shown by Batchelor (1967), this inequality leads to the following conditions that must be satisfied in order to assume that the Boussinesq approximation is valid: (1) the Mach number, defined as $M = U_0/a$, where a is the speed of sound, has the requirement that $M^2 \ll 1$. For reference air, at 15°C, $a = 340.6$ m/sec and, for water at 15° C, $a = 1470$ m/sec; (2) the ratio, $\omega L/a \ll 1$ as well, where ω is the frequency of a possible periodic flow. In this case, if $\omega \sim U_0/L$, then this ratio is simply that for the Mach number. Of course, when $\omega L/a = 1$, with L the wave length of sound waves with frequency ω, then compressibility cannot be irrelevant. (Sonar does work in the ocean!) A last condition, (3) $gL/a^2 \ll 1$, is more sensitive for the atmosphere than the ocean since this inequality leads to what is known as a scale height. The consequences of all these conditions will be understood as we develop the necessary linear perturbation equations that govern the motion of disturbances in geophysical flows.

For a homogeneous, isentropic fluid that is in equilibrium with respect to the rotating Earth, a reference state can be obtained from a hydrostatic balance, or summation of the forces on the fluid. Thus, neglecting viscous effects,

$$\rho_r \nabla G + \nabla p_r = 0. \tag{7.1}$$

And, if we recognize that the relative gravitational potential, G, is only in the vertical z-direction with $G = gz$, this relation becomes

$$\rho_r g + \frac{\partial p_r}{\partial z} = 0, \tag{7.2}$$

where standard gravity is $g = 9.8 \text{ m/s}^2$. The solution to (7.2) is

$$\rho_r = \rho_0 e^{-g \int_0^z dz/a^2}, \tag{7.3}$$

where $dp_r/d\rho_r = a^2$ has been used; ρ_0 is the value of the density at $z = 0$. For the ocean, this would mean the value at the free surface or the mean oceanic density is usually referred to as standard conditions. For pressure, this would be referred to as the ambient pressure. Note this relation for the reference state is characterized by the scale height (e.g., altitude or depth). Call the value $H = a^2/g$, say, and it is clear that we must require that any variation of the perturbation field in this direction or ℓ_z must satisfy the inequality $\ell_z/H \ll 1$.

The set of governing equations for such flows is obtained by subtracting the hydrostatic reference values from the momentum equations (2.34)–(2.37), yielding

$$\rho \frac{Du}{Dt} = -\nabla(p - p_r) - (\rho - \rho_r)\nabla G \tag{7.4}$$

when, again, viscous effects are neglected.

Now, the difference between the actual and the reference states is small and typically $(\rho - \rho_r)/\rho_r \approx O(10^{-3}) \ll 1$. This suggests replacing of ρ by ρ_r in the inertia term of the equations but not in the force due to gravity because the acceleration due to gravity, g, is $O(10^3)$. In fact, with $\ell_z/H \ll 1$, we can further assume that $\rho_r \cong \rho_0$. Finally, the equations of motion, under these conditions, are

$$\frac{Du}{Dt} = -\frac{1}{\rho_0}\nabla\bar{p} - \frac{(\rho - \rho_0)}{\rho_0}\nabla G, \tag{7.5}$$

with $\bar{p} = (p - p_r)$. Along with (7.5) we have the continuity equation

$$\nabla \cdot u = 0 \tag{7.6}$$

(or essentially zero) that is a result of the Boussinesq approximation. This means the relation for the conservation of mass uncouples and

$$\frac{D\rho}{Dt} = 0 \tag{7.7}$$

is also true in order to completely define a problem within this environment.

The system can now be perturbed and the necessary linear equations can be found for determining the stability. However, it should be noted that it is no longer necessary to maintain the notation with respect to the reference values. In short, density variations only appear together with the force of gravity and the uncoupled equations for the conservation of the density for a fluid particle. Thus, we assume a mean velocity in the ocean or atmosphere in the x-direction,

$U(z)$, along with a mean pressure and a mean density varying in the vertical z-direction, and we then write

$$
\begin{aligned}
u(x,y,z,t) &= U(z) + \tilde{u}(x,y,z,t), \\
v(x,y,z,t) &= \tilde{v}(x,y,z,t), \\
w(x,y,z,t) &= \tilde{w}(x,y,z,t), \\
\rho(x,y,z,t) &= \overline{\rho}(z) + \tilde{\rho}(x,y,z,t), \\
p(x,y,z,t) &= P(z) + \tilde{p}(x,y,z,t).
\end{aligned}
\right\}
\tag{7.8}
$$

After substituting (7.8) into (7.5)–(7.7), the following linearized disturbance equations result:

$$
\tilde{u}_x + \tilde{v}_y + \tilde{w}_z = 0,
\tag{7.9}
$$

$$
\tilde{u}_t + U\tilde{u}_x + U'\tilde{w} = -\tilde{p}_x/\rho_0,
\tag{7.10}
$$

$$
\tilde{v}_t + U\tilde{v}_x = -\tilde{p}_y/\rho_0,
\tag{7.11}
$$

$$
\tilde{w}_t + U\tilde{w}_x = -\tilde{p}_z/\rho_0 - g\tilde{\rho}/\rho_0,
\tag{7.12}
$$

$$
\tilde{\rho}_t + U\tilde{\rho}_x + \overline{\rho}'\tilde{w} = 0,
\tag{7.13}
$$

with $()' = d()/dz$ and all variables denoted with tilde being the linear perturbations.

Although not obvious in this form, these equations can be reduced to a set for the stability analysis that is akin to those of the non-stratified problem. We can easily show that the three momentum equations (7.10) to (7.12) can be reduced to the following equation by taking the divergence and using (7.9) and thus

$$
\nabla^2 \tilde{p}/\rho_0 = -2U'\tilde{w}_x - g\tilde{\rho}_z/\rho_0.
\tag{7.14}
$$

Due to the variation in density, there is now a term for the source of pressure in addition to that which is prevalent for the constant density problem. Then, operating on (7.12) with ∇^2, the three-dimensional Laplacian, and using (7.14) and (7.13) to eliminate the pressure and density, the governing equation becomes

$$
\left(\frac{\partial}{\partial t} + U\frac{\partial}{\partial x}\right)^2 \nabla^2 \tilde{w} - \left(\frac{\partial}{\partial t} + U\frac{\partial}{\partial x}\right)U''\tilde{w}_x + N^2\left(\tilde{w}_{xx} + \tilde{w}_{yy}\right) = 0, \quad (7.15)
$$

where

$$
N^2 = N^2(z) = -\frac{g}{\rho_0}\overline{\rho}'
\tag{7.16}
$$

is known as the "Brunt–Väisälä[1] frequency." Using the normal mode assumption for the solution,

$$\tilde{w}(x,y,z,t) = \hat{w}(z)e^{i(k_1 x + k_2 y - \omega t)}, \tag{7.17}$$

equation (7.15) can be reduced to an ordinary differential equation, which is

$$\hat{w}'' + (k_1^2 + k_2^2)\left\{ \frac{N^2 - (k_1 U - \omega)^2 - (k_1 U - \omega)U'' k_1/k^2}{(k_1 U - \omega)^2} \right\}\hat{w} = 0, \tag{7.18}$$

where $k^2 = k_1^2 + k_2^2$.

One remark should be made here, namely that the Squire transformation is still quite valid for this problem. This can easily be seen by using the same normal mode decomposition for the set of equations (7.9)–(7.13) and then recognizing the following:

$$\left. \begin{aligned} k^2 &= k_1^2 + k_2^2, \\ \tan\theta &= k_2/k_1, \\ k\bar{u} &= k_1\hat{u} + u_2\hat{v}, \\ k\bar{v} &= k_2\hat{u} - k_1\hat{v}. \end{aligned} \right\} \tag{7.19}$$

Substitution of these expressions into the three-dimensional set of equations that have been decomposed using normal modes will result in an equivalent two-dimensional system. In other words, no z-dependence and no $\hat{w}$ component of the perturbation velocity in this direction as would be for the two-dimensional problem at the outset.

In order to illustrate some of the essential features of this kind of flow, let us consider the problem in the absence of a mean shear or $U = 0$. In this case, equation (7.15) reduces to

$$\nabla^2 \tilde{w}_{tt} + N^2\left(\tilde{w}_{xx} + \tilde{w}_{yy} \right) = 0 \tag{7.20}$$

and the ordinary equation is obtained by substituting the solution form of (7.17) into (7.20) to find

$$\hat{w}'' + k^2\left[\frac{N^2 - \omega^2}{\omega^2} \right]\hat{w} = 0. \tag{7.21}$$

[1] Sir David Brunt (1886–1965) English meteorologist. First full-time professor of meteorology at Imperial College, 1934–1952. His textbook *Physical and Dynamical Meteorology* (Cambridge University Press, London, 1934; 2nd ed., 1939) was one of the first modern unifying accounts of meteorology.
Vilho Väisälä (1899–1969) Finnish meteorologist. Developed a number of meteorological instruments, including a version of the radiosonde in which readings of temperature, pressure and moisture are telemetered in terms of radio frequencies. The modern counterpart of this instrument (manufactured by Väisälä Oy/Ltd.) recently won recognition as one of Finland's most successful exports.

In this limit, the governing system is hyperbolic so long as the coefficient of $\hat{w}$ is positive. This result is that of internal gravity waves. The general properties for these waves are significant and will now be reviewed. From the point of stability, this is the same as a neutral dynamical system since there is neither amplification nor decay.

A local vertical length scale for the wave motion can be defined as

$$\ell_v = \left(\frac{\hat{w}}{\hat{w}''} \right)^{1/2}. \tag{7.22}$$

The horizontal length is $\ell_h = k^{-1}$ and the ratio of the two is

$$\frac{\ell_h}{\ell_v} = \frac{1}{k} \left(\frac{\hat{w}''}{\hat{w}} \right)^{1/2} = \left| \frac{N^2(z) - \omega^2}{\omega^2} \right|^{1/2}. \tag{7.23}$$

For low frequency internal gravity waves, $\omega/N \ll 1$ and, in fact

$$\frac{\ell_h}{\ell_v} \to \infty \qquad \text{as} \qquad \omega/N \to 0. \tag{7.24}$$

The horizontal length scale is therefore much larger than the vertical length scale, i.e., low-frequency waves always have a horizontally elongated structure.

Suppose we now only consider wave lengths small compared to the scale of variation of $N(z)$. This input allows for the use of the WKBJ method for solving $\hat{w}(z)$. In fact, take $N(z)$ to be constant (e.g., the mean density varies exponentially in the vertical) and then a local solution can be found as

$$\tilde{w} = ae^{i(\alpha_1 x + \alpha_2 y + \alpha_3 z - \omega t)}. \tag{7.25}$$

By substituting (7.25) into (7.20), a dispersion relation for the waves results and is

$$\omega = \pm N \frac{k}{\alpha} = \pm N \cos\theta, \tag{7.26}$$

where $\alpha = \sqrt{\alpha_1^2 + \alpha_2^2 + \alpha_3^2}$, $k = \sqrt{\alpha_1^2 + \alpha_2^2}$ and θ is the angle between the two- and the three-dimensional wave vectors. It is clear that the waves are anisotropic since the frequency depends upon the direction of the propagation and not just the magnitude of the wavenumber. Here, in the limit of $\omega \to 0$, $\theta \to \pi/2$. At the same time, for $\omega \to N$, the wavenumber approaches the horizontal. When $\omega > N$, the wavenumber is no longer real and wave motion ceases altogether.

An alternative form of the derivation of the dispersion relation can be given and thereby lends additional insight into this type of perturbation dynamics. Consider a plane of fluid, with an angle θ to the vertical, and displaced a distance ξ parallel to itself, as shown in Fig. 7.1. Because of the density gradient in the vertical, each element of fluid finds itself more dense than that of its surroundings, so that it has an excess in mass per unit volume of $d\overline{\rho}/dz$ times the vertical displacement $\xi \cos \theta$. There is an excess weight per unit volume of this times g, which acts vertically downwards or this multiplied by $\cos \theta$ in the plane of the motion. This excess weight must be balanced by the acceleration and is given by Newton's second law of motion. Hence, in these terms,

$$\rho_0 \ddot{\xi} = -\left(g\overline{\rho}' \cos^2 \theta \right) \xi.$$

The result is a linear oscillator with frequency:

$$\omega - \left(\frac{g}{\rho_0} \overline{\rho}' \cos^2 \theta \right)^{1/2}.$$

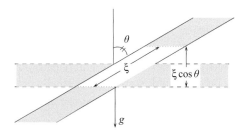

Figure 7.1 Mechanical analogue of dispersion relation.

There are more salient points to this wave motion. The phase velocity, $\underline{C}_p$, is

$$\underline{C}_p = \pm \frac{N \cos \theta}{\alpha^2} \underline{\alpha} = \pm \frac{N}{\alpha} \left(\frac{\alpha_1}{\alpha} \hat{i}, \frac{\alpha_2}{\alpha} \hat{j}, \frac{\alpha_3}{\alpha} \hat{k} \right). \tag{7.27}$$

The group velocity, $\underline{C}_g$, is defined in the usual manner and is

$$\underline{C}_g = \nabla_\alpha(n) = \frac{N}{\alpha^2} \left[\left(\alpha_1 - \frac{k\alpha_1}{\alpha} \right) \hat{i}, \left(\alpha_2 - \frac{k\alpha_2}{\alpha} \right) \hat{j}, \frac{k\alpha_3}{\alpha} \hat{k} \right], \tag{7.28}$$

where k is the wave vector in the plane or $k = (k_1, k_2, 0)$. It can be seen that $\underline{C}_p \cdot \underline{C}_g = 0$, and hence these vectors are perpendicular. This is the consequence of the transverse nature of waves in an incompressible fluid. More specifically, for a single Fourier disturbance, the velocity vector must be perpendicular to the wave vector in order to satisfy the solenoidal property of the vector that is related to the condition of incompressibility. Thus the energy flux, which is the product of the pressure with the velocity, is perpendicular to the same wave vector. In turn, the energy flux is proportional to the group velocity and therefore this phenomenon.

The just-described combination of results leads to curious effects that are common in anisotropic wave propagation, and examples should be mentioned to illustrate same. First, consider the effect that ensues when internal waves reflect at a sloping bottom, say. Figure 7.2 depicts this situation.

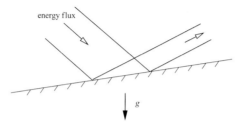

Figure 7.2 Scaling change due to sloping bottom.

The basic rule governing the reflection is that the frequencies of the incident and reflected waves are the same. This implies that the inclination of the wavenumbers to the *vertical* must be the same and the angle of incidence of the waves is not the same as the angle of reflection. The width of the reflected beam is less than that of the incident beam, and the energy is concentrated and the scale is reduced. In spectral terms this corresponds to a transfer of energy to higher wavenumbers and then it can be dissipated easier by viscous action. This kind of reflection is termed *anomalous* and it does not occur at a horizontal surface when the angle of incidence and reflection are the same. In this case, the wavenumber magnitude is conserved on reflection at a horizontal surface.

Second, suppose there is a source of disturbance at a constant frequency $\omega < N$ in a continuously stratified fluid. More specifically, consider the situation as in Fig. 7.3. Since the frequency is fixed, the wavenumbers of the disturbance produced are all at a fixed angle to the horizontal. As a result, the group velocity is at a fixed angle to the vertical since $\underline{C}_g$ is perpendicular to $\underline{C}_p$ and is determined by the frequency. Since the group velocity provides the direction of the energy flux, it follows that the energy radiated from the disturbance is confined to a set of beams whose angle to the vertical is fixed by the frequency. The energy radiates away in these directions only and any disturbance in the dividing wedges is much less.

A last example is known as Brunt–Väisälä trapping. Suppose the buoyancy frequency has a local maximum. A wave, whose frequency is less than this value, will have a group velocity that is inclined to the vertical. As the wave

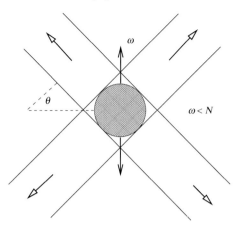

Figure 7.3 Energy flux in a stratified medium.

propagates upwards, the local value of N decreases until it becomes equal to the wave frequency. As this happens, the direction of the group velocity turns toward the vertical and its magnitude decreases. And, when the local value of the frequency tends to be $\omega = N$ exactly, the group velocity is directed vertically but is vanishingly small. An accumulation of energy takes place at this region, and the energy is reflected and returns downwards to be reflected again at a lower level where $\omega = N$. The net result is a trapping of energy in the layer where $\omega < N$. The energy propagates horizontally but is restricted vertically. This is depicted in Fig. 7.4 and is tantamount to a wave guide. Again, since viscosity is neglected, the reflection points here are cusps, and hence all energy is contained within the bounds where $\omega = N$.

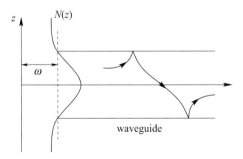

Figure 7.4 Brunt–Väisälä trapping; waveguide.

Whether the full three-dimensional Fourier solution is used, or only the de-

composition in the horizontal plane, there is still the question of proper boundary conditions. In any direction that extends to infinity, the Fourier modes are
quite correct since these are bounded in these directions. In the z-direction of
variation for the mean density, though, there is more than one possibility. First,
there could be confinement between two solid barriers in z, indicating that
$\hat{w} = 0$ at these locations. As a result, the solution is the same as any two-point-
boundary-value, initial-value problem, and there are an infinite number of discrete Fourier modes possible for the solution. The case where there are are no
solid boundaries or only one in the vertical, discrete modes are no longer the
sole basis for the problem, and one must resort to the use of a continuous spectrum as well. This result is no different from that of the non-stratified problem.
And certainly problems that deal with free surface waves will require somewhat different boundary conditions. This is especially true if surface tension is
to be incorporated into the analysis.

Now, these waves do have vorticity and, for three-dimensional perturbations
all components can, in principle, be present. But, since the fluid is taken as
incompressible and there is no mean motion, then the component, ω_z, will
be nonzero only if it is so specified initially, and then it will remain constant
in time without further generation or diffusion, since viscous effects are neglected. The other components, ω_x and ω_y, will be generated even if not given
initial values but, like ω_z, cannot diffuse. This appraisal will be demonstrated
more fully by examining the governing equations for the perturbation vorticity.

The definitions for the three vorticity components are, respectively,

$$\omega_x = w_y - v_z, \quad \omega_y = u_z - w_x, \quad \text{and} \quad \omega_z = v_x - u_y. \tag{7.29}$$

When there is no mean motion, the vorticity rests solely with the perturbations,
and the governing linear equations, derived from (7.10)–(7.12), are

$$\frac{\partial \tilde{\omega}_x}{\partial t} = -\frac{g}{\rho_0} \tilde{\rho}_y, \tag{7.30}$$

$$\frac{\partial \tilde{\omega}_y}{\partial t} = \frac{g}{\rho_0} \tilde{\rho}_x, \tag{7.31}$$

$$\frac{\partial \tilde{\omega}_z}{\partial t} = 0. \tag{7.32}$$

As can be seen, generation of vorticity is due to the fact that there is a variation
in the density field rather than a gradient in the velocity. This is known as
baroclinic torque and, because of the assumptions on the mean density and
pressure, the only generation possible is in the plane perpendicular to the line
marking the stratification.

Consider the mean motion restored to the flow. It is now that competition in the dynamics emerges and, in fact, the flow has the possibility to be unstable. The equation that is equivalent to (7.15) for $\hat{w}$ is, with $U = U(z)$, just (7.18) or

$$(k_1 U - \omega)^2 (\hat{w}'' - k^2 \hat{w}) - k_1 (k_1 U - \omega) U'' \hat{w} + k^2 N^2 \hat{w} = 0 \qquad (7.33)$$

when rewritten, where $k^2 = k_1^2 + k_2^2$. In essence, this is just equation (7.15) written in terms of the modal form for solution. As (7.33) stands, however, it is not in the familiar format from the standpoint of stability. Instead of the Brunt–Väisälä frequency as the parameter, it is the Richardson number that is more meaningful here: it is defined as

$$\overline{Ri} = -g \frac{\overline{\rho}'}{\overline{\rho}} \frac{L^2}{U_0^2} = \frac{N^2 L^2}{U_0^2}. \qquad (7.34)$$

This parameter can be interpreted as the ratio of the stabilizing influence due to the vertical stratification and the destabilizing influence due to the mean velocity and is a parameter of bulk form. Thus, after nondimensionalizing (7.33), the following equation becomes the center of the analysis:

$$(k_1 U - \omega)^2 (\hat{w}'' - k^2 \hat{w}) - k_1 (k_1 U - \omega) U'' \hat{w} + k^2 \overline{Ri} \hat{w} = 0, \qquad (7.35)$$

where all quantities are nondimensional with respect to the mean length and velocity scales L and U_0. Now, unlike the case with no mean flow, the time behavior no longer has to be that of neutral wave motion. Such determination depends critically on the Richardson number, which is often expressed in terms of the local Richardson number rather than the bulk value. If the mean velocity profile is approximated locally as a linear function of the z-coordinate, then the local shear value is a constant. In these terms, the local Richardson number is

$$Ri_{\text{loc}} = N^2 / (U')^2. \qquad (7.36)$$

In many ways, this definition is more meaningful from the standpoint of stability since the mean shear is tantamount to vorticity. This point will now be put into perspective.

As has already been shown, the use of the energized form of the perturbation equations can be beneficial to indicating the means whereby there are sources for amplification of the disturbances. This can be done here as well. Define the integrated value for the total energy to be

$$E(t) = \iiint \frac{\rho_0}{2} (\tilde{u}^2 + \tilde{v}^2 + \tilde{w}^2) dV \qquad (7.37)$$

and it is possible to show that, if the volume boundaries are passive, then the temporal balance for the perturbation energy is

$$\frac{dE}{dt} = \int -\rho_0 \tilde{u}\tilde{w}U'dz + \int -g\tilde{\rho}\tilde{w}dz. \tag{7.38}$$

Thus, from previous considerations it is known that the Reynolds stress acting on the mean shear can, if the phasing is correct, lead to an increase in the energy. Counteracting such generation is the stabilizing effect of buoyancy because the fluid is stratified. Likewise, there can now be neutrality because the two mechanisms can be in balance.

In a similar fashion, the differential equations can be investigated directly as was done by Rayleigh (1883). Return to (7.33) for $\hat{w}$, let $U = 0$, multiply by the complex conjugate of $\hat{w}$, and integrate over the region of the flow variation. These operations will lead to

$$c^2 \int_{z_1}^{z_2} \{|\hat{w}'|^2 + k^2|\hat{w}|^2\} dz = \int_{z_1}^{z_2} N^2|\hat{w}|^2 dz, \tag{7.39}$$

where $c^2 = \omega^2/k^2$ has been used. First, the flow is stable so long as $N > 0$ everywhere in the flow regime. Second, for the case where $N < 0$ everywhere, then the flow is unstable. In fact, such a flow is even unstable statically due to the fact that the more dense fluid is above that which is lighter.

Now, when $U \neq 0$, then Howard (1961) was able to extract a general stability criterion for two-dimensional stratified shear flow. Define a new variable, F, as $F = \hat{w}/(U - c)^{1/2}$, and, in these terms, (7.35) is

$$\frac{d}{dz}\left\{(U - c)F'\right\} - \left\{k^2(U - c) + \frac{1}{2}U'' + \left(\frac{1}{4}U' - \overline{Ri}\right)\Big/\left(U - c\right)\right\}F = 0. \tag{7.40}$$

After multiplying (7.40) by the complex conjugate of F and integrating over the flow regime, then

$$\int_{z_1}^{z_2}\left[(U - c)\left\{|F'|^2 + k^2|F|^2\right\} + \frac{1}{2}U''|F|^2 + \frac{(1/4U' - \overline{Ri})}{(U - c)}|F|^2\right]dz = 0 \tag{7.41}$$

follows. The imaginary part of this expression is

$$-c_i \int_{z_1}^{z_2}\left[|F'|^2 + k^2|F|^2 + \left(\overline{Ri} - \frac{1}{4}U'\right)|F|^2/|U - c|^2\right]dz = 0 \tag{7.42}$$

and, if $c_i \neq 0$, then the local Richardson number, Ri_{loc}, has the requirement that

$$Ri_{loc} = \frac{\overline{Ri}}{U'} < \frac{1}{4} \tag{7.43}$$

somewhere within the flow regime. As a result, the flow is stable if the local Richardson number is everywhere $Ri_{loc} > 1/4$. For most flows of this type, the frequency of the internal gravity waves is usually much less than N, the Brunt–Väisälä frequency, and hence are probably in a stable environment.

One example for such flow should be given. Specifically, the flow that was perturbed was defined by $U = \tanh z$ and $\ln \overline{\rho} = b \tanh z$, and is a stratified mixing layer. This choice leads to $N^2 = \mathrm{sech}^2 z$ and the flow is defined for $-\infty < z < \infty$. In terms of the neutral stability locus, the stability boundary was computed by Hazel (1972) for this flow and is shown in Fig. 7.5. As can be seen, the requirement for the Richardson number is confirmed.

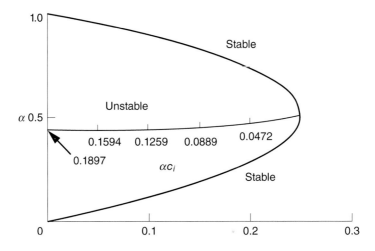

Figure 7.5 Neutral stability boundary for stratified shear layer (after Hazel, 1972, reproduced with permission).

When treating the fluctuations inviscidly, the fundamental equation for w can be considered in much the same way as was done for v in the Rayleigh equation (1.43) when there is no stratification. From kinematics,

$$\nabla^2 w = \frac{\partial \omega_x}{\partial y} - \frac{\partial \omega_y}{\partial x}.$$

In wave space, this is $ik\hat{\Omega}_\varphi = ik_2\hat{\omega}_x - ik_1\hat{\omega}_y$ and equation (7.15) can be written as

$$\left(\frac{\partial}{\partial t} + ik_1 U\right)^2 \hat{\Omega}_\varphi = \left(\frac{\partial}{\partial t} + ik_1 U\right)\cos\varphi U''\hat{w} + ikN^2\hat{w}. \tag{7.44}$$

There are now two possible generating terms for the vorticity. First, the interaction with the mean vorticity and second, the baroclinic torque due to stratification. Now, similarly there is an equation for the component of the vorticity in the z-direction, just as done for the y-component in the constant density fluid.

This equation has become known as the Squire equation and, in this case, is

$$\left(\frac{\partial}{\partial t} + ik_2 U\right)\hat{\omega}_z = ik_2 U'\hat{w}. \tag{7.45}$$

Unlike (7.44) for $\hat{\Omega}_\varphi$, this relation is exactly as that for the constant density fluid where a mean shear and three-dimensionality are crucial to the full dynamical behavior.

More details of interest in flows that have variable density can be revealed by the use of an all-encompassing simplified example in the manner of what is now known as the Kelvin–Helmholtz model. This technique was used in Chapter 2 as well. In other words, flows joined at an interface that has discontinuous mean properties. Consider the prototype as shown in Fig. 7.6.

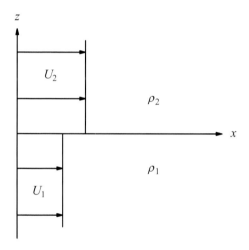

Figure 7.6 Discontinuous stratified mean flow model.

In this case, both above and below the line of discontinuity, (7.33) reduces to

$$(k_1 U - \omega)^2 (\hat{w}'' - k^2 \hat{w}) = 0. \tag{7.46}$$

And, unless there is an initial specification of vorticity, solutions of (7.46) are tantamount to a velocity field that is harmonic. Thus, the velocity can be replaced by $\nabla\phi$ where ϕ is the scalar potential and satisfies $\nabla^2\phi = 0$, just as w does in (7.46). The far-field boundary conditions for $z \to \pm\infty$ are easily satisfied by the solutions for this equation and, in fact, the perturbations must vanish for these limits.

At the interface, $z = 0$, both a kinematic relation and a balance in the pressure (normal stress) must be insured. If $F = z - h(x,y,t) = 0$ defines the interface, then $DF/Dt = 0$ is the kinematic requirement. In linearized form this is

$$\frac{\partial h}{\partial t} + U\frac{\partial h}{\partial x} = w = \frac{\partial \phi}{\partial z}, \tag{7.47}$$

and must be satisfied at $z = 0$. To do this, solutions for ϕ are obtained separately in the two regions $z > 0$ and $z < 0$. In so doing, the far-field boundary conditions can be satisfied and the net results used in (7.47).

A relation for the pressure is derived from one of the forms of the Bernoulli equation. Since the perturbation field is harmonic, this means a Bernoulli equation that can be written for the pressure is

$$p/\rho + \frac{1}{2}|\nabla\phi|^2 + gh + \frac{\partial \phi}{\partial t} = c,$$

where g is the acceleration due to gravity. The balance at the interface $z = 0$ requires the pressure difference to be a consequence of the force due to surface tension. Simply stated, this requires $\Delta p = (p_2 - p_1) = T\left(\frac{\partial^2 h}{\partial x^2} + \frac{\partial^2 h}{\partial y^2}\right)$ where T is the coefficient for surface tension. Linearized, the expression for the pressure becomes

$$p/\rho + U\frac{\partial \phi}{\partial x} + gh + \frac{\partial \phi}{\partial t} = 0. \tag{7.48}$$

This formulation results in three homogeneous equations for the unknown coefficients, or simply an eigenvalue problem. In fact, assuming $h(x,y,t) = He^{ik_1x+ik_2y-i\omega t}$ and $\phi = Ae^{-kz}e^{ik_1x+ik_2y-i\omega t}$ for $z > 0$; $\phi = Be^{kz}e^{ik_1x+ik_2 2y-i\omega t}$ for $z < 0$, then the eigenvalues are given by the determinant of the coefficients A, B, and H. This is

$$\begin{vmatrix} i(k_1 U_2 - \omega) & 0 & k \\ i(k_1 U_1 - \omega) & -k & 0 \\ (\rho_2 - \rho_1)g + k^2 T & -i\rho_1(k_1 U_1 - \omega) & i\rho_2(k_1 U_2 - \omega) \end{vmatrix} = 0. \tag{7.49}$$

When expanded, (7.49) is

$$k\left[\rho_2(k_1 U_2 - \omega)^2 + \rho_1(k_1 U_1 - \omega)^2 + k(\rho_2 - \rho_1)g - k^3 T\right] = 0. \tag{7.50}$$

The eigenvalue relation (7.50) is to be solved for the frequency ω ($k = 0$ is not of any interest) in order to determine whether or not there is an instability and, if the imaginary part, $\omega_i > 0$, the flow is unstable. The general result for the frequency is

$$(1+s)\overline{\omega} = \cos\theta(1+s\overline{U}) \pm i\cos\theta\left[s(\overline{U}-1)^2 - \frac{(1+s)\overline{T}+(1-s^2)\overline{g}}{\cos^2\theta}\right]^{1/2}$$
(7.51)

where all the quantities have now been nondimensionalized and are defined as

$$\overline{\omega} = \frac{\omega}{kU_1}, \qquad s = \rho_2/\rho_1, \qquad \overline{U} = U_2/U_1,$$

$$\cos\theta = k_1/k, \qquad \overline{g} = g/kU_1^2, \qquad \overline{T} = \frac{kT}{\rho_1 U_1^2}.$$

The first observation should be directed to the fact that, if $\rho_1 = \rho_2|_{s=1}$, then the force of gravity is no longer present. Second, the maximum amplification rate, when there is instability, is for $\cos\theta = 1$ or $\theta = 0$. Historically, if $\overline{T} = 0$, the general result is due to Kelvin; $\overline{T} \neq 0$ was considered by Stokes. More specific cases are:

(a) $s = 1, \overline{T} = 0$:

$$\overline{\omega} = \frac{(1+\overline{U})}{2}\cos\theta \pm i\frac{(\overline{U}-1)}{2}\cos\theta.$$

This is the classical Helmholtz model where, unless $\overline{U} = 1$, the flow is always unstable.

(b) $\overline{U} = 1, \overline{T} = 0$:

$$\overline{\omega} = \cos\theta \pm \frac{[(1-s^2)\overline{g}]^{1/2}}{(1+s)}.$$

In this case, there is no relative shearing motion but instability depends upon whether or not the heavier or the lighter fluid is in the upper location, i.e., $s \leq 1$.

(c) $\overline{U} = 1, \overline{T} \neq 0$:

Here, it is a question of the balance of the force of gravity with surface tension with the neutral focus, $\overline{\omega}_i = 0$, being given by

$$g/T = \frac{(1+s)}{(s^2-1)}\frac{k^2}{\rho_1}$$

when expressed in the original dimensional variables. But, since there is no natural length scale in a discontinuous model, it is the polar wave length, k, that is of importance. As such, large values of k simply mean very small spatial scales and are nonphysical.

(d) With all effects considered, full neutrality is given by

$$(1 - s^2)\overline{g} = s(\overline{U} - 1)^2 \cos^2 \theta - (1+s)\overline{T},$$

or a balance of all forces in the problem. Generally speaking, gravity and surface tension tend to be stabilizing, whereas relative motion (shear) destabilizes.

7.3 Effects of Rotation

The basis for this problem concerns motions that are on the surface of the Earth, in the atmosphere or perhaps on a rotating table in a laboratory. As such, the analyses have been made using what is known as the beta-plane approximation for the fundamental model, and it provides an excellent description for understanding such motions in the northern or southern latitudes of the Earth where the value of the rotation varies as one moves from the equator to the poles. In this case, three-dimensionality is excluded because, in order to do this properly, stratification of the mean density must be incorporated in order to correctly model the full physics (cf. Drazin, 1978, among others, for the full analysis). Viscous effects can, however, be neglected to high order and the fluid is incompressible.

The perturbation equations for this problem can be written as

$$u_x + v_y + w_z = 0, \tag{7.52}$$

$$u_t + 2\Omega(w\cos\varphi - v\sin\varphi) = -\frac{1}{\rho}p_x, \tag{7.53}$$

$$v_t + 2\Omega u \sin\varphi = -\frac{1}{\rho}p_y, \tag{7.54}$$

$$w_t - 2\Omega u \cos\varphi = -\frac{1}{\rho}p_z - g, \tag{7.55}$$

for a constant density fluid and motion where the Earth's rotation, Ω, plays a central role. The angle, φ, is measured from the equator in the northerly or southerly directions. Mean motion is neglected for the moment and, typically, $u, v \approx 200$ cm/sec, $w \approx 1$ cm/sec; $g = 10^3$cm/sec and $2\Omega = 1.456 \times 10^{-4}$/sec. Thus, rotation is essentially relegated to the (x, y)-plane under these conditions. Once a location is established, in either the northern or southern latitude and not too near the equator, i.e., $\varphi > \pm 4°$, then it is also true that $|w\cos\varphi| \ll |v\sin\varphi|$ and $|\Omega u\cos\varphi| \ll g$.

Define $f = 2\Omega \sin \varphi$ as the Coriolis parameter and the reduced set of equations for the momenta are

$$u_t - fv = -\frac{1}{\rho} p_x, \tag{7.56}$$

$$v_t + fu = -\frac{1}{\rho} p_y \tag{7.57}$$

and

$$w_t = -\frac{1}{\rho} p_z - g. \tag{7.58}$$

For hydrostatic balance to be valid, then $|w_t| \ll g$ with the centrifugal effect included in the definition of the pressure. In fact, if $w = \hat{w}e^{i\sigma t}$, then $|\sigma \hat{w}/g| \ll 1$ must follow or $|\sigma a/g| \ll 1$ with a the amplitude of any vertical motion. Hence, this assumption results in a condition on the frequency σ of the vertical motion. Under these conditions, (7.58) immediately reduces to

$$p_z \approx -\rho g \tag{7.59}$$

and can be integrated to give

$$p = p_a + \int_z^h \rho g dz, \tag{7.60}$$

where p_a is the atmospheric pressure. If $\rho = \rho_0$, then $p = p_a + \rho_0 g(h - z)$ with h the height of the sea surface, say. The result is that hydrostatic balance allows for u and v to be independent of the coordinate z.

Continuity is expressed by the fact that the velocity must be solenoidal or (7.52). This equation can now be integrated in the vertical direction over the height of the fluid column and, in so doing, the pressure variable can be replaced by h, the surface height. Similarly, w can be replaced in the same way since there are boundary conditions both above and below that are given in terms of w. The combination of these operations leads to the new set of governing equations or

$$(u\bar{h})_x + (v\bar{h})_y + h_t = 0, \tag{7.61}$$

$$u_t - fv = -gh_x, \tag{7.62}$$

$$v_t + fu = -gh_y, \tag{7.63}$$

where $\bar{h} = h + s$; $z = s$ is the location of the bottom and taken independent of time; the atmospheric pressure is taken as constant. It is interesting to note that these changes have now changed the set of linear equations to ones that are again nonlinear. However, the approximation

$$\bar{h} = h + s = s\left(1 + \frac{h}{s}\right) \approx s \qquad \text{for} \qquad \frac{h}{s} \ll 1, \qquad (7.64)$$

must be used for consistency, and then (7.61) becomes

$$(us)_x + (vs)_y = -h_t, \qquad (7.65)$$

with $s = s(x,y)$ in general and is a known function in order to solve any problem.

There are several noted examples from the field of oceanography that use this set of equations:

(1) $s = s_0$ is a constant: If rotation and the v-component of the motion are neglected, then the governing equation in terms of h as the dependent variable is nothing more than a one-dimensional wave equation, and $\sqrt{gs_0}$ is just the shallow water speed for these waves:

$$\frac{\partial^2 h}{\partial t^2} = s_0 g \frac{\partial^2 h}{\partial x^2}. \qquad (7.66)$$

(2) $s = s(x)$ only and constant rotation and in a narrow channel: Solutions can be assumed of the form $h(x,t) = a(x)e^{i\sigma t}$, $u = U(x)e^{i\sigma t}$, $v = V(x)e^{i\sigma t}$ and the net equation will read

$$\frac{d^2}{dx^2}(sgU) + (\sigma^2 - f_0^2)U = 0. \qquad (7.67)$$

As can be seen, wave motion is possible and depends upon the relative frequency when compared to the value of the rate of rotation.

(3) Long progressive waves: Here, $s = s_0$ and $f = f_0$, both constants. The more common solution for this problem is taken as

$$\{h,u,v\} = \{H,A,B\}e^{i(kx-\omega t)+\ell y} \qquad (7.68)$$

and leads to a dispersion relation by substituting (7.68) into the set of governing equations (7.62), (7.63) and (7.65). This is:

$$\omega\left[\omega^2 - f_0^2 + gs_0(\ell^2 - k^2)\right] = 0. \qquad (7.69)$$

Thus, either (a) $\omega = 0$ and is just the steady motion and is geostrophic or, (b) $\omega^2 = f_0^2 + gs_0(k^2 - \ell^2)$. In this case ℓ can be real or complex. If purely real, then Kelvin waves result but require $v = 0$ in order to satisfy the boundary conditions on the walls of the channel. If ℓ is imaginary, then the waves are known as Poincaré waves and the channel can be broad.

(4) Rossby waves: For this case, $s = s_0$, constant but $f \cong f_0 + \beta y$ and the assumption is known as the beta-plane approximation. After eliminating u and v from the equations, an equation for h under these circumstances can be found. Furthermore, the temporal behavior remains periodic or $h(x,y,t) = \hat{h}(x,y)e^{-i\omega t}$, and the result becomes

$$(\omega^2 - f^2)\hat{h} + (gs_0)\nabla^2\hat{h} - i\frac{\beta}{\omega}(gs_0)\hat{h}_x = 0, \qquad (7.70)$$

with ∇^2 the two-dimensional Laplace operator in x, y variables. The fact that this wave equation has a term involving an extra derivative with respect to x is the heart of Rossby waves, namely these waves have a westerly propagation since x denotes the east–west orientation. This point can be illustrated by using the WKBJ form for solution. To this end, let

$$\hat{h}(x,y) = a(x,y)e^{i\theta(x,y)} \qquad (7.71)$$

with $i\theta = i\ell x + iny + \lambda x + \mu y$. And, for plane waves, λ/ℓ and μ/n are both $\ll 1$. To lowest order, the dispersion relation for these waves is

$$(\ell + \ell_0)^2 + n^2 = \ell_0^2 - \frac{(f_0^2 - \omega^2)}{gs_0} = r^2, \qquad (7.72)$$

where

$$\ell_0 = \frac{1}{2}\frac{(f_0^2 + \omega^2)}{(f_0^2 - \omega^2)}\frac{\beta}{\omega}.$$

Again, the westerly nature of the Rossby wave is quite evident from (7.72).

Return to the set of equations (7.52) and (7.56) to (7.57) and insert the beta- (or spanwise-) plane approximation for f along with the restoration of the mean flow, $U(y)$. Elimination of the pressure from these equations will lead to the following nondimensional equation for v:

$$\left(\frac{\partial}{\partial t} + U\frac{\partial}{\partial x}\right)\nabla^2 v - \left(U'' - \frac{1}{Ro}\right)v_x = 0, \qquad (7.73)$$

where all quantities have been nondimensionalized with respect to U_0 and L, mean flow values for the velocity and spatial scales. This governing equation is the analogue for that of internal gravity waves with shear except that the new parameter is known as the Rossby number, defined as

$$Ro = U_0/\beta L^2, \qquad (7.74)$$

and is the ratio of the relative values of inertia to the Coriolis force in the problem. In the limit, $Ro \to 0$, the flow is in geostrophic balance as noted previously.

Equation (7.73), even with the addition of the Rossby number, is mathematically the same type of equation as that used by Rayleigh (1880) for the study of stability without rotation in the inviscid limit (see 1.43). This basis was recognized by Kuo (1949), Fjørtoft (1950) and Tung (1981), among others. Consequently it has led to a result that has become known as the Rayleigh–Kuo theorem for flows in this environment. The theorem is directly related to the necessity for an inflection point in the mean flow distribution. This can be demonstrated in a straightforward manner just as was done for stratified flow.

Normal-mode solutions can be found for v as has been done throughout, or $v = \hat{v}(y)e^{i\alpha x - i\omega t}$. Upon substitution of this form into (7.73), an ordinary differential equation for $\hat{v}$ is obtained:

$$\hat{v}'' + \left[\frac{1/Ro - U''}{(U - c)} - \alpha^2 \right] \hat{v} = 0, \tag{7.75}$$

where $c = \omega/\alpha$, the complex wave speed. Now multiply (7.75) by the complex conjugate of $\hat{v}$ and integrate the result over the range of the flow or from $y = y_1$ to $y = y_2$, say. With the boundary conditions, such that $\hat{v} = 0$ or $\hat{v} \to 0$ for the y_1, y_2 values and, after integration by parts, this operation yields

$$\int_{y_1}^{y_2} \left[\frac{1/Ro - U''}{(U - c)} - \alpha^2 \right] |\hat{v}|^2 dy = \int_{y_1}^{y_2} |\hat{v}'|^2 dy. \tag{7.76}$$

The imaginary part of (7.76) is

$$c_i \int_{y_1}^{y_2} (Ro^{-1} - U'') \frac{|\hat{v}|^2}{|U - c|^2} dy = 0. \tag{7.77}$$

Thus, a necessary condition for instability where $c_i > 0$ is that $(Ro^{-1} - U'') = 0$, the mean vorticity gradient ($\beta - U'' = 0$, dimensionally) must change sign somewhere within the flow domain. In short, a generalized inflection point criterion and a result due to Kuo (1949). In like manner, the real part of (7.76) was used by Fjørtoft (1950) to demonstrate that the generalized inflection point should be located exactly as that of the critical layer, $(U - c_r) = 0$, when there is a neutral disturbance, $c_i = 0$. These two results mimic those established by Rayleigh when there is no rotation ($\beta = 0$) and are particularly relevant to the bases of barotropic instability in zonal flows. Tung (1981) has further provided an extensive description of this type of flow and its instabilities.

There is still more to this problem other than the analogy to the non rotating case. In part, such inferences are directly related to the net value and sign of the coefficient $(\beta - U'')$. One extreme is for the case when there is no mean flow and it is only the effect due to the beta-plane that dictates the perturbation motion. All relevant possibilities have been given, and these are tantamount to linear wave motion. Then there can be a finite mean flow but it varies only linearly and thus the term U'' still vanishes. For the most part such flows are stable but, for flows in the geophysical environment, it is not just the mean velocity variation that dictates the salient nature of the problem but the specific boundary conditions that must be satisfied as well. The barotropic description is that of perturbations on the beta-plane in the presence of a mean shear with $U'' \neq 0$, and the perturbation velocity is required to vanish at the boundaries. As has been shown, this flow can be unstable. As a final comment, the extension of the beta-plane approximation that includes the effect of free surface waves was recently investigated by Shivamoggi & Rollins (2001).

7.4 Baroclinic Flow

Another important problem deals with the stability of what is known as baro-clinic flow. In this case the boundary conditions are physically connected to the perturbation pressure instead of the velocity. Of course these conditions are, by use of the governing perturbation equations, easily set in terms of the velocity but such conditions are not simply the requirement that the velocity vanish and the results lead to far different conclusions. Progress in this area was pioneered by Charney (1947) and Eady (1949). Both of these efforts considered only a linear variation for the mean velocity but, as has been mentioned, this is not a requirement, and instability is still possible.

The Eady problem is also described by Drazin & Reid (1984) in fine detail and the underlying physics are presented, making note that there is an inter-esting mathematical equivalent of this problem to that for the stability of plane Couette flow of an inviscid constant density fluid where the boundary condi-tions require the pressure to be constant on the upper and lower boundaries in lieu of the vanishing of the velocity. For this purpose, however, the perturba-tion problem should be cast in a convective coordinate reference frame in order to fully make the analogy. In this way, Criminale & Drazin (1990) examined the Eady problem, among other examples, in order to demonstrate an alter-native method for solving the complete dynamics for perturbations as well as determine the stability. A thorough presentation of this technique will be given in Chapter 8 but it is useful to apply the technique to the Eady problem now. And, since the inviscid plane Couette flow problem is perhaps the simplest of all modeled flows, the impact of the change in the boundary conditions is profound.

Consider the flow defined between $y = \pm H$ with $U(y) = \sigma y$. The inviscid perturbation equation for the $\tilde{v}$-velocity component is

$$\left(\frac{\partial}{\partial t} + U\frac{\partial}{\partial x}\right)\nabla^2\tilde{v} = 0 \tag{7.78}$$

and the boundary conditions are that the pressure be constant for $y = \pm H$.

The convective coordinate transformation is defined by the change of variables

$$T = t; \qquad \xi = x - U(y)t; \qquad \eta = y; \qquad \zeta = z.$$

In these terms, equation (7.78) becomes

$$\frac{\partial}{\partial T}\nabla^2\tilde{v} = 0, \tag{7.79}$$

where $\nabla^2 = \frac{\partial^2}{\partial\eta^2} - 2\sigma T\frac{\partial^2}{\partial\eta\partial\xi} + \sigma^2 T^2\frac{\partial^2}{\partial\xi^2} + \frac{\partial^2}{\partial\xi^2} + \frac{\partial^2}{\partial\zeta^2}$. It can be immediately seen that there are two possible solutions for v. First, $\nabla^2\tilde{v} = F(\xi,\eta,\zeta)$. Second, $\nabla^2\tilde{v} = 0$. These solutions correspond physically to whether or not there is an initial perturbation vorticity, since all solutions for $\nabla^2\tilde{v} = 0$ are those of a velocity field has neither a divergence nor a curl and thus harmonic.

Again, since the ξ and ζ variables are unbounded in both directions, a Fourier transformation is used, and (7.79) takes the form

$$\frac{\partial}{\partial T}\Delta\check{v} = 0 \tag{7.80}$$

where

$$\check{v} = \int_{-\infty}^{+\infty}\int_{-\infty}^{+\infty}\tilde{v}(\xi,\eta,\zeta,T)e^{i\alpha\xi+i\gamma\zeta}d\xi d\zeta$$

is the Fourier transformation, and

$$\Delta = \frac{\partial^2}{\partial\eta^2} + i2\sigma T\alpha\frac{\partial}{\partial\eta} - (\tilde{\alpha}^2 + \sigma^2 T^2\alpha^2)$$

with $\tilde{\alpha}^2 = \alpha^2 + \gamma^2$. A further change of variables where

$$\check{v}(\alpha;\eta;\gamma;T) = V(\alpha;\eta;\gamma;T)e^{-i\sigma T\alpha\eta}$$

and then

$$\Delta\check{v} \equiv \left(\frac{\partial^2 V}{\partial\eta^2} - \tilde{\alpha}^2 V\right)e^{-i\sigma T\alpha\eta}.$$

The solution takes the form $V = A(T)e^{-\tilde{\alpha}\eta} + B(T)e^{\tilde{\alpha}\eta} + V_p$ with V_p depending upon the choice for F, the initial vorticity.

An expression for the pressure in terms of the velocity can be obtained from the transformed equations for u and w. This is

$$-\tilde{\alpha}^2 \check{p} = \frac{\partial^2 \check{v}}{\partial T \partial \eta} + i\sigma T \alpha \frac{\partial \check{v}}{\partial T} + 2i\sigma \alpha \check{v}$$

$$= \left(\frac{\partial^2 V}{\partial \eta \partial T} + i\alpha\sigma V - i\alpha\sigma\eta \frac{\partial V}{\partial \eta} \right) e^{-i\sigma T \alpha \eta} \qquad (7.81)$$

and must be constant for $\eta = \pm H$. By applying these conditions two coupled first-order ordinary equations for A and B result and can be written as

$$\begin{pmatrix} \tilde{\alpha} e^{-\tilde{\alpha} H} & -\tilde{\alpha} e^{\tilde{\alpha} H} \\ \tilde{\alpha} e^{\tilde{\alpha} H} & -\tilde{\alpha} e^{-\tilde{\alpha} H} \end{pmatrix} \begin{pmatrix} \overset{\circ}{A} \\ \overset{\circ}{B} \end{pmatrix}$$

$$= i\sigma\alpha \begin{pmatrix} (1+\tilde{\alpha} H)e^{-\tilde{\alpha} H} & (1-\tilde{\alpha} H)e^{\tilde{\alpha} H} \\ (1-\tilde{\alpha} H)e^{\tilde{\alpha} H} & (1+\tilde{\alpha} H)e^{-\tilde{\alpha} H} \end{pmatrix} \begin{pmatrix} A \\ B \end{pmatrix} + \begin{pmatrix} F_1 \\ F_2 \end{pmatrix}, \quad (7.82)$$

where the pressure constant at the boundaries has been taken to be zero. Here, for notation purposes,

$$\overset{\circ}{A} = \frac{dA}{dT} \qquad \text{and} \qquad \overset{\circ}{B} = \frac{dB}{dT}.$$

The functions F_1 and F_2 are defined as

$$F_1 = \tilde{\alpha}^2 \check{p}_p \Big|_{\eta=H} e^{i\sigma T \alpha H} \qquad \text{and} \qquad F_2 = \tilde{\alpha}^2 \check{p}_p \Big|_{\eta=-H} e^{-i\sigma T \alpha H},$$

with $\check{p}_p$ is the pressure resulting from V_p. Inverting the matrix on the left-hand side of (7.82) produces a standard form or the system:

$$\frac{d}{dT} \begin{pmatrix} A \\ B \end{pmatrix} = \frac{i\sigma \cos\varphi}{2} \begin{pmatrix} 2 - q\coth q & q/\sinh q \\ -q/\sinh q & -2 + q\coth q \end{pmatrix} \begin{pmatrix} A \\ B \end{pmatrix}$$

$$+ \frac{H}{q\sinh q} \begin{pmatrix} e^{\tilde{\alpha} H} F_2 - e^{-\tilde{\alpha} H} F_1 \\ e^{-\tilde{\alpha} H} F_2 - e^{\tilde{\alpha} H} F_1 \end{pmatrix}, \qquad (7.83)$$

where $q = 2\tilde{\alpha} H$ and $\alpha = \tilde{\alpha}\cos\varphi$ have been used.

The homogeneous solutions for (7.83) are those of normal modes and, with the solutions proportional to $e^{\omega T}$, the eigen-frequencies are

$$\omega = \pm \frac{i\sigma \cos\varphi}{2} (4 - 4q\coth q + q^2)^{1/2}. \qquad (7.84)$$

The system is dynamically neutral so long as $q > 2.4$ since ω remains purely imaginary. A and B are pure constants for $q = 0, 2.4$ or $\varphi = \pi/2$, since $\omega = 0$ for these values. For values $0 < q < 2.4$ the flow is unstable. These results are in sharp contrast to those of plane Couette flow where there are no normal modes for the inviscid problem and $A = B = 0$ is the only possibility, since it is the velocity that must vanish at the two boundaries in lieu of the pressure.

When an initial vorticity has been specified, the forced problem offers much more to either the Eady or the plane Couette flow problem. This is mathematically the result due to the continuous spectrum, as discussed in Chapter 3. For inviscid plane Couette flow, the forced problem was done by Criminale, Long & Zhu (1991). The exact behavior for either problem will depend upon the choice for the forcing, but it is significant to note that algebraic dependence in time evolves and dominates the early transient period of the dynamics.

In the case treated by Eady for baroclinic instability, the results for the homogeneous unforced problem are as the expression in (7.84) save for the fact that the relevant physical parameter is the Burger number, a number that involves the density variation, perturbation temperature, the acceleration due to gravity as well as the value of the rotation. The exact result for the forced problem will depend upon the initial input but, as for plane inviscid Couette flow, this part of the solution will involve the continuous spectrum.

7.5 The Ekman Layer

The perturbation problem for the Ekman layer is remarkable for several reasons. First of all, the functions that define the mean flow that Ekman (1905) determined are known to be an exact set of solutions for the full Navier–Stokes governing equations. In fact, this result is true for both the steady and the unsteady solutions for this flow. Ekman in essence considered this problem as one that approximated the planetary boundary layer and, to this end, it was assumed that the problem was local in that the surface was taken to be flat and the Coriolis parameter constant. Then, because of the important role that the rotation plays, this boundary layer is truly parallel. The mean flow vector is not, however, simply in one specified direction. On the contrary, it has two components that combine to form what is known as the Ekman spiral. If $\underline{U} = (U, V, W)$ defines the mean velocity vector, then the solution in dimensional terms is

$$\left.\begin{array}{rcl} U(z) & = & U_g \left[1 - e^{-z/\sqrt{f/2v}} \cos(z/\sqrt{f/2v})\right], \\ V(z) & = & U_g e^{-z/\sqrt{f/2v}} \sin(z/\sqrt{f/2v}), \\ W & = & 0, \end{array}\right\} \qquad (7.85)$$

where U_g is the geostrophic value of the velocity outside the boundary layer. Here, the z-coordinate is maintained for the geophysical flow examples and is the vertical coordinate perpendicular to the boundary. The velocity lies entirely in the (x, y)-plane, is zero at the boundary and approaches the geostrophic value in the freestream $(z \to \infty)$. The referenced spiral is defined in the (x, y)-plane and can be evaluated by the ratio of U and V as a function of z and is shown in Fig. 7.7. The existence of the spiral nature also is the cause of notable features, such as the fact that the transport and the stress on the fluid at the boundary are opposed to each other, when viewed from the orientation of the vector $\underline{U}$. Figure 7.8 shows the velocity component solutions within the Ekman layer.

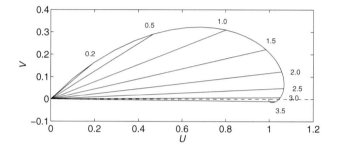

Figure 7.7 Ekman spiral (after Ekman, 1905).

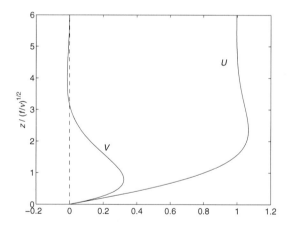

Figure 7.8 Ekman profile (after Ekman, 1905).

In nondimensional terms, the natural length scale is $\sqrt{f/\nu}$. And, based on the considerations of the geophysical flows that have been presented, two parameters should be present in the governing equations, namely the Rossby number and one that assesses the effects of viscosity. The latter is known as the Ekman number in this context but is akin to the Reynolds number of non-rotating flows. In this case, however, the Rossby number fails to appear in the solution because there is no nonlinearity for an exact solution. More specifically, the Ekman solution is a direct consequence of a balance of the viscous force and the Coriolis acceleration and this is a linear problem.

This flow is now perturbed in a full three-dimensional manner. The linearized perturbation equations are

$$\tilde{u}_x + \tilde{v}_y + \tilde{w}_z = 0, \tag{7.86}$$

$$\tilde{u}_t + U\tilde{u}_x + V\tilde{u}_y + U'\tilde{w} - f\tilde{v} = -\tilde{p}_x/\rho_0 + \nu\nabla^2\tilde{u}, \tag{7.87}$$

$$\tilde{v}_t + U\tilde{v}_x + V\tilde{v}_y + V'\tilde{w} + f\tilde{u} = -\tilde{p}_y/\rho_0 + \nu\nabla^2\tilde{v}, \tag{7.88}$$

$$\tilde{w}_t + U\tilde{w}_x + V\tilde{w}_y = -\tilde{p}_z/\rho_0 + \nu\nabla^2\tilde{w}. \tag{7.89}$$

Modal solutions combined with Fourier decomposition can be used as before to reduce this system to that of ordinary differential equations. Assume that all variables are a sum of modes written as

$$\{\tilde{u}, \tilde{v}, \tilde{w}, \tilde{p}\} = \{\hat{u}, \hat{v}, \hat{w}, \hat{p}\}(z)e^{i(\alpha_1 x + \alpha_2 y - \omega t)}. \tag{7.90}$$

Then, with this ansatz, the pressure can be eliminated from (7.87)–(7.89), and the following pair of coupled equations must be solved as an eigenvalue problem.

$$(\alpha\overline{U} - \omega)\Delta\hat{w} - \alpha\overline{U}''\hat{w} + i\nu\Delta^2\hat{w} = f(\alpha\overline{v})', \tag{7.91}$$

$$(\alpha\overline{U} - \omega)(\alpha\overline{v}) + i\nu\Delta(\alpha\overline{v}) = i\alpha\overline{V}'\hat{w} + f\hat{w}', \tag{7.92}$$

where again

$$\Delta = \frac{d^2}{dy^2} - \alpha^2$$

and full use of the Squire transformation has been used, namely

$$\alpha\overline{U} = \alpha_1 U + \alpha_2 V, \qquad \alpha\overline{u} = \alpha_1\hat{u} + \alpha_2\hat{v}, \tag{7.93}$$

$$\alpha\overline{V} = \alpha_2 U - \alpha_1 V, \qquad \alpha\overline{v} = \alpha_2\hat{u} - \alpha_1\hat{v}, \tag{7.94}$$

in order to put the equations in this form. Here, $\alpha^2 = \alpha_1^2 + \alpha_2^2$. Originally, this was done by Lilly (1966) and later by Spooner & Criminale (1982) for a different purpose. But, it should be noticed that, even though (7.91) and (7.92) bear a strong resemblance to the classical Orr–Sommerfeld and Squire equations (exactly these equations if $f = 0$), this system is fully coupled. As such, it is now a sixth-order system if one elects to obtain one governing equation.

Equations (7.91) and (7.92) can be better scrutinized in a nondimensional form. The basic velocity is taken to be U_g, the freestream geostrophic value and the Ekman length scale is chosen to be $L = \sqrt{\nu/f}$. In this measure, the Ekman number becomes unity and the dominant parameter is again the Reynolds number, $Re = U_g L/\nu$. In this way, the governing equations are

$$\Delta^2 \hat{w} - iRe(\alpha \overline{U} - \omega)\Delta \hat{w} + i\alpha Re\overline{U}'' \hat{w} = -i\alpha \overline{v}', \qquad (7.95)$$

$$\Delta(i\alpha \overline{v}) - iRe(\alpha \overline{U} - \omega)(i\alpha \overline{v}) = i\alpha Re\overline{V}' \hat{w} + \hat{w}', \qquad (7.96)$$

where all quantities are now nondimensional. It is this set of equations that was numerically integrated by Lilly (1966), Faller & Kaylor (1967) and Spooner (1980) in order to determine the discrete eigenvalues for this flow. Unlike the Blasius boundary layer, where the instability requires viscosity to exist and is manifested by Tollmien–Schlichting waves, the Ekman layer has still other instabilities. Figure 7.9 displays the results in the conventional (α, Re)-plane. Figures 7.10 to 7.12 depict surfaces in the Fourier space or cross sections of the data from Fig. 7.9 in order to better illustrate the possibilities. Such additional diagrams are the direct results from the fact that the system is a fully coupled sixth-order system and the existence of an important new parameter that defines the orientation of the Ekman spiral. Figure 7.9 already reveals other possibilities for instability, particularly as $Re \rightarrow \infty$. It can also be seen that the phase speed of the perturbation can change sign. In Fig. 7.10, two maxima for the perturbation amplification are now possible. Figures 7.11 and 7.12 display such details in other ways.

Lilly explained the results that were obtained in terms of three distinct designations. First, the parallel mode is viscous and exists because the flow is rotating. This instability vanishes as $Re \rightarrow \infty$ and is more prevalent at low values of the Reynolds number. Second, as the Reynolds number is increased, dual regions of instability emerge. This is due to the lingering of the parallel mode and the viscous-induced instability of the Tollmien–Schlichting type that is more common in the boundary layer. Third, there is an inviscid instability, or the type that requires the mean profile to have an inflection. In terms of the physics, an inflection point in the velocity profile implies an extremum in the vorticity distribution (Lin, 1954) and follows exactly the bases established in Chapter 3. This can be made more lucid if (7.95) and (7.96) are considered at $Re = \infty$. Immediately there is an uncoupling, and (7.95) is effectively the Rayleigh equation save for the particular distribution of the mean velocity.

The parallel instability that Lilly suggested was substantiated by a series of approximations to the set of governing equations. Although the approximations did demonstrate this type of instability, it was better done by a somewhat milder approximation made by Stuart (cf. Greenspan, 1969). Here, the mean flow was taken to be $\overline{V}' = $ constant and $\overline{U} = \overline{U}'' = 0$ (and hence the designation of parallel mode), and then solutions of the approximate set of equations was sought by requiring that the solutions be oscillatory in z. This is possible for the resulting sixth-order system is one of constant coefficients. Thus, if the solutions are proportional to $e^{-i\beta z}$, then the following dispersion relation can be found:

$$\omega = iRe^{-1}\left[-(\beta^2 + \alpha^2) \pm \left(\frac{\beta\alpha Re\overline{V}' - \beta^2}{\beta^2 + \alpha^2}\right)^{1/2}\right]. \qquad (7.97)$$

Consequently, the perturbations will be unstable if

$$Re > \left[\beta^2 + (\beta^2 + \alpha^2)^3\right]/\beta\alpha\overline{V}'.$$

Again, the dependence on the parallel component of the mean velocity is essential for this mechanism. And, as mentioned, it is a viscous instability for, as $Re \to \infty$, the instability vanishes. The other viscous instability is that which was discussed in Chapter 3 for the flat-plate boundary layer.

It should be mentioned that Spooner (1980) has thoroughly examined all the modes for the Ekman layer and, not only were the eigenvalues that were obtained by others confirmed, but the details of the eigenfunctions, the phasing and the Reynolds stress distributions were given in addition. Such intense details were needed so that he could do an initial-value problem for the Ekman boundary layer, which was done by Spooner & Criminale (1982).

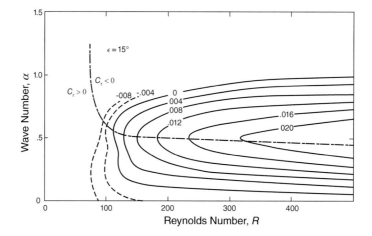

Figure 7.9 Neutral stability of the Ekman layer. (After Lilly, 1966, ©American Meteorological Society. Used with permission.)

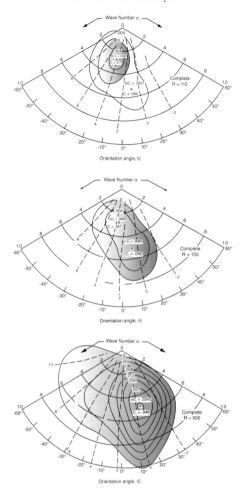

Figure 7.10 Ekman eigenvalue display for fixed wavenumber as a function of spiral angle and Reynolds number. (After Lilly, 1966, ©American Meteorological Society. Used with permission.)

In this chapter, we have discussed the class of flows that are termed geophysical. Stratified flows with a mean density variation were introduced and shown that the body force due to gravity has a significant role and can lead to internal gravity waves. Next, we showed that the effects of the Earth's rotation have a dominant role on the generation of waves. Finally, we looked at the case where, with a geophysical boundary layer where the rotation is present but taken as

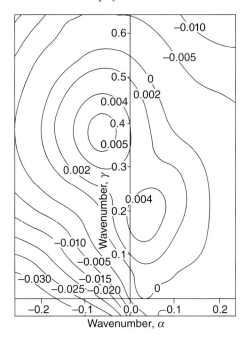

Figure 7.11 Contours of constant amplification for the Ekman layer at Reynolds number, $Re = 105$ (after Spooner & Criminale, 1982, reprinted with permission).

constant, the surface is assumed flat. In this case, viscous shear is important and the flow is known as the Ekman layer. The resulting mean flow solution also happens to be an exact solution of the full Navier–Stokes equations. For this perturbation problem, the flow can be inviscidly unstable as well as having a viscous instability.

7.6 Exercises

Exercise 7.1 Analyze qualitatively and quantitatively the stability of a flow that has a mean velocity, $\underline{U} = (\sigma z, 0, 0)$ with σ constant and is in an environment that is rotating with a value, f = constant about the z-axis. In making this analysis, answer the following questions:

(a) What is the mean flow solution?
(b) What can be said of the mean pressure?
(c) What are the relevant scales and parameters if the flow is (i) viscous, (ii) inviscid?

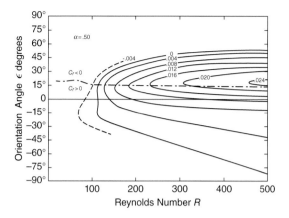

Figure 7.12 Unstable regions for the Ekman layer as a function of wavenumber and spiral angle. (After Lilly, 1966, ©American Meteorological Society. Used with permission.)

(d) Solve the problem when all fluctuations are independent of x and comment on the viscous versus inviscid solutions.

(e) Solve the problem by two different means and describe the type of instability – if one is present.

(f) Determine the energy and vorticity for both the mean and perturbation quantities.

(g) Compare to the Ekman–Blasius problem.

Exercise 7.2 Investigate the stability of the flow diagramed in Fig. 7.13.

Exercise 7.3 Derive equation (7.14).

Exercise 7.4 Derive equation (7.15).

Exercise 7.5 Derive the perturbation energy equation (7.38).

Exercise 7.6 State the linearized equations that result in (7.73).

Exercise 7.7 Derive the coupled pair of instability equations (7.91)–(7.92), beginning with the disturbance form of the Navier–Stokes equations (7.86)–(7.89).

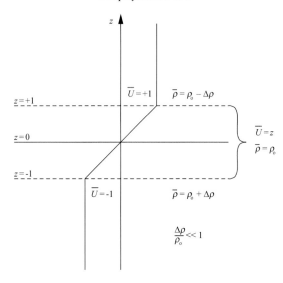

Figure 7.13 Sketch of profile for Exercise 7.2.

8

Transient Dynamics

The history of science teaches only too plainly the lesson that no single method is absolutely to be relied upon, that sources of error lurk where they are least expected, and that they may escape the notice of the most experienced and conscientious worker.

– Sir John William Strutt (Lord Rayleigh) (1842–1919)

8.1 The Initial-Value Problem

The fundamental needs for specifying an initial-value problem for stability investigations are not in any way different from those that have long since been established in the theory of partial differential equations. This is especially true in view of the fact that the governing equations are linear. Thus, by knowing the boundary conditions as well as the particular initial specification, the problem is, in principle, complete. Unfortunately, in this respect, classical theory deals almost exclusively with second-order systems and, as such, few problems in this area can be cast in terms of well-known orthogonal functions. For the equations that are the bases of shear flow instability, however, it is only the inviscid problem that is second order (Rayleigh equation) and even this limiting equation does not have a detailed set of known functional solutions. The more serious case where viscous effects are retained, then the minimum requirement is an equation that is fourth order (Orr–Sommerfeld equation) and even this, as previously noted, is fortuitous. An a priori inspection would have led one to believe that the full three-dimensional system should be sixth order, such as that discussed for the case of the Ekman boundary layer, for example. The net result is one where there are neither known closed solutions nor mutual orthogonality. It is only the accompanying Squire equation, where the solutions are coupled to those of the Orr–Sommerfeld equation, that eventually makes for sixth order. These facts already form a sufficient basis for the necessity to inquire as to the early time as well as the long time behavior in the dynamics but, depending upon the particular flow under investigation, there are other considerations that must be included as well. All of these facts combine to strongly influence the initial temporal response and, in fact, under certain circumstances, can make the asymptotic behavior quite moot by comparison. And, although work dealing with this aspect of the problem did not make any significant progress in the early years, both Kelvin (1887a) and Orr (1907a,b) had already recognized that the early time period contained significant information.

As has been demonstrated, there are three main categories for parallel shear flows, namely those (i) in enclosed channels, (ii) boundary layers and (iii) free shear flows that do not have any solid boundary influence in the flow. Physically, viscosity is critical to the understanding of the results for (i) and (ii) but not essential to (iii). Mathematically, (i) has only discrete eigensolutions and the set is complete (DiPrima & Habetler, 1969), whereas (ii) and (iii) must include the continuous as well as the discrete spectrum when one wishes to investigate the complete fate of an arbitrary initial disturbance. Grosch & Salwen (1978) and Salwen & Grosch (1981) have lucidly presented the details for these conclusions. In short, merely ascertaining that there may be at least one positive eigenvalue, and therefore the flow is unstable, is not sufficient. For example, Mack (1976) has long since shown that there are only a finite number of normal modes for the Blasius boundary layer and, as such, it would be impossible to represent any arbitrary initial distribution. A similar argument follows for free shear flows.

From the description of the mathematical complexities cited, it can be concluded that there are algebraic as well as exponential solutions in time. Specifically, there are three reasons for this. First, the nonorthogonality of the eigenfunctions is a reason for such behavior. Second, a possible resonance between the Orr–Sommerfeld and Squire solutions can lead to algebraic dependence. For channel flows, this is all that is possible for the viscous problem. With the boundary layer and free shear flows, then the continuous spectrum contributes to such behavior as well. Of the three, the nonorthogonality is ubiquitous and inherent in all of these flows when normal mode solutions are used for solving the equations. Resonance has been demonstrated to be possible for channel flow by Gustavsson & Hultgren (1980) and Benney & Gustavsson (1981) but, as the latter authors have further shown, this does not occur for the boundary layer. These conclusions are independent of any particular initial disturbance specification and are the consequences of the nature of the homogeneous set of governing equations and the respective boundary conditions.

Regardless of the underlying source that is the cause, the algebraic behavior translates to a linear dependence in the time variable. But, because there is viscous dissipation, any increase during the early period will eventually decay after reaching a maximum in finite time. Then, any exponentially growing mode, if one exists, will prevail as time increases beyond this point. If there is no growing mode, then the normal conclusion is that the flow is stable even though the initial algebraic growth may have increased to an amplitude that violates the assumption of linearity for the amplitude. This point will be elaborated more fully as the methods are presented. And, toward this end, it will be seen that certain specific initial conditions can be made that will result in algebraic behavior that is stronger than just linear in time. In turn, initial input depends critically upon the physics of the problem. This thesis will be demonstrated in due course but, first, the fundamentals of the proper method for describing the initial-value problem should be reviewed.

The two major equations that are needed for determining the stability of any parallel or almost parallel flow are, in partial differential equation form (cf. Chapter 1):

$$\left(\frac{\partial}{\partial t} - i\alpha U\right)\Delta \check{v} + i\alpha U''\check{v} = \varepsilon \Delta\Delta\check{v}, \tag{8.1}$$

$$\left(\frac{\partial}{\partial t} - i\alpha U\right)\check{\omega}_y - \varepsilon\Delta\check{\omega}_y = i\gamma U'\check{v}, \tag{8.2}$$

and are the Orr–Sommerfeld and Squire equations respectively in the two-dimensional Fourier (α, γ) space; i.e., after taking the Fourier transformation in (x, z). In this way (8.1) and (8.2) can be used in a very general way in order to understand the needs of the temporal initial-value, boundary value problem. The Fourier decomposition assures the fact that all dependent variables are bounded in the x-, z-directions. Within the framework of the model assumptions, this is consistent in that these two variables range from $-\infty$ to $+\infty$. The remaining spatial variable, y, also requires boundary conditions to be met but the specifics are bound to the mean flow, $U(y)$, that is in question. In every case, the boundary conditions are requirements for the velocity that must be met. There is no reason, however, that initial data must be given in terms of the velocity. On the contrary, not only can it be given in terms of the vorticity mathematically, for example, but there is strong justification to do this in terms of the physics of the problems. Moreover, if the problem was one where the fluid is compressible, then the pressure might be yet another alternative. For the incompressible medium, however, this is not appropriate.

One of the original attempts to do an initial-value problem was made by Criminale & Kovasznay (1962). In this work, a localized disturbance was given in terms of the disturbance velocity in a Blasius boundary layer and the ensuing dynamics inferred from but a few normal modes. A similar effort was made by Gaster (1968) for the Blasius boundary layer, and Spooner & Criminale (1982) considered the Ekman planetary boundary layer. In none of these cases was any consideration given to the need of a continuous spectrum. As a result, the results for these examples were limited and, in some ways, the work pointed more to the needs for the assumptions that were required then that are not, fortunately, needed today. For example, Breuer & Haritonidis (1990) and Breuer & Kuraishi (1994) have since provided a modern treatment for the Blasius boundary layer. Still, even this work has assumptions that are needed in order to make the necessary numerical calculations but, generally speaking, it is a vast improvement over the earlier attempts. The noticeable weakness for these calculations has to do with the particular set of functions used to represent the vertical spatial variation. Chebyshev polynomials were chosen and these decay quite rapidly in the far-field, a constraint that is too severe and loses the important physics that are related to the continuous spectrum.

The correct means of formulation is best given by the works of Grosch & Salwen (1978) and Salwen & Grosch (1981) as discussed in Chapter 3. In this work a completeness argument was shown that, any solution to (8.1), together with an initial-condition, can be written as

$$\check{v}(\alpha, y, \gamma, t) = \sum_{j=1}^{N} A_j e^{i\omega_j t} \bar{v}_j(y) + V_c(y, t), \qquad (8.3)$$

where N is finite and the number of discrete modes, $\bar{v}_j(y)$, are the eigenfunctions, and $V_c(y,t)$ is the continuum. The amplitude factors A_j and the frequencies ω_j are in general functions of $\tilde{\alpha} = (\alpha^2 + \gamma^2)^{1/2}$, $\phi = \tan^{-1}(\gamma/\alpha)$ and the Reynolds number, $Re = \varepsilon^{-1}$. Once a choice for the initial value has been prescribed, use of the orthogonality principle between the eigenfunctions and the adjoint eigenfunctions allows for the determination of the amplitude factors and the continuum to be found. Unfortunately, although this procedure is mathematically correct, it is of limited use when actually studying transient behavior because of the underlying difficulties in the expansion process per se and the eigenfunctions for the system have been obtained numerically. Still, there are some important observations for this approach that should be noted.

For angles of obliquity, $\phi < 90°$, then N, the number of discrete modes, is finite with the number depending upon the value of the Reynolds number. For $\phi = 90°$, there are no discrete modes whatsoever ($N = 0$) and therefore only a continuum remains, regardless of the specific problem under investigation.

The basic equations, (8.1) and (8.2), when $\alpha = 0$ ($\phi = 90°$) are

$$\frac{\partial}{\partial t}\Delta\check{v} = \varepsilon\Delta\Delta\check{v},\tag{8.4}$$

$$\frac{\partial\check{\omega}_y}{\partial t} - \varepsilon\Delta\check{\omega}_y = i\gamma U'\check{v},\tag{8.5}$$

where $\check{\omega}_y = i\gamma\bar{w}$ and $\Delta = \dfrac{\partial^2}{\partial y^2} - \gamma^2$ have been used. Equation (8.4) no longer has any dependence on the mean flow, $U(y)$. As a result, Salwen & Grosch (1981) have shown that the solution for $\check{v}$ can be written as

$$\check{v}(y,t) = \int_0^\infty a_k(\gamma)\bar{v}_k(y)e^{-\varepsilon(\gamma^2+k^2)t}dk.\tag{8.6}$$

The eigenfunctions are given by

$$\bar{v}_k(y) = \left(\frac{2}{\pi}\right)^{1/2}\frac{k}{(\gamma^2+k^2)}\left[e^{-|\gamma|y} - \cos(ky) + |\gamma|k^{-1}\sin(ky)\right].\tag{8.7}$$

The coefficients in (8.6) are found by use of initial data or

$$a_k(\gamma) = \int_0^\infty \check{v}(y,0)\bar{v}_k(y)dy.\tag{8.8}$$

Salwen & Grosch further showed that this solution decays in time as

$$\check{v} \sim t^{-1/2}e^{-\varepsilon\gamma^2 t} \quad \text{for} \quad t \to \infty.\tag{8.9}$$

Thus, in these terms, the Blasius boundary layer is asymptotically stable for all wavenumbers and Reynolds numbers. Still, there is early transient algebraic growth and it is due entirely to the fact that there is an inhomogeneous term in the Squire equation, (8.5). This result is not due to resonance, however, (cf. Benney & Gustavsson, 1981) and there is no contribution from non-normality of the Orr–Sommerfeld and Squire operators for this extreme angle of obliquity. It is only when the bounded channel flow is in question do these additional sources contribute to this kind of behavior.

Although our outline provides a rationale for treating the initial-value problem, the actual details that are needed for the calculations are lacking, and it suggests that an alternative means for analysis might provide a more expedient means for accomplishing this task.

8.2 Laplace Transforms

Traditionally, when one wishes to solve an initial-value problem, a very powerful tool for this purpose is the Laplace transform in time. This is true even if

the governing equations are partial differential equations and this method has
been used in the study of shear flows as well. Of note are the contributions of
Gustavsson (1979) and Hultgren & Gustavsson (1980). Earlier contributions
are due to Eliassen, Høiland & Riis (1953), Case (1960a,b) and Dikii (1960).

The general problem using Laplace transforms can be applied directly to
(8.1) and (8.2). With the definition that

$$\bar{\bar{v}}(y,s) = \int_0^\infty \check{v}(y,t)e^{-st}dt, \qquad \overline{\overline{\omega}}_y(y,s) = \int_0^\infty \check{\omega}_y(y,t)e^{-st}dt, \qquad (8.10)$$

then these two equations become

$$(s - i\alpha U)\Delta\bar{\bar{v}} + i\alpha U''\bar{\bar{v}} - \varepsilon\Delta\Delta\bar{\bar{v}} = \Delta\check{v}|_{t=0} \qquad (8.11)$$

and

$$(s - i\alpha U)\overline{\overline{\omega}}_y - \varepsilon\Delta\overline{\overline{\omega}}_y = i\gamma U'\bar{\bar{v}} + \check{\omega}_y(y,0). \qquad (8.12)$$

It can be noted that the original partial differential equations have now been
transformed to ones that are ordinary and, more significantly, both equations
contain terms that are not homogeneous. For (8.12), the counter part for the
Squire equation, the term denoting the coupling to Orr–Sommerfeld remains
but now there is a term that can be interpreted as that of vorticity or the polar
velocity, as defined by the Squire transform. For (8.11), it is not the initial
value of the normal component of the velocity but rather the initial value of the
Laplace operator on v. In Fourier wave space, this is tantamount to vorticity.
From kinematics,

$$\nabla^2\tilde{v} = \frac{\partial\tilde{\omega}_z}{\partial x} - \frac{\partial\tilde{\omega}_x}{\partial z} \qquad (8.13)$$

in real space. And, when Fourier transformed, this is

$$\Delta\check{v} = -i\alpha\check{\omega}_z + i\gamma\check{\omega}_x \qquad (8.14)$$

or the vorticity in that direction in wave space. This result is not at all surprising
for it was reviewed in Chapter 1, and the two major governing equations are
the result of taking the curl of the governing linearized equations in order to
eliminate the pressure. Thus, in any form, these are equations for components
of the vorticity, strictly speaking. This observation further substantiates the
claim that initial value specifications can be given for either physical variable.
Regardless, solutions for (8.11) and (8.12) have the same requirements and
consequences as those of the homogeneous equations that were obtained by
the assumption of normal modes. In short, the equations in Laplace space do
contain more information but are just as complex when consequences are to be
determined. It will be expedient, then, to consider some limiting problems in
this sense in order to better understand the needs for the initial-value problem.

The simplest problem of all the prototypical flows is that of inviscid plane Couette flow. In this case, (8.11) and (8.12) reduce to

$$(s - i\alpha U)\Delta\bar{\bar{v}} = \Delta\check{v}|_{t=0} \tag{8.15}$$

and

$$(s - i\alpha U)\bar{\bar{\omega}}_y = i\gamma U'\bar{\bar{v}} + \check{\omega}_y(y,0), \tag{8.16}$$

with $U(y) = \sigma y$. Here, there is no problem for solutions to these reduced equations but, it must be remembered, there are boundary conditions that must be satisfied at the upper and lower boundaries, namely $\bar{\bar{v}} = 0$ at these locations. If no initial value was given, then the solution of (8.15) that meets this requirement is $\bar{\bar{v}} \equiv 0$. This result is identical to that which would come from normal modes. On the other hand, when there is an initial vorticity, there is now a nonzero solution. In fact, this solution will have the factor $(s - i\alpha U)$ in the denominator. Consequently, from the inversion of the Laplace transform, this implies the existence of a branch cut in the complex plane and therefore a continuous spectrum of eigenvalues that are related to the zeros of $(s + i\alpha U)$. From the basis of the differential equation, this is the same as a singularity, that is a condition where the coefficient of the highest derivative vanishes. It is for this reason that, when viscosity is added, the singularity is no longer present in the differential equation and only poles will be present in the complex plane. These become exponential solutions in time or normal discrete modes. The same argument is true for inviscid plane Poiseuille flow when compared to the viscous problem. More details can be found in Case (1960a, 1961), who was the principal author to analyze inviscid problems in incompressible flow in this way. Reference should also be made to Lin (1961) for a discussion of mathematical problems involved in the work of Case.

Similar conclusions can be made for the solution of (8.16), except that the real time inversion for $\bar{\bar{\omega}}_y$ can lead to a power of time greater than that of linearity if $\bar{\bar{v}}$ has a nonzero solution and already is proportional to time, for example. Otherwise the result mimics that of $\bar{v}$ with a nonzero initial value of the vorticity. A short review for this problem and its solution can be found in Drazin & Reid (1984) where the singular solution is given in terms of a Green's function and the Green's function needed for $\bar{\bar{v}}$ is provided explicitly. The Squire equation is not included in any of these references but the general implications can be inferred.

Use of the Laplace transform has been used in the analysis for the stability of the Blasius boundary layer by Gustavsson (1979). The important aspect of this effort is again the fact that there is the existence of the continuum as well as the poles in the complex Laplace plane for the discrete spectrum that must be assessed when making the inversion of the transform to real time. Since one of the boundaries is at infinity, then this fact manifests itself in the evaluation of singularities in the complex plane with the results being that such disturbances (a) move with the value of the freestream velocity, (b) vary as waves in the vertical spatial variable and hence are bounded rather than decay exponentially in this direction and (c) all eventually decay in time because there is viscous dissipation. No specific initial input data was analyzed in this work and, again, the Squire equation was ignored.

As can be seen, the use of the Laplace transform is only marginally better than the traditional separation of variables using normal modes. And, unfortunately, since it is the consequence of specific initial data in the dynamics that is sought, then other strategies must be invoked or more succinct numerical treatment is required in order to arrive at this goal.

8.3 Moving Coordinates and Exact Solutions

Explicit unsteady solutions for perturbations in plane Couette flow were initially found by Kelvin (1887a) and further developed by Orr (1907a,b). In addition to the class of exact solutions given by Craik & Criminale (1986), the history of such effort, together with many references of note in the development, are given by these authors. More recent work that uses the same or equivalent bases for solving initial-value problems in this manner are due to Criminale & Drazin (1990, 2000), Criminale, Long & Zhu (1991), Bun & Criminale (1994) and Criminale, Jackson & Lasseigne (1995). The particular problems analyzed range from free shear flows to the boundary layer.

In terms of the physics to this approach, the fundamental mechanism is that the vorticity of the disturbance is advected by the basic flow and the basic vorticity is advected by the disturbance. When viscous effects are retained, then the vorticity of the disturbance can also be diffused. If the basic flow is linear or piecewise-linear in the appropriate spatial variable, then the basic vorticity is constant or piecewise-constant. It is this last condition that ensures the fact that the solutions are exact for the full Navier–Stokes equations, and the full ramifications have been elucidated by Craik & Criminale (1986) and later in a more general fashion by Criminale (1991), namely that a set of basic solutions can be found for the linear perturbation problem that are (i) of closed form; (ii) contain both the discrete and continuous spectra allowing for arbitrary disturbances; (iii) the complications of critical layers or singular perturbation analysis are no longer required; (iv) even the near and far-fields can be determined as well as the early and asymptotic temporal behavior; (v)

Lagrangian descriptions can also be ascertained by using the solutions obtained for the velocity. This is a fringe benefit and allows for insight into mixing and vorticity physics.

The premise for this approach is to decompose the motion into a mean and a fluctuation part as

$$\underline{u} = \underline{U} + \underline{\tilde{u}}, \qquad p = P + \tilde{p}, \tag{8.17}$$

where the underbar denotes a vector quantity. In this case the perturbations are not necessarily small and thus the governing Navier–Stokes equations are

$$\nabla \cdot \underline{\tilde{u}} = 0 \tag{8.18}$$

and

$$\frac{\partial \underline{\tilde{u}}}{\partial t} + \underline{U} \cdot \nabla \underline{\tilde{u}} + \underline{\tilde{u}} \cdot \nabla \underline{U} + \underline{\tilde{u}} \cdot \nabla \underline{\tilde{u}} = -\nabla(\tilde{p}/\rho_0) + \varepsilon \nabla^2 \underline{\tilde{u}}, \tag{8.19}$$

where it is assumed that the fluid is incompressible and $\underline{U}$ satisfies its own equations. A footnote here should be added: incompressibility is required in the sense that the velocity field is solenoidal and not necessarily constant density (cf. Chapter 7).

In the special case for which the perturbation velocity can be written in the form

$$\underline{\tilde{u}}(\underline{x}, t) = f(\underline{x}, t)\underline{\hat{u}}(t), \tag{8.20}$$

it follows that

$$\underline{\hat{u}} \cdot \nabla f = \nabla \cdot \underline{\tilde{u}} = 0. \tag{8.21}$$

As a result, the nonlinear terms in the Navier–Stokes equations that are due to the fluctuations vanish identically for reasonable functions f and $\underline{\hat{u}}$. Thus, the perturbation problem can be solved by a linear set of equations. If the velocity is initially in the same direction everywhere so that $\underline{\tilde{u}}(\underline{x}, 0) = f(\underline{x}, 0)\underline{\hat{u}}(0)$, then the linearized problem ensues and the results are exact solutions to the full governing equations.

For a steady basic flow, the linearized problem is customarily solved by the method of normal modes. This form is the same as the assumption of separation of variables and represents traveling waves so that

$$\left. \begin{aligned} f(\underline{x}, t) &= F(\xi), \\ \underline{\hat{u}} &= e^{\int \sigma dt}, \\ \xi &= \underline{\alpha} \cdot \underline{x} + \int \omega dt. \end{aligned} \right\} \tag{8.22}$$

Hence, along with the proper boundary conditions, this problem can in principle be solved as an eigenvalue problem where ω, σ, $\underline{\alpha}$ satisfy the eigenvalue relation. When $\underline{U}$ is not a function of time and the parameters are constants, the F is the exponential function and the eigenvalue relation is the traditional type in stability theory. This formulation also allows for a superposition of normal modes if $\tilde{u}$ has the form given by (8.20). And, for an initial-value problem for $\tilde{u}$ in the same direction, then all $\underline{\alpha}_N$ lie in the plane perpendicular to $\hat{\underline{u}}$. This point is significant because, although one Fourier mode is clearly a solution to the Navier–Stokes equations, a sum of modes is not necessarily but such is true under these constraints.

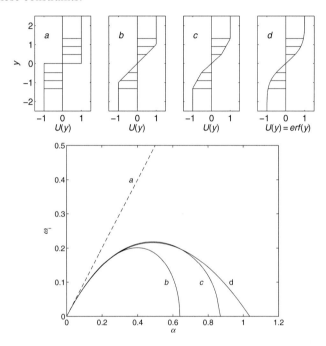

Figure 8.1 Stability as a function of piecewise linear mean mixing layer profile (after Criminale, 1991).

One problem should be given for illustration and demonstration of the benefits of the technique. A good one for this purpose that demonstrates all the benefits is that of the mixing layer. The essential need for the method is a coordinate transformation that changes the set of linear partial differential equations to ones where the coefficients are at most functions of time. The general procedure is given by Criminale (1991); for various mean unidirectional flows that can be modeled as piecewise-linear, reference is made to Criminale & Drazin (1990). In this manner, Bun & Criminale (1994) modeled the incompressible mixing layer as shown in Fig. 8.1 and considered the perturbation problem inviscidly.

Here, $\underline{U} = (\sigma y, 0, 0)$ in any of the regions. The moving coordinate transformation for such a parallel flow is

$$T = t, \quad \xi = x - \sigma yt, \quad \eta = y, \quad \zeta = z, \qquad (8.23)$$

which is just a subset of the general transformation that can be used (cf. Criminale, 1991) if the mean flow can be written as

$$U_i = \sigma_{ij}(t)x_j + U_i^0(t). \qquad (8.24)$$

With (8.23), the governing equation (cf. (7.78)) reduces to

$$\frac{\partial}{\partial T}\Delta \check{v} = 0, \qquad (8.25)$$

where

$$\check{v}(\alpha, \eta, \gamma, T) = \int\!\!\int_{-\infty}^{+\infty} \tilde{v}(\xi, \eta, \zeta, T)e^{i\alpha\xi + i\gamma\zeta}\,d\xi d\zeta$$

and

$$\Delta \check{v} = \frac{\partial^2 \check{v}}{\partial \eta^2} + i2\alpha\sigma T\frac{\partial \check{v}}{\partial \eta} - (\tilde{\alpha}^2 + \alpha^2\sigma^2 T^2)\check{v}, \qquad \tilde{\alpha}^2 = \alpha^2 + \gamma^2.$$

It should be noted that this transformation is not one of Eulerian to one of Lagrangian coordinates but should be thought of as a change to a moving set of coordinates. In this form, the far-field conditions in the respective spatial variables are satisfied by finiteness of the dependent variables. Boundedness in the y- (or η-) direction is also required but, in view of the form the governing equations take, this will automatically be met and, instead, matching conditions will replace such at the locations where the mean velocity changes from one linear variation to another or to a constant value. This means that $\check{v}$ and the pressure are continuous at these locations in the inviscid limit.

The Squire transformation for the velocity components in wave space is just as valid here. Hence, if $\tilde{\alpha}\check{w} = -\gamma\check{u} + \alpha\check{w}$, then the Squire equation becomes

$$\frac{\partial \overline{w}}{\partial T} = \sigma \sin\phi\,\check{v}. \qquad (8.26)$$

Although (8.25) and (8.26) are written using velocity components as the dependent variables, it is again stressed that these equations are those for vorticity. More specifically it should be noted that

$$\begin{aligned}
\check{\omega}_y &= \check{\omega}_\eta = i\tilde{\alpha}\overline{w}, \\
\check{\omega}_{\tilde{\alpha}} &= \left(\frac{\partial}{\partial \eta} + i\alpha\sigma T\right)\overline{w}, \\
\check{\omega}_\varphi &= -\frac{i}{\tilde{\alpha}}\Delta\check{v}.
\end{aligned} \right\} \qquad (8.27)$$

In other words, the vorticity components in wave space can be transformed to polar dependent variables in the same manner as the velocity.

Every vector field can be decomposed into its solenoidal, rotational and harmonic parts. Since the velocity is divergence free, there is no solenoidal component and thus only the rotational and harmonic parts are possible. And, as a result, both $\breve{\omega}_\eta$ and $\breve{\omega}_{\tilde{\alpha}}$ are in the η-direction normal to the (α, γ)-plane or $(\tilde{\alpha}, \varphi$-plane). The solution of (8.25) can thus be interpreted as an initial source of vorticity if $\Delta\breve{v} \neq 0$ at time $t = 0$. When $\Delta\breve{v} = 0$, there is no vorticity and the resulting vector field is harmonic. In fact, the solution for the harmonic field is

$$\breve{v}_H = A(T)e^{\tilde{\alpha}\eta - i\alpha\sigma T_\eta} + B(T)e^{-\tilde{\alpha}\eta - i\alpha\sigma T_\eta}. \tag{8.28}$$

It should be especially noted that the coefficients can be functions of time and are proportional to an oscillatory factor that reflects the fact that these solutions are in the moving coordinate system.

The complete solution will then be the sum of (8.28) and a particular rotational component, $\breve{v}_R$, say. General considerations were made as to either initial velocity or vorticity by Criminale & Drazin (1990).

In the case of the mixing layer, modeled with a three-section piecewise-linear mean profile, Bun & Criminale (1994) treated the problem by prescribing the vorticity initially, and selected this as a combination of an oblique wave in the (x, z)-plane and a localized pulse in the y-variable and located within the inner shear zone. Not only does this choice provide important physics but also makes for ease in inverting the double Fourier transforms to real space. In both the upper and lower non-shear regions, no initial value was given and the flow is irrotational in the outer flow. The solutions in the respective regions are established so that the conditions as $y = \eta \to \pm\infty$ are met. From (8.28) this means the solutions will exponentially decay in this regions. The remaining solutions then must allow for continuity of $\breve{v} = \breve{v}_H + \breve{v}_R$ and the pressure, $\breve{p}$, or

$$-\tilde{\alpha}^2\breve{p} = \frac{\partial^2\breve{v}}{\partial\eta\partial T} + i\alpha\sigma T\frac{\partial\breve{v}}{\partial T} + i2\alpha\sigma\breve{v}$$

when written in terms of $\breve{v}$, at the two locations where the mean velocity changes. This matching leads to a linear system of equations for the coefficients of the unknown harmonic part of the velocity field and can be written in vector form as

$$\dot{\underline{x}} = A\underline{x} + \underline{f}, \tag{8.29}$$

where $\underline{x}$ are the coefficients of the irrotational components and $\underline{f}$ is due to the initial input of vorticity.

The system (8.29) has two solutions. First, the homogeneous problem and there are eigenvalues of A. The eigenvalues are those of the normal modes. The second solutions are the ones that are forced and can result in algebraic behavior in time and this is due to the continuous spectrum. The display of Fig. 8.1 shows the traditional results as that of normal modes. And it is clear that the piecewise-linear model contains all of the essential features. And, although not shown, the inviscid problem has a damped mode for every growing one in the range of wavenumbers where possible growth is to be found.

The initial-value is completed in this way by further recognizing that, for the conditions given, this translates to $\underline{x}(0) = 0$ for (8.29). In other words, there are no normal modes at time $t = 0$ but, as time goes on, such a mode will ultimately dominate asymptotically. Initially, it is the transient motion that is prominent. It was also demonstrated that three-dimensionality is strongly influential.

Another benefit from this means of solution for the problem comes from the fact that all the velocity components can be written as closed form functions of all spatial and temporal variables. Consequentially one can make a Lagrangian representation as

$$\frac{dx}{dt} = U + \tilde{u}, \qquad \frac{dy}{dt} = \tilde{v}, \qquad \frac{dz}{dt} = \tilde{w},$$

and then actually trace particle paths. The results of Bun & Criminale using these relations are shown in Figs. 8.2 and 8.3.

The classical roll-up process is equally robust under pure algebraic dynamics as it is for that combined with the maximum for normal mode growth. In addition, algebraic growth can lead to nonlinearity on a time scale shorter than the rate of growth for a normal mode. Then, the inclusion of three-dimensionality allows for the so-called cross-ribbing effect, as reported experimentally.

Criminale, Jackson & Lasseigne (1995) examined the case of the jet and wake mean flows when modeled in this manner, and compared the results to the integration of the full linear equations with continuous mean profiles. And, although no specific feedback mechanism was determined, it was demonstrated how the disturbances could be delayed or enhanced.

Criminale, Long & Zhu (1991) considered the complete three-dimensional disturbance problem for plane inviscid Couette flow. As has already been noted, there are no normal modes for this limiting problem. And clearly the linear mean profile is no longer an approximation for the mean velocity. The results show that rapid algebraic growth can evolve and, again, three-dimensionality should not be neglected in the initial-value problem.

The viscous boundary layer was modeled in this way was done by Criminale & Drazin (2000). The piecewise-linear mean profile is used so that the governing equations can be evaluated but now viscous effects are included, and the problem is then solved by the method of matched asymptotic expansions for large Reynolds number. In this way the complete boundary conditions for

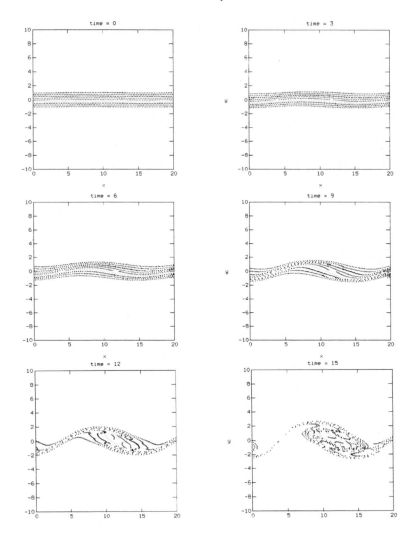

Figure 8.2 Material particles with initial data $\varphi = 0$, amplitude 0.1 at $t = 0$, $\tilde{\alpha} = 0.4$ (normal mode maximum) (after Bun & Criminale, 1994, reproduced with permission).

no slip at the flat plate can be satisfied. The solutions are remarkably explicit although somewhat complicated. Various initial conditions are employed using both the velocity and the vorticity, and the results substantiate once again that linear disturbances may grow so much transiently as to excite nonlinear growth. The equivalent nondimensional Orr–Sommerfeld and Squire equations

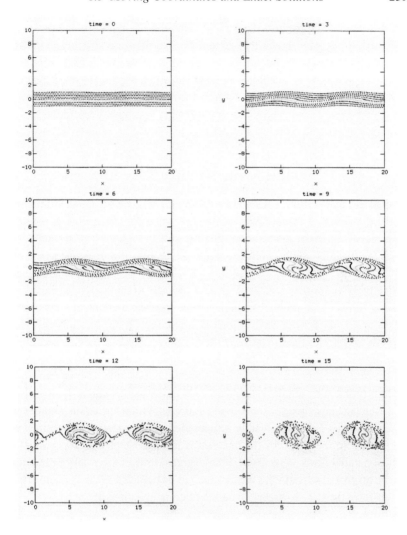

Figure 8.3 Material particles with initial data $\varphi = 0$, amplitude 0.1 at $t = 0$, $\tilde{\alpha} = \tilde{\alpha}_s$ (zero growth normal mode) (after Bun & Criminale, 1994, reproduced with permission).

for this case are

$$\frac{\partial}{\partial T}\Delta\check{v} = \varepsilon\Delta\Delta\check{v} \tag{8.30}$$

and

$$\frac{\partial\overline{w}}{\partial T} = \varepsilon\Delta\overline{w} + \sin\varphi\,\check{v}, \tag{8.31}$$

where ε is the reciprocal of the Reynolds number. Both (8.30) and (8.31) are analyzed and it was shown that streamwise vortices are strongly amplified. The analytics also corroborates the use of the Laplace transform in time used by Gustavsson (1979) where it was established that algebraic growth was due to the continuous spectrum that stemmed from the branch cut in the complex plane and even complements numerical results that have been found for the evolution of specific initial perturbations.

8.4 Multiple Scale, Multiple Time Analysis

In spite of meaningful results that can be obtained vis-à-vis the determination of the dynamics of prescribed disturbances when the mean flow is modeled by a piecewise-linear distribution, there is still the question of the influence when the mean profile is continuous. In other words, how does one assess any errors that the approximation may invoke when the profile has discontinuous derivatives? A most recent work that is devoted to an analytical means for solving initial-value problems has been able to make a creditable response to this concern and it is due to Criminale & Lasseigne (2002). In summary: (1) it exploits the physics fully in terms of the vorticity; (2) uses a moving coordinate transformation in order to simplify the governing equations; (3) uses the partial differential equations directly for solution; (4) one is able to examine a fully viscous problem by regular rather than by singular perturbations. It also has all of the advantages cited for the moving coordinate transformation, and has the additional asset that it can readily be extended to investigation of nonlinearity.

Once more the basis for the analysis is as that established for parallel or almost parallel flows. As a result, Fourier transforms in the x- and z-directions can be taken and all variables are assumed to be bounded in the far-field. Thus, in this form the Orr–Sommerfeld and the Squire equations are

$$\left(\frac{\partial}{\partial t} - i\alpha U \right) \Delta \check{v} + U'' i\alpha \check{v} = \varepsilon \Delta \Delta \check{v} \qquad (8.32)$$

and

$$\left(\frac{\partial}{\partial t} - i\alpha U \right) \check{\omega}_y - \varepsilon \Delta \check{\omega}_y = i\gamma U' \check{v}, \qquad (8.33)$$

where $\check{v}$ and $\check{\omega}_y$ are the respective dependent variables for the velocity and vorticity components in Fourier space. Now in polar coordinates, $\alpha = \tilde{\alpha} \cos \varphi$, $\gamma = \tilde{\alpha} \sin \varphi$. Moreover,

$$\Delta \check{v} = \frac{\partial^2 \check{v}}{\partial y^2} - \tilde{\alpha}^2 \check{v}.$$

Define this as $\check{\Omega}$, which has already been shown to be the vorticity component in the φ-direction in wave space.

With these quantities substituted in (8.32) and (8.33), there will now be three equations or

$$\Delta \check{v} = \check{\Omega},$$ (8.34)

$$\frac{\partial \check{\Omega}}{\partial t} - \varepsilon \frac{\partial^2 \check{\Omega}}{\partial y^2} = i\tilde{\alpha} \cos \varphi U \check{\Omega} - i\tilde{\alpha} \cos \varphi U'' \check{v} - \varepsilon \tilde{\alpha}^2 \check{\Omega},$$ (8.35)

and

$$\frac{\partial \check{\omega}_y}{\partial t} - \varepsilon \frac{\partial^2 \check{\omega}_y}{\partial y^2} = i\tilde{\alpha} \cos \varphi U \check{\omega}_y + i\tilde{\alpha} \sin \varphi U' \check{v} - \varepsilon \tilde{\alpha}^2 \check{\omega}_y.$$ (8.36)

Now, for purposes of illustration, consider the case of the Blasius boundary layer. Define new dependent variables as

$$\begin{aligned} \hat{\Omega} &= e^{-\varepsilon\tilde{\alpha}^2 t} e^{i\tilde{\alpha}\cos\varphi t} \overline{\Gamma}(y,t;\tilde{\alpha},\varphi), \\ \hat{v} &= e^{-\varepsilon\tilde{\alpha}^2 t} e^{i\tilde{\alpha}\cos\varphi t} \overline{v}(y,t,\tilde{\alpha},\varphi), \\ \hat{\omega}_y &= e^{-\varepsilon\tilde{\alpha}^2 t} e^{i\tilde{\alpha}\cos\varphi t} \overline{\omega}_y(y,t;\tilde{\alpha},\varphi). \end{aligned} \right\}$$ (8.37)

Equations (8.34) to (8.36) can now be written as

$$\nabla^2 \overline{v} = \frac{\partial^2 \overline{v}}{\partial y^2} - \tilde{\alpha}^2 \overline{v} = \overline{\Gamma},$$ (8.38)

$$\frac{\partial \overline{\Gamma}}{\partial t} - \varepsilon \frac{\partial^2 \overline{\Gamma}}{\partial y^2} = i\tilde{\alpha} \cos \varphi (U-1)\overline{\Gamma} - i\tilde{\alpha} \cos \varphi U'' \overline{v}$$ (8.39)

and

$$\frac{\partial \overline{\omega}_y}{\partial t} - \varepsilon \frac{\partial^2 \overline{\omega}_y}{\partial y^2} = i\tilde{\alpha} \cos \varphi (U-1)\overline{\omega}_y - i\tilde{\alpha} \sin \varphi U' \overline{v}.$$ (8.40)

The change given by (8.37) is one that shifts all quantities to move with the value of the freestream velocity and is a special case of the more general moving coordinate transformation; the other factor is merely an indicator of the net effect of viscosity in wave space.

In the new form, (8.39) and (8.40) have important implications when the question of the far-field boundary condition in y is considered. In fact, as $y \to \infty$, then the right-hand sides of these two equations will vanish, since $U \to 1$ and $U'' \to 0$ in this limit. In turn, this means that the two equations are just those of classical heat diffusion in the freestream and are readily solvable.

As has been noted previously, the case when $\varphi = \pi/2$ reduces the system to one that can be solved explicitly.

The third observation of note deals with the moving coordinate transformation discussed in Section 8.3 as compared to just having movement with respect to the freestream. For the more general transformation, there would be terms as $T(t = T)$, $\tilde{\alpha}T$ and $\tilde{\alpha}^2 T^2$ in the Laplace operator. This fact has already been noted by Lasseigne, Joslin, Jackson & Criminale (1999) when investigating the early period dynamics for disturbances in the boundary layer. To supplement this is the established fact that the value of $\tilde{\alpha} < 1$, based on numerical computations of instability. Large values of $\tilde{\alpha}$ would imply very small spatial scales and these are heavily damped. It is also known that even the growth factors for normal modes are $O(10^{-3})$. These points suggest that there are multiple times as well as multiple scales in this problem. Indeed, for other mean flows as well. Thus, let

$$\overline{\Gamma}(y, t; \tilde{\alpha}, \varphi) = \overline{\Gamma}(y, \tilde{\alpha}y, t, \tilde{\alpha}t, \tilde{\alpha}^2 t^2; \tilde{\alpha}, \varphi)$$

$$= \overline{\Gamma}(y, Y, t, \tau, T; \tilde{\alpha}, \varphi) \tag{8.41}$$

and similarly for $\overline{v}$, $\overline{\omega}_y$, indicating two spatial and three temporal scales are well suited for this problem. As a result, $\tilde{\alpha}$ is now the small parameter and viscous terms can be retained at the outset making for a straightforward means for satisfying boundary conditions in the y-direction. This is a regular perturbation problem.

Assume that the respective dependent variables can be expanded as

$$\left. \begin{aligned} \overline{\Gamma} &= \overline{\Gamma}_0 + \tilde{\alpha}\overline{\Gamma}_1 + \tilde{\alpha}^2\overline{\Gamma}_2 + \cdots, \\ \overline{v} &= \overline{v}_0 + \tilde{\alpha}\overline{v}_1 + \tilde{\alpha}^2\overline{v}_2 + \cdots, \\ \overline{\omega}_y &= \overline{\omega}_{y0} + \tilde{\alpha}\overline{\omega}_{y1} + \tilde{\alpha}^2\overline{\omega}_{y2} + \cdots, \end{aligned} \right\} \tag{8.42}$$

with

$$\overline{\Gamma}(y, Y, 0, 0, 0; \tilde{\alpha}, \varphi) = \overline{\Gamma}_0(y; \tilde{\alpha}, \varphi),$$

$$\overline{\Gamma}_1 = \overline{\Gamma}_2 = \cdots = 0 \text{ at } t = 0, \tag{8.43}$$

and

$$\left. \begin{aligned} \overline{\omega}_y(y, Y, 0, 0, 0; \tilde{\alpha}, \varphi) &= \overline{\omega}_{y0}(y; \tilde{\alpha}, \varphi), \\ \overline{\omega}_{y1} &= \overline{\omega}_{y2} = \cdots = 0 \text{ at } t = 0. \end{aligned} \right\}$$

The respective equations are

$$\left. \begin{aligned} \frac{\partial^2 \overline{v}_0}{\partial y^2} &= \overline{\Gamma}_0, \\ \frac{\partial^2 \overline{v}_1}{\partial y^2} &= -2\frac{\partial^2 \overline{v}_0}{\partial y \partial Y} + \overline{\Gamma}_1, \\ \frac{\partial^2 \overline{v}_2}{\partial y^2} &= -2\frac{\partial^2 \overline{v}_1}{\partial y \partial Y} - \frac{\partial^2 \overline{v}_0}{\partial Y^2} + \overline{v}_0 + \overline{\Gamma}_2, \end{aligned} \right\} \tag{8.44}$$

along with

$$
\left.
\begin{aligned}
&\bar{v}(0,0,t,\tau,T;\tilde{\alpha},\varphi) = 0, \\
&\bar{v} \quad \text{bounded as} \quad y \to \infty, \\
&\frac{\partial \bar{\Gamma}_0}{\partial t} - \varepsilon \frac{\partial^2 \bar{\Gamma}_0}{\partial y^2} = 0, \\
&\bar{\Gamma}_0(y,Y,0,0,0;\tilde{\alpha},\varphi) = \bar{\Gamma}_0(y;\tilde{\alpha},\varphi), \\
&\bar{\Gamma}_0(\infty,\infty,t,\tau,T;\tilde{\alpha},\Omega) = 0 \text{ or bounded.}
\end{aligned}
\right\}
\qquad (8.45)
$$

The condition where the vorticity is zero precludes any value in the freestream. Then,

$$
\left.
\begin{aligned}
\frac{\partial \bar{\Gamma}_1}{\partial t} - \varepsilon \frac{\partial^2 \bar{\Gamma}_1}{\partial y^2} &= -\frac{\partial \bar{\Gamma}_0}{\partial \tau} + 2\varepsilon \frac{\partial \bar{\Gamma}_0}{\partial y \partial Y} + i\cos\varphi(U-1)\bar{\Gamma}_0 - i\cos\varphi U'' \bar{v}_0, \\
\frac{\partial \bar{\Gamma}_2}{\partial t} - \varepsilon \frac{\partial^2 \bar{\Gamma}_2}{\partial y^2} &= -\frac{\partial \bar{\Gamma}_1}{\partial \tau} - \frac{\partial \bar{\Gamma}_0}{\partial T} + 2\varepsilon \frac{\partial^2 \bar{\Gamma}_1}{\partial y \partial Y} + \varepsilon \frac{\partial^2 \bar{\Gamma}_0}{\partial Y^2} \\
&\quad + i\cos\varphi(U-1)\bar{\Gamma}_1 - i\cos\varphi U'' \bar{v}_1,
\end{aligned}
\right\}
$$
$$
(8.46)
$$

and

$$
\left.
\begin{aligned}
\frac{\partial \overline{\omega}_{y0}}{\partial t} - \varepsilon \frac{\partial^2 \overline{\omega}_{y0}}{\partial y^2} &= 0, \\
\frac{\partial \overline{\omega}_{y1}}{\partial t} - \varepsilon \frac{\partial^2 \overline{\omega}_{y1}}{\partial y^2} &= -\frac{\partial \overline{\omega}_{y0}}{\partial \tau} + 2\varepsilon \frac{\partial^2 \overline{\omega}_{y0}}{\partial y \partial Y} - i\sin\varphi U' \bar{v}_0 \\
&\quad + i\cos\varphi(U-1)\overline{\omega}_{y0}, \\
\frac{\partial \overline{\omega}_{y2}}{\partial t} - \varepsilon \frac{\partial^2 \overline{\omega}_{y2}}{\partial y^2} &= -\frac{\partial \overline{\omega}_{y1}}{\partial \tau} - \frac{\partial \overline{\omega}_{y0}}{\partial T} + 2\varepsilon \frac{\partial^2 \overline{\omega}_{y1}}{\partial y \partial Y} + \varepsilon \frac{\partial^2 \overline{\omega}_{y0}}{\partial Y^2} \\
&\quad - i\sin\varphi U' \bar{v}_1 + i\cos\varphi(U-1)\overline{\omega}_{y1}, \\
\overline{\omega}_y(y,Y,0,0,0;\tilde{\alpha},\varphi) &= \overline{\omega}_{y0}(y;\tilde{\alpha},\varphi), \\
\overline{\omega}_{y0}(0,0,t,\tau,T;\tilde{\alpha},\varphi) &= 0, \qquad \overline{\omega}_{y0} \text{ bounded as } y \to \infty.
\end{aligned}
\right\}
$$
$$
(8.47)
$$

After the initial input of vorticity, the system of perturbation equations for the vorticity components become a series of forced heat equations, whereas the equation for the vertical component of the velocity is forced at the outset. All equations can be solved in a most general fashion.

The work of Lasseigne *et al.* (1999) essentially used this method to examine the transient motion for disturbances in the boundary layer and then compared the results to direct numerical integration of the linear partial differential equations. The agreement is very good, even to low orders of the expansion.

8.5 Numerical Solution of Governing Partial Differential Equations

The partial differential equations (8.1) and (8.2) can be solved numerically by the method of lines (see e.g., Ames, 1977). This choice is a convenient numerical method; other techniques are possible. Given a mean flow with appropriate boundary conditions, the spatial derivatives are first centered-differenced on a uniform grid. By specifying arbitrary initial conditions, the resulting system can then be integrated in time by a fourth-order Runge–Kutta scheme. As with all our codes, calculations are to be carried out using 64-bit precision. Selected results for channel flows and for the Blasius boundary layer are briefly presented in the next subsection.

8.5.1 Channel Flows

In this section we consider both plane Poiseuille flow where

$$U(y) = 1 - y^2 \tag{8.48}$$

and plane Couette flow where

$$U(y) = y, \tag{8.49}$$

defined over the domain $-1 < y < 1$; see Criminale, Jackson, Lasseigne & Joslin (1997). The system (8.1) and (8.2) can now be solved by the numerical strategy described in this section. To validate the numerical procedure, numerical solutions were compared to those obtained from an Orr–Sommerfeld solver. For unstable modes, the two should agree. Table 8.1 shows the numerically computed temporal growth rate for plane Poiseuille flow as a function of grid points for $Re = 10^4$, $\tilde{\alpha} = 1$ and $\phi = 0°$. Growth rates were computed by integrating the equations forward in time beyond the transient until the growth rate, defined as $\omega_i = \ln |E(t)|/2t$ where $|E|$ is the amplitude of the perturbation energy defined later in the section, asymptotes to a constant value. The value given by an Orr–Sommerfeld solver is 0.00373967 (Orszag, 1971). Figure 8.4

Table 8.1 *Numerically computed temporal growth rate for plane Poiseuille flow as a function of grid points. $Re = 10^4$, $\tilde{\alpha} = 1$ and $\phi = 0°$.*

GRID POINTS	GROWTH RATE
500	0.003726
1000	0.003736
2000	0.003739

shows the growth rates as a function of wavenumber obtained from the numerical solution (shown as circles) and those obtained from the Orr–Sommerfeld equation (shown as the solid curve) for $Re = 10^4$ and $\phi = 0°$. The agreement is excellent. The corresponding real and imaginary parts of the eigenfunctions from both the numerical solution (solid) and the Orr–Sommerfeld solution (dashed) are displayed in Fig. 8.5 for the case $\tilde{\alpha} = 1$ and $\phi = 0°$. Note that the two curves essentially lie on top of each other. Similar results are obtained at higher values of the Reynolds number, as well as for plane Couette flow.

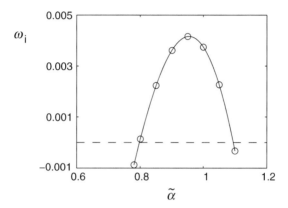

Figure 8.4 Plot of growth rates as a function of wavenumber $\tilde{\alpha}$. The circles corresponds to the numerically computed values from the partial differential equation, and the solid curve corresponds to the growth rate computed using the Orr–Sommerfeld equation. Results for plane Poiseuille flow with $\phi = 0°$ and $Re = 10^4$.

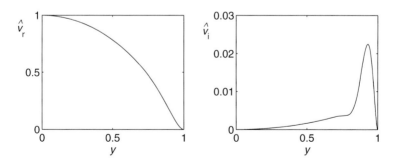

Figure 8.5 The real and imaginary parts of the eigenfunction as a function of y for $\tilde{\alpha} = 1$. Results for plane Poiseuille flow with $\phi = 0°$ and $Re = 10^4$.

Table 8.2 *Numerically computed temporal growth rate of a Tollmien–Schlichting wave as a function of grid points. Results for Blasius flow with Re* $= 10^3$, $\tilde{\alpha} = 0.24$ *and* $\phi = 0°$.

GRID POINTS	GROWTH RATE
500	0.00285574
1000	0.00285181
2000	0.00284947
4000	0.00284961

8.5.2 Blasius Boundary Layer

We now consider the Blasius boundary layer, given by

$$2f''' + ff'' = 0, \tag{8.50}$$

subject to the conditions

$$f(0) = f'(0) = 0; \quad f'(\infty) = 1, \tag{8.51}$$

with $U(y) = f'(y)$; see Lasseigne, Joslin, Jackson & Criminale (1999). As before, we validate the numerical procedure by comparing the numerical solutions to those obtained from an Orr–Sommerfeld solver. For a Tollmien–Schlichting wave, the two should agree. Table 8.2 shows the numerically computed temporal growth rate as a function of grid points for $Re = 10^3$, $\tilde{\alpha} = 0.24$ and $\phi = 0°$. The value obtained from an Orr–Sommerfeld solver is 0.00284962. No effort was made to optimize the number of grid points by employing non-uniform meshes. If this were done, far fewer grid points would be needed. Figure 8.6 shows the growth rates obtained from the numerical solution (shown as circles) and those obtained from the Orr–Sommerfeld equation (shown as the solid curve) for three different values of the Reynolds number and $\phi = 0°$. As was for the case of plane Poiseuille flow, the agreement is excellent.

We have shown, both for channel flows and Blasius flow, that the numerical method is capable of determining the temporal growth rate for any given value of wavenumber, angle of obliquity and Reynolds number. If one is searching for the most unstable mode, this method works quite well and can be used in lieu of an Orr–Sommerfeld solver. Unfortunately, the numerical method can only select that mode which has the largest growth rate, or if the flow is stable, the least stable mode. If higher modes exist this method is not the proper choice. In this case it is best to resort back to the Orr–Sommerfeld solver. However, as will be shown in the next section, the numerical method can also be used to investigate transient behavior for subcritical Reynolds numbers. Thus,

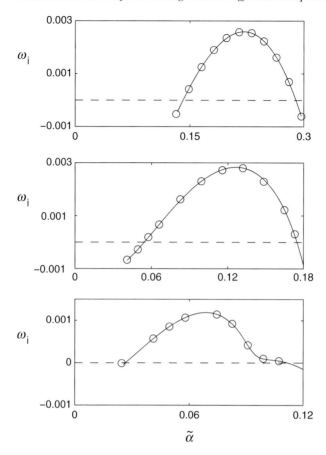

Figure 8.6 Plot of the growth rate versus wavenumber for two-dimensional disturbances, $\phi = 0$. Circles represent growth rates calculated by numerical integration, and solid curves are growth rates obtained from the Orr–Sommerfeld equation (normal mode solution). Reynolds numbers: (top) $Re = 10^3$, (middle) $Re = 10^4$, (bottom) $Re = 10^5$.

such a numerical method has significant advantages for understanding the flow characteristics in those regions of the (wavenumber, Reynolds number)-plane where the Orr–Sommerfeld equation is not applicable.

8.6 Optimizing Initial Conditions

8.6.1 Perturbation Energy

Of particular interest is the effects of various initial conditions and their subsequent transient behavior at subcritical Reynolds numbers. In order to examine
the evolution of various initial conditions, the energy density in the $(\tilde{\alpha}, \phi)$-
plane as a function of time is computed. The energy density is defined as

$$E(t; \tilde{\alpha}, \phi, Re) = \int_{-1}^{1} \left[|\breve{u}^2| + |\breve{v}^2| + |\breve{w}^2| \right] dy \tag{8.52}$$

for channel flows, or

$$E(t; \tilde{\alpha}, \phi, Re) = \int_{0}^{\infty} \left[|(\breve{u} - \breve{u}_\infty)^2| + |(\breve{v} - \breve{v}_\infty)^2| + |(\breve{w} - \breve{w}_\infty)^2| \right] dy \tag{8.53}$$

for Blasius flow. The subscript denotes the value at infinity, which may be a
function of time. Note that, in using (8.53), we have allowed for disturbances
that are bounded at infinity. The total energy of the perturbation can be found
by integrating E over all $\tilde{\alpha}$ and ϕ. A growth function can be defined in terms
of the normalized energy density, namely

$$G(t; \tilde{\alpha}, \phi, Re) = \frac{E(t; \tilde{\alpha}, \phi, Re)}{E(0; \tilde{\alpha}, \phi, Re)}, \tag{8.54}$$

and effectively measures the growth of the energy at time t for a prescribed initial condition at $t = 0$. For cases where it is known that the solution is asymptotically stable for large time, we use the bases that, if $G > 1$ for some time
$t > 0$, then the flow is said to be algebraically unstable; if $G = 1$ for all time,
the flow is algebraically neutral; and, if $G < 1$ for all time, the flow is algebraically stable.

Various initial conditions can be specified to explore transient behavior at
subcritical Reynolds numbers with the important issue here being the ability to
make, in a simple manner, arbitrary specifications. For channel flows, a normal
mode decomposition provides a complete set of eigenfunctions, and thus it is
true that any arbitrary specification can (theoretically) be written in terms of an
eigenfunction expansion. For the boundary layer, one must include the continuum to form a complete set; see (8.3). However, there is nothing special about
the eigenfunctions when it comes to specifying initial conditions, but they do
represent the most convenient means of specifying the long-time solution. In
addition, the use of the (non-orthogonal) eigenfunctions in the attempt to make
any truly arbitrary specification introduces unnecessary mathematical complications that actually involve tedious numerical calculations. For channel flows,
it would seem physically plausible that the natural issues affecting the initial

specification is whether the disturbances being considered are (a) symmetric or antisymmetric and, (b) whether or not they are local or more diffuse across the channel. The cases analyzed by Criminale, Jackson, Lasseigne & Joslin (1997) satisfy both of these needs and uses functions that can be definitively employed to represent any arbitrary initial distribution. This approach, of course, offers a complete departure from the specification of the initial conditions using normal mode decomposition, but, as previously stated, the use of an eigenfunction expansion to address the natural issues affecting arbitrary initial disturbances is mathematically infeasible. For the boundary layer there does not exist a complete discrete set of orthogonal functions, square integrable over $(0, \infty)$ with the function and its first derivative zero at the plate, that can completely describe all possible initial conditions. Thus, Lasseigne, Joslin, Jackson & Criminale (1999) considered three different subspaces, with different characteristics at infinity, to examine the effect of initial conditions.

The remaining salient question is whether or not the large optimal transient growth previously determined can be at all realized by an arbitrarily specified initial condition (which are the only kind of disturbances that occur naturally in an unforced environment). The analysis presented by Criminale *et al.* (1997) and Lasseigne *et al.* (1999) clearly show what properties that the initial conditions must have in order to produce significant growth. Furthermore, it should be asked whether or not the basis functions used for the expansion of the initial conditions has any predictive properties that can be exploited a priori in determining the optimal conditions. Because of the temporal dependence of the eigenfunctions, any eigenfunction taken individually does not provide any clue as to its importance in a calculation of optimal conditions; furthermore, some of the eigenfunctions are nearly linearly dependent in a spatial sense, which can further cloud the issue of their importance. After identifying the initial conditions that are the most relevant to transient growth, a straightforward optimization procedure can be used to show that the results of the transient calculations performed do indeed have a strong predictive property.

We begin by examining channel flows subject to specified initial conditions. Figure 8.7 plots the growth function G as a function of time for the initial condition

$$\check{v}(y,0) = \frac{\Omega_0}{\beta^2}[\cos\beta - \cos(\beta y)], \quad \check{\omega}_y(y,0) = 0, \quad (8.55)$$

where $\beta = n\pi$, and for various values of n and $\Omega_0 = 1$. Note that this set is complete over the domain $[-1,1]$, and so any arbitrary initial condition can be expressed as a linear combination of these modes. Other sets can be chosen, as shown by Criminale *et al.* (1997). For plane Poiseuille flow and with $n = 7$ the maximum value is 12, and for plane Couette flow and with $n = 3$ the maximum value of G is 4.8. In both cases, moderate transient growth is observed, with the maximum growth being lower than that obtained by Butler & Farrell (1992). For Couette flow, these authors have shown that the maximum optimal energy growth for this choice of $\tilde{\alpha}$ and ϕ occurs at $t = 8.7$. Here, we observe that the largest growth is for the initial condition with $n = 3$ and the maximum occurs at time $t = 7.8$. The same can be said of Poiseuille flow. Butler & Farrell have shown that the optimal initial conditions for Poiseuille flow produce a maximum at time $t = 14.1$ and the largest growth here is for $n = 7$ that has a maximum at time $t = 14.4$. It is easy to see how these solutions for different values of n can be combined to produce an optimal solution. This issue is explored further in the next section.

Figure 8.8 plots the growth function as a function of time for plane Poiseuille flow with $\tilde{\alpha} = 2.044$, $\phi = 90°$ and a Reynolds number of $Re = 5,000$, and for plane Couette flow with $\tilde{\alpha} = 1.66$, $\phi = 90°$ and $Re = 1,000$. In both cases the initial condition (8.55) is used with $\Omega_0 = n = 1$. The parametric values again correspond to those from Butler & Farrell (1992); the choice of $\tilde{\alpha}$ corresponds to the streamwise vortex with largest growth. In the case of plane Poiseuille flow, the global optimal coincides with a streamwise vortex ($\phi = 90°$) but not so for Couette flow, where the global optimal was shown to be at $\phi \approx 88°$.

Comparing Figs. 8.7 and 8.8, we see that the transient growth is significantly larger for three-dimensional disturbances than it is for two-dimensional disturbances. For plane Poiseuille flow, the maximum is within 90% of the global maximum reported by Butler & Farrell. They point out that the presence of streamwise vorticity, while passive to nonlinear dynamics (Gustavsson, 1991) can cause the development of streaks, which may themselves be unstable to secondary instabilities or possibly produce transient growth of other types of perturbations. For plane Couette flow, the maximum is within 97% of the maximum reported by Butler & Farrell. Thus, any initial condition with $\check{v}$ velocity symmetric and no initial vorticity will give near optimum results when three-dimensionality is considered. This easily explains the growth observed by Gustavsson when only a limited normal mode initial condition was employed.

For Blasius boundary layer flow, we show similar results for the initial condition given by a Gaussian, namely

$$\check{v} = V_0 \left(\frac{y}{y_0} \right)^2 e^{-(y-y_0)^2/\sigma}, \quad \check{\omega}_y = 0, \tag{8.56}$$

centered at y_0 with width σ. The value of y_0 and σ are chosen so that the boundary conditions at $y = 0$ are essentially satisfied. Figure 8.9 is a plot of the normalized energy $G(t)$ for $\hat{\alpha} = 0.24$, $\phi = 0$, $Re = 1,000$, $\sigma = 0.25$ and

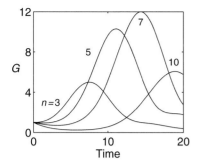 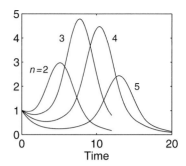

Figure 8.7 The growth function G as a function of time for various values of n. Left: plane Poiseuille flow with $\tilde{\alpha} = 1.48$, $\phi = 0$, and $Re = 5,000$. Right: plane Couette flow with $\tilde{\alpha} = 1.21$, $\phi = 0$, and $Re = 1,000$.

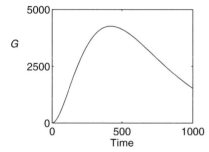 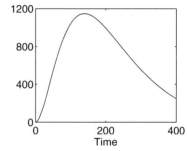

Figure 8.8 The growth function G as a function of time for $n = 1$. Left: plane Poiseuille flow with $\tilde{\alpha} = 2.044$, $\phi = 90°$, and $Re = 5,000$. Right: plane Couette flow with $\tilde{\alpha} = 1.66$, $\phi = 90°$, and $Re = 1,000$.

for various values of y_0; these parametric values corresponds to an unstable Tollmien–Schlichting wave. From this plot we see that for each value of y_0, the perturbation energy grows exponentially in accordance with classical stability theory. However, the time at which the eventual exponential growth sets in depends upon the location of the Gaussian initial condition. That is, as the location moves further out of the boundary layer (recall that the boundary layer edge is at $y \approx 5$), the time at which exponential growth occurs also increases. What might not be expected is that even for $y_0 = 12$, which lies well outside the boundary layer, a Tollmien–Schlichting wave is still generated. This can be explained by appealing to the general solution given by (8.3). Since any solution can be expanded in terms of the normal modes and the continuum, and since the initial condition is not entirely contained within the continuum, there must be a nonzero coefficient for the normal modes, no matter how small, which eventually gives rise to the observed exponential growth. This observation is further explained in the work of Hill (1995), and the reader is referred to that work for more details. And, more importantly, this point can be completely missed if one ignores the effect of initial conditions, and relies entirely on the classical stability framework, which might have critical consequences in the areas of receptivity and flow control. Also shown are results for $\phi = 90°$ (Fig. 8.9; right). Recall that at this angle only the continuum exists and there are no normal modes. We plot the maximum of G in time (denoted by $G_{\max}$) for a given fixed wavenumber; the maximum does not increase without bound. The largest value of $G_{\max} = 124$ occurs at $\tilde{\alpha} = 1.5$. The idea of determining the largest possible value of $G_{\max}$ for a set of initial conditions is explored further in Lasseigne *et al.* (1999).

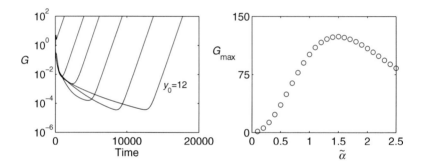

Figure 8.9   Left: Normalized energy G versus time with $Re = 10^3$, $\sigma = 0.25$, $\tilde{\alpha} = 0.24$, $\phi = 0$ and $y_0 = 2, 4, 6, 8, 10, 12$ (increasing from left to right). Right: Maximum over time of normalized energy G versus wavenumber $\tilde{\alpha}$ with $Re = 10^3$, $\sigma = 0.25$, $\phi = 90°$ and $y_0 = 2$.

8.6.2 Optimization Scheme

A mechanism for rapid transient growth when the initial condition is expressed as a sum of the eigenfunctions has been given by Reddy & Henningson (1993). The concept is that a group of eigenfunctions are nearly linearly dependent so that, in order to represent an arbitrary disturbance (say), then it is possible that the coefficients can be quite large. Now, since each one of these nearly linearly dependent eigenfunctions has differing decay rates, the exact cancellations that produce the given initial disturbance might not persist in time and thus significant transient growth can occur. The mechanism can be (and is) taken a step further in order to determine the optimal initial condition (still expressed as a sum of the non-orthogonal eigenfunctions) that produces the largest relative energy growth for a certain time period. When this procedure is completed, it can be seen to have the feature that the nearly linearly dependent eigenfunctions are multiplied by coefficients three orders of magnitudes greater than the others. This optimal initial condition produces a growth factor of about 20 for the two-dimensional disturbance in Poiseuille flow. However, this optimal growth is nearly destroyed by not including the first eigenfunction (growth drops to a factor of 6 rather than 20) which seems to indicate that the prior explanation of (initial) exact cancellations by the nearly linearly dependent eigenfunctions is not the entire mechanism. Butler & Farrell (1992) also calculated optimal initial conditions in terms of a summation of the eigenfunctions (although they put no particular emphasis on the importance of using this approach) and reiterated the importance of near linear dependence of the modes to the transient growth. This work also explained the transient growth of the optimal initial conditions in terms of the vortex-tilting mechanism and the Reynolds stress mechanism, since these (physical) arguments apply no matter what the solution method.

By the method that we have been following, an optimization procedure can be determined without resorting to a variational procedure. We shall describe the optimization procedure for channel flows (Criminale *et al.*, 1997); the procedure for Blasius flow follows similarly and can be found in Lasseigne *et al.* (1999). A closer inspection of initial conditions (8.55) suggests that each of these disturbances is in essence a single Fourier mode of an arbitrary initial condition. If one were to consider an arbitrary odd function for the $\breve{u}$ velocity satisfying the boundary conditions written in terms of a Fourier sine series, then the initial condition in the $\breve{v}$ velocity is given by (8.55). Thus, if one wished to determine an optimal initial disturbance, a maximization procedure could be applied to an arbitrary linear combination of these modes, all of which are initially orthogonal and linearly independent. Clearly, if one wanted to also include nonzero initial vorticity in such an optimization scheme, it would not be difficult to include (and these initial conditions are of course very important when modeling real disturbances as opposed to optimal disturbances). The results presented by Criminale *et al.* (1997) show that, if included in the optimization procedure, the initial vorticity modes would not contribute to the optimal solution for the cases considered.

To start the optimization scheme, we consider the total solution $\underline{\breve{u}} = (\breve{u}, \breve{v}, \breve{w})$ to be the sum

$$\underline{\breve{u}}(y,t) = \sum_{k=1}^{N} (a_k + ib_k)\underline{\breve{u}}_k(y,t), \tag{8.57}$$

where each of the vectors $\underline{\breve{u}}_k(y,t)$ represents a solution to equations (8.1) and

(8.2), subject to the initial conditions

$$\underline{\check{u}}(y,0) = \left\{ \begin{array}{c} \cos\phi\,\sin k\pi y \\[4pt] \dfrac{i\tilde{\alpha}}{k\pi}(\cos k\pi - \cos k\pi y) \\[4pt] \sin\phi\,\sin k\pi y \end{array} \right\}. \tag{8.58}$$

In order to maximize the growth function, it is sufficient to maximize the energy,

$$E(t) = \int_{-1}^{1} \underline{\check{u}}(y,t)\cdot\underline{\check{u}}^*(y,t)\,dy, \tag{8.59}$$

subject to the constraint

$$E(0) = 1, \tag{8.60}$$

where $*$ denotes the complex conjugate. Therefore, we use Lagrange multipliers to maximize the function

$$\overline{G}(t) = \int_{-1}^{1} \underline{\check{u}}(y,t)\cdot\underline{\check{u}}^*(y,t)\,dy - \lambda\left(\int_{-1}^{1}\underline{\check{u}}(y,0)\cdot\underline{\check{u}}^*(y,0)\,dy - 1\right), \tag{8.61}$$

that requires

$$\frac{\partial\overline{G}}{\partial a_k} = 0, \qquad \frac{\partial\overline{G}}{\partial b_k} = 0, \qquad k = 1,2,\dots,N. \tag{8.62}$$

The set of equations thus derived produces a $2N \times 2N$ generalized eigenvalue problem, and can be solved by any standard eigenvalue solver. A search over the eigenvectors gives the initial condition with initial unit energy that maximizes the function $\overline{G}$ at time t.

To illustrate the optimization procedure, we perform the calculations for the two cases reported by Butler & Farrell (1992). Both cases correspond to plane Poiseuille flow. The first is the computation of the two-dimensional optimal for $\tilde{\alpha} = 1.48$, $\phi = 0°$ and $Re = 5,000$. In Fig. 8.10 we show the growth factor at $t = 14.1$ for each individual mode as well as for the optimal solution for various values of N. The convergence as $N \to \infty$ is well illustrated; compare with Reddy & Henningson (1993), e.g. Also shown in the figure are the magnitudes of the coefficients that produce the optimum with $N = 20$. There are no surprises. Each coefficient is of reasonable size, with the largest coefficient being a factor of ten greater than the first coefficient, and not a factor of 1,000, as is the case when using eigenfunction expansions where a group of eigenfunctions is nearly linearly dependent and is not orthogonal. The magnitudes peak for $k = 6, 7$ and 8, which could be easily predicted from the previous graphs for the responses to each individual mode. The second calculation is for the optimal three-dimensional disturbance. The parameters chosen are $Re = 5{,}000$, $\tilde{\alpha} = 2.044$ and $\phi = 90°$. The initial conditions that produce a maximum growth at $t = 379$ are found. The results are also shown in Fig. 8.10, and the composition of the initial conditions in terms of the modes chosen here could be easily determined from the individual responses of each mode.

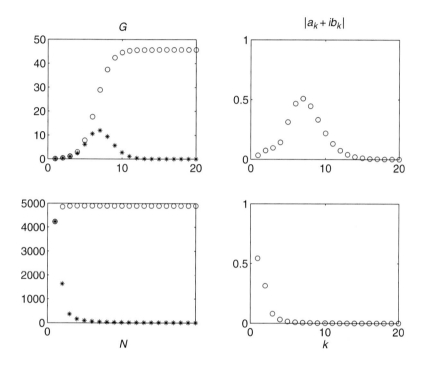

Figure 8.10 Top: Left: The growth function G at $t = 14.1$; individual mode results denoted by *; cumulative results from optimization procedure denoted by o. Right: The magnitude of the coefficients $a_k + ib_k$ from optimization procedure for $N = 20$. For plane Poiseuille flow with $\tilde{\alpha} = 1.48$, $\phi = 0$, and $Re = 5{,}000$. Bottom: Same as above except $t = 379$, $\tilde{\alpha} = 2.044$, $\phi = 90°$, and $Re = 5{,}000$.

It must be reiterated that, although it is possible and conceptually easy to reproduce the optimal initial conditions that have been previously found, the maximum transient growth is only a measure of what is possible and not what will actually occur as has been the difficulty in devising experiments. It is at least as important to investigate whether such large growth is possible for arbitrary initial conditions. In this regard, the results presented by Criminale *et al.* (1997) produce a mostly negative answer to this question. For two-dimensional disturbances in Poiseuille flow, the transient growth observed for arbitrarily chosen initial conditions using this approach is, at best, only 25% of the optimal. When considering a fixed wavelength $\tilde{\alpha}$ and a fixed obliqueness ϕ, it is seen that very large relative energy growth of the perturbation can be observed in Poiseuille flow for oblique disturbances with arbitrary velocity profiles restricted to having zero initial normal vorticity, but the relative energy growth quickly decreases when arbitrary disturbances are combined with initial normal vorticity. Similar results are found for Couette and Blasius flows.

8.6.3 Concluding Remarks

Plane Poiseuille and plane Couette flows in an incompressible viscous fluid have been investigated subject to the influence of small perturbations (Criminale *et al.*, 1997). In lieu of using the techniques of classical stability analysis or the more recent techniques involving eigenfunction expansions, the approach here has been to first Fourier transform the governing disturbance equations in the streamwise and spanwise directions only and then solve the resulting partial differential equations numerically by the method of lines. Unlike traditional methods, where traveling wave normal modes are assumed for solution, this approach offers another means whereby arbitrary initial input can be specified. Thus, arbitrary initial conditions can be imposed and the full temporal behavior, including both early-time transients and the long-time asymptotics, can be determined. All of the stability data that are known for such flows can be reproduced. In addition, an optimization scheme is presented using the orthogonal Fourier series and all previous results using variational techniques and eigenfunction expansions are reproduced. However, it was shown that the transient growth of the perturbation energy density is very sensitive to the presence of an initial normal vorticity perturbation.

The benefit of this approach is clear, for it can be applied to other classes of problems where only a finite number of normal modes exist. For unstable conditions, a localized initial velocity disturbance always excites the unstable Tollmien–Schlichting wave, even when the disturbance lies far outside the boundary layer. For stable conditions, the degree of transient growth was found to depend on the location of the localized disturbance, with localized disturbances within the boundary layer showing greater transient growth. For fixed disturbances, the transient growth was seen to depend greatly on the wavenumber and angle of obliquity. Since no complete infinite set of functions span the flow, three subspaces were defined, with different flow characteristics in the freestream, for the optimization procedure. Using these orthogonal sets, it was determined what type of initial conditions were necessary to produce almost the same maximum growth that the governing equations allow; the contribution of the continuum to the initial conditions producing the maximum transient growth is properly included. Disturbances that are nonzero and nonlocalized in the freestream were found to produce the greatest transient growth. However, large transient growth was found only in response to disturbances with zero initial normal vorticity. When nonzero normal vorticity was included in the initial conditions, the transient growth either diminished or was eliminated.

Finally, this numerical approach was recently been successfully applied to free shear flows in an inviscid fluid (Criminale, Jackson & Lasseigne, 1995).

8.7 Exercises

Exercise 8.1   Derive the disturbance equations equivalent to (8.25) and (8.26) when the mean profile U is a continuous function of y. Assume three-dimensional disturbances, incompressibility and include viscous effects. In addition, derive the corresponding vorticity components.

Exercise 8.2   Consider the baroclinic plane Couette flow of Section 7.4. Redo the problem using the normal mode approach, and compare the eigensolution to that found using the moving coordinate transformation.

Exercise 8.3   Consider the inviscid boundary layer flow given by

$$U(y) = \begin{cases} U_0 & y \geq H, \\ \sigma y & 0 < y < H. \end{cases}$$

Compute the solutions using the normal mode approach and the moving coordinate approach, and compare the two.

Exercise 8.4   Consider an inviscid shear flow given by Fig. 8.11. Compute the solutions using both the normal mode approach, and using the moving coordinate transformation. Compare the two methods.

Exercise 8.5   Consider inviscid plane Couette flow. Solve this limiting case by use of the moving coordinate transformation and with initial distribution $\tilde{v}_0(x, y, 0) = \Omega e^{-i(\alpha_0 x + \beta_0 y)}$, which corresponds to a plane wave in the x-, y-directions with wavenumbers α_0 and β_0, respectively.

Exercise 8.6   Redo the inviscid plane Couette flow with the boundary conditions requiring the perturbation pressure to vanish at $y = \pm H$.

Exercise 8.7   An idealized problem for internal gravity waves in the upper ocean can be modeled as show in Fig. 8.12. With the additional assumption that the flow is inviscid, determine the fate of linear perturbations when the

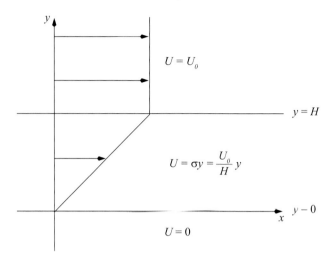

Figure 8.11 Figure for Exercise 8.4.

full equations for $W(z)$ is used and, above and below $z = -\frac{(d \pm \varepsilon)}{2}$, $N = N_0 = 0$. Otherwise, $N^2 = -g\frac{\bar{\rho}'}{\rho} = -g\lambda$.

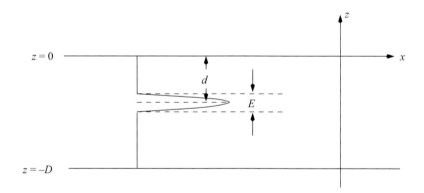

Figure 8.12 Figure for Exercise 8.7.

Exercise 8.8   Write a numerical code that solves the two-dimensional partial differential equation (8.1) by the method of lines. Verify the code by computing the temporal growth rate for plane Poiseuille flow for $\alpha = 1$ and $Re = 10^4$ and compare to Table 8.1. Also, reproduce Figs. 8.4 and 8.5.

9

Nonlinear Stability

The sciences do not try to explain, they hardly even try to interpret, they mainly make models. By a model is meant a mathematical construct which, with the addition of certain verbal interpretations, describes observed phenomena. The justification of such a mathematical construct is solely and precisely that it is expected to work – that is, correctly to describe phenomena from a reasonably wide area.

– John von Neumann (1903–1957)

In this chapter, because energy can only be expressed as a nonlinear product, we examine the energy equation that is associated with disturbances that can decay, be neutral or amplify, depending upon the attributes of the equation for the energy. Next, we extend the investigation by discussing weakly nonlinear theory, secondary instability theory and resonant wave interactions. The chapter will close with a presentation of an all-encompassing theory that enables direct solutions for linear, secondary and nonlinear instabilities within a single theoretical framework. This theory is now referred to as parabolized stability equation theory, denoted as "PSE theory".

9.1 Energy Equation

For the linear regime of stability theory, the Orr–Sommerfeld equation (2.28) provides a reasonable basis to describe the characteristics for the stability of the flow, particularly two-dimensional disturbances. Moreover, instability theory based on an energy equation is useful to help us understand the physical processes that lead from linear to a nonlinear instability basis, and this has been demonstrated on several occasions.

Just as was done by Mack (1984), we define the kinetic energy for a two-dimensional disturbance per unit density of an incompressible fluid[1] as

$$e = \frac{1}{2}\left(u^2 + v^2\right).$$ (9.1)

Note that the absence of nonlinear terms is not a weakness for this description. Specifically, the nonlinear terms only serve to shift energy between velocity components and can neither increase nor decrease the *total* energy. To obtain the equation for the kinetic energy, first multiply the x-momentum equation (2.7) by u, the y-momentum equation (2.8) by v, and add the two to give

$$e_t + U e_x + uvU' = -up_x - vp_y + Re^{-1}(u\nabla^2 u + v\nabla^2 v),$$ (9.2)

where only a parallel mean flow is considered. For temporally amplifying disturbances, (9.2) is integrated over the range of mean flow, from $y = a$ to $y = b$, say, and then averaged over a single wavelength in x. This leads to the equation

$$E_t = \int_a^b -\langle uv\rangle U' dy - Re^{-1}\int_a^b \langle(v_x - u_y)^2\rangle dy,$$ (9.3)

where E is now the total kinetic energy per wavelength per unit density as defined by the integration of e in (9.1) over x and y. The right-hand side of equation (9.3) involves a total energy production component (over a wavelength) and a dissipation component (over a wavelength) due to viscosity. The production term consists of a Reynolds stress $\langle uv\rangle$, and the dissipation term is the product of viscosity and the disturbance vorticity ω_z, since $\omega_z = v_x - u_y$.

For a fixed mean flow, the term that can change sign is that due to the production term or the Reynolds stress $\langle uv\rangle$. For disturbance amplification, the Reynolds stress must be such that it overcomes the dissipation; otherwise, the disturbance will either be neutrally stable (production balances dissipation) or decay (dissipation overcomes production).

From inviscid stability theory we have seen that a flow with a convex velocity profile has only damped instability waves. On the other hand, one of the most important convex profiles is the Blasius similarity solution for the boundary layer. In this case, the Reynolds stress associated with inviscid instabilities for the Blasius profile has $U'' < 0$ and can support amplifying disturbances provided the disturbance phase velocity is less than the freestream velocity in the region $y < y_c$, where y_c is the location of the critical layer. The critical layer is the distance from the wall to the point away from the wall where the freestream velocity matches the disturbance phase speed, as

---

[1] The assumption of two-dimensional disturbances is not critical and, in fact, the full three-dimensional case follows easily.

shown by Rayleigh's Theorem (see Result 2.1 of Chapter 2). Therefore, amplifying disturbances can be present only if viscosity causes sufficient positive Reynolds stress near the wall. The mechanical analogue for such production was presented in Section 1.6, in that even a linear spring force that has a time delay can lead to instability. In short, it is a question of phasing.

Because the Blasius boundary layer is a similarity profile governed by a single parameter (the Reynolds number) and there is no inflection point in the profile, the only possible convective instability is the viscous traveling wave instability as described by normal mode solutions. Hence, we can study the viscous instability without potentially competing mechanisms. For the remainder of this section, we will use the Blasius profile and the associated viscous instability to study nonlinear effects.

For completeness and before we proceed with the next topic in this chapter, let us look at the energy equation for spatially amplifying disturbances. From the work of Hama, Williams & Fasel (1979) we find that, with the parallel flow assumption and averaging over one period in time, the energy equation will be

$$\frac{d}{dx}\int_0^\infty \langle e \rangle U \, dy = \int_0^\infty -\langle uv \rangle U' dy - Re^{-1}\int_0^\infty \langle (v_x - u_y)^2 \rangle dy$$

$$-\frac{d}{dx}\int_0^\infty \langle pu \rangle dy + Re^{-1}\frac{d}{dx}\int_0^\infty \langle v(v_x - u_y) \rangle dy. \quad (9.4)$$

Not surprisingly, the production term (first term on right-hand side of equation) is the dominant process in determining whether a disturbance is amplified or decays. The second term on the right-hand side of (9.4) is the dissipation. The net total transfer of kinetic energy is governed by the last two terms of (9.4). Unlike temporal stability, for spatially growing disturbances, the production and dissipation terms do not balance. Moreover, for spatial disturbances, the $\langle pu \rangle$ correlation plays a significant role in the energy balance, and is always opposite to the trend of the fluctuations. When disturbances are amplified, it suppresses the energy; when disturbances are decaying, energy is supplied.

9.2 Weakly Nonlinear Theory

Squire's theorem states that, for every unstable three-dimensional mode, there is an unstable two-dimensional mode at a lower Reynolds number (see the discussion in Subsections 2.1.5 and 2.1.6). However, Squire's theorem is only applicable for linear disturbances and is not valid when the disturbances are nonlinear. This fact was not recognized in early nonlinear studies, where the theory was restricted to two-dimensional perturbations. Fortunately, as is usually the case, beginning with the simpler two-dimensional approach can often shed significant light on our understanding of a subject before we approach the full three-dimensional system. A more careful nonlinear analysis must take into account three-dimensional perturbations.

Toward developing a framework for nonlinear stability, Watson (1960) and Stuart (1960) expanded the Navier–Stokes equations in powers of a temporal disturbance amplitude, $A(t)$. This amplitude must satisfy the following equation

$$A_t = \sigma(Re, \alpha)A - \sum_{n=1}^{N} l_n(Re, \alpha)A^{2n+1}, \tag{9.5}$$

where $\sigma(Re, \alpha)$ is the small linear growth rate of a wave at a near-critical value of Reynolds number (Re) for wavenumber α. $l_n(Re, \alpha)$ are referred to as Landau coefficients (Landau, 1944). This weakly nonlinear theory is often referred to as the Stuart–Watson expansion and is often truncated at $N = 1$. As such, equation (9.5) becomes

$$A_t = \sigma(Re, \alpha)A - l_1(Re, \alpha)A^3. \tag{9.6}$$

Solutions of the Stuart–Watson equation depend on both the linear growth rate σ of the disturbance and the Landau constant l_1. For $\sigma > 0$ and $l_1 > 0$, linear disturbances amplify, but the amplitude, A, reaches a stable state $A \simeq \sigma/l_1$ as $t \to \infty$. Near the critical point, this is referred to as supercritical bifurcation. For $\sigma < 0$ and $l_1 > 0$, both linear and nonlinear disturbances are damped and $A = 0$ as $t \to \infty$. For $\sigma > 0$ and $l_1 < 0$, the disturbances are linearly unstable and grow unbounded. This type of behavior is referred to as subcritical bifurcation. For $\sigma < 0$ and $l_1 < 0$, small amplitude disturbances decay; however, when the threshold amplitude ($A_0 \simeq \sqrt{\sigma/l_1}$) is exceeded, the disturbances grow unbounded. The amplitude A_0 itself is a finite-amplitude stable equilibrium condition. Note that this predicted unbounded growth results from truncating (9.5).

As the amplitude increases, the higher-order terms are no longer negligible, and should alter the solution behavior. For an infinite Stuart–Watson expansion, the solutions should tend to the full Navier–Stokes solutions. However, no true minimum exists for equation (9.5).

For Poiseuille mean flow, $U(y) = 1 - y^2$, results from the truncated series (9.6) compared with full Navier–Stokes solutions indicate that subcritical bifurcation should exist and that both linear and nonlinear neutral curves are possible.

For the study of instabilities in flows that are more complicated than the Poiseuille or Blasius flows, the Stuart–Watson expansion is replaced by an alternate equation. Based on the reaction-diffusion work of Kuramoto (1980) and the flame propagation modeling work of Sivashinsky (1977), the Kuramoto–Sivashinsky equation was identified as playing an important role in studying the linear and nonlinear instability (and potentially chaotic behavior) of non-traditional fluid mechanics problems.

The one-dimensional Kuramoto–Sivashinsky equation is given by

$$u_t + u u_x + u_{xx} + u_{xxxx} = 0, \tag{9.7}$$

with appropriate boundary conditions. Subsequent to their work, the equation has been used to study the thermal diffusive instabilities in laminar flame fronts, interfacial instabilities between concurrent viscous fluids, viscous film flow down vertical or included planes, the interfacial stress from adjacent gas flow and the drift waves in plasmas.

We will not discuss the many uses of the Kuramoto–Sivashinsky equation, but have listed it for completeness sake.

9.3 Secondary Instability Theory

As we have discussed in Chapter 3, the viscous Tollmien–Schlichting wave instabilities of a boundary layer begins to amplify at rather long wavelengths at branch I (low frequency) of the neutral curve, amplify until branch II (high frequency) is reached downstream, and then once again decay. The transition from a laminar flow in boundary layers can begin with these seemingly harmless Tollmien–Schlichting waves. However, at some point, these two-dimensionally dominant waves begin to develop a three-dimensional, short-wavelength structure reminiscent of a boundary layer transition to turbulent flow. The theory of secondary instabilities provides one understanding of this two-dimensional process becoming three-dimensional and of the long wavelength dominant instabilities developing short scales.

The experimental observations and documentation of what we now know as secondary instabilities began with the early publications of Klebanoff, Tidstrom & Sargent (1962), who found that the Tollmien–Schlichting wave in the boundary layer evolves from a two-dimensional wave into an aligned arrangement of Λ (Lambda) vortices. Boundary layer transition soon followed the appearance of these Λ vortices. This alignment of Λ vortices, as shown in Fig. 9.1, is also referred to as peak-valley splitting, whereby the peak of a wave is aligned with the peak of an adjacent wave and the valley of a wave is aligned with the valley of an adjacent wave. Later, Kachanov, Kozlov & Levchenko (1979) and Kachanov & Levchenko (1984) presented experimental results that showed two-dimensional Tollmien–Schlichting waves evolving to a staggered arrangement of Λ vortices and evidence of an energy gain forming a peak in the spectrum at near-subharmonic wavelengths. This staggered alignment of Λ vortices, as shown in Fig. 9.2, is referred to as peak-valley alignment because the peak of a wave is aligned with the valley of an adjacent wave.

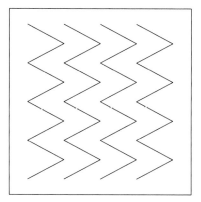

Figure 9.1 Sketch of peak-valley splitting secondary instabilities.

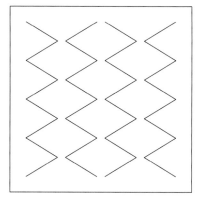

Figure 9.2 Sketch of peak-valley alignment secondary instabilities.

The theoretical work of Herbert (1983) and Orszag & Patera (1980, 1981) led to the common state of the art in what is now known as *secondary instability theory*. This theory explains the process of how a two-dimensional Tollmien–Schlichting wave evolves into either the peak-valley splitting (Fig. 9.1) or the peak-valley alignment (Fig. 9.2). The theory proposes that the three-dimensional disturbances originate from parametric excitation in the periodic streamwise flow that arises from the Tollmien–Schlichting wave as it reaches finite amplitude. Whether peak-valley splitting or alignment results from this excitation depends upon a threshold amplitude for the Tollmien–Schlichting wave. For low finite amplitudes, subharmonic resonance and peak-valley alignment are predicted and verified in the experiments. As the threshold amplitude of the Tollmien–Schlichting wave is increased, the peak-valley splitting is predicted and observed in the experiments. The theory proposes that the growth of three-dimensional disturbances arises from vortex tilting and vortex stretching. A redistribution of energy in the spanwise vorticity near the critical layer causes the growth of secondary instabilities.

Similar to linear stability theory, secondary instability decomposes the velocity field and pressure field into basic state velocities $\underline{u}_2 = (u_2, v_2, w_2)$ and pressure p_2 and disturbance velocities $\underline{u}_3 = (u_3, v_3, w_3)$ and pressure p_3. The sum of these basic values plus disturbance quantities are substituted into the Navier–Stokes equations and, by subtracting the basic state equations, which are assumed exactly satisfied, we find

$$\nabla \cdot \underline{u}_3 = 0, \tag{9.8}$$

and

$$\left(Re^{-1}\nabla^2 - \frac{\partial}{\partial t} \right) \underline{u}_3 - (\underline{u}_2 \cdot \nabla)\underline{u}_3 - (\underline{u}_3 \cdot \nabla)\underline{u}_2 = \nabla p_3, \tag{9.9}$$

where $\nabla = (\partial/\partial x + \partial/\partial y + \partial/\partial z)$. The pressure can be eliminated by taking the curl of the momentum equations (9.9). In this way, equations for the vorticity result, namely

$$\left[Re^{-1}\nabla^2 - \frac{\partial}{\partial t} \right]\underline{\Omega}_3 - (\underline{u}_2 \cdot \nabla)\underline{\Omega}_3 + (\underline{\Omega}_2 \cdot \nabla)\underline{u}_3$$

$$- (\underline{u}_3 \cdot \nabla)\underline{\Omega}_2 + (\underline{\Omega}_3 \cdot \nabla)\underline{u}_2 = 0; \tag{9.10}$$

the streamwise, normal and spanwise vorticity components, $\underline{\Omega} = \{\xi, \omega, \zeta\}$, are given by

$$\xi = w_y - v_z, \qquad \omega = u_z - w_x, \qquad \zeta = v_x - u_y, \tag{9.11}$$

for the basic state vorticity $\underline{\Omega}_2$ and the disturbance vorticity $\underline{\Omega}_3$. If we consider three-dimensional locally parallel boundary layers, the basic flow becomes a composite of the mean profile (U_b, W_b) and a two-dimensional or oblique Tollmien–Schlichting wave component (u, v, w), or

$$\begin{rcases} u_2(x, y, z, t) &= U_b(y) + Au(x, y, z, t), \\ v_2(x, y, z, t) &= Av(x, y, z, t), \\ w_2(x, y, z, t) &= W_b(y) + Aw(x, y, z, t). \end{rcases} \tag{9.12}$$

The mean flow (U_b, W_b) may be a similarity solution such as the two-dimensional Blasius solution or the three-dimensional Falkner–Skan–Cooke solution. The Tollmien–Schlichting solution is obtained from the Orr–Sommerfeld and Squire equations. The primary amplitude A is an input to the solution procedure and would directly reflect the maximum streamwise root-mean-square fluctuations in the flow, provided the profiles are normalized by the streamwise disturbance component. The mean basic state (U_b, W_b) is based on the choice of reference frame in which the solutions are are obtained. Hence, for a coordinate system moving with the freestream direction, $U_b = U$ and $W_b = 0$, or just the Blasius solution. For oblique primary waves in a Blasius flow, $U_b = U \cos \theta$ and $W_b = -U \sin \theta$, where θ is the angle the traveling wave is aligned relative to the freestream direction.

The primary instability is moving with a phase speed, defined with streamwise and spanwise components, as

$$c_x = \omega_r/(\alpha_r \cos \theta) \qquad \text{and} \qquad c_z = \omega_r/(\alpha_r \sin \theta), \qquad (9.13)$$

where $\alpha_r = \sqrt{\alpha_r^2 + \beta_r^2}$ and α_r and β_r are the real parts of the streamwise and spanwise wavenumbers, respectively. In general, for flat-plate boundary layers, the dominant primary mode is a two-dimensional Tollmien–Schlichting wave. However, for compliant (or flexible) walls that are used for drag reduction (see Chapter 12), the dominant primary instability is an oblique wave. This chapter will preserve the more general three-dimensional nature of the primary instability.

The equations are now transformed to a system moving with the primary wave according to Floquét (1883) theory. Hence, flow visualization of the Blasius flow together with Tollmien–Schlichting wave solutions would show a periodic state basic flow. Further, by assuming the primary wave as part of the basic flow solution for the secondary instability analysis, we are inherently assuming that the secondary instability amplifies much faster than the primary instability. This is a fundamental notion for the solution of the secondary instability equations as will be discussed later in this section.

By substituting (9.12) into the vorticity equation (9.10) and using the transformed reference frame, the normal vorticity for the secondary instability mode takes the form

$$Re^{-1}[\omega_{3,xx} + \omega_{3,yy} + \omega_{3,zz} - \omega_{3,t} - (U_b - c_x)\omega_{3,x} - (W_b - c_z)\omega_{3,z}]$$
$$- U_b' v_{3,z} + W_b' v_{3,x} + A\{-u\omega_{3,x} - v\omega_{3,y} - w\omega_{3,z}$$
$$- u_3 \omega_x - v_3 \omega_y - w_3 \omega_z + (\xi + v_z)v_{3,x}$$
$$+ (\zeta - v_x)v_{3,z} - v_z u_{3,y} + v_x w_{3,y}\} = 0. \qquad (9.14)$$

By taking

$$\frac{\partial}{\partial z}(\text{streamwise vorticity equation}) - \frac{\partial}{\partial x}(\text{spanwise vorticity equation})$$

for the secondary mode, the final equation, in the moving reference frame, is

$$
\left[Re^{-1}\nabla^2\frac{\partial}{\partial t} - (U_b - c_x)\frac{\partial}{\partial x} - (W_b - c_z)\frac{\partial}{\partial z} \right]\nabla^2 v_3
$$

$$
+ U_b''v_{3,x} + W_b''v_{3,z} + A\left\{ -\left[u\frac{\partial}{\partial x} + v\frac{\partial}{\partial y} + w\frac{\partial}{\partial z} \right]\nabla^2 v_3 \right.
$$

$$
- \nabla^2 v_y v_3 - (u_{xx} + u_{zz} - u_{yy})v_{3,x} - 2v_{xy}v_{3,x}
$$

$$
+ (v_{zz} - v_{xx} + v_{yy} + 2u_{xy})v_{3,y} + (w_{yy} - w_{xx} + u_{xz} - v_{yz})v_{3,z}
$$

$$
+ (v_y + 2u_x)(v_{3,zz} - v_{3,xx} + v_{3,yy}) - 2v_x v_{3,xy} - 2(u_z + w_x)v_{3,xz}
$$

$$
- v_z v_{3,yz} - \nabla^2 v_x u_3 + 2(v_{zz} - v_{xx} + v_{yy} + 2u_{xy})u_{3,x}
$$

$$
+ 2(w_{xy} - v_{xz})u_{3,z} - v_x(u_{3,xx} + u_{3,zz} - u_{3,yy})
$$

$$
+ 2(v_y + 2u_x)u_{3,xy} + 2w_x u_{3,yz} + v_z u_{3,xz} - \nabla^2 v_z w_3
$$

$$
\left. + 2(u_{yz} - v_{xz})w_{3,x} - v_z(w_{3,xx} - w_{3,yy}) + 2u_z w_{3,xy} \right\} = 0. \qquad (9.15)
$$

This equation involves the secondary velocity components. To reduce the equation to normal velocity and vorticity representation only (similar to the Orr–Sommerfeld and Squire equations), normal modes are introduced. Additionally, the primary amplitude, A, is a parameter in the equations and is assumed to be locally non-varying. As $A \to 0$ the Orr–Sommerfeld and Squire equations result. For the case of interest where $A \neq 0$, the primary eigenfunctions (u, v, w) appear in the equations as coefficients.

To solve the secondary system, an appropriate normal mode representation is sought similar to the linear stability normal mode assumption, and this is

$$v_3(x, y, z, t) = V(x, y, z)e^{\sigma t + i\beta(z\cos\phi - x\sin\phi)}, \qquad (9.16)$$

where $\beta = 2\pi/\lambda_z$ is a specified spanwise wavenumber, and $\sigma = \sigma_r + i\sigma_i$ is a temporal eigenvalue or is a specified real number for spatial analyses. $V(x, y, z)$ is a function that represents the class of secondary modes. Floquét theory suggests the form of solution for periodic systems and, for the present problem, this may be written as

$$V(x, y, z) = \tilde{V}(x, y, z)e^{\gamma(x\cos\phi + z\sin\phi)}, \qquad (9.17)$$

where $\gamma = \gamma_r + i\gamma_i$ is the characteristic exponent and $\tilde{V}(x, y, z)$ is periodic in the (x,z)-plane and may be represented by Fourier decomposition. Thus the representation of the secondary instability for a three-dimensional basic flow is

$$\{v_3, \omega_3\}(x,y,z,t) = Be^{\sigma t + i\beta(z\cos\phi - x\sin\phi) + \gamma(x\cos\phi + z\sin\phi)}$$

$$\times \sum_{n=-\infty}^{\infty} \{v_n(y), \underline{\omega}_n(y)\} e^{i(n/2)\alpha_r(x\cos\phi + z\sin\phi)}, \quad (9.18)$$

where B is the amplitude of the secondary instability mode. This suggests a form of solution for the secondary disturbance based on a coordinate system oriented at an angle ϕ with respect to the mean flow and moving with the primary wave. If the coordinate system is aligned with the primary wave, or $\phi = 0°$, then the solution for the secondary disturbance would follow the presentation presented by Herbert, Bertolotti, & Santos (1985), who considered the case for two-dimensional primary wave.

If the secondary disturbance form (9.18) is substituted into (9.14) and (9.15), an infinite system of ordinary differential equations results. The dynamic equations are determined by collecting terms in the governing equations with like exponentials. The system consists of two distinct classes of solution because the even and odd modes decouple. Even modes correspond to the fundamental mode of secondary instability, and the odd modes are the subharmonic mode. Only a few terms of the Fourier series are retained since, as shown by Herbert, Bertolotti, & Santos (1985), this provides a sufficiently accurate approximation for a two-dimensional disturbance.

The fundamental modes, v_f, and subharmonic modes, v_s, would satisfy

$$v_f(x+\lambda,y,z) = v_f(x,y,z) \quad \text{and} \quad v_s(x+2\lambda,y,z) = v_s(x,y,z). \quad (9.19)$$

Thus, the fundamental modes are associated with primary resonance in the Floquet system and subharmonic modes originate from principal parametric resonance.

This form of solution indicates two complex quantities, σ and γ, that lead to an ambiguity similar to that found with the Orr–Sommerfeld–Squire problem. There are four unknowns, $\sigma_r, \sigma_i, \gamma_r, \gamma_i$. Two can be determined while two must be chosen in some other way. For brevity in this text, only temporally growing tuned modes are considered. The temporal growth rate is σ_r, and σ_i can be interpreted as a shift in frequency. In this case, $\gamma_r = \gamma_i = 0$. If $\sigma_i = 0$, then the secondary disturbance is traveling synchronously with the basic flow.

The rigid wall boundary conditions for the secondary disturbance are given as

$$v_n, v_n', \underline{\omega}_n \to 0 \quad \text{as} \quad y \to \infty, \quad (9.20)$$

Table 9.1 *Spectral convergence of temporal eigenvalues for the subharmonic mode of secondary instability for $R_\delta = 880$, $F = \omega/Re \times 10^6 = 58.8$, $A = 0.00695$, $\beta = 0.214$ and $\alpha = 0.15488 - i0.005504$.*

N	σ_1	$\sigma_{2,3}$
20	0.0039769	0.0007067 ± 0.010675
25	0.0041494	0.0011586 ± 0.010510
30	0.0041667	0.0011565 ± 0.010456
35	0.0041714	0.0011640 ± 0.010448
40	0.0041713	0.0011628 ± 0.010448
45	0.0041713	0.0011628 ± 0.010448

and

$$v_n, v'_n, \underline{\omega}_n = 0 \quad \text{at} \quad y = 0, \tag{9.21}$$

which are the same as those for the Orr–Sommerfeld and Squire equations.

Solutions to the secondary instability equations can be determined by using whatever methods that are used for the primary Orr–Sommerfeld and Squire equations. For example, Joslin (1990), Joslin, Morris & Carpenter (1991) and Joslin & Morris (1992) have used Chebyshev polynomials and the Tau method together with the Gramm–Schmidt orthonormalization approaches to solve both the primary Orr–Sommerfeld and Squire equations and the secondary disturbance equations.

The primary and secondary disturbance equations are nondimensionalized using the freestream velocity U_∞, kinematic viscosity v and a boundary layer scale. Here, the boundary layer displacement thickness δ^* is used and results in the Reynolds number $Re_{\delta^*} = 1.7207 Re_x^{1/2}$, where $Re_x = U_\infty x/v$ for the boundary layer similarity solution. An alternate Reynolds number often used in secondary instability analysis is $Re_\delta = 1.4 Re_{\delta^*}$.

Convergence of the subharmonic secondary instability eigenvalues is shown in Table 9.1 for the first three eigenmodes for a Reynolds number $Re_\delta = 880$, frequency $F = \omega Re^{-1} \times 10^6 = 58.8$, or $\omega = 0.051744$, and spanwise wavenumber $\beta = 0.214$. Here, spectral methods was used to compute the eigenvalues. For this test case, the primary wave has an amplitude $A = 0.00695$ and spatial wavenumber and growth rate of $\alpha = 0.15488 - i0.005504$.

To demonstrate how well the secondary instability theory models the physics of the true boundary layer transition problem, Fig. 9.3 shows a comparison of the velocity profiles from the shooting and spectral methods compared with experimental data of Kachanov & Levchenko (1984) for $Re_\delta = 608$, $F = 124$, and $b = \beta Re^{-1} \times 10^3 = 33$. Equation (9.22) shows the relationship between the disturbance profiles for the subharmonic mode and the eigenfunctions, namely

$$u_3 = B\cos\beta z[u_1 u_1^* + u_{-1} u_{-1}^*]^{1/2}, \tag{9.22}$$

where $*$ indicates a complex conjugate. Note that the secondary instability theory matches quite well the experimental data.

To compute the amplification of disturbances with downstream location, we use the following relationships

$$A = A_o e^{\int_{x_0}^{x_N} \alpha_i dx} \quad \text{and} \quad B = B_o e^{\int_{x_0}^{x_N} \sigma_r dx}. \tag{9.23}$$

The amplification of both the primary and secondary instabilities, as a function of the downstream distance, are shown in Fig. 9.4 for comparison of the theory versus the experiments.

Note that A_o and B_o are somewhat arbitrarily chosen to match the initial values of the experiments. Again, secondary instability theory shows that the assumption of rather insignificant amplification of the primary mode (shape assumption) with the explosive secondary disturbance amplification is justified and that the theory agrees well with the experiments.

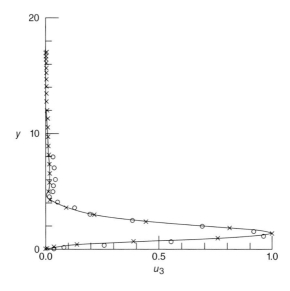

Figure 9.3 Comparison of the u_3 distribution (solid, x) of a subharmonic disturbance at $Re = 608$, $F = 124$, $b = 0.33$ with experimental data of Kachanov & Levchenko (1984).

The analysis is repeated for the fundamental or peak-peak alignment mode. Again, the Reynolds number $Re_\delta = 880$, frequency $F = \omega Re^{-1} \times 10^6 = 58.8$

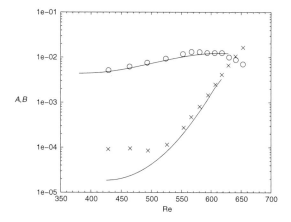

Figure 9.4 Amplitude growth with Reynolds number of the subharmonic mode (x) of a two-dimensional primary wave (0) over a rigid wall at $F = 124$, $A_o = 0.0044$, $B_o = 1.86 \times 10^{-5}$ and $b = \beta Re^{-1} = 0.33$. Symbols, Kachanov & Levchenko (1984); solid, theory.

Table 9.2 *Spectral convergence of temporal eigenvalues for the fundamental mode of secondary instability for $Re_\delta = 880$, $F = 58.8$, $A = 0.00695$, $\beta = 0.214$ and $\alpha = 0.15488 - i0.005504$.*

N	σ_1
20	0.00091129
25	0.00087073
30	0.00090107
35	0.00091794
40	0.00091743
45	0.00091831

and spanwise wavenumber $\beta = 0.214$ are selected. The primary wave has an amplitude $A = 0.00695$ and spatial wavenumber and growth rate of $\alpha = 0.15488 - i0.005504$. With the spectral method, convergence of the fundamental secondary instability eigenvalues are shown in Table 9.2 for the dominant mode.

In summary, the secondary instability theory discussed in this section has been shown to describe well the experimental amplification as a function of the downstream distance (Fig. 9.4). This theory represents a link between the two-dimensional Tollmien–Schlichting wave and the three-dimensional flow fields sketched in Figs. 9.1 and 9.2. The theory does tend to be less valid in the later nonlinear stages of breakdown. The difference between the experiments and theory in the highly nonlinear stages results from the assumptions made needed to obtain the simplified ordinary differential equations used to solve for the secondary instability modes. Specifically, the equations were linearized in the sense that all terms with A^2, AB and B^2 were neglected. Hence, secondary instability theory is a pseudo-nonlinear theory.

9.4 Resonant Wave Interactions

Numerous attempts have been made to explain the three-dimensional nature of boundary layer transition where the origins are with the evolution of predominately two-dimensional traveling wave instabilities. Nonresonant models have been proposed that attempt to link a composite of Orr–Sommerfeld modes into a rational model for the onset of the three-dimensional experimental observations. Prior to the discovery of secondary instability theory, Benney & Lin (1960) proposed a link between the two-dimensional mode $(\alpha, 0)$ and a pair of oblique waves $(\alpha, \pm\beta)$ that would explain the observation of streamwise vortices (Klebanoff, Tidstrom & Sargent, 1962) in a laminar boundary layer downstream of the two-dimensional amplifying traveling wave. The two-dimensional wave was of the form

$$u(x, y, t) = A u_{2D}(y) e^{i\alpha(x - c_{2D}t)} + c.c. \tag{9.24}$$

and the three-dimensional waves had the form

$$u(x, y, z, t) = \left[B u_{3D}(y) e^{i\alpha(x - c_{3D}t)} + c.c. \right] \cos(\beta z), \tag{9.25}$$

where A and B are the constant wave amplitudes, c_{2D} and c_{3D} are the phase velocities of the waves and were assumed to be equal. Benney & Lin (1960) found that a secondary flow was generated by this wave interaction and was proportional to $BA \cos(\beta z)$, $BA \sin(\beta z)$, $BB \cos(2\beta z)$ and $BB \sin(2\beta z)$. The terms that form the secondary flow make up streamwise vorticity and qualitatively show viability as to how streamwise vortices develop as found in experimentally observed boundary layers.

Stuart (1960) argued against this theory by noting a flaw in the assumptions made by Benney & Lin (1960) for describing the streamwise vortices. Whereas the Benney–Lin model assumed the two- and three-dimensional wave speeds were equivalent, Stuart noted that, for the Blasius flow, the real parts of the wave speeds differ by as much as 15%. A component of the flow generated by a nonlinear interaction of the fundamental modes make for a frequency of 1/6th or 1/7th of the fundamental. Stuart then argued that the oscillatory terms associated with $c_{2D} \neq c_{3D}$ undergo a slow phase change relative to the fundamental mode. This phase change can have a reinforcing effect on the fundamental modes and streamwise vorticity.

Later, Craik (1971) proposed a model that considered that a triad resonance occurs between the Tollmien–Schlichting wave $(\alpha, 0)$ and a pair of oblique subharmonic waves $(\alpha/2, \pm\beta)$, where the oblique waves have twice the wavelength of the two-dimensional Tollmien–Schlichting wave. Resonance occurs only if the phase velocity of the two-dimensional Tollmien–Schlichting wave matches the phase velocity of the oblique wave pair. Resonant triads of this type have the same critical layer and hence the potential for powerful interactions can lead to amplification. A main feature of Craik's analysis was the inclusion of the nonlinear terms of the Navier–Stokes equations simplified by using the linear analysis. This system is weakly nonlinear. Craik showed that the weakly nonlinear system can undergo an explosive (infinite) growth. Here we will use the more compact derivation of the weakly nonlinear theory of Craik (1971) as summarized somewhat differently by Nayfeh and Bozatli (1979, 1980) and Nayfeh (1987).

For the purpose of this text, the equations for the amplitudes of the resonant triad system follows Nayfeh and Bozatli (1979, 1980). For simplicity, we assume a two-dimensional mean flow field. The two-dimensional mean flow is assumed to be "slightly" nonparallel and is expressed as

$$U = U_0(x_1, y), \quad V = \varepsilon V_0(x_1, y), \quad P = P_0(x_1), \qquad (9.26)$$

where ε is a small dimensionless parameter characterizing the growth of the mean flow and $x_1 = \varepsilon x$. The instantaneous velocities and pressure are

$$
\left.
\begin{aligned}
u(x,y,z,t) &= U_0(x_1,y) + u'(x,y,z,t), \\
v(x,y,z,t) &= \varepsilon V_0(x_1,y) + v'(x,y,z,t), \\
w(x,y,z,t) &= w'(x,y,z,t), \\
p(x,y,z,t) &= P_0(x_1) + p'(x,y,z,t).
\end{aligned}
\right\}
\qquad (9.27)
$$

Substituting (9.26) into the Navier–Stokes equations (2.34) to (2.37) yields the following equations that include nonlinear and nonparallel effects

$$\frac{\partial u'}{\partial x} + \frac{\partial v'}{\partial y} + \frac{\partial w'}{\partial z} = 0, \qquad (9.28)$$

$$\frac{\partial u'}{\partial t} + U_0 \frac{\partial u'}{\partial x} + v' \frac{\partial U_0}{\partial y} + \frac{\partial p'}{\partial x} - Re^{-1}\left(\frac{\partial^2 u'}{\partial x^2} + \frac{\partial^2 u'}{\partial^2 y} + \frac{\partial^2 u'}{\partial^2 z}\right)$$

$$= -\varepsilon\left(u' \frac{\partial U_0}{\partial x_1} + V_0 \frac{\partial u'}{\partial y}\right) - u' \frac{\partial u'}{\partial x} - v' \frac{\partial u'}{\partial y} - w' \frac{\partial u'}{\partial z}, \qquad (9.29)$$

$$\frac{\partial v'}{\partial t} + U_o \frac{\partial v'}{\partial x} + \frac{\partial p'}{\partial y} - Re^{-1}\left(\frac{\partial^2 v'}{\partial x^2} + \frac{\partial^2 v'}{\partial^2 y} + \frac{\partial^2 v'}{\partial^2 z}\right) - \varepsilon^2 u' \frac{\partial V_o}{\partial x_1}$$

$$= -\varepsilon\left(V_o \frac{\partial v'}{\partial y} + v' \frac{\partial V_o}{\partial y}\right) - u' \frac{\partial v'}{\partial x} - v' \frac{\partial v'}{\partial y} - w' \frac{\partial v'}{\partial z}, \qquad (9.30)$$

and

$$\frac{\partial w'}{\partial t} + U_o \frac{\partial w'}{\partial x} + \frac{\partial p'}{\partial z} - Re^{-1}\left(\frac{\partial^2 w'}{\partial x^2} + \frac{\partial^2 w'}{\partial^2 y} + \frac{\partial^2 w'}{\partial^2 z}\right)$$

$$= -\varepsilon V_o \frac{\partial w'}{\partial y} - u' \frac{\partial w'}{\partial x} - v' \frac{\partial w'}{\partial y} - w' \frac{\partial w'}{\partial z}, \qquad (9.31)$$

with boundary conditions

$$u',v',w' = 0 \quad \text{at} \quad y = 0 \qquad \text{and} \qquad u',v',w' \to 0 \quad \text{at} \quad y \to \infty. \qquad (9.32)$$

Introducing the parameter ε_1 to characterize the amplitude of the small but finite disturbance, both the nonlinear and the nonparallel contributions can be accounted for in one form with $\varepsilon_1 = \gamma\varepsilon$. If $\varepsilon_1 \ll \varepsilon$, the nonlinear effects are small compared with the nonparallel effects; if $\varepsilon_1 \gg \varepsilon$ the opposite is true. γ is taken as unity. The method of multiple scales is used to expand the disturbance quantities. Retaining only the first two terms of the expansion gives

$$\begin{aligned} u'(x,y,z,t) &= \varepsilon u_1(x_o,x_1,y,z_o,z_1,t_o,t_1) + \varepsilon^2 u_2(x_o,x_1,y,z_o,z_1,t_o,t_1), \\ v' &= \varepsilon v_1 + \varepsilon^2 v_2 \quad w' = \varepsilon w_1 + \varepsilon^2 w_2 \quad p' = \varepsilon p_1 + \varepsilon^2 p_2, \end{aligned} \right\} \qquad (9.33)$$

where $x_o = x$, $x_1 = \varepsilon x$, $z_o = z$, $z_1 = \varepsilon z$, $t_o = t$ and $t_1 = \varepsilon t$. Note the following derivative relationship (van Dyke, 1975)

$$\frac{\partial u'}{\partial x} = \varepsilon\left[\frac{\partial u_1}{\partial x_o} + \varepsilon \frac{\partial u_1}{\partial x_1}\right] + \varepsilon^2\left[\frac{\partial u_2}{\partial x_o} + \varepsilon \frac{\partial u_2}{\partial x_1}\right]. \qquad (9.34)$$

Substituting the multiple scale approximation (9.33) into (9.28)–(9.32) and equating coefficients of ε, we obtain the following

$O(\varepsilon)$: First-order (quasi-parallel) equations:

$$\mathbf{L_1}(u_1,v_1,w_1) = \frac{\partial u_1}{\partial x_o} + \frac{\partial v_1}{\partial y} + \frac{\partial w_1}{\partial z_o} = 0, \qquad (9.35)$$

$$\mathbf{L_2}(u_1,v_1,w_1,p_1) = \frac{\partial u_1}{\partial t_o} + U_o \frac{\partial u_1}{\partial x_o} + v_1 \frac{\partial U_o}{\partial y}$$

$$+ \frac{\partial p_1}{\partial x_o} - Re^{-1}\left(\frac{\partial^2 u_1}{\partial^2 x_o} + \frac{\partial^2 u_1}{\partial^2 y} + \frac{\partial^2 u_1}{\partial^2 z_o}\right) = 0, \quad (9.36)$$

$$\mathbf{L_3}(u_1,v_1,w_1,p_1) = \frac{\partial v_1}{\partial t_0} + U_0\frac{\partial v_1}{\partial x_0}$$

$$+\frac{\partial p_1}{\partial y} - Re^{-1}\left(\frac{\partial^2 v_1}{\partial^2 x_0} + \frac{\partial^2 v_1}{\partial^2 y} + \frac{\partial^2 v_1}{\partial^2 z_0}\right) = 0, \quad (9.37)$$

$$\mathbf{L_4}(u_1,v_1,w_1,p_1) = \frac{\partial w_1}{\partial t_0} + U_0\frac{\partial w_1}{\partial x_0}$$

$$+\frac{\partial p_1}{\partial z_0} - Re^{-1}\left(\frac{\partial^2 w_1}{\partial^2 x_0} + \frac{\partial^2 w_1}{\partial^2 y} + \frac{\partial^2 w_1}{\partial^2 z_0}\right) = 0. \quad (9.38)$$

The boundary equations for the quasi-parallel equations are

$$u_1,v_1,w_1 = 0 \quad \text{at} \quad y = 0 \quad \text{and} \quad u_1,v_1,w_1 \to 0 \quad \text{as} \quad y \to \infty. \quad (9.39)$$

The quasi-parallel problem is simply that described by the Orr–Sommerfeld and Squire equations.

$O(\varepsilon^2)$: Second-order equations:

$$\mathbf{L_1}(u_2,v_2,w_2) = -\frac{\partial u_1}{\partial x_1} - \frac{\partial w_1}{\partial z_1} = 0, \quad (9.40)$$

$$\mathbf{L_2}(u_2,v_2,w_2,p_2) = -\frac{\partial u_1}{\partial t_1} - U_0\frac{\partial u_1}{\partial x_1} - \frac{\partial p_1}{\partial x_1} + 2Re^{-1}\frac{\partial^2 u_1}{\partial x_0\partial x_1} + 2Re^{-1}\frac{\partial^2 u_1}{\partial z_0\partial z_1}$$

$$-u_1\frac{\partial U_0}{\partial x_1} - V_0\frac{\partial u_1}{\partial y} - u_1\frac{\partial u_1}{\partial x_0} - v_1\frac{\partial u_1}{\partial y} - w_1\frac{\partial u_1}{\partial z_0}$$

$$= 0, \quad (9.41)$$

$$\mathbf{L_3}(u_2,v_2,w_2,p_2) = -\frac{\partial v_1}{\partial t_1} - U_0\frac{\partial v_1}{\partial x_1} + 2Re^{-1}\frac{\partial^2 v_1}{\partial x_0\partial x_1} + 2Re^{-1}\frac{\partial^2 v_1}{\partial z_0\partial z_1}$$

$$-v_1\frac{\partial V_0}{\partial y} - V_0\frac{\partial v_1}{\partial y} - u_1\frac{\partial v_1}{\partial x_0} - v_1\frac{\partial v_1}{\partial y} - w_1\frac{\partial v_1}{\partial z_0}$$

$$= 0, \quad (9.42)$$

$$\mathbf{L_4}(u_2,v_2,w_2,p_2) = -\frac{\partial w_1}{\partial t_1} - U_0\frac{\partial w_1}{\partial x_1} - \frac{\partial p_1}{\partial z_1} + 2Re^{-1}\frac{\partial^2 w_1}{\partial x_0\partial x_1} + 2Re^{-1}\frac{\partial^2 w_1}{\partial z_0\partial z_1}$$

$$-V_0\frac{\partial w_1}{\partial y} - u_1\frac{\partial w_1}{\partial x_0} - v_1\frac{\partial w_1}{\partial y} - w_1\frac{\partial w_1}{\partial z_0}$$

$$= 0, \quad (9.43)$$

The boundary equations for the second-order problem are

$$u_2,v_2,w_2 = 0 \quad \text{at} \quad y = 0 \quad \text{and} \quad u_2,v_2,w_2 \to 0 \quad \text{as} \quad y \to \infty. \quad (9.44)$$

We can use a linear summation of Tollmien–Schlichting waves to represent
these disturbances. Here, we use three waves generalized as three-dimensional
waves. Of course, for two-dimensional waves, the corresponding spanwise
wavenumber is zero. Thus

$$
\left.
\begin{aligned}
u_1 &= A_1(x_1,z_1,t_1)\hat{u}_1(x_1,y)e^{i\theta_1} + A_2(x_1,z_1,t_1)\hat{u}_2(x_1,y)e^{i\theta_2} \\
&\quad + A_3(x_1,z_1,t_1)\hat{u}_3(x_1,y)e^{i\theta_3} + cc, \\
v_1 &= A_1\hat{v}_1 e^{i\theta_1} + A_2\hat{v}_2 e^{i\theta_2} + A_3\hat{v}_3 e^{i\theta_3} + c.c., \\
w_1 &= A_1\hat{w}_1 e^{i\theta_1} + A_2\hat{w}_2 e^{i\theta_2} + A_3\hat{w}_3 e^{i\theta_3} + c.c., \\
p_1 &= A_1\hat{p}_1 e^{i\theta_1} + A_2\hat{p}_2 e^{i\theta_2} + A_3\hat{p}_3 e^{i\theta_3} + c.c.,
\end{aligned}
\right\} \qquad (9.45)
$$

where *c.c.* is the complex conjugate and

$$
\theta_n = \int \alpha_n dx_0 - \omega_n t_0 + \beta_n z_0,
$$

$$
\frac{\partial \theta_n}{\partial x_0} = \alpha_n, \quad \frac{\partial \theta_n}{\partial z_0} = \beta_n \quad \text{and} \quad \frac{\partial \theta_n}{\partial t_0} = -\omega.
$$

As before, α_n are complex streamwise wavenumbers, β_n are complex spanwise
wavenumbers and ω_n are frequencies.

Substituting (9.45) into (9.35)–(9.39) leads to

$$
\left.
\begin{aligned}
\mathbf{M_1}(\hat{u}_n, \hat{v}_n, \hat{w}_n) &= i\alpha_n \hat{u}_n + \frac{\partial \hat{v}_n}{\partial y} + i\beta_n \hat{w}_n = 0, \\
\mathbf{M_2}(\hat{u}_n, \hat{v}_n, \hat{w}_n) &= i(\alpha_n U_0 - \omega_n)\hat{u}_n + \hat{v}_n \frac{\partial U_0}{\partial y} + i\alpha_n \hat{p}_n \\
&\quad - Re^{-1}\left(\frac{\partial^2}{\partial^2 y} - \alpha^2 - \beta^2\right)\hat{u}_n = 0, \\
\mathbf{M_3}(\hat{u}_n, \hat{v}_n, \hat{w}_n) &= i(\alpha_n U_0 - \omega_n)\hat{v}_n + \frac{\partial \hat{p}_n}{\partial y} \\
&\quad - Re^{-1}\left(\frac{\partial^2}{\partial^2 y} - \alpha^2 - \beta^2\right)\hat{v}_n = 0, \\
\mathbf{M_4}(\hat{u}_n, \hat{v}_n, \hat{w}_n) &= i(\alpha_n U_0 - \omega_n)\hat{w}_n + i\beta_n \hat{p}_n \\
&\quad - Re^{-1}\left(\frac{\partial^2}{\partial^2 y} - \alpha^2 - \beta^2\right)\hat{w}_n = 0
\end{aligned}
\right\}
$$

with boundary conditions

$$
\hat{u}_n, \hat{v}_n, \hat{w}_n = 0 \quad \text{at} \quad y = 0 \quad \text{and} \quad \hat{u}_n, \hat{v}_n, \hat{w}_n \to 0 \quad \text{as} \quad y \to \infty. \quad (9.46)
$$

Similar to the linear stability equations used in the previous chapters, A_n are
indeterminable because the first-order equations are linear; however, using the
second-order equations with a solvability condition, these amplitudes can be
established.

For the second-order equations, (9.45) and a wave form for (u_2, v_2, w_2, p_2) are substituted into the second-order equations (9.40)–(9.44). The right-hand side of the resulting equations contain terms with

$$e^{\pm 2i\theta_n}, \qquad e^{\pm i(\theta_n + \theta_m)}, \qquad e^{\pm i(\theta_n - \theta_m)}, \qquad e^{\pm i\theta_n}.$$

A combination of resonances can occur for these waves with

$$\theta_n - \theta_m = \theta_o, \qquad \theta_n + \theta_m = \theta_o$$

for some combination of n, m, o waves.

Assume the triad wave for the second-order problem for

$$
\begin{aligned}
u_2(x_1, y_1, z_1, t_1) &= \tilde{u}_1(x_1, y_1, z_1, t_1)e^{i\theta_1} + \tilde{u}_2(x_1, y_1, z_1, t_1)e^{i\theta_2} \\
&\quad + \tilde{u}_3(x_1, y_1, z_1, t_1)e^{i\theta_3} + c.c., \\
v_2 &= \tilde{v}_1 e^{i\theta_1} + \tilde{v}_2 e^{i\theta_2} + \tilde{v}_3 e^{i\theta_3}, \\
w_2 &= \tilde{w}_1 e^{i\theta_1} + \tilde{w}_2 e^{i\theta_2} + \tilde{w}_3 e^{i\theta_3}, \\
p_2 &= \tilde{p}_1 e^{i\theta_1} + \tilde{p}_2 e^{i\theta_2} + \tilde{p}_3 e^{i\theta_3},
\end{aligned}
\right\} \tag{9.47}
$$

After substituting (9.45) and (9.47) into the second-order equations (9.40)–(9.44), the following equations are found:

$$
\begin{aligned}
\mathbf{M_1}(\hat{u}_n, \hat{v}_n, \hat{w}_n) &= a_n, \\
\mathbf{M_2}(\hat{u}_n, \hat{v}_n, \hat{w}_n) &= b_n, \\
\mathbf{M_3}(\hat{u}_n, \hat{v}_n, \hat{w}_n) &= c_n, \\
\mathbf{M_4}(\hat{u}_n, \hat{v}_n, \hat{w}_n) &= d_n.
\end{aligned}
\right\} \tag{9.48}
$$

The right-hand sides, a_n, b_n, c_n, d_n, will be derived as part of the exercise at the end of the chapter. The boundary conditions are

$$\tilde{u}_n, \tilde{v}_n, \tilde{w}_n = 0 \quad \text{at} \quad y = 0 \qquad \text{and} \qquad \tilde{u}_n, \tilde{v}_n, \tilde{w}_n \to 0 \quad \text{as} \quad y \to \infty. \tag{9.49}$$

The second-order system (9.48)–(9.49) have a nontrivial solution. These inhomogeneous equations have a solution if the inhomogeneous parts are orthogonal to every solution of the adjoint homogeneous problem, or

$$\int_0^\infty (a_n \hat{u}_n^* + b_n \hat{v}_n^* + c_n \hat{w}_n^* + d_n \hat{p}_n^*)\,dy = 0, \tag{9.50}$$

where the adjoint equations are

$$
\begin{aligned}
\mathbf{M_1}(\hat{u}_n^*, \hat{v}_n^*, \hat{w}_n^*) &= i\alpha_n \hat{u}_n^* + \frac{\partial \hat{v}_n^*}{\partial y} + i\beta_n \hat{w}_n^* = 0, \\
\mathbf{M_2}(\hat{u}_n^*, \hat{v}_n^*, \hat{w}_n^*) &= i(\alpha_n U_0 - \omega_n)\hat{u}_n^* + \hat{v}_n^* \frac{\partial U_0}{\partial y} + i\alpha_n \hat{p}_n^* \\
&\quad - Re^{-1}\left(\frac{\partial^2}{\partial^2 y} - \alpha^2 - \beta^2 \right)\hat{u}_n^* = 0, \\
\mathbf{M_3}(\hat{u}_n^*, \hat{v}_n^*, \hat{w}_n^*) &= i(\alpha_n U_0 - \omega_n)\hat{v}_n^* + \frac{\partial \hat{p}_n^*}{\partial y} \\
&\quad - Re^{-1}\left(\frac{\partial^2}{\partial^2 y} - \alpha^2 - \beta^2 \right)\hat{v}_n^* = 0, \\
\mathbf{M_4}(\hat{u}_n^*, \hat{v}_n^*, \hat{w}_n^*) &= i(\alpha_n U_0 - \omega_n)\hat{w}_n^* + i\beta_n \hat{p}_n^* \\
&\quad - Re^{-1}\left(\frac{\partial^2}{\partial^2 y} - \alpha^2 - \beta^2 \right)\hat{w}_n^* = 0,
\end{aligned}
\right\} \tag{9.51}
$$

with boundary conditions

$$
\hat{u}_n^*, \hat{v}_n^*, \hat{w}_n^* = 0 \quad \text{at} \quad y = 0 \qquad \text{and} \qquad \hat{u}_n^*, \hat{v}_n^*, \hat{w}_n^* \to 0 \quad \text{as} \quad y \to \infty. \tag{9.52}
$$

Finally, use

$$
\hat{A}_n = A_n e^{-\int \alpha_{n,i} dx_0 - \beta_{n,i} dz_0} \tag{9.53}
$$

to obtain the final system of equations that can be used to obtain various resonant conditions. This lengthy derivation of the final system will be left to the student in the exercises.

Craik showed that, in finite time, the amplitudes become infinite. The experimental results do not, however, show this explosive instability amplification for perfectly tuned triad resonance. Unlike the Benney–Lin model, the current resonant triad model estimates a preferred spanwise wavelength and streamwise vorticity.

9.5 PSE Theory

Herbert (1991, 1997) and Bertolotti (1992) developed a theory, now called the Parabolized Stability Equations (PSE), that sought approximate solutions to the unsteady Navier–Stokes equations by invoking a parabolic nature to the equations. Then, when parabolic, numerical solutions can be obtained by an efficient marching procedure. Generally, there are a number of ways to parabolize the Navier–Stokes equations. However, any acceptable approximation must be able to capture the physics of instability waves. The underlying notion of the PSE approach is to first decompose the disturbance into an oscillatory wave and a shape function. By properly choosing a streamwise wave number to resolve the wave motion, the governing equations reduce to a set of partial differential equations for the shape functions which vary slowly in the streamwise direction and their second derivatives are assumed negligible. These partial differential equations can be parabolized by neglecting the dependence of convected disturbances on downstream events and by neglecting the second derivatives $(\partial^2/\partial x^2)$ of the shape functions. Since most of the oscillatory wave motion is absorbed in the streamwise wavenumber and the terms neglected in the shape function equations are of order Re^{-2}, the resulting system provides the desired results. A brief discussion of the theory is as follows.

For disturbances that are present in the flow field, periodicity is assumed both in time and in the spanwise direction. The total disturbance can then be described by the following Fourier-series expansion:

$$\left\{\begin{array}{c} u \\ v \\ p \end{array}\right\}(x,y,z,t) = \sum_{n=-N_z}^{N_z}\sum_{m=-N_t}^{N_t} \left\{\begin{array}{c} \hat{u}_{m,n} \\ \hat{v}_{m,n} \\ \hat{p}_{m,n} \end{array}\right\}(x,y)e^{i(n\beta z - m\omega t)}, \qquad (9.54)$$

where N_z and N_t are the numbers of modes retained in the truncated series, ω is an imposed frequency and β is an imposed spanwise wavenumber. The disturbance form (9.54) is substituted into the Navier–Stokes equations so that a set of elliptic equations for the transformed variables $\{\hat{u}_{m,n}, \hat{v}_{m,n}, \hat{p}_{m,n}\}$ results. Owing to the wave nature of these transformed variables, a further decomposition is made into a fast oscillatory wave part and a slowly varying shape function:

$$\left\{\begin{array}{c} \hat{u}_{m,n} \\ \hat{v}_{m,n} \\ \hat{p}_{m,n} \end{array}\right\} = \left\{\begin{array}{c} u_{m,n} \\ v_{m,n} \\ p_{m,n} \end{array}\right\}(y)e^{i\int_{x_0}^{x}\alpha_{m,n}dx}. \qquad (9.55)$$

In (9.55), the fast-scale variation along the streamwise direction x is represented by the streamwise wavenumber $\alpha_{m,n}$ and therefore the second-order variation of the shape function in x is negligible. In turn, this observation leads to the desired parabolized stability equations for the shape functions $\{u_{m,n}, v_{m,n}, p_{m,n}\}$. These equations are obtained by neglecting all second derivatives in the streamwise direction and the terms associated with upstream influence. Similar to the Orr–Sommerfeld and Squire equations, pressure can be eliminated by taking the curl of the Navier–Stokes equations. The resulting governing equations take the form of two equations that are given in matrix notation as

$$\mathbf{L_u}u_{m,n} + \mathbf{L_v}v_{m,n} + \mathbf{M_u}\frac{du_{m,n}}{dx} + \mathbf{M_v}\frac{dv_{m,n}}{dx} + \frac{d\alpha_{m,n}}{dx}(\mathbf{F_u}u_{m,n} + \mathbf{F_v}v_{m,n}) = 0,$$
$$(9.56)$$

where $\alpha_{m,n}$ is the complex wavenumber for mode m,n, composed of a real part describing the growth rate and an imaginary part describing the wavenumber, the operators $\mathbf{L},\mathbf{M},\mathbf{F}$ depend on $\alpha_{m,n}$, the frequency ω and containing derivatives only in y. The operator $\mathbf{L}$ contains the Orr–Sommerfeld and Squire operators that are well known in parallel flow stability theory. The term $F_{m,n}$ is the convolution that stems from the nonlinear products.

The matrices for the first equation are given by

$$\mathbf{L_u} = -2\alpha_{m,n}U_{xy} - 2\alpha_{m,n}U_x\frac{d}{dy},$$

$$\mathbf{L_v} = -Re^{-1}\frac{d^4}{dy^4} + V\frac{d^3}{dy^3} + [\alpha_{m,n}U - i\omega - \frac{2}{Re}(\alpha_{m,n}^2 - \beta^2) - U_x]\frac{d^2}{dy^2},$$

$$\mathbf{M_u} = -2U_x\frac{d}{dy} - 2U_{xy},$$

$$\mathbf{M_v} = -4\alpha Re^{-1}\frac{d^2}{dy^2} + 2\alpha V\frac{d}{dy} + U\frac{d^2}{dy^2} - 2i\alpha\omega + 3\alpha^2 U - \beta^2 U$$
$$-4\alpha Re^{-1}(\alpha^2 - \beta^2) - U_y + 2\alpha U_x,$$

$$\mathbf{F_u} = 0,$$

$$\mathbf{F_v} = -2iRe^{-1}\frac{d^2}{dy^2} + iV\frac{d}{dy} + \omega + i\alpha U - 6iRe^{-1}(\alpha^2 - \beta^2) - 2\alpha + iU_x.$$

For the second equation, the matrices take the form

$$\mathbf{L_u} = -Re^{-1}(\alpha^2 - \beta^2)\frac{d^2}{dy^2} + V(\alpha^2 - \beta^2)\frac{d}{dy}$$
$$+ (\alpha^2 - \beta^2)[\alpha U - i\omega - \frac{1}{Re}(\alpha^2 - \beta^2) - V_y],$$

$$\mathbf{L_v} = -\alpha Re^{-1}\frac{d^3}{dy^3} + \alpha V\frac{d^2}{dy^2} + \alpha[\alpha U - i\omega$$
$$- \frac{1}{Re}(\alpha^2 - \beta^2) - V_y]\frac{d}{dy} - U'\beta^2,$$

$$\mathbf{M_u} = -2\alpha Re^{-1}\frac{d^2}{dy^2} + 2\alpha V\frac{d}{dy} - 2i\alpha\omega + 3\alpha^2 U - U\beta^2$$
$$-4\alpha Re^{-1}(\alpha^2 - \beta^2) - 2\alpha V_y,$$

$$\mathbf{M_v} = -Re^{-1}\frac{d^3}{dy^3} + V\frac{d^2}{dy^2} + [2\alpha U - i\omega - Re^{-1}(3\alpha^2 - \beta^2) - V_y - U_y\beta^2]\frac{d}{dy},$$

$$\mathbf{F_u} = -iRe^{-1}\frac{d^2}{dy^2} + iV\frac{d}{dy} + \omega + 3i\alpha U - iRe^{-1}(5\alpha^2 - \beta^2) - iV_y,$$

$$\mathbf{F_v} = iU\frac{d}{dy} - Re^{-1}2i\alpha.$$

In order to solve the nonlinear problem, the nonlinear convection terms are placed on the right-hand side of the governing equation

$$\underline{F}(x,y,z,t) = (\underline{u}\cdot\nabla)\underline{u}. \tag{9.57}$$

For the PSE approach, the governing equations are solved in wavenumber space. The Fourier coefficients, obtained from the corresponding Fourier transform of $\underline{F}$ in (9.57), provide a nonlinear forcing for each of the linearized shape function equations. These inhomogeneous equations for the shape functions are solved by applying a marching procedure along the streamwise direction for each Fourier mode. If $\alpha_{n,m}$ are chosen (or computed) properly, the evolution of disturbances can be described by the parabolized equations for the shape functions. Equations (9.57) can then be marched in x using an Euler differencing for the x-derivative terms.

For the PSE theory boundary conditions, Bertolotti & Joslin (1995) showed that either asymptotic, Dirichlet, Neumann or mixed boundary conditions can be imposed at various distances from the wall. These are

Dirichlet conditions:

$$u_{m,n} = 0, \qquad v_{m,n} = 0, \qquad \frac{\partial v_{m,n}}{\partial y} = 0; \tag{9.58}$$

Neumann conditions:

$$\frac{\partial u_{m,n}}{\partial y} = 0, \qquad \frac{\partial v_{m,n}}{\partial y} = 0, \qquad \frac{\partial^2 v_{m,n}}{\partial y^2} = 0; \tag{9.59}$$

Mixed conditions:

$$\left.\begin{array}{c} \dfrac{\partial u_{m,n}}{\partial y} + \alpha_{m,n}u_{m,n} = 0, \qquad \dfrac{\partial v_{m,n}}{\partial y} + \alpha_{m,n}v_{m,n} = 0, \\[2mm] \dfrac{\partial^2 v_{m,n}}{\partial y^2} + \alpha_{m,n}\dfrac{\partial v_{m,n}}{\partial y} = 0; \end{array}\right\} \tag{9.60}$$

Asymptotic conditions:

$$\underline{x}_{m,n}\cdot\underline{B}^T\underline{e}_i = \underline{c}_{m,n}\cdot\underline{e}_i, \qquad i = 1,2,3, \tag{9.61}$$

where the conditions for the highest derivative of $\hat{v}$ with respect to y are derived from the continuity equation. The mixed boundary conditions and the asymptotic boundary conditions are altered for the mean flow distortion term (i.e., $m = 0, n = 0$) to form

$$u_{0,n} = 0, \qquad \frac{\partial v_{0,n}}{\partial y} = 0, \qquad \frac{\partial^2 v_{0,n}}{\partial y^2} = 0. \qquad (9.62)$$

To close the problem, a relationship for updating α must be obtained. One such relationship is given by

$$\alpha_{m,n}^{k+1} = \alpha_{m,n}^k - i \int u u_x^* dy \bigg/ \int u u^* dy, \qquad (9.63)$$

where $*$ refers to complex conjugates and k implies the level of iteration.

The solution sequence is as follows:

(1) The x-derivative terms in equations (9.56) with the appropriate boundary conditions are first-order backward Euler differenced (e.g., $\partial u / \partial x = (u_{i+1} - u_i)/dx$, where the i solutions are known and dx is the step size in the x-direction).

(2) To start the solution procedure, the initial solutions (profiles and wavenumber α) at $i = 1$ are obtained from linear stability theory.

(3) The solutions at $i + 1$ are obtained by iterating on (9.56) and (9.63) until the solution no longer changes with continued iteration. Then, depending on the algorithm, convergence can be obtained in three to four iterations.

Some sample two-dimensional PSE theory results are taken from Bertolotti & Joslin (1995). Here, the reference length is $\delta(x_0) = \sqrt{v x_0 / U_\infty}$ that is defined at the streamwise location x_0. The corresponding Reynolds number at x_0 is $Re_0 = U_\infty \delta(x_0)/v = 400$. The nondimensional frequency (Strouhal number) of the two-dimensional Tollmien–Schlichting wave is $F = 2 \times 10^6 \pi f v / U_\infty^2 = 86$, and leads to $\omega = 0.0344$.

The initial condition was composed of the single Fourier mode $m = 1$ obtained from the Orr–Sommerfeld equation. The arbitrary initial amplitude was selected to be 0.25% rms, based on the maximum of the u component of velocity.

At the wall, no-slip boundary conditions (i.e., homogeneous Dirichlet conditions) are enforced for the disturbance equations used by direct numerical simulations (DNS; see Chapter 11) and the Fourier-coefficient equations used by PSE theory. So, both DNS and PSE incorporate the same boundary conditions at the wall.

In the far-field, homogeneous Dirichlet boundary conditions are imposed for the DNS computations. This far-field condition is exact at infinity but, to computationally solve the system using DNS, the semi-infinite domain is truncated. For the PSE approach, homogeneous Dirichlet boundary conditions are used for all Fourier-coefficient equations except for the mean flow distortion equations. This nonzero component arises from assumptions of PSE theory. Unlike the DNS that solves the full Navier–Stokes equations, PSE theory reduces the equations to a simplified parabolic system in a manner as described in the previous subsection. As a result of the PSE simplification, the mean flow distortion equations are essentially of the boundary layer type. With boundary layer equations, the wall-normal velocity component approaches a constant in the far-field. Similarly, the mean flow distortion equation in PSE theory, that is the boundary layer equation type, incorporates a Neumann boundary condition for the wall-normal velocity component. This Neumann condition allows the total normal velocity (mean flow + mean flow correction) in the far-field as predicted by PSE theory to vary at infinity. Thus, the far-field boundary conditions, used by both the DNS and PSE approaches, are approximate. The spatial DNS far-field boundary conditions cannot be changed to mimic the PSE approach because the DNS cannot accommodate an a priori Fourier modal boundary condition treatment as is present in PSE theory. When the Neumann condition is changed to that of Dirichlet, numerical instabilities in the PSE approach are generated. As a result, the far-field boundary conditions for PSE theory are different from the boundary conditions employed for the DNS approach for the mean flow distortion equation. With the present boundary conditions, the PSE theory approximate far-field conditions should prove more accurate for a far-field boundary fixed close to the wall. The DNS conditions should prove more realistic for the boundary far from the wall.

Figure 9.5 shows the evolution of the disturbance amplitude based on the u component of velocity for the Fourier modes $F = 1$, $F = 2$ and their steady component $F = 0$ with a Reynolds number of $Re = U_\infty \delta(x)/v = \sqrt{xRe_0}$ for results calculated by both the PSE and DNS codes. Both codes enforced the Dirichlet boundary conditions (9.58) at $y_{max} = 130$. The results agree well, indicating a reasonable equivalence of the two procedures for the flat-plate boundary layer problem.

Computations were conducted with PSE theory to compare the maximum amplitudes of $F = 0$ and $F = 1$ modes as functions of far-field boundary locations. At the downstream location, that corresponds to $R = 940$ (near the maximum amplification amplitude of the $F = 1$ mode), Table 9.3 shows the variation in the results by simply altering the boundary conditions. The exact values are 0.595% for $F = 0$ and 2.843% for $F = 1$. The results indicate that asymptotic and mixed boundary conditions yield the most accurate mean

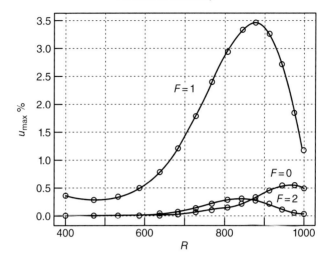

Figure 9.5 Amplitude of $F = 0$, $F = 1$ and $F = 2$ modes from PSE (solid) and DNS (symbols) with Dirichlet boundary conditions (after Bertolotti & Joslin, 1995).

Table 9.3 *Modal maximums (in percent of U_∞) at Re = 940.*

BC Type	$u_0(\text{max})$	$u_1(\text{max})$
Asymptotic	0.596	2.844
Mixed	0.598	2.858
Neumann	0.684	3.807
Dirichlet	0.298	1.895

flow distortion and unsteady instability modes in comparison with the results obtained with either Dirichlet or Neumann conditions.

The use of a finite domain in y plus Dirichlet and Neumann boundary conditions eliminates some coding difficulties in direct Navier–Stokes simulation codes, but at the cost of error introduction. As in the case considered here, the errors are small when the truncation location y_{max} is located well into the region of exponential decay of the disturbances. Exceptions are for the steady component $F = 0$, that does not decay in the freestream and for which the error introduced by the use of Dirichlet conditions does not vanish as y_{max} is increased. A similar error is also expected for three-dimensional steady disturbances because they decay slowly (i.e., as in $\exp(-\beta^2 y)$) in the freestream. On the other hand, the errors introduced in the calculation of traveling modes by either Dirichlet or Neumann conditions are negligible if a truncation location y_{max} is chosen sufficiently far from the plate. By contrast, asymptotic boundary conditions and mixed boundary conditions yield accurate results when imposed beyond the 99.99% definition of the boundary layer edge. The asymptotic conditions are exact but require a significantly greater amount of coding to implement.

9.6 Exercises

Exercise 9.1   Derive (9.3) and state all necessary conditions and assumptions.

Exercise 9.2   Derive (9.4) and state all necessary conditions and assumptions.

Exercise 9.3   Derive the secondary instability theory equations (9.14) and (9.15) using normal modes (9.18) of velocity (v_3) and vorticity (ω_3).

(a) Substitute the secondary disturbance form (9.18) into the continuity equation (9.8) and definition of vorticity (9.11).
(b) Substitute the secondary disturbance form (9.18) into the secondary instability equations (9.14) and (9.15).
(c) Use the normal mode equations obtained from the normal mode continuity equation and definitions of vorticity (part (a)) to reduce the equations (part (b)) to normal velocity and vorticity only.
(d) If $A = 0$ in the secondary instability equations, how do the terms that remain compare with the Orr–Sommerfeld and Squire equations?

Exercise 9.4   In the discussion of the secondary instability results of Section 9.3, why are the frequency $F = \omega/R \times 10^6$ and spanwise wavenumber $b = \beta/R \times 10^3$ used?

Exercise 9.5   By referring to equation (9.22) for the subharmonic profiles, derive the relationship between disturbance profile and eigenfunction for the fundamental mode profiles. (Note, use the $n = 0, \pm 2$ modes.)

Exercise 9.6   From Section 9.4, use equation (9.46) to show that you can obtain the Orr–Sommerfeld and Squire equations.

Exercise 9.7   Substituting equations (9.45) and (9.47) into the second-order equations (9.40)–(9.43) leads to equations (9.48). Determine the form of the right-hand sides a_n, b_n, c_n, d_n.

Exercise 9.8   Derive the PSE equations for two-dimensional disturbances evolving in a two-dimensional basic flow.

Exercise 9.9   Use the relation for $\hat{A}_n$ in equation (9.53) and the equation derived in the previous exercise to determine the final system from which resonant analysis can be performed.

Exercise 9.10   Write a numerical solver for the linear PSE system and verify that the code works for the following initial conditions and basic flows.

(a) Reynolds number based on displacement thickness ($R = 900$) and nondimensional frequency ($F = 86 = \omega/R \times 10^6$) and the profiles from the Orr–Sommerfeld equation and Blasius basic flows are used for the initial conditions.
(b) Compare the PSE results for parallel and nonparallel basic flows with the Orr Sommerfeld solution.
(c) Duplicate the results in Table 9.3.

Exercise 9.11   Write a numerical solver for the secondary instability equations assuming a two-dimensional mean flow. Validate the code for the problem of Table 9.1.

Exercise 9.12   Discuss the process you would follow using secondary instability theory to duplicate the results of Fig. 9.4.

10
Transition and Receptivity

*Under certain circumstances the steady motion becomes unstable, so
that an indefinitely small disturbance may lead to a change to the sinuous motion – it changes from steady to eddying motion.*

– Sir George Gabriel Stokes (1819–1903)

10.1 Introduction

In this chapter, we discuss the breakdown of hydrodynamic stability, a theory
that is initially characterized by a system of linear equations, as discussed in
great detail in Chapters 1–8. Breakdown thus implies that the linear assumption is becoming invalid and the flow now has several modes interacting and
amplifying. This interaction can then transfer energy to modes not yet dominant in the flow. The culmination of this breakdown process is a turbulent flow.
One might suppose that the characteristics of the breakdown stage depends on
the initial conditions – as receptivity – as well as freestream conditions such
as vorticity and freestream turbulence. Today, we understand much about this
initial stage and the linear amplification stage but have only limited knowledge
for the nonlinear processes of many flows (cf. Chapter 9) because the complete Navier–Stokes equations must be solved, and tracing measurements in
this stage back to their origin to ascertain the cause and effect is challenging.

The major goal of this text has been to present the subject of hydrodynamic
instability processes for many different engineering problems. The initial chapters demonstrated that this understanding can most often be achieved with linear systems. However, as was somewhat evident in Chapters 8 and 9, the transition from a laminar to turbulent flow is extremely complicated. This chapter
and the next will expose the reader to issues effecting hydrodynamic instabilities and the nonlinear breakdown of modes after linear growth, and we will

summarize a condensed history of methods that have been used to predict loss of laminar flow and onset of transition to turbulence.

10.2 Influence of Free Stream Turbulence and Receptivity

Both experiments and computations in the field of transitional flows are plagued by uncertainty, which comes primarily from external conditions. Consider a box or volume of space where hydrodynamic instabilities are under study, such as a test section for either experiments or computations. This test section has a laminar flow that is subject to disturbances. The boundary layer may ingest disturbances and, as a result, hydrodynamic instabilities may be induced or triggered. Such disturbances can take the form of acoustics, turbulent fluctuations or organized vorticity. Furthermore, the plate may have small roughness discontinuities or joints that stem from the many parts to form a single plate. Any and all of these can induce instability modes. In addition, computations must deal with one additional mode, namely, numerical round-off or loss of conservation in the governing equations. This artifact is inherent to various degrees in all computations. Furthermore, when attempting to compare numerical or theoretical solutions with the seemingly experimental counterpart, uncertainty in the differences between these external conditions must always be assessed.

Morkovin (1969) is usually given the credit for coining the term "receptivity." By this, it is meant to describe the process by which freestream turbulence interacts with the boundary layer, and it is believed by many to be a significant part in the transition process. Reshotko (1984) put forth a description of transition and the role of receptivity by stating: "In an environment where initial disturbance levels are small, the transition Reynolds number of a boundary layer is very much dependent upon the nature and spectrum of the disturbance environment, the signatures in the boundary layer of these disturbances and their excitation of the normal modes (receptivity), and finally the linear and nonlinear amplification of the growing modes."

Receptivity prediction tools provide the disturbance spectrum and initial amplitudes to be used by the linear or nonlinear evolution module, whether it be linear stability theory, PSE theory, etc. to predict the transition location or to provide a means to correlate the transition location. Such capabilities already exist for the simplest of disturbance initiation processes, as shown by Bertolotti & Crouch (1992).

Leehey & Shapiro (1980), Kachanov & Tararykin (1990), Saric, Hoos & Radeztsky (1991) and Wiegel & Wlezien (1993) have conducted receptivity experiments, and Kerschen (1987), Tadjfar & Bodonyi (1992), Fedorov & Khokhlov (1993), Choudhari & Streett (1994), Choudhari (1994) and Crouch (1994) have conducted theoretical studies of receptivity in order to extend the knowledge base and capability for predicting the receptivity process. Acoustic noise, turbulence and vorticity are freestream influences that couple with single and distributed roughness, steps and gaps, surface waviness, etc., to produce disturbances in the viscous boundary layer flow. For example, as shown in Fig. 10.1, the influence of the transition location with changes in Reynolds number is displayed. Here, the receptivity mechanism involves a roughness element.

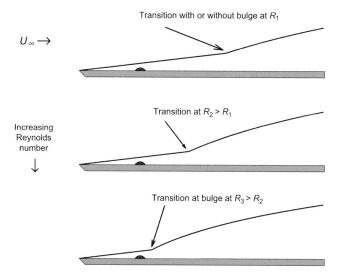

Figure 10.1 Effects of two-dimensional surface imperfection on laminar flow extent (after Holmes *et al.*, 1985a).

These ingestion mechanisms are referred to as natural receptivity. In addition, there is forced as well as natural receptivity. Because the dominant instabilities in a boundary layer flow are of a small scale, receptivity initiation must input energy into this part of the spectrum in order to result in the most efficient excitation of disturbances. As Kerschen (1989) pointed out, forced receptivity usually involves the intentional generation of instability waves by supplying energy to the flow at finite and selected wavelengths and frequencies that match the boundary layer disturbance components. Examples of forced receptivity include unsteady wall suction and blowing or heating and cooling, such as used for active flow control.

Forced theoretical and computational receptivity are linked to linear stability theory through a forcing boundary condition. This is done by introducing the boundary condition for the generation of a disturbance due to suction and blowing through a single orifice in the wall (or computational boundary) as

$$v = v_w f(x) e^{-i\omega t}, \tag{10.1}$$

where ω is the frequency of the disturbance that one desires to initiate, $f(x)$ is the shape of the suction or blowing distribution over a discrete region of the boundary and v is the resulting wall-normal velocity component at the wall. Similar techniques can be used for unsteady thermal forcing in a numerical simulation and can be used to excite disturbances in a wind tunnel experiment.

Natural receptivity is more complicated in that freestream acoustics, turbulence and vorticity have wavelengths that are much longer than that of the boundary layer disturbance. In addition, complicating the matter is the fact that the freestream disturbance in nature has a well defined propagation speed and the energy concentrated at specific wavelengths. Hence, the freestream disturbance has no energy in wavelengths that correspond to the boundary layer disturbance. As a result, any mechanism must effectively and efficiently be able to transfer energy from the long wavelength range to that of the short scale wavelengths. Mechanisms to accomplish this transfer include the leading edge of a plate and wing and surface discontinuities, such as bugs, surface roughness, rivets or other surface distortion.

To determine the process of length-scale conversion, Goldstein (1983, 1985, 1987) put forth a theory that showed that the primary means of conversion was through nonparallel mean flow effects. Hence, the two locations where nonparallel effects are strongest are:

(1) regions of rapid boundary layer growth such as at the leading edge where the boundary layer is thin and rapidly growing, and
(2) downstream at a surface discontinuity, such as a bump on the wall.

To determine the receptivity of the boundary layer in the leading edge region of a particular geometry to freestream disturbances, solutions of the linearized unsteady boundary layer equations are required. These solutions match downstream with the Orr–Sommerfeld equation that governs the linear instability and serves to provide a means for determining the amplitude of the viscous boundary layer disturbance.

Finally, the second class of natural receptivity involves the interaction of long wavelength, freestream disturbances with local mechanisms, such as wall roughness, suction or steps, for example, to generate boundary layer disturbances. In this case, adjustments made to the mean flow cannot be obtained with standard boundary-layer equations and, instead, the triple-deck asymptotic approximation to the Navier–Stokes equations is used. The triple deck produces an interactive relationship between the pressure and the displacement thickness due to matching of the requirements between the three decks. The middle deck or main deck responds inviscidly to the short-scale wall discontinuities. The viscous layer or lower deck between the main deck and the surface is required to ensure a no-slip boundary condition at the wall. Finally, the rapid change in displacement thickness at the surface discontinuity induces a correction to the outer potential flow. This correction takes place in the upper deck. The mean-flow gradients, due to the discontinuity, serve as forcing terms for the disturbance equations. So, although much understanding about receptivity has been gained, continued research must be conducted, especially in the three-dimensional effects and in supersonic flows before the tools can become widely used as transition prediction tools.

10.3 Tollmien–Schlichting Breakdown

The breakdown of Tollmien–Schlichting waves was discussed in great detail in Section 9.3 as an instrument for secondary instability. This section will briefly summarize the process so that crossflow vortex, oblique-wave and Görtler vortex breakdown can be better appreciated.

The viscous Tollmien–Schlichting wave instabilities of a boundary layer begin to amplify at rather long wavelengths at branch I of the neutral curve, amplify until branch II is reached downstream and then decay. In a real flow, many other modes exist as well making it impractical to design many laminar configurations. However, the transition from a laminar flow to turbulent flow in boundary layers can begin with the Tollmien–Schlichting waves. At some point these two-dimensional dominant waves tend to be that of three-dimensional, short-wavelength structure. The theory of secondary instabilities provides for understanding of this two-dimensional process becoming three dimensional and of the long wavelength dominant instabilities developing short scales. When these secondary instabilities amplify, as discussed in Chapter 9, they amplify with a larger growth rate than that of the Tollmien–Schlichting wave. As the secondary mode amplitude approaches that of the Tollmien–Schlichting amplitude, the spectrum rapidly fills in a short spatial distance and breakdown occurs. This is shown in Fig. 10.2. Clearly at the end of the simulation the solution is under-resolved; however, the results show the cascade of energy to many additional modes and, as the amplitudes increase, the process must be described by a nonlinear system of equations.

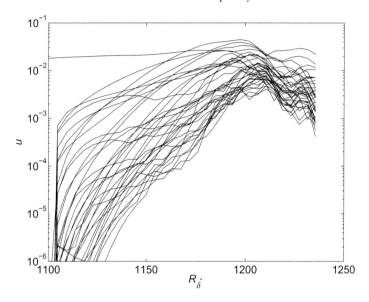

Figure 10.2 Amplification of modes associated with subharmonic breakdown process. (Because the simulation is under-resolved, the results simply show a qualitative cascade of energy to higher frequency/lower wavenumbers.)

10.4 Oblique Wave Breakdown

The oblique-wave breakdown procedure is due to the linear or nonlinear interactions of a pair of oblique waves. This process is more likely to occur in high speed flows because, for low speed flows, the two-dimensional Tollmien–Schlichting mode dominates or has the largest growth rate, while for high speed flows, it is the three-dimensional Tollmien–Schlichting modes that dominate. We will briefly discuss this breakdown process because the modes interact in an unusual manner and induce stationary vortex structures. Due to nonlinearity, no adequate formal theory is available to explain the breakdown process. However, similar mechanisms have been studied by Hall & Smith (1991) using asymptotic methods. Specifically, Hall & Smith discussed vortex wave interactions within a large wavenumber and Reynolds-number limit. To further quantify the mechanisms of interest in the finite Reynolds-number range, DNS (Chapter 11) and possibly PSE theory (Chapter 9) are just some options available to study the wave interactions.

Schmid & Henningson (1992a) studied bypass transition by introducing a pair of large amplitude oblique waves into channel flow. The evolution of the disturbances was computed with temporal DNS. They found that the development of the oblique waves was dominated by a preferred spreading of the energy spectra into low streamwise wavenumbers and led to the rapid development of streamwise-elongated structures. Schmid & Henningson (1992b) also looked at small-amplitude wave pairs over a variety of parameters and suggested that the mechanism of energy transfer is primarily linear. Fasel & Thumm (1991) and Bestek, Thumm & Fasel (1992) computed such breakdown structure in a compressible boundary layer and described the physical structure as honeycomb-like in order to identify a distinction from the secondary instability Λ-like structures discussed in Chapter 9. Chang & Malik (1992) used PSE theory to examine the breakdown of supersonic boundary layers because the dominant first mode is an oblique wave in supersonic flow. Chang & Malik found that even waves with amplitudes as small as 0.001% that are initiated at the lower branch can lead to transition in this breakdown scenario and depend on the frequency of the induced oblique waves. Here, we focus on the nonlinear flow breakdown process. A brief discussion is also presented in Chapter 11 because these results were obtained using DNS.

The profiles for the oblique-wave pair are obtained from linear stability theory for the Reynolds number, $Re_{\delta_0^*} = 900$, frequency, $\omega = 0.0774$ and spanwise wavenumbers, $\beta = \pm 0.2$. Details of the spatial DNS computations are included here in the event that they may be used as a test case. The grid consists of 901 uniformly spaced streamwise nodes, 61 wall-normal collocation points and 10 symmetric spanwise modes. In the streamwise direction, the outflow boundary is $465\delta_0^*$ from the inflow boundary; the far-field or freestream boundary is $75\delta_0^*$ from the wall; and the spanwise boundary consists of a length equal to one half of the spanwise wavelength, or $\lambda_z/2 = \pi/\beta$. For the time-marching scheme, the disturbance period is divided into 320 time-steps. For the PSE computational approach, 100 wall-normal grid points are used; 7 frequency modes and 7 spanwise Fourier modes are used; and the far-field boundary is $58\delta_0^*$ from the wall.

The input modes consist of a pair of oblique traveling waves that were obtained from linear stability theory. The disturbance forcing consists of modes $(1,1)$ and $(1,-1)$, or (ω, β) and $(\omega, -\beta)$, and their complex conjugates $(-1,1)$ and $(-1,-1)$. Theoretically, if these modes self-interact initially, then only certain higher modes are likely to be excited. These higher modes are: $(0,0)$, $(0,2)$, $(2,0)$ and $(2,2)$, etc.

The oblique waves are introduced with larger amplitudes $A_{1,1}^o = 0.01$. The computed primary disturbance $(1,1)$ and higher modes are shown in Fig. 10.3. Again, the modes predicted by PSE theory are shown to be in agreement with the DNS results. The small wavenumber modes gain initial energy. The vortex mode $(0,2)$ is clearly dominant. The self-interaction of the wave pairs and the interaction with the streamwise vortex lead to a rapid cascade of energy to the other modes.

Rather than the meager growth and downstream decay, as occurs if the initial amplitudes are too small, these higher modes now grow with growth rate characteristics that are similar to the vortex mode. The vortex and harmonics rapidly overtake the introduced waves $(1, 1)$ and $(1, -1)$ and subsequent breakdown occurs. At breakdown, the spectrum is filled, and both the DNS and PSE computations are under-resolved near the downstream end of Fig. 10.3. Further evidence that the onset of transition from laminar to turbulent flow has begun, and the skin-friction curve begins to rise.

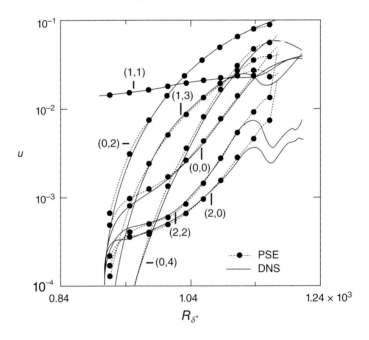

Figure 10.3 Amplitude growth with downstream distance for a pair of oblique waves. (Reprinted by permission from Springer Nature: Joslin, Streett & Chang, 1993.)

10.5 Crossflow Vortex Breakdown

In Section 6.5, crossflow vortex modes were introduced by way of the rotating disk flow problem. As discussed in Chapter 6, the crossflow instability occurs due to the existence of an inflection in the profile of a three-dimensional velocity profile as found in rotating disks and on boundary layer associated with a swept wing. For the boundary layer of a wing, the Tollmien–Schlichting process dominates until the wing is swept to approximately $25°$–$30°$. At that point, the inflectional-profile properties cause the crossflow vortex mode to dominate with transition occurring very near the leading edge of a wing in most cases, depending on the pressure gradient and wing sweep.

Here, we discuss the breakdown process from results given by Joslin & Streett (1994) and Joslin (1995a). Again, because the breakdown process is of interest and is inherently nonlinear, the DNS code (Chapter 11) is used to obtain results for discussion. For the simulations, no surface imperfections, such as particulates, weather condition effects, noise or spanwise inhomogeneities, exist. Surface curvature is neglected to simplify the numerics and because the simulation is conducted on a chordwise region of the wing that corresponds to a relatively flat portion of a laminar-flow airfoil. Note that all of these effects must be considered when analyzing the transitional process in a wind tunnel experiment or flight test. The base flow and most of the parameters used in the initial study by Joslin & Streett (1994) are used here to enhance the understanding of the transition process on swept wings.

The vortex packets are forcibly imposed into the boundary layer by steady suction and blowing through the wedge surface in the same manner as described by Joslin & Streett (1994). Suction and blowing techniques may be used because, as demonstrated by Kachanov & Tararykin (1990), the results from suction and blowing and roughness-element disturbance generators correlate well and lead to disturbances that graphically coincide.

Consider the base flow given by the Falkner–Skan–Cooke profile. The pressure field used by Müller & Bippes (1988) is given by the following linear equation

$$c_p(x_c) = 0.941 - 0.845x_c. \qquad (10.2)$$

The first simulation (SIM-I) is the case of Joslin & Streett (1994). This simulation has a grid of 901 streamwise, 61 wall-normal and 32 spanwise grid points. The far-field boundary is located 50δ from the wedge, the streamwise distance is 857δ from the inflow, with the spanwise distance 108δ. For time marching, a time-step size of 0.2 is chosen for the three-stage Runge–Kutta method. For all simulations, crossflow vortex packets are generated through a periodic strip of steady suction and blowing holes that are equally spaced on the wing surface, and the shape of the wall-normal velocity profiles at the wall have a half-period sine wave in the chordwise direction and a full-period sine wave in the spanwise direction. This mode of disturbance generation would correspond to an isolated roughness element within the computational domain. Stationary crossflow vortex packets are generated by steady suction and blowing with a wall-normal velocity component at the wall with an amplitude of $v_w = 1 \times 10^{-5}$. The holes for SIM-I have a chordwise length of 8.572δ and a spanwise length of 16.875δ.

The second simulation (SIM-II) has a grid of 901 streamwise, 81 wall-normal and 48 spanwise grid points. The far-field boundary is located 50δ from the wedge, the streamwise distance is 550δ and the spanwise distance is 108δ. A time-step size of 0.2 is chosen for time marching. Suction and blowing with a wall-normal velocity amplitude of $v_w = 1 \times 10^{-4}$ is used to generate stationary crossflow vortices. The holes for SIM-II have a chordwise length of 8.5δ and a spanwise length of 36δ and are aligned side by side in the spanwise direction.

The final simulation (SIM-III) has a grid of 721 streamwise, 81 wall-normal and 64 spanwise grid points. The far-field boundary is located 50δ from the wedge, the streamwise distance is 440δ and the spanwise distance is 108δ. A time-step size of 0.2 is chosen for time marching. Stationary crossflow vortex packets are generated with steady suction and blowing with a wall-normal velocity component at the wall with an amplitude $v_w = 1 \times 10^{-3}$. The holes for SIM-III have a chordwise length of 5.5δ and a spanwise length of 36δ and are aligned side by side in the spanwise direction.

Distinct stages of disturbance evolution are found for crossflow evolution, amplification and breakdown. If the disturbances are generated by a local means (e.g., roughness element), then the initial growth of individual disturbance packets occurs in isolation from adjacent packets. The individual packets coalesced at a chordwise location downstream, depending upon the distance between the suction holes, the directions of the disturbance evolution and the spreading rate. When the vortex packets reach sufficiently large amplitude in the later stages of transition, the disturbance field becomes dominated by nonlinear interactions and vortex rollover. These stages are shown in Fig. 10.4, where spanwise planar views of chordwise velocity contours, viewed from the trailing edge toward the leading edge, are shown. Note that the wing tip is to the left and the wing root is to the right. For each simulation, the contour results show that, immediately downstream of the disturbance initialization point, a distinct vortex packet evolves that is isolated from nearby disturbances. As the disturbance evolves and spreads, additional vortices fill the span as a result of the adjacent vortex superposition. This superposition process leads to apparent rapid increase in the disturbance amplitudes and phase adjustments. In the later nonlinear stage of breakdown, the contours indicate that low-speed fluid is dragged out and over the high-speed fluid, which is then drawn toward the surface. Dagenhart & Saric (1994) observed this same phenomenon in their experiments.

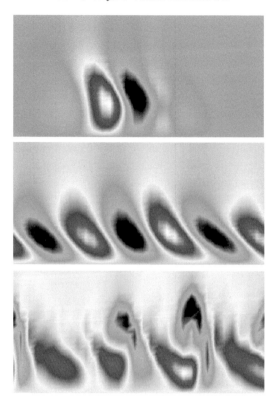

Figure 10.4 Spanwise planes of disturbance velocity (u) contours at chordwise locations for swept wedge flow of SIM-I; $x_c = 0.25, 0.34, 0.45$.

The breakdown sequence of SIM-I may be typical for isolated roughness elements where initial energy resides in many instability modes, but the evolution sequence can be more generalized by the following description. Instead of describing the first stage as a region of isolated growth that is specific to an isolated roughness, the initial growth stage could be described as linear or has an exponential growth. In both SIM-II and SIM-III, the initial amplitude levels of the disturbances are much larger than SIM-I, and the disturbance initiation process imitates distributed roughness where initial energy resides in a single dominant mode. Still, all of the simulations have this linear growth stage. The second stage can be generically described as coalescence although, unlike SIM-I that is a linear superposition, the process may be nonlinear in SIM-II or SIM-III because of the much larger disturbance amplitudes. The final stage can be typically described as a nonlinear interaction because all of the simulations have very large amplitudes in this region.

In this nonlinear interaction region, inflectional velocity profiles are observed in all of the simulations. Figure 10.5 shows the instantaneous chordwise velocity profiles $(U + u)$ for each simulation at a chordwise station that corresponds to the nonlinear vortex stage of rollover. The various profiles at each station correspond to adjacent spanwise locations. Across the span, the flow is accelerated in regions near the wedge surface and is retarded in other areas out in the boundary layer. The characteristic inflectional profiles have been observed in experiments by both Müller & Bippes (1988), Dagenhart *et al.* (1989) and Dagenhart & Saric (1994). Dagenhart & Saric noted that the appearance of inflectional profiles was rapidly followed by the discovery of a high-frequency instability and, subsequently, by transition.

The theoretical studies of Kohama, Saric & Hoos (1991) and Balachandar, Streett & Malik (1990) indicated that this high-frequency instability in the experiments is reminiscent of secondary instabilities that spawn from these inflectional velocity profiles. Thus, the late stages of crossflow breakdown, the most likely cause of transition, is the appearance of this secondary instability mode.

10.6 Dean–Taylor–Görtler Vortex Breakdown

Recall from Chapter 6, centrifugal instabilities were discussed using an analysis of a linear system of equations. Among others, centrifugal instability occurs for shear flows over concave surfaces. Rayleigh (1916b) determined the necessary and sufficient conditions for the existence of an inviscid instability. This instability is referred to as the Rayleigh circulation criteria and is a function of the circulation. As Saric (1994a) notes, there are three centrifugal instabilities that can occur with each sharing the same physical mechanism of generation. The Taylor (1923) instability occurs in flow between co-rotating cylinders; the Dean (1928) instability occurs in curved channel flows; and the Görtler (1940) instability occurs for open, curved plate boundary-layer flows. Take note that, although the instabilities may have similar generation mechanisms, the mean flow states are quite different. However, these common vortex disturbances arise when the surface geometry becomes concave and are reminiscent of counter-rotating vortex structures (see Fig. 6.4). Note the changes that occur in the flow due to the presence of these vortices. Low-speed fluid near the wall is transported up and high-speed fluid is transported down toward the wall. Such changes can then induce additional instability that effect a transition from a laminar to turbulent state.

The nonlinear breakdown process for these vortex-based flows can be quite complicated because additional traditional instabilities may become unstable. As discussed by Coles (1965) for the case of Taylor–Couette flow, with the

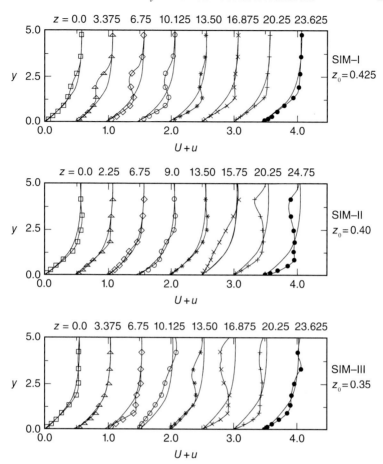

Figure 10.5 Chordwise (base + disturbance) velocity profiles at various chord-wise and spanwise locations for swept wedge flow.

inner cylinder having a larger angular velocity than the outer cylinder, the expected Taylor motion of periodic axial vortices developed, followed by a secondary pattern of traveling circumferential waves. As these dual modes amplify, energy is observed in the harmonic modes. This process can be referred to as cascade process whereby other modes gain energy from their parental modes until they have sufficient energy to interact nonlinearly. A catastrophic transition process may be observed when the outer cylinder has a larger angular velocity than the inner cylinder. Figure 6.2 shows the bounds between stable and unstable regions; however, the problem can become more complicated than simply stable-unstable regions. With both catastrophic and doubly-period transition processes, more regions can be added. For doubly periodic, a region can be identified in the first quadrant with a region where only singly-period flow is maintained. Also, for catastrophic transition, bands of turbulence mingle with the laminar flow. These turbulent regions can appear and disappear in a random manner or may form in a regular pattern in the form of a spiral turbulent pattern. To represent this appearance of turbulence, a boundary can be added to Fig. 6.2 above the Taylor bound. Above this line, the flow is turbulent and between these lines the flow is characterized as transitional.

Among others, Ligrani *et al.* (1994) examined a breakdown process for Dean vortex flow. Since fluctuations in the mean flow can trigger vortex initiation, the vortices meander in the spanwise direction as they are convected downstream with the flow. A splitting and merging of vortices is observed in the flow presumably because the flow supports a spanwise dominant instability. The symmetric counter-rotating vortex-pair system encounters an instability. After this encounter, the vortex pair experiences a strong up-wash and the flow pattern now appears as an asymmetric mushroom like structure. This mushroom structure has been observed by Saric (1994a) as shown by Fig. 10.6.

Prior to this mushroom characteristic, vortex pairs can merge. As two pairs of vortices incur a decrease in radial extent, these pairs are observed to merge and form a single pair. This merging phenomenon is quite common. Additionally, two vortex pairs can split and result in three vortex pairs. The dynamics of this process involves an average preferred spanwise wavenumber. As a final note for the Dean-vortex problem, the observations for the vortex splitting and merging process are very dependent on the freestream and initial conditions similar to crossflow vortex problem. Roughness induces more of a stationary vortex system, whereas fluctuations induce more of an unsteady vortex system.

The explanation by Saric (1994a) for Görtler vortex breakdown provides a consistent description of the formation of the symmetric counter-rotating vortex pairs. The actions of the vortex pair induces up-wash, and down-wash regions have been discussed. In the down-wash region, high-momentum fluid is drawn toward the wall and shear is increased while in the up-wash region the low-momentum fluid is ejected and the region has decreased shear. As the vortices amplify, the mean flow now encounters regions of large distortions similar to crossflow breakdown. The now highly distorted and inflectional mean flow is susceptible to inviscid Rayleigh or secondary instabilities. From computations and the experiments of Swearingen & Blackwelder (1986, 1987), it was concluded that Görtler breakdown was caused by a secondary instability as-

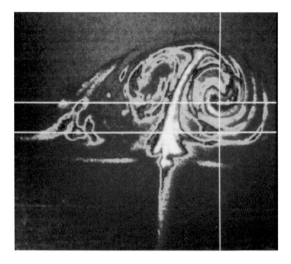

Figure 10.6 Transport of low momentum fluid by a stationary vortex structure (after Peerhossaini, 1987; courtesy of W. Saric).

sociated with the spanwise velocity gradient and the strongly distorted mean velocity profiles. As measured by Winoto, Zhang & Chew (2000), transition is observed to start and become turbulent earlier in the up-wash region when compared with the down-wash region.

10.7 Transition Prediction

Because the performance of a configuration is directly tied to the amount of laminar and turbulent flow present on a configuration, it is imperative to be able to accurately predict and design for the transition location. This section reviews the transition prediction methodologies and focuses on the theoretical and computational aspects of the transition prediction. More detailed reviews of currently used approaches are provided by Cousteix (1992) and Reed & Saric (1989).

The reason why laminar flow is usually more desirable than the turbulent counterpart for external aerodynamic vehicles lies with the reduction of the viscous drag penalty and noise source. Do we have a sufficient understanding of the fundamental flow physics for the problem to design an optimal, reliable, cost effective system to control the flow? The answer is encouraging at low speed!

As discussed in Chapters 1 and 2, the first major theoretical contributions to the study of boundary layer transition were made by Helmholtz (1868),

Kelvin (1880), Reynolds (1883) and Rayleigh (1879, 1880, 1887). Although these early investigations neglected the effects of viscosity, the second derivative of the mean velocity proved to be of key physical importance in explaining boundary layer instabilities. These fundamental studies proved to be the basis for future progress in the theoretical development. Viscous effects were added by Orr (1907a,b) and Sommerfeld (1908), who developed an ordinary differential equation that governs the linear instability of two-dimensional disturbances in channel flows. This was later extended by others to the incompressible boundary layer. Later, Squire (1933) accounted for three-dimensional waves by introducing a transformation from two to three dimensions. Tollmien (1929) and Schlichting (1932) provided the basis for convective traveling wave instabilities that are now termed Tollmien–Schlichting instabilities. Liepmann (1943) and Schubauer & Skramstad (1947) experimentally confirmed the existence and amplification of these instabilities in the boundary layer. One can visualize this disturbance by remembering the image of water waves created by dropping a pebble into a still lake or puddle. In this image, the waves that are generated decay as they travel from the source. Such is the case in boundary layer flow, except when certain critical flow parameters such as the Reynolds number are reached and the waves will grow in strength and lead to turbulent flow.

The improvements in aerodynamic efficiency directly scale with the amount of laminar flow that can be achieved. Hence, the designer must be able to accurately predict the location of boundary layer transition on complex three-dimensional geometries as a function of suction distribution and suction level or the accurate prediction of the suction distribution for a given target transition location. Pressure gradients, surface curvature and deformation, wall temperature, wall-mass transfer and unit Reynolds number are known to influence the stability of the boundary layer and transition location and must be reviewed.

This section describes the conventional and advanced transition prediction tools, some of which include the prediction of perturbations in the laminar boundary layer, the spectrum and amplitudes of these perturbations and the linear and nonlinear propagation of these perturbations that ultimately lead to transition. For literature focusing on the theoretical and computational aspects of transition prediction, refer to Cousteix (1992), Arnal, Habiballah & Coustols (1984) and Arnal (1994).

10.7.1 Granville Criterion

Granville (1953) reported a procedure for calculating viscous drag on bodies of revolution, and developed an empirical criterion for locating the transition location associated with low turbulence flows. Low-(or zero-) turbulence characteristics of flight (or low-turbulence wind tunnels) and high-turbulence characteristics of most wind tunnels are the two problems considered relative to a transition criterion. The low-turbulence case assumed that transition was Tollmien–Schlichting disturbance dominated and began with infinitesimally small amplitude disturbances. Granville (1953) showed that a variety of flight and low-turbulence wind tunnel data collapsed onto a criterion curve based on $(Re_{\theta,T} - Re_{\theta,N})$, the difference between the momentum-thickness Reynolds number at transition and at the neutral point, versus $\overline{(\theta^2/\nu)(dU/dx)}$, which is the average pressure gradient parameter. This correlation was demonstrated for two-dimensional flows and is shown in Fig. 10.7. Granville used a transformation to convert this information to a body-of-rotation problem. The data was also correlated with the turbulence level in the freestream as shown in Fig. 10.8. Extrapolation of the criteria does work for a two-dimensional airfoil that is dominated by Tollmien–Schlichting waves (see Holmes *et al.*, 1983), and the existing database included this form of transition. However, when the design configuration begins to significantly differ from the existing database, this transition prediction criteria fails.

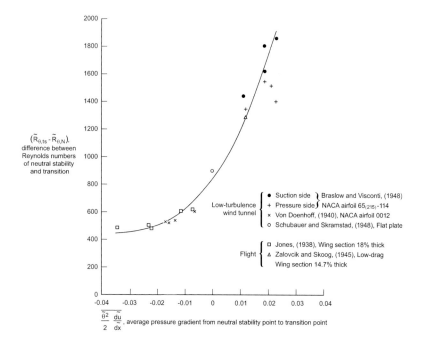

Figure 10.7 Transition location as a function of average pressure gradient (after Granville, 1953).

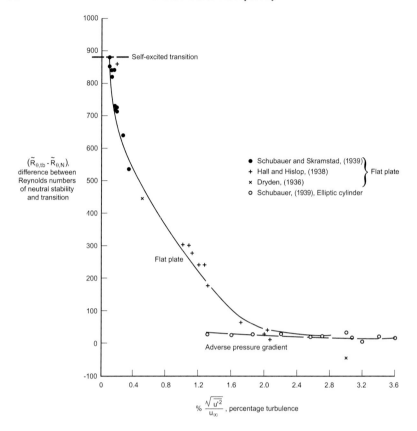

Figure 10.8 Transition location as a function of turbulence level (after Granville, 1953).

10.7.2 *C1 and C2 Criteria*

At ONERA,[1] Arnal, Juillen & Casalis (1991) performed *N*-factor correlations with wind tunnel experimental results of a LFC suction infinite swept wing. The motivation for the study was to gain fundamental understanding of the transition process with suction and to test the methodologies developed at ONERA/CERT[2] for three-dimensional flows. The streamwise instability criteria was based on an extension of Granville (1953). Two crossflow transition criteria have been developed by Arnal, Habiballah & Coustols (1984) at ONERA and are referred to as C1 and C2. The C1 criterion involves a correlation of transition onset integral values of the crossflow Reynolds number and the

---

[1] Office National d'Etudes et de Recherches Aérospatiales, www.onera.fr/en
[2] Centre d'Etudes et de Recherches de Toulouse

streamwise shape factor. The C2 criterion is a correlation of transition onset with a Reynolds number computed in the direction of the most unstable wave, the streamwise shape factor and the freestream turbulence level. The results demonstrate that the transition criteria cannot be applied in regions where the pressure gradient is mild because there is a large range of unstable directions. In that region, one cannot only look at pure streamwise or crossflow instabilities. The C1 criterion leads to bad results with wall suction present whereas the C2 criterion correctly accounts for wall suction.

10.7.3 Linear Stability Theory and e^N

Although the growth or decay of small-amplitude disturbances in a viscous boundary layer can be predicted by the Orr–Sommerfeld and Squire equations within the quasi-parallel approximation, the ability to predict transition using these results was first achieved in the 1950s with the semiempirical method of Smith (1953), denoted as the e^N or N-factor method that correlates the predicted disturbance growth with measured transition locations. Although limited to empirical correlations of available experimental data, it has been the main tool that has been used throughout the 1990s. Moreover, linear stability theory represents the current state of the art for transition location prediction for three-dimensional subsonic, transonic and supersonic flows. To begin a transition prediction analysis, the steady laminar mean flow must first be obtained directly from Navier–Stokes solutions or the boundary layer equations. Then, the three-dimensional boundary layer stability equations are solved in order to determine the amplification rate at each point along the surface.

Significant advances have been made in the understanding of the fundamentals of two- and three-dimensional unsteady viscous boundary layer flow physics associated with transition (cf. reviews by Reshotko, 1976; Herbert, 1988; Bayly, Orszag & Herbert, 1988; Reed & Saric, 1998; Kachanov, 1994), CFD mean flow capabilities in complex geometries, turbulence modeling efforts and in the direct numerical simulation of the unsteady flow physics (Kleiser & Zang, 1991). However, the devised transition–prediction methodology is considered state of the art and has been used by industry for LFC related design. This transition prediction methodology, termed the e^N method, is semiempirical and relies on experimental data to determine the N-factor value at transition.

The disturbance evolution and transition-prediction tools require an accurate representation of the mean flow velocity profiles. Either the velocity profiles can be extracted from Navier–Stokes solutions or are derived from solutions of a coupled Euler and boundary layer equation solver. Harris, Iyer & Radwan (1987) and Iyer (1990, 1993, 1995) have provided a solver for the Euler and boundary layer equations. Harris *et al.* (1987) demonstrated the accuracy of a fourth-order finite-difference method for a Cessna airplane fuselage forebody flow, flat-plate boundary layer, flow around a cylinder on a flat plate, a prolate spheroid and the flow over a NACA0012 swept wing. In terms of computational efficiency, the Euler and boundary layer approach for obtaining accurate mean flows will be the solution of choice for most of preliminary design stages. Navier–Stokes solvers can be used for LFC design. A limiting factor for the Navier–Stokes mean flows is the demanding convergence required for the suitability of the results in the boundary layer stability codes.

For linear stability theory that makes use of the quasi-parallel flow assumption, the mean flow $U(y), W(y)$ are functions of the distance from the wall only and $V = 0$. Then, writing the velocities and pressure as a mean part plus a fluctuating part $\{\tilde{u}, \tilde{v}, \tilde{w}, \tilde{p}\}$, substituting into the Navier–Stokes equations and linearizing, the following linear system results:

$$\frac{\partial \tilde{u}}{\partial x} + \frac{\partial \tilde{v}}{\partial y} + \frac{\partial \tilde{w}}{\partial z} = 0, \tag{10.3}$$

$$\frac{\partial \tilde{u}}{\partial t} + U\frac{\partial \tilde{u}}{\partial x} + \tilde{v}\frac{dU}{dy} + W\frac{\partial \tilde{u}}{\partial z} = -\frac{\partial \tilde{p}}{\partial x} + Re^{-1}\left[\frac{\partial^2 \tilde{u}}{\partial x^2} + \frac{\partial^2 \tilde{u}}{\partial y^2} + \frac{\partial^2 \tilde{u}}{\partial z^2}\right], \tag{10.4}$$

$$\frac{\partial \tilde{v}}{\partial t} + U\frac{\partial \tilde{v}}{\partial x} + W\frac{\partial \tilde{v}}{\partial z} = -\frac{\partial \tilde{p}}{\partial y} + Re^{-1}\left[\frac{\partial^2 \tilde{v}}{\partial x^2} + \frac{\partial^2 \tilde{v}}{\partial y^2} + \frac{\partial^2 \tilde{v}}{\partial z^2}\right], \tag{10.5}$$

and

$$\frac{\partial \tilde{w}}{\partial t} + U\frac{\partial \tilde{w}}{\partial x} + \tilde{v}\frac{dW}{dy} + W\frac{\partial \tilde{w}}{\partial z} = -\frac{\partial \tilde{p}}{\partial z} + Re^{-1}\left[\frac{\partial^2 \tilde{w}}{\partial x^2} + \frac{\partial^2 \tilde{w}}{\partial y^2} + \frac{\partial^2 \tilde{w}}{\partial z^2}\right], \tag{10.6}$$

where the Reynolds number, $Re = U_o\delta/\nu$.

According to the conventional normal mode assumption used to derive the Orr–Sommerfeld equation, the eigensolutions take the form

$$\{\tilde{u}, \tilde{v}, \tilde{w}, \tilde{p}\}(x, y, z, t) = \{\hat{u}, \hat{v}, \hat{w}, \hat{p}\}(y)e^{i(\alpha x + \beta z - \omega t)} + c.c., \tag{10.7}$$

where α and β are the nondimensional wavenumbers in the streamwise and spanwise directions, ω is the frequency and $\{\hat{u}, \hat{v}, \hat{w}, \hat{p}\}$ are the eigenfunctions in y. Thus, using (10.7) in the linear equations (10.3)–(10.6), the Orr–Sommerfeld

$$\left[\left[\frac{d^2}{dy^2} - \alpha^2 - \beta^2\right]^2 \hat{v} - iRe^{-1}\left[-\alpha\frac{d^2U}{dy^2} - \beta\frac{d^2W}{dy^2}\right.\right.$$

$$\left.\left. + (\alpha U + \beta W - \omega)\left(\frac{d^2}{dy^2} - \alpha^2 - \beta^2\right)\right]\hat{v} = 0, \tag{10.8}$$

and Squire

$$\frac{d^2\hat{\omega}_y}{dy^2} - \left[\alpha^2 + \beta^2 - iRe^{-1}(\alpha U + \beta W - \omega)\right]\frac{d\hat{\omega}_y}{dy}$$
$$= iRe^{-1}\left(\alpha\frac{dU}{dy} + \beta\frac{dW}{dy}\right)\hat{v}, \qquad (10.9)$$

equations result, where $\hat{\omega}_y$ is the perturbation vorticity in the y-direction. The wall boundary conditions are

$$\hat{v}, \hat{v}', \hat{\omega}_y = 0 \quad \text{at} \quad y = 0, \qquad (10.10)$$

and, in the freestream,

$$\hat{v}, \hat{v}', \hat{\omega}_y \to 0 \quad \text{as} \quad y \to \infty. \qquad (10.11)$$

Either spatial or temporal stability analysis may be performed, with the temporal analysis less expensive and the spatial analysis more physical. In addition to the Reynolds number that must be prescribed, a stability analysis requires that the mean flow and its first and second wall-normal derivatives be determined very accurately. A small deviation in the mean flow can cause significant changes in the second derivative and contaminate the stability calculations. Once the mean flow is obtained, a stability problem has to determine six unknowns, namely $\{\alpha_r, \alpha_i, \beta_r, \beta_i, \omega_r, \omega_i\}$, the streamwise wavenumber, streamwise (spatial) growth rate, spanwise wavenumber and growth rate, wave frequency and temporal growth rate, respectively. For the temporal formulation, α and β are real numbers and ω is a complex number that is determined through an eigenvalue solver. For the spatial approach, α and β are complex and ω is the real wave frequency.

Because the spatial formulation is more representative of the real boundary layer instability physics and the temporal-to-spatial conversion is only valid on the neutral curve, the remaining transition–prediction methodologies will be described via the spatial approach. For the temporal approach see Srokowski & Orszag (1977) and Malik (1982). Malik also included the effect of compressibility in the equations. For the spatial approach in three-dimensional flows, the frequency (ω_r) is fixed, $\omega_i = 0$ and $\{\alpha_r, \alpha_i, \beta_r, \beta_i\}$ are parameters to be determined. While an eigenvalue analysis will provide two of these values, the main issue with the application of the e^N methodology to three-dimensional flows is the specification or determination of the remaining two parameters. Figure 10.9 illustrates the instability concept within linear stability theory. A certain parameter range exists whereby a certain combination of wavenumbers and frequencies characterize disturbances that decay at low Reynolds numbers, amplify over a range of Reynolds numbers and then decay as the Reynolds number is increased. The Reynolds numbers nondimensionally represents the spatial chordwise location on a wing for example. Although by now the reader should be extremely familiar with illustration, we restate that the boundary between regions of amplification (unstable) and decay (stable) is termed the neutral curve (location where disturbances neither amplify nor decay).

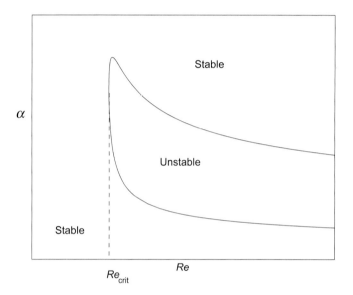

Figure 10.9 Illustration of neutral curve for linear stability theory.

By assuming a method is available to determine the two remaining free parameters, the N-factor correlation with experiments can now be made. By integrating from the neutral point with arbitrary disturbance amplitude A_0, the amplification of the disturbance is tracked until the maximum amplitude A_1 is reached at which decay ensues. Since this is a linear method, the amplitudes A_0 and A_1 are never really used. Instead, the N-factor relation of interest is defined as

$$N = \ln \frac{A_1}{A_0} = \int_{s_0}^{s_1} \gamma ds, \qquad (10.12)$$

where s_0 is the point at which the disturbance first begins to grow, s_1 is the point at which transition is correlated and γ is the characteristic growth rate of the disturbance. Figure 10.10 illustrates the amplification and decay of four disturbances (wavenumber, frequency combinations) leading to four N-values. The inclusion of all individual N-values leads to the N-factor curve. By correlating this N-factor with many transition cases, the amplification factor for which transition is likely or expected for similar flow situations can be inferred. The resulting N-factor is correlated with the location of transition for a variety of experimental databases and is traced in Fig. 10.10. This information is then used to determine the laminar flow extent. Hence, this methodology is critically dependent on the value of the experimental databases and the translation of the N-factor value to a new design.

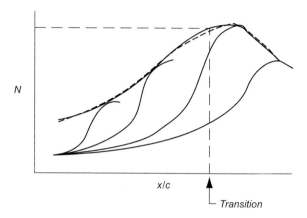

Figure 10.10 Amplification of four waves of different frequency to illustrate the determination of the N-factor curve.

The saddle point, fixed wave angle and fixed spanwise wavelength methods are three means that have been devised in order to determine the two free parameters for three-dimensional flows.

Strictly valid only in parallel flows, the saddle point method suggests that the derivative of $(\alpha x + \beta z)$ with respect to β equals zero. As noted by Nayfeh (1980) and Cebeci & Stewartson (1980), carrying out this derivative implies that $d\alpha/d\beta$ must be real, or

$$\frac{\partial \alpha_i}{\partial \beta_r} = 0. \tag{10.13}$$

The group velocity angle, ϕ_{g}, is given by

$$\phi_{\mathrm{g}} = \tan^{-1}\left(\partial \alpha_r / \partial \beta_r\right). \tag{10.14}$$

The final condition needed to close the problem requires that the growth rate be maximized along the group velocity trajectory. Then, the N-factor or integrated growth would be

$$N = \int_{s_0}^{s_1} \gamma ds \quad \text{where} \quad \gamma = \frac{-\left(\alpha_i - \beta_i \frac{\partial \alpha_r}{\partial \beta_r}\right)}{\sqrt{1 + \left(\frac{\partial \alpha_r}{\partial \beta_r}\right)^2}} \tag{10.15}$$

and where s_0 is the location where the growth rate γ is zero and s_1 is the distance along the tangent of the group velocity direction.

For the second method, developed by Arnal, Casalis & Juillen (1990), the fixed wave angle approach sets $\beta_i = 0$ and the N-factors are computed with a fixed wave orientation, or

$$N = \int_{s_0}^{s_1} -\alpha ds. \tag{10.16}$$

Many calculations have to be made over the range of wave angles to determine the highest value of N.

Finally, the fixed spanwise wavelength approach proposed by Mack (1988) sets $\beta_i = 0$ and β_r is held fixed over the N-factor calculation, computed with

$$N = \int_{s_0}^{s_1} -\alpha ds. \tag{10.17}$$

Many calculations have to be made over the range of β_r in order to determine the maximum value of N. It is not clear what significance holding β_r to a constant has in three-dimensional flows.

A major obstacle in validating or calibrating current and future transition prediction tools results from insufficient information in both wind tunnel and flight test databases. For example, Rozendaal (1986) correlated N-factor tools for Tollmien–Schlichting and cross flow disturbances on a Cessna Citation III business jet flight test database. The database consisted of transition locations measured with hot film devices for points that varied from 5 to 35% chord on both upper and lower wing surfaces for Mach numbers that ranged from 0.3 to 0.8 and altitudes ranging from 10,000 to 43,000 ft. The results showed that crossflow and Tollmien–Schlichting disturbances may interact and that crossflow disturbances probably dominated. Crossflow N-factors were scattered around the value 5 and Tollmien–Schlichting N-factors varied from 0 to 8. The stability analysis showed no relationship between Mach number and disturbance amplification at transition. Rozendaal noted that the quality of the results was suspect because no information on the surface quality existed, an unresolved shift in the pressure data occurred and an inadequate density of transition sensors on the upper wing surface was used. Furthermore, the impact of the engine placement relative to the wing could be added as a potential contributing factor. The Rozendaal analysis reinforced that the N-factor method relies on creditable experimental data.

In a discussion of the application of linear stability theory and e^N method in LFC, Malik (1987) described the methodology for both incompressible and compressible flows and presented a variety of test cases. In situations where transition occurs near the leading edge of wings, the N-factors can be quite large compared to the N = 9 to 11 range applicable for transition in the later portion of a wing. Malik makes an important contribution to this understanding by noting that the linear quasi-parallel stability theory normally does not account for surface curvature effects. However, for transition near the leading edge of a wing, the stabilizing effects of curvature are significant and must be included to achieve N-factors of 9 to 11. The remainder of this subsection documents samples of the extended use of the N-factor method for predicting laminar flow extent.

Schrauf, Bieler & Thiede (1992) indicated that transition prediction is a key problem of laminar flow technology. They presented a description of the N-factor code developed and used at Deutsche Airbus that documents the influence of pressure gradient, compressibility, sweep angle and curvature during calibrations with flight tests and wind tunnel experiments.

Among others, Vijgen *et al.* (1986) used N-factor linear stability theory to ascertain the influence of compressibility on disturbance amplification. They compared Tollmien–Schlichting disturbance growth for incompressible flow over a NLF fuselage with the compressible formulation and noted that compressibility has a stabilizing influence on the disturbances (first mode). For the NLF and LFC, an increase in Mach number (enhanced compressibility) is stabilizing to all instabilities for subsonic and low supersonic flow.

Nayfeh (1987) used the method of multiple scales to account for the growth of the boundary layer (nonparallel effects). The nonparallel results showed increased growth rates compared with the parallel flow assumption. These results indicate that nonparallel flow effects are destabilizing to the instabilities. Singer, Choudhari & Li (1995) attempted to quantify the effect of nonparallelism on the growth of stationary crossflow disturbances in three-dimensional boundary layers using the multiple scales analysis. The results indicate that multiple scales can accurately represent the nonparallel effects when nonparallelism is weak; however, as the nonparallel effects increase, multiple scales results diminish in accuracy.

Finally, Hefner & Bushnell (1980) investigated the status of linear stability theory and the N-factor methodology for predicting transition location. They noted that the main features lacking in the methodology are the inability to account for the ingestion and characterization of the instabilities entering the boundary layer (the receptivity problem).

10.7.4 Parabolized Stability Equations Theory

Because the N-factor methodology is based in linear stability theory, it has limitations. Other methods must be considered that account for nonparallelism, curvature effects and ultimately nonlinear interactions. The final method considered relative to the evolution of disturbances in boundary layer flow is the PSE theory or method. Unlike the Orr–Sommerfeld equation N-factor method that assumes a parallel mean flow, the PSE method enables disturbance evolution computations in a growing boundary layer. As first suggested by Herbert (1991) and Bertolotti (1992), PSE theory assumes that the dependence of the convective disturbances on downstream development events is negligible and that no rapid streamwise variations occur in the wavelength, growth rate and mean velocity and disturbance profiles. At present, the disturbance $\tilde{\Phi} = (\tilde{u}, \tilde{v}, \tilde{w}, \tilde{p})$ in the PSE formulation assumes periodicity in the spanwise direction (uniform spanwise mean flow) and time (temporally uniform) and takes the form

$$\tilde{\Phi} = \sum_{m=-N_z}^{N_z} \sum_{n=-N_t}^{N_t} \Phi_{m,n}(x,y) e^{i\left(\int_{x_0}^x \alpha_{m,n} dx + m\beta z - n\omega t \right)}, \tag{10.18}$$

where N_z and N_t are the total numbers of modes kept in the truncated Fourier series. The convective or streamwise direction has decomposition into a fast oscillatory wave part and a slow varying shape function. Since the disturbance profile $\hat{\Phi}$ is a function of x and y, partial differential equations result and describe the shape function. These equations take the matrix form

$$\left[\mathbf{L} \right] \Phi + \left[\mathbf{M} \right] \frac{d\Phi}{dx} + \left[\mathbf{N} \right] \frac{d\alpha}{dx} = f. \tag{10.19}$$

The fast variations of the streamwise wavenumber mean that the second derivatives in the shape function are negligible. By the proper choice of $\alpha_{n,m}$, the above system can be solved by marching in x. For small-amplitude disturbances, $f = 0$ while, for finite amplitude disturbances, f, in physical space, stems from the nonlinear terms of the Navier–Stokes equation, or

$$\underline{F} = (\underline{u} \cdot \nabla)\underline{u}. \tag{10.20}$$

After the initial values of $\alpha_{n,m}$ are selected, a sequence of iterations is required during the streamwise marching procedure to satisfy the shape function equations at each streamwise location.

Joslin, Streett & Chang (1992, 1993) and Pruett & Chang (1995) have shown that the PSE solutions agree with direct numerical simulation results for the case of incompressible flat-plate boundary layer transition and for compressible transition on a cone.

Haynes & Reed (1996) investigated the nonlinear evolution of stationary crossflow disturbances over a $45°$ swept wing by computing with nonlinear PSE theory, and compared their results with the experiments of Reibert *et al.* (1996). The nonlinear computational results agree with the experiments in that the stationary disturbances reach a saturation state, also confirmed with DNS by Joslin & Streett (1994) and Joslin (1995a), whereas the linear N-factor-type results suggest that the disturbances continue to grow. Hence, the linear predictions inadequately predict the behavior of the disturbances.

Finally, theoretical and computational tools are being developed to predict a rich variety of instabilities that could be growing along the attachment line of a swept wing. Lin & Malik (1994, 1995, 1996) describe a two-dimensional eigenvalue method that predicts symmetric and asymmetric disturbances about incompressible and compressible attachment-line flows that are growing along the attachment line. Such methodologies could provide important parametric information for the design of NLF and LFC swept wings. However, the costly eigenvalue approach has been superceded by an exact ordinary differential equation theory by Theofilis (2002). This new theory leads to identical results as those due to DNS and eigenvalue approaches.

10.7.5 Transition Prediction Coupled to Turbulence Modeling

In this final subsection, a relatively recent concept will be outlined that involves coupling transition prediction methodology with a two equation turbulence model approach. Warren & Hassan (1996, 1997) posed the transition prediction problem within a nonlinear system of equations involving the kinetic energy and enstrophy. The exact governing equations provided a link between the laminar boundary layer flow instabilities, the nonlinear transitional flow state and the fully turbulent flow fluctuations. By assuming the breakdown is initiated by a disturbance with a frequency reminiscent of the dominate growing instability, the simulations are initiated. The influence of freestream turbulence and surface roughness on the transition location were accounted for by a relationship between turbulence level and roughness height with initial amplitude of the disturbance. The initial comparisons with flat-plate, swept flat-plate and infinite swept-wing wind-tunnel experiments suggest a good correlation between the computations and experiments for a variety of freestream turbulence levels and surface conditions.

Then, building on the intermittency accomplishments of Dhawan & Narasimha (1958), the transitional flow region was modeled with an intermittency function Γ. The function ranges from $\Gamma = 0$ for laminar flow to $\Gamma = 1$ for fully developed turbulent flow. In between, the flow is intermittent and one function describing this process is

$$\Gamma = 1 - e^{-0.412\varepsilon^2}, \tag{10.21}$$

where

$$\varepsilon = \max(x - x_t, 0)/\lambda,$$

x_t is the location where turbulent spots begin to form and λ is the extent of the transition region.

The viscosity can then be modeled using the following relationship between laminar and equivalent turbulent viscosity or

$$\mu = \mu_l + \Gamma \mu_t. \tag{10.22}$$

Warren & Hassan (1996) extended this model to include additional information on the fluctuation level where they introduce an expression for μ given by

$$\mu = \mu_l + [(1 - \Gamma)\mu_{lt} + \Gamma \mu_t], \tag{10.23}$$

where μ_{lt} is the contribution of the laminar fluctuations. Originally, the expression for μ_{lt} was determined using correlations from linear stability theory. Warren & Hassan (1997) extended this model so that μ_{lt} can be solved as part of the solution when the onset of transition is known (i.e., minimum skin friction, or some user specified criteria). Such a link then removes the need for linear stability theory analysis altogether. As a result the individual instability modes do not directly play a role in the transition process. Further, the governing equations for the turbulent flow are not closed and stress-strain law modeling assumptions are required to close the system.

Finally, other researchers, such as Liou & Shih (1997), have been approaching the transition prediction and modeling process by a similar intermittency process. A significant difference results from the direct inclusion of freestream turbulence magnitudes and our understanding of the development of turbulent spots in flat plate turbulent boundary layers.

The discussion on transitional models will close out at this point because it is beyond the scope of this text to proceed too far beyond the original focus of hydrodynamic instability theory; however, the potential for these models to support the turbulence modeling efforts continue to be an active area of research.

10.8 Exercises

By now the reader should be extremely familiar with the linear processes associated with hydrodynamic instability theory. Because this chapter is certainly more complicated and is a culmination of the amplification of linear modes, modal interactions and subsequent breakdown, the exercises will primarily focus on essay types of assignments requiring you to think in a cause-and-effect manner. This cause-and-effect hypothesizing is consistent with what is required daily from the research engineer.

Exercise 10.1 Outline all of the potential external factors that may impact or induce instabilities. Characterize which factors induce Tollmien–Schlichting waves versus other modes such as crossflow vortex modes.

Exercise 10.2 Describe the similarities and differences between the Görtler, Dean and Taylor vortices.

Exercise 10.3 Describe the similarities and differences between the Görtler and crossflow breakdown processes.

Exercise 10.4 Generate a main routine to loop through frequency and Reynolds numbers so that you can make e^N calculations. For $Re_{\delta^*} = 2240$, calculate the range of N values for each frequency over the neutral curve. What N-value do you use to assess whether transition will occur? This is in fact tracking/predicting a wave of a given physical frequency as it either decays or amplifies downstream.

11

Direct Numerical Simulation

Perhaps someday in the dim future it will be possible to advance the computations faster than the weather advances and at a cost less than the saving to mankind due to the information gained. But that is a dream.

– Lewis Fry Richardson (1881–1953)

11.1 Introduction

Throughout this text, various assumptions have been employed to simplify this mathematical system in order to extract theoretical insights into the physics of the problems or applications of interest. In this chapter, we return to the complete mathematical system to seek solutions without (or with minimal) a priori assumptions about the flow physics of the problem. The term "direct numerical simulation (DNS)" will be used hereafter to denote direct solutions of the Navier–Stokes equations; sometimes referred to as Direct Navier–Stokes solutions. Inherently, high-order methods for spatial and temporal discretization of the equations are employed and the grids and time stepping are such that all relevant scales of the flows of interest are sufficiently and accurately resolved. For linear hydrodynamic instability there might be one or a few scales of interest whereas, for fully turbulent flows, there are two to three decades in the scales that must be resolved for a computation. In this chapter, we review aspects of the equations relative to the problem of hydrodynamic stability and provide some solution methodologies.

316

11.2 Governing Equations

The Navier–Stokes equations can be written with the primitive variables using velocity and pressure, velocity and vorticity, streamfunctions and vorticity or streamfunctions. For incompressible two-dimensional flows, the primitive-variable formulation leads to

$$u_t + uu_x + vu_y = -p_x + Re^{-1}\nabla^2 u, \tag{11.1}$$

$$v_t + uv_x + vv_y = -p_y + Re^{-1}\nabla^2 v, \tag{11.2}$$

and

$$u_x + v_y = 0. \tag{11.3}$$

Here, the velocity vector is (u, v) and the scalar pressure is p in the Cartesian coordinate system (x, y). Unlike the compressible equations, the pressure in the above equations implicitly adjusts itself to satisfy the divergence-free condition associated with incompressible flow. Hence, no initial conditions or boundary conditions are required for the pressure.

Vorticity is denoted as the curl of the velocity, or $(\xi, \omega, \zeta) = \nabla \times \underline{u}$. For two-dimensional flows, only the spanwise or z component of vorticity exists and is defined as $\zeta = v_x - u_y$. The dynamic equations governing vorticity can then be found by taking the curl of the momentum equations and thus

$$\zeta_t + u\zeta_x + v\zeta_y = Re^{-1}\nabla^2\zeta. \tag{11.4}$$

Note that the pressure is now absent from the dynamic equation of interest and that boundary conditions on vorticity must now be applied. However, there are no physical boundary conditions for vorticity. Such boundary conditions must be approximated.

For the streamfunction–vorticity formulation, a relationship between streamfunction (ψ) and velocity (u, v) and vorticity (ζ) is introduced. This is

$$u = \psi_y, \quad v = -\psi_x, \quad \text{and} \quad -\zeta = \nabla^2\psi. \tag{11.5}$$

This definition of streamfunction inherently satisfies conservation of mass in the continuity equation (11.3). The velocity–streamfunction relationships (11.5) are substituted into equation (11.4), leading to the streamfunction–vorticity equation or

$$\zeta_t + \psi_y\zeta_x - \psi_x\zeta_y = Re^{-1}\nabla^2\zeta. \tag{11.6}$$

The boundary conditions for the streamfunction become homogeneous Dirichlet and Neumann conditions.

Finally, by substituting the streamfunction relationship with vorticity into equation (11.6), the streamfunction formulation of the Navier–Stokes equations is obtained, or

$$\nabla^2 \psi_t + \psi_y \nabla^2 \psi_x - \psi_x \nabla^2 \psi_y = Re^{-1} \nabla^2 (\nabla^2 \psi), \qquad (11.7)$$

and two second-order momentum equations become one fourth-order equation with one dependent variable, ψ.

Each of the four formulations have distinct advantages and disadvantages when it comes to numerically solving the respective systems of equations. Most often, either the primitive variables or the velocity–vorticity formulation is used to represent conservation of mass and momentum for hydrodynamic stability computations. For the remainder of this chapter, the discussion of direct numerical simulation will focus on solutions of the primitive variable formulation, or (11.1)–(11.3).

In applications of hydrodynamic stability, it is customary to first obtain a time-independent (mean or basic state) solution, and then compute the evolution of perturbations to that basic flow state. The final instantaneous solutions are then a composite of the basic mean and disturbance solutions. The instantaneous velocities and pressure are given by

$$\{u,v\}(x,y,t) = \{U,V\}(x,y) + \{u,v\}(x,y,t),$$

$$p(x,y,t) = P(x,y) + p(x,y,t). \qquad (11.8)$$

After substituting the instantaneous values (11.8) into the Navier–Stokes equations (11.1)–(11.3) and subtracting the mean values, the conservation of mass and momentum equations for the disturbances result. These are

$$u_t + (U+u)u_x + (V+v)u_y + uU_x + vU_y = -p_x + Re^{-1}\nabla^2 u, \qquad (11.9)$$

$$v_t + (U+u)v_x + (V+v)v_y + uV_x + vV_y = -p_y + Re^{-1}\nabla^2 v, \qquad (11.10)$$

and

$$u_x + v_y = 0. \qquad (11.11)$$

This disturbance formulation has a distinct numerical advantage over directly solving equations (11.1)–(11.3) for the instantaneous quantities. Because the basic state is typically many orders of magnitude larger than the disturbance values (initially at least), round-off errors can be avoided by solving for the basic state and disturbance solutions separately.

Before we proceed to solution methodologies for the Navier–Stokes equations and for completeness' sake a summary of some comments by Zang (1991) and Vasilyev (2000) will be presented here relative to the form of the Navier–Stokes equations used in numerical simulations. Zang discussed potential errors with the different forms of the Navier–Stokes equations. While equations (11.1)–(11.3) are referred to as the convective form, additional forms include skew-symmetric, rotational and divergence. The skew-symmetric form is given by

$$u_t + \frac{1}{2}\left[(U+u)u_x + (V+v)u_y + uU_x + vU_y\right]$$
$$+ \frac{1}{2}\left[(2Uu + uu)_x + (Uv + Vu + uv)_y\right]$$
$$= -p_x + Re^{-1}\nabla^2 u, \tag{11.12}$$

and

$$v_t + \frac{1}{2}\left[(U+u)v_x + (V+v)v_y + uV_x + vV_y\right]$$
$$+ \frac{1}{2}\left[(Uv + Vu + uv)_x + (2Vv + vv)_y\right]$$
$$= -p_y + Re^{-1}\nabla^2 v. \tag{11.13}$$

The rotational form of the Navier–Stokes equations is

$$u_t - v\zeta = -p_x + Re^{-1}\nabla^2 u, \tag{11.14}$$

and

$$v_t + u\zeta = -p_y + Re^{-1}\nabla^2 v, \tag{11.15}$$

where p is the total pressure in the rotational equations. Finally, the divergence form of the equations is

$$u_t + (2Uu + uu)_x + (Vu + Uv + uv)_y = -p_x + Re^{-1}\nabla^2 u, \tag{11.16}$$

and

$$v_t + (Vu + Uv + uv)_x + (2Vv + vv)_y = -p_y + Re^{-1}\nabla^2 v. \tag{11.17}$$

Zang (1991) concluded from his work, using a series of detailed numerical simulations, that the rotational form produces more aliasing errors, which, in turn, contaminate the solutions compared with the convective, skew symmetric or the divergence form of the equations. However, all forms converge to the same solution as the grids are refined. Vasilyev (2000) conducted a renewed look at the different forms of the Navier–Stokes equations on uniform and non uniform grids, discussing the conservative/nonconservative properties with the different forms. Vasilyev summarized that uniform versus non-uniform meshes and the order of the numerical scheme (e.g., second-order) can impact conservation of mass, momentum and kinetic energy properties, depending on the form of the equations. So some attention should be paid to the numerical formulations, results and interpretation of the results.

In the remaining sections of this chapter, the Navier–Stokes equations for the disturbances will be used to describe two fundamentally different solution approaches for hydrodynamic stability applications. Although the convective form of the equations (11.9)–(11.11) are used hereafter, all forms of the equations are quite similar and, in practice, it is quite easy to code all forms by employing a simple switching method with if-then loops. For example, NFORM is an integer that the user prescribes at the beginning of a computation, whereby

$$\text{if NFORM} = 1, \quad \text{then use convective form;}$$
$$\text{if NFORM} = 2, \quad \text{then use divergence form;}$$
$$\text{else,} \quad \text{use skew-symmetric form.}$$

This provides a very simple means of evaluating changes in conservation properties while performing a simulation.

11.3 Temporal DNS Formulation

The temporal formulation for DNS parallels and complements the discussions in Chapters 2 and 3. A temporal formulation implies that the disturbances amplify or decay in time yielding complex frequencies. The spatial representation of disturbances are real and periodic. For most hydrodynamic stability applications of relevance, one or more spatial directions are nonhomogeneous. If one direction is nonhomogeneous, the temporal formulation can effectively and efficiently be used to compute the instability evolution. For two or three directions of nonhomogeneity, the temporal approach will give trends of behavior, but with some quantitative deficiency because some important spatial features of the flow would be neglected. If quantitative accuracy is required, the spatial formulation may be required and is described in the next section. The temporal formulation may be well suited to the study of absolute instabilities where the disturbance amplifies in time and is fixed in space. In addition, if the spatial changes are of minor importance, then the temporal approach may yield results quantitatively close to the true physical results.

Consider here the case with only one nonhomogeneous direction. For example, channel flow between parallel plates or pipe flow are good examples.

Let the nonhomogeneous direction be the y-coordinate; in this direction a variety of discretization approaches are available. Spectral collocation, finite-difference and compact-difference methods have all been successfully applied to hydrodynamic stability. In the homogeneous direction (x or z coordinates), it is customary to use Fourier series for approximation of the flow because of the efficiency of the fast Fourier transform methods. Second-order Adams–Bashforth with implicit Crank–Nicolson for the y diffusion terms, or a low storage Runge–Kutta approach, have all been used for time advancement. For the temporal formulation, these discretization approaches can reduce to the system for each spatial wavenumber, $k\alpha$, given by

$$\mathbf{L}(\underline{u}^{n+1}, p)\Big|_k = \mathbf{F}(\underline{u}^n, p^n)\Big|_k, \qquad (11.18)$$

where $\mathbf{L}$ and $\mathbf{F}$ are matrix operators and n are known solutions at current time $n \cdot t$ and $n+1$ are desired (to be computed) solutions at the next time step. The system can be solved with either a direct or an iterative approach.

To visualize what is occurring with this temporal system, imagine a computational grid that has one or two temporal periods of the dominant instability mode and a grid in the y-direction. The x coordinate would be the streamwise direction. The Fourier series in the x coordinate suggests that this computational box moves downstream with time advancement. Effectively, the periodicity of the Fourier series means that the outflow in x winds back to become new inflow conditions. Hence, after the initial conditions are applied at the beginning of the simulation, the system becomes a self-sustaining system, which has only boundary conditions to enforce.

Because the temporal formulation has been used in the study of both hydrodynamic stability and turbulence since the 1970s, the reader can easily find additional solution strategies in the literature. Refer to Canuto *et al.* (1988), among others, for numerous references using the temporal formulation.

11.4 Spatial DNS Formulation

The spatial formulation for DNS parallels and complements the discussions in Chapters 4 and 5. A spatial formulation implies that the disturbances amplify or decay in space and require complex wavenumbers for description. This formulation can be done for both two- or three-dimensional directions. As such, truly nonhomogeneous features of the flow, such as nonparallel effects, are captured in the computation.

However, the simulation of spatially evolving disturbances is available with a cost penalty. For this formulation, a sufficiently refined grid is required for all of the spatial length scales and for all of time. This means that, if the evolution of a disturbance over a length equal to 40 wavelengths is of interest, then the entire 40 wavelengths must be resolved throughout the computation. To defer some of this penalty, two approaches have successfully been applied in hydrodynamic stability. The first involves the use of a short domain (4–5 wavelengths) to compute the first few periods of disturbance evolution. Then, additional regions of the grid are gradually added in the downstream direction as the disturbance travels further in this direction. This step saves considerable computational cost, but care must be taken so that the disturbance does not reach the outflow of the computational domain prior to adding additional grid in the downstream direction. The second approach (as described by Huai, Joslin & Piomelli, 1999) involves saving all temporal information near the end of one shorter computational domain and using this information for the forcing of a second computational domain positioned downstream of the first. This is schematically shown in Fig. 11.1. This approach is optimal when the station in the first box, where temporal information is being stored for the second computation, has only a few modes of interest (i.e., there are only a few dominant modes in the frequency spectrum). Otherwise, a huge temporal database must be stored so that all of the energy containing spectrum is continued as the inflow of the second computational box.

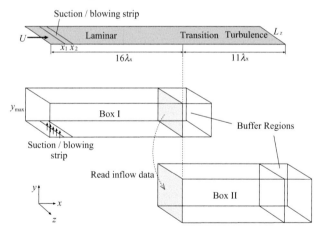

Figure 11.1 Spatial and temporal DNS computational approaches.

Because of the cost penalties associated with the spatial formulation, this approach should be reserved for the situations where the spatial nonhomogeneities are important, or for the quantitative validation of proposed theoretical methods. For example, a two-dimensional flat-plate boundary layer has two nonhomogeneous directions and a swept-wedge flow has three directions of nonhomogeneity. And, as shown in Chapter 9, the spatial DNS approach was invaluable in validating PSE theory. Additional PSE validation is demonstrated in Section 11.6. Since it is beyond the scope of this text to summarize all possible numerical methods of solution, we will focus on a combined use of spectral and high-order finite-difference techniques to formulate a spatial DNS code for three nonhomogeneous flow directions. The questions at the end of this chapter will contain some problems which will require the same kinds of procedures as outlined here but with different numerical techniques and with the less complex one or two nonhomogeneous directions.

11.4.1 Boundary and Initial Conditions

Once the basic state or mean flow solution is obtained, a simulation for the disturbance equations can begin at time $t = 0$ with no initial conditions. Then, at time $t = t + \Delta t$, the disturbance(s) can be forced at the inflow using a theoretical solution(s) or with some unsteady wall condition. See Fig. 11.2 for different forcing regions and the outflow buffer domain sketched.

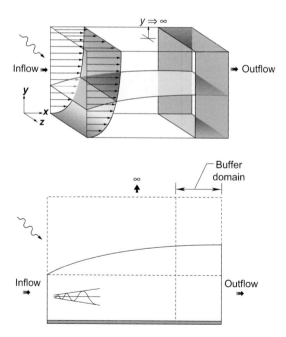

Figure 11.2 Sketch of forcing and buffer regions for computational domain.

Because all of the applications in this chapter are of a boundary layer type, the boundary conditions that will be used at the wall and in the far-field are

$$u, v, w = 0 \quad \text{at} \quad y = 0 \quad \text{and} \quad u, v, w \to 0 \quad \text{as} \quad y \to \infty. \tag{11.19}$$

Thus, the far-field boundary conditions used by the DNS approach are approximate. The mean-flow distortion component of a nonlinearly amplifying disturbance should change in the freestream, but DNS cannot accommodate an a priori Fourier modal boundary condition treatment as the PSE method of Chapter 9.

Time-dependent inflow or wall boundary conditions may be used to generate or force disturbances. Solutions to either the Orr–Sommerfeld or PSE theory equations may be used for inflow forcing. As such, the inflow conditions for the spatial DNS approach appear as

$$\underline{u}_\text{o} = \underline{U}_\text{o} + \sum_{n=-N_z}^{N_z} \sum_{m=-N_t}^{N_t} A_{m,n}^\text{o} \cdot \underline{u}_{m,n}^\text{o}(y) e^{i(n\beta z - m\omega t)}, \tag{11.20}$$

where $\underline{U}_\text{o}$ is the inflow basic components; $A_{m,n}^\text{o}$ the two- and three-dimensional disturbance amplitudes; β is an imposed spanwise wavenumber; and ω is an imposed disturbance frequency. The terms $\underline{u}_{m,n}^\text{o}(y)$ are complex eigenfunctions found either by solving the Orr–Sommerfeld and Squire equations or obtained from a secondary instability theory (see Herbert, 1983). The eigenfunctions $\underline{u}_{m,n}^\text{o}(y)$ are normalized with respect to the maximum streamwise velocity component such that the initial amplitudes of the induced disturbances are prescribed by $A_{m,n}^\text{o}$.

Disturbances can be forcibly imposed into the boundary layer by unsteady suction and blowing with the wall-normal velocity component through the wall (harmonic source generators). An equal amount of mass injected by blowing is extracted by suction so that zero-net-mass is added to the boundary layer. Although the disturbances may be generated by random frequency input, disturbances of interest can be forced with known frequencies. Essentially, this disturbance generator is an alteration to the no-slip boundary conditions that are conventionally used for the wall condition in a viscous flow problem. An example of a common boundary condition is

$$v(x, y = 0, t) = A \cdot \sin^2(\omega \cdot t) \sin(\pi x / \Delta x) \quad x_1 < x < x_2, \tag{11.21}$$

where ω is the frequency of the disturbance of interest, and $\Delta x = x_2 - x_1$ is a somewhat arbitrarily selected forcing region of the boundary. Typically, $\Delta x = \lambda/5$ is an effective means to initiate Tollmien–Schlichting waves. The half-period sine wave for the shape of the forcing is somewhat arbitrary as well.

The buffer-domain technique (see Figs. 11.1 and 11.2) introduced by Streett

& Macaraeg (1989) is used for the outflow condition. Essentially, the elliptic equations (11.9)–(11.11) are parabolized in the streamwise direction using an attenuation function that varies smoothly from one at the beginning of the buffer region (end of physical region of interest) to zero at the outflow boundary.

This method is motivated by the recognition that, for incompressible flow, the ellipticity of the Navier–Stokes equations and thus their potential for upstream feedback, comes from two sources, namely the viscous terms and the pressure field. Examination of earlier unsuccessful attempts at spatial simulations indicated that upstream influence occurred through the interaction of these two mechanisms; strong local velocity perturbations would interact with the condition imposed at the outflow boundary, producing a pressure pulse that was immediately felt everywhere in the domain, especially at the inflow boundary. Therefore both mechanisms for ellipticity have to be treated. To deal with the first source of upstream influence, the streamwise viscous terms are smoothly reduced to zero by multiplying by an appropriate attenuation function in a buffer region, that is appended to the end of the computational domain of interest. The viscous terms are unmodified in the domain of interest. To reduce the effect of pressure-field ellipticity to acceptable levels, the source term of the pressure Poisson equation is multiplied by the attenuation function in the buffer domain. This is akin to the introduction of an artificial compressibility in that region and locally decouples the pressure solution from the velocity computation in the time-splitting algorithm. Thus, in effect the boundary layer equations that determine the solution at outflow are parabolic and hence do not require an added condition at this location. Finally, the advection terms are linearized about the imposed mean or base-flow solution in order that the effective advection velocity that governs the direction of disturbance propagation is strictly positive at outflow even in the presence of large disturbances.

The attenuation function can be defined as

$$s_j = \frac{1}{2}\left[1 + \tanh\left(4 - 8\,\frac{(j - N_b)}{(N_x - N_b)}\right)\right], \qquad (11.22)$$

where N_b marks the beginning of the buffer domain and N_x marks the outflow boundary location. As will be shown later, a buffer domain length of about three streamwise wavelengths is adequate to provide a smooth enough attenuation function to avoid upstream influence for most practical problems.

Alternatively, a moving downstream boundary may be employed to prevent wave reflections. For this boundary-condition treatment, the domain is forever being increased in the downstream direction until sufficient results are obtained in order that the computations may be discontinued.

11.4.2 Time-Marching Methods

For time-marching, a time-splitting procedure can be used with implicit Crank–Nicolson differencing for normal diffusion terms along with either an explicit Adams–Bashforth second-order method or an explicit Runge–Kutta type of method for the remaining terms. Although the Adams–Bashforth approach is easier to code than the Runge–Kutta approach, an order of magnitude more time steps are required to reach the same point in time. This increase in the number of time steps for the Adams–Bashforth approach results from numerical instability of the method if the time-step size is too large. Here, a third-order three-stage Runge–Kutta method (Williamson, 1980) is used for the remaining terms. The pressure is omitted from the momentum equations for the fractional Runge–Kutta stage, leading to

$$\frac{\underline{u}^* - \underline{u}^m}{\Delta t^m} = C_1^m H^m(\underline{u}) + C_2^m Re^{-1} D^2(\underline{u}^* + \underline{u}^m), \qquad (11.23)$$

where

$$H^m(\underline{u}) = L(\underline{u})^m + C_3^m H^{m-1}(\underline{u}), \qquad (11.24)$$

and the operator is given by

$$L(\underline{u}) = (\underline{U} \cdot \nabla)\underline{u} + (\underline{u} \cdot \nabla)\underline{U} + (\underline{u} \cdot \nabla)\underline{u} - Re^{-1}(\underline{u}_{xx} + \underline{u}_{zz}). \qquad (11.25)$$

Here, $\underline{u}^*$ are disturbance velocities at the intermediate Runge–Kutta stages, $\underline{u}^m$ are velocities at previous Runge–Kutta stages ($m = 1, 2$ or 3), $\underline{u}^o$ are velocities at the previous time step, Δt is the time-step size and D is the wall-normal spectral derivative operator. For a full Runge–Kutta stage, the momentum equations with the pressure are

$$\frac{\underline{u}^{m+1} - \underline{u}^m}{\Delta t^m} = C_1^m H^m(\underline{u}^m) + C_2^m Re^{-1} D^2(\underline{u}^{m+1} + \underline{u}^m) - \nabla \wp^{m+1}. \qquad (11.26)$$

After subtracting (11.23) from (11.26) we have

$$\frac{\underline{u}^{m+1} - \underline{u}^*}{\Delta t^m} = -\nabla \wp^{m+1}. \qquad (11.27)$$

By taking the divergence of (11.27) and imposing zero divergence of the flow field at each Runge–Kutta stage, a pressure equation is obtained or

$$\nabla^2 \wp^{m+1} = \frac{1}{\Delta t^m}(\nabla \cdot \underline{u}^*), \qquad (11.28)$$

which is subject to homogeneous Neumann boundary conditions. This boundary condition is justified in the context of a time-splitting scheme, as discussed by Streett & Hussaini (1991). The solution procedure is as follows: The intermediate Runge–Kutta velocities $(\underline{u}^*)$ are determined by solving equation (11.23). The pressure correction $(\wp^{m+1})$ is found by solving (11.28). Then, the full Runge–Kutta stage velocities $(\underline{u}^{m+1})$ are obtained from (11.27). Upon solving the system three consecutive times and full time step, $(n+1)$ velocities are determined, where $\underline{u}^{n+1} = \underline{u}^{m=3}$. The Runge–Kutta coefficients and time steps given by Williamson (1980) are

$$\begin{pmatrix} C_1^1 & C_2^1 & C_3^1 \\ C_1^2 & C_2^2 & C_3^2 \\ C_1^3 & C_2^3 & C_3^3 \end{pmatrix} = \begin{pmatrix} 1 & 1/2 & 0 \\ 9/4 & 1/2 & -4 \\ 32/15 & 1/2 & -153/32 \end{pmatrix}, \quad (11.29)$$

and

$$\begin{pmatrix} \Delta t^1 \\ \Delta t^2 \\ \Delta t^3 \end{pmatrix} = \begin{pmatrix} 1/3\Delta t \\ 5/12\Delta t \\ 1/4\Delta t \end{pmatrix}, \quad (11.30)$$

where the sum of the three Runge–Kutta time steps equals the full time step (Δt). For details of the time-marching procedure, refer to Joslin, Streett & Chang (1993). The much simplified Adams–Bashforth time-marching approach can be used in place of the Runge–Kutta approach, as documented in Joslin, Streett & Chang (1992).

11.4.3 Spatial Discretization Methods

For discretization in the streamwise direction, fourth-to sixth-order central and compact differences have been demonstrated in numerous studies to be sufficiently accurate. Here, fourth-order central finite differences are used for the pressure equation. At boundary and near boundary nodes, third- or fourth-order differences can be used. For first and second derivatives in the momentum equations, sixth-order compact differences are used. As described by Lele (1992), the difference equations are

$$\frac{1}{3}f'_{i-1} + f'_i + \frac{1}{3}f'_{i+1} = \frac{7}{9\Delta x}(f_{i+1} - f_{i-1}) + \frac{1}{36\Delta x}(f_{i+2} - f_{i-2}), \quad (11.31)$$

and

$$\frac{2}{11}f''_{i-1} + f''_i + \frac{2}{11}f''_{i+1} = \frac{12}{11\Delta x^2}(f_{i+1} - 2f_i + f_{i-1})$$

$$+ \frac{3}{44\Delta x^2}(f_{i+2} - 2f_i + f_{i-2}), \quad (11.32)$$

where Δx is the uniform streamwise step size, and f is an arbitrary function whose derivatives are sought. At boundary and near-boundary nodes, explicit fifth-order finite differences are used (e.g., Carpenter, Gottlieb, & Abarbanel, 1993). The discretization yields a pentadiagonal system for the finite difference scheme and a tridiagonal system for the compact-difference scheme, where both can be solved efficiently by LU decomposition with appropriate backward and forward substitutions.

In both the wall-normal (y) and spanwise (z) directions, Chebyshev series are used to approximate the disturbances at Gauss–Lobatto collocation points. A Chebyshev series is used in the wall-normal direction because it provides good resolution in the high-gradient regions near the boundaries. Furthermore, the use of as few grid points as possible results in significant computational cost savings. In particular, the use of the Chebyshev series enables an efficient pressure solver. Since this series and its associated spectral operators are defined on $[-1,1]$ and the physical problem of interest has a truncated domain $[0, y_{max}]$ and $[-z_{max}, z_{max}]$, transformations are employed. Furthermore, stretching functions are used to cluster the grid near both the wall and the attachment line. Here an algebraic mapping is used, namely

$$ y = \frac{y_{max} s_p (1 + \bar{y})}{2 s_p + y_{max}(1 - \bar{y})} \quad \text{or} \quad \bar{y} = \frac{(2 s_p + y_{max}) y - y_{max} s_p}{y_{max}(s_p + y)}, \tag{11.33} $$

where $0 \le y \le y_{max}$ and $-1 \le \bar{y} \le 1$; y_{max} is the wall-normal distance from the wall to the far-field boundary in the truncated domain; and s_p controls the grid stretching in the wall-normal direction.

The solution is determined on a staggered grid. The intermediate Runge–Kutta velocities are determined on Gauss–Lobatto points. To avoid the use of pressure boundary conditions, the pressure is found by solving the Poisson equation on Gauss points and is then spectrally interpolated onto Gauss–Lobatto points. Then, the full Runge–Kutta stage velocities are obtained from on Gauss–Lobatto points with the updated pressure. The above system found in (11.23), (11.26) and (11.28) is solved three consecutive times to obtain full time-step velocities.

To satisfy global mass conservation, an influence matrix method is employed and is described in some detail by Streett & Hussaini (1991), Danabasoglu, Biringen & Streett (1990, 1991) and Joslin, Streett & Chang (1992, 1993). For boundary layer flow, four Poisson–Dirichlet problems are solved for the discrete mode that correspond to the zero eigenvalue of the system and single Poisson–Neumann problems are solved for all other modes.

11.4.4 Influence Matrix Method

An influence matrix method is employed to solve for the pressure. Streett & Hussaini (1991) used it for the Taylor–Couette problem and later Danabasoglu,

Biringen & Streett (1990) used the method for the two-dimensional channel flow problem. Instead of solving a Poisson–Neumann problem, two Poisson–Dirichlet problems are solved.

The solution of the following Poisson–Dirichlet problem, which is the pressure like equation, is sought:

$$\nabla^2 \wp = F \quad \text{in} \quad \wp, \qquad \wp_n = 0 \quad \text{on} \quad \partial\Gamma, \tag{11.34}$$

where Γ is the computational domain, $\partial\Gamma$ is the computational boundary and $\wp_n$ indicates a derivative of the pressure-like quantity normal to the boundary, $\partial\Gamma$. To accomplish this, a sequence of solutions to the following problem is first determined:

$$\nabla^2 \wp^j = 0 \quad \text{in} \quad \Gamma, \qquad \wp^j = \delta_{ij} \quad \text{on} \quad \partial\Gamma \tag{11.35}$$

for each discrete boundary point, x_j. The δ_{ij} is the Dirac delta function defined as $\delta_{ij} = 1$ for $i = j$, and $\delta_{ij} = 0$ for $i \neq j$. Upon computing the vector of normal gradients $\wp_n^j$ at all of the boundary points, these vectors are then stored in columns, yielding a matrix that is referred to as the influence matrix, or

$$I_{NF} = [\wp_n^1, \wp_n^2, \dots, \wp_n^{N_B}], \tag{11.36}$$

where N_B is the number of boundary points.

For two-dimensional problems, the influence matrix, which is dense, is of order $N_B \times N_B$ and of order $N_B \times N_B \times N_z$ for three-dimensional problems; it is dependent on the computational mesh only. Because of this dependence, it needs to be calculated only once for a given geometry. The memory requirements for the influence matrix for a three-dimensional problem can quickly become overbearing and thereby eliminate the possibility of performing simulations into later stages of transition as a result of such insufficient memory.

The composed influence matrix gives the residuals of $\wp$ as a result of the unit boundary condition influence, or

$$[I_{NF}]\wp = \text{residual}. \tag{11.37}$$

The value of one boundary condition is temporarily relaxed so that the problem is not over-specified. This is done by setting one column of the influence matrix to zero, except for the boundary point of interest, which is set to unity. The corresponding residual in (11.37) is exactly zeroed.

The Poisson equation with Neumann boundary conditions is equivalent to the following solution of a Poisson problem and a Laplace problem (or Helmholtz problems) with Dirichlet boundary conditions. First, solve

$$\nabla^2 \wp^J = F \quad \text{in} \quad \Gamma, \qquad \text{and} \qquad \wp^J = 0 \quad \text{on} \quad \partial\Gamma. \tag{11.38}$$

Again, compute the gradients normal to the boundary, $\wp_n^J$. This gives the influence of the right-hand side, F, on the boundary. Then, solve

$$\nabla^2 \wp^{JI} = 0 \quad \text{in} \quad \Gamma \tag{11.39}$$

subject to the boundary constraint

$$\wp^{JI} = I_{NF}^{-1} \cdot \wp_n^J \quad \text{on} \quad \partial\Gamma. \tag{11.40}$$

The final solution that satisfies the original problem and boundary conditions is $\wp = \wp^J - \wp^{JI}$.

Since the gradient or boundary condition at one discrete boundary point was relaxed in the influence matrix formulation, the desired condition ($\wp_n = 0$) may not hold at that boundary point. And, this may result since the discrete compatibility relation may not hold for the pure Neumann problem. In order to regain this boundary condition, the pressure problem (11.38)–(11.40) is resolved, but this time adding a nonzero constant (say 0.01) to the right-hand side of (11.38). A pressure correction ($\overline{\wp}$) results. The composite solution satisfies the boundary conditions at all discrete nodes and then consists of a linear combination of $\wp$ and $\overline{\wp}$. This combination is found by satisfying the following two equations:

$$a_1 \wp_n + a_2 \overline{\wp}_n = 0 \quad \text{on} \quad \partial\Gamma_i \quad \text{and} \quad a_1 + a_2 = 1. \tag{11.41}$$

The final pressure correction ($\wp^{m+1}$) is then given by

$$\wp^{m+1} = a_1 \wp + (1 - a_1)\overline{\wp} \quad \text{with} \quad a_1 = \overline{\wp}_n / (\overline{\wp}_n - \wp_n). \tag{11.42}$$

As a note, the corner points are not included in the discretization and are used in the tangential slip velocity correction only. The pressure at the corners are of minor significance and interpolations are sufficient to compute pressures used for the two-dimensional or zeroth wavenumber mode for three-dimensional problems.

To efficiently solve the resulting Poisson problem, the tensor product method of Lynch, Rice & Thomas (1964) is used in addition to the influence matrix method. The discretized form of the Poisson equation for the pressure is

$$\left(L_x \otimes I \otimes I + I \otimes L_y \otimes I + I \otimes I \otimes L_z \right) \wp = R, \tag{11.43}$$

where $\wp$ is the desired pressure solution; R results from the time splitting procedure; I is the identity matrix; L_x is the streamwise directed central finite difference operator; L_y and L_z are the wall-normal directed and spanwise directed spectral operators; and $\otimes$ implies a tensor product. By decomposing the operators L_y and L_z into their respective eigenvalues and eigenvectors, we find

$$L_y = Q\Lambda_y Q^{-1} \quad \text{and} \quad L_z = S\Lambda_z S^{-1}, \tag{11.44}$$

where Q and S are the eigenvectors of L_y and L_z, Q^{-1} and S^{-1} are inverse matrices of Q and S, and Λ_y and Λ_z are the eigenvalues of L_y and L_z. The solution procedure reduces to the following sequence of operations to determine the pressure $\wp$:

$$\left.\begin{aligned}
\wp^* &= (I \otimes Q^{-1} \otimes S^{-1})R, \\
\wp^\dagger &= (L_x \otimes I \otimes I + I \otimes \Lambda_y \otimes I + I \otimes I \otimes \Lambda_z)^{-1}p^*, \\
\wp &= (I \otimes Q \otimes S)p^\dagger.
\end{aligned}\right\} \qquad (11.45)$$

Because the number of grid points in the attachment-line direction is typically an order of magnitude larger than the wall-normal and flow acceleration directions, the operator L_x is much larger than both L_y and L_z. Since L_x is large and has a sparse pentadiagonal structure and since Λ_y and Λ_z influence the diagonal only, an LU decomposition is performed for the second stage of equation (11.45) once, and forward and backward solvers are performed for each time step of the simulation. The first and third steps to solve for the pressure from equation (11.45) involve matrix multiplications.

To obtain the attachment-line-directed operator L_x, central finite differences are used. To find the wall-normal L_y and flow acceleration L_z operators, the following matrix operations are required:

$$L_y = I_{\text{GL}}^{\text{G}} D_y \tilde{D}_y I_{\text{G}}^{\text{GL}} \qquad \text{and} \qquad L_z = I_{\text{GL}}^{\text{G}} D_z \tilde{D}_z I_{\text{G}}^{\text{GL}}, \qquad (11.46)$$

where D_y is a spectral wall-normal derivative operator for the stretched grid, D_z is the spectral derivative operator that is grid clustered in the spanwise region and $\tilde{D}_y$ and $\tilde{D}_z$ are the derivative operators with the first and last rows set to 0. The interpolation matrix I_{GL}^{G} operates on variables at Gauss–Lobatto points and transforms them to Gauss points; the interpolation matrix I_{G}^{GL} performs the inverse operation. The spectral operators are described in detail by Streett & Chang (1993), Canuto *et al.* (1988) and Joslin, Streett & Chang (1993). The operators $\{L_x, L_y, L_z\}$, the eigenvalue matrices $\{\Lambda_y, \Lambda_z\}$, the eigenvector matrices $\{Q, Q^{-1}, S, S^{-1}\}$ and the influence matrix are all mesh-dependent matrices and must be calculated only once.

The description in this section of a DNS methodology is a rather general but complex code that permits solutions of hydrodynamic instability studies in three-dimensional, nonhomogeneous directions. Very often one or more of the directions are homogeneous and the solution approach can be simplified. For example, if the spanwise z-direction was homogeneous, as is the case for the Blasius boundary layer, then Fourier series can be used to approximate disturbances in that direction. Then, one can essentially solve a set of two-dimensional equations for each mode of the Fourier series. Equation (11.28) simply becomes

$$\nabla^2 p^{m+1}(x,y) - \beta^2 p^{m+1}(x,y) = \frac{1}{h_t^m}(\nabla \cdot \underline{u}^*),\qquad (11.47)$$

which is now a two-dimensional equation in Fourier transform space. Upon obtaining a solution, a Fourier transform yields the pressure needed to update the velocity in (11.27).

11.5 Large Eddy Simulation

Here, the Large Eddy Simulation (LES) methodology is discussed primarily for completeness sake with its relevance to hydrodynamic stability. LES implies that the large scales present in the flow are directly computed on a sufficiently fine grid and the scales smaller than this grid are modeled. Inherent in the approach is the assumed existence of a range of relevant scales and an adequate model for the smaller scales. Although the LES approach is most appropriate for the study of turbulent flows, the method has been used in hydrodynamic stability for the study of the highly nonlinear disturbance region associated with the transition from laminar to turbulent flow. Therefore, the basis of LES will be outlined here and some limited results will be presented in the next section.

While the application of LES to turbulent flows dates back to the 1960s, where it was used to study atmospheric turbulence, this technique has only recently been used for the study of transitional flow. Piomelli *et al.* (1990), Piomelli & Zang (1991) and Germano *et al.* (1991) computed the transition in temporally developing boundary layer and plane channel flow. Their results indicate that, at the early stages of transition, the eddy viscosity must be inactive to allow the correct growth of the perturbations. The dynamic model (Germano *et al.*, 1991) achieves this result without the *ad hoc* corrections required by other models; e.g., the Smagorinsky (1963) model.

As with the DNS approach, the present technique begins with the disturbance form of the Navier–Stokes equations (11.9)–(11.11). In LES, the large-scale (i.e., grid-resolved) components of the velocity and pressure are calculated and the effects of the small, unresolved scales are modeled. By applying the filtering operation,

$$\bar{f}(\underline{x}) = \int_\Gamma f(\underline{x}')G(\underline{x},\underline{x}')d\underline{x}' \qquad (11.48)$$

to (11.9)–(11.11), where G is the filter function and Γ is the entire domain, the governing equations for the large-scale velocity and pressure can be obtained:

$$\frac{\partial \bar{u}_i}{\partial t} + \bar{u}_j \frac{\partial \bar{u}_i}{\partial x_j} + U_j \frac{\partial \bar{u}_i}{\partial x_j} + \bar{u}_j \frac{\partial U_i}{\partial x_j} = -\frac{\partial \bar{p}}{\partial x_i} - \frac{\partial \tau_{ij}}{\partial x_j} + Re^{-1} \frac{\partial^2 \bar{u}_i}{\partial x_j \partial x_j}, \quad (11.49)$$

and

$$\frac{\partial \bar{u}_i}{\partial x_i} = 0, \quad (11.50)$$

where τ_{ij} is the subgrid scale (SGS) stress tensor given by $\tau_{ij} = \overline{u_i u_j} - \bar{u}_i \bar{u}_j$, which must be modeled.

The modeling of τ_{ij} remains a topic of considerable research for both transitional and turbulent flows. We will proceed no further with our discussion of LES since it goes beyond the scope of this text, but the student should take note of the similarities between the LES equations (11.49)–(11.50) and the DNS equations. As an LES computation uses a finer and finer grid, the term with the τ_{ij} component tends to zero and the computation becomes a DNS approach.

11.6 Applications

In this section, some sample applications are explored using the spatial DNS approach described in Section 11.4 and the LES approach described in Section 11.5. However, the mean flow can be obtained using computational fluid dynamics (CFD). Because the first and second derivatives of the mean flow are required for theoretical investigations, significant computational cost results and it becomes more practical to use known analytical mean or basic states. With analytical basic states, the DNS and the theoretical analysis begin with the same mean flows. Here, disturbance evolution in a two-dimensional flat-plate boundary layer flow and an attachment-line flow are reviewed.

11.6.1 Tollmien–Schlichting Wave Propagation

By using the Blasius similarity profile to represent the flat-plate boundary layer, we can quantify various transitional flow mechanisms of interest from the linear region to the nonlinear breakdown stage and provide a critical comparison of results for the spatial DNS and PSE theory. To date, this comparison offers the most rigorous test of the PSE approach for accuracy and the main focus will be to point out strengths and potential weaknesses of PSE theory as well as the impact of these weaknesses on the overall flow field prediction. To accomplish this goal, three test cases are computed by spatial DNS and then compared to PSE theory:

(1) two-dimensional Tollmien–Schlichting wave propagation;
(2) subharmonic breakdown; and
(3) oblique-wave breakdown.

The equations are nondimensionalized with respect to the freestream velocity U_∞, the kinematic viscosity ν and displacement thickness δ_0^* at the inflow as the length scale. A Reynolds number is then be defined as $Re_{\delta_0^*} = U_\infty \delta_0^* / \nu$.

A Tollmien-Schlichting disturbance with a root mean squared (rms) amplitude $A_{1,0}^o = 0.0025$ is introduced into the boundary layer by a forcing at the inflow for the DNS as well as for the PSE calculations. For reference, recall the definition of $A_{m,n}^o$ as given in (11.20). Through nonlinear interactions, all other harmonic waves including the mean flow distortion are generated for both DNS and PSE. Calculations are made with an inflow Reynolds number $Re_{\delta_0^*} = 688.315$ and frequency $F = 86$. To generate resolved benchmark data to test the PSE theory, the spatial DNS was computed on a grid of 2,041 uniformly spaced streamwise nodes (60 nodes per disturbance wavelength) and 81 wall-normal collocation points. The outflow boundary is $442\delta_0^*$ from the inflow boundary, and the far-field (or freestream) boundary is $75\delta_0^*$ from the wall. The DNS parameters were chosen based on convergence studies by Joslin, Streett & Chang (1992). For the time-marching scheme, the disturbance period is divided into 320 time steps. For the PSE computational approach, several numerical experiments have been performed by varying the grid, far-field boundary location and the number of Fourier modes. These numerical experiments led to the choice of 100 wall-normal grid points, 5 frequency modes of series (11.20) ($N_t = 6$) and a far-field boundary located $58\delta_0^*$ from the wall.

Figure 11.3 shows the maximum streamwise amplitudes for the mean flow distortion u_0, fundamental wave u_1 and first harmonic u_2 predicted by PSE theory and compared to the DNS results with the downstream distance. Both the fundamental waves ($F = 1$) and the first harmonics ($F = 2$) are in good quantitative agreement throughout the initial linear region and the later weakly nonlinear region. The mean flow distortion components ($F = 0$) are in good agreement throughout the initial linear region. Later, however, a discrepancy begins to occur downstream at an apparent "notch" in the results at $Re_{\delta^*} = 1,400$. At the local streamwise location $Re_{\delta^*} = 1,519$, Fig. 11.4 shows comparisons of the streamwise velocity component. The fundamental (Tollmien–Schlichting) wave and harmonics are in good quantitative agreement, even in regions of high gradients. A comparison of the streamwise velocity (u_0) profiles illustrates a comparable difference. The discrepancy in the mean flow distortion results identified in Fig. 11.3 arises from the change of profile contributions. This discrepancy is due to the homogeneous Neumann boundary conditions used in the far-field for the mean flow distortion equations in PSE theory. As in the traditional boundary layer equations approach, this boundary condition leads to a nonzero, wall-normal mean flow velocity component in the far-field, as discussed in some detail in Chapter 9.

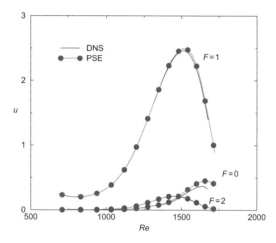

Figure 11.3 Plot of the amplitude growth with downstream distance for the Tollmien–Schlichting mode ($F = 1$), the first harmonic ($F = 2$) and the mean flow distortion ($F = 0$).

In this first test problem of Tollmien–Schlichting wave propagation, the results from DNS and PSE theory agree very well for the fundamental and harmonic waves. However, a discrepancy exists in the mean flow distortion component. This discrepancy is a result of the difference in the treatment of the far-field boundary condition of the two approaches.

11.6.2 Subharmonic Breakdown

A well-understood breakdown scenario in an incompressible boundary layer on a flat plate begins with a predominantly two-dimensional disturbance that emerges downstream into aligned and staggered three-dimensional distinct vortex structures through spanwise vortex stretching and tilting or some indistinct, non-unique combination of vorticities in the later stages. These vortex patterns

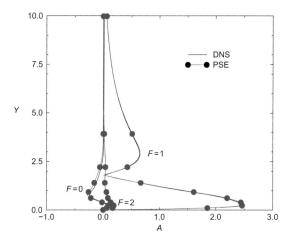

Figure 11.4 Streamwise velocity profile for the Tollmien–Schlichting mode ($F = 1$), the first harmonic ($F = 2$) and the mean flow distortion ($F = 0$) at $Re_{\delta^*} = 1519$.

are referred to as fundamental, subharmonic and combination resonant modes, respectively, and may be described by secondary instability theory and PSE theory (Chapter 9). For the present study, the subharmonic mode of secondary instability breakdown will be computed and the results compared with the experiments of Kachanov & Levchenko (1984).

For spatial DNS, computations are performed on a grid of 1,021 uniformly spaced streamwise nodes, 81 wall-normal collocation points and 5 symmetric-spanwise nodes. In the streamwise direction, the outflow boundary is $442\delta_o^*$ from the inflow boundary, the far-field boundary is $75\delta_o^*$ from the wall and the spanwise boundary consists of a length equal to one half of the spanwise wavelength, or $\lambda_z/2 = \pi/\beta$. Note that the spanwise computational length would be λ_z for the general, nonsymmetric computation. For the time-marching scheme, the disturbance period is divided into 320 time steps and time is advanced using a three stage Runge–Kutta method. For the PSE computational approach, 100 wall-normal grid points are used; 7 frequency modes and 3 spanwise modes from series (9.54) are used and the far-field boundary is $58\delta_o^*$ from the wall.

The prescribed primary and subharmonic disturbances are obtained at the Reynolds number $Re_{\delta^*} = 732.711$ and the primary frequency $F = 124$; values that correspond to the experiments. The primary wave has an inflow rms amplitude of $A_{2,0}^o = 0.0048$. The subharmonic mode has an inflow rms amplitude of $A_{1,1}^o = 0.145 \times 10^{-4}$ and corresponds to a mode with spanwise wavenumber $\beta = 0.2418$.

Figure 11.5 compares the maximum amplitudes predicted by PSE theory with the results from DNS. For this test case, note the extremely good quantitative agreement between PSE theory and DNS for the growth rates of the fundamental and the dominant harmonic modes. Even the mean flow distortion components are in good agreement. The DNS profiles are compared with the experiments in Fig. 11.6 at the local downstream Reynolds number $Re_{\delta*} = 1,049$. For this test case, it can be seen that DNS results agree with both the PSE predictions and the experiments.

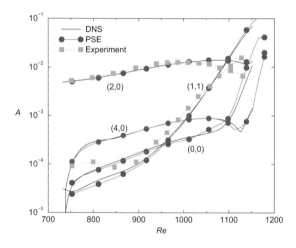

Figure 11.5 Plot of the amplitude growth with downstream distance for the Tollmien–Schlichting wave (2,0), subharmonic (1,1), first harmonic (4,0) and mean flow distortion (0,0).

11.6.3 Oblique-Wave Breakdown

The final test case that uses both DNS and PSE theory is that of oblique-wave breakdown. Oblique-wave breakdown is due to the nonlinear interactions of a pair of oblique waves. Because of this nonlinearity, no adequate formal theory is available to explain the breakdown process. However, similar mechanisms have been studied by Hall & Smith (1991) by use of asymptotic methods. Hall and Smith discussed the vortex wave interactions within a large wavenumber

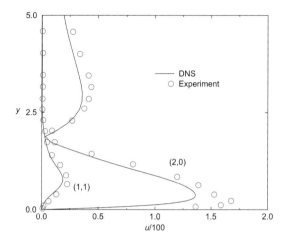

Figure 11.6 Streamwise velocity profile of the Tollmien–Schlichting wave (2,0) and subharmonic (1,1) at $Re_{\delta^*} = 1049$.

and Reynolds number limit. To quantify the mechanisms of interest in the finite Reynolds number range, DNS and possibly PSE theory are available to study the wave interactions.

Because of this alternative route to, or mechanism of, nonlinear transition, limited research has been done for oblique-wave breakdown. It is worth citing a few of the interesting papers on this topic that are available. Schmid & Henningson (1992a) studied bypass transition by introducing a pair of large-amplitude oblique waves into channel flow. The evolution of disturbances was computed with temporal DNS. They found that the development of the oblique waves was dominated by a preferred spreading of the energy spectra into low streamwise wavenumbers, and this led to the rapid development of streamwise elongated structures. Schmid & Henningson (1992b) also investigated small-amplitude wave pairs over a variety of parameters. They suggested that the mechanism of energy transfer is primarily linear.

For this example problem, the profiles for the oblique-wave pair are obtained from linear stability theory for the Reynolds number $Re_{\delta_0^*} = 900$, frequency $\omega = 0.0774$ and spanwise wavenumbers $\beta = \pm 0.2$. Spatial DNS computations are performed on a grid of 901 uniformly spaced streamwise nodes, 61 wall-normal collocation points and 10 symmetric-spanwise modes. In the streamwise direction, the outflow boundary is $465\delta_0^*$ from the inflow boundary.

The far-field, or freestream boundary, is $75\delta_0^*$ from the wall, and the spanwise boundary consists of a length equal to one half of the spanwise wavelength, or $\lambda_z/2 = \pi/\beta$. For the time-marching scheme, the disturbance period is divided into 320 time steps. For the PSE computational approach, 100 wall-normal grid points are used, seven frequency modes and seven spanwise modes of series (9.54) are used and the far-field boundary is $58\delta_0^*$ from the wall.

The input modes are represented by (11.20) that are truncated to four terms. The disturbance forcing consists of modes $(1,1)$ and $(1,-1)$, or (ω,β) and $(\omega,-\beta)$, and their complex conjugates $(-1,1)$ and $(-1,-1)$. Theoretically, if these modes self-interact initially, then only certain higher modes are likely to be excited (supplied energy). These higher modes are: $(0,0)$, $(0,2)$, $(2,0)$ and $(2,2)$, etc.

The oblique waves each have the small amplitude $A_{1,1}^0 = 0.001$. In Fig. 11.7, the primary disturbance $(1,1)$ and the higher modes that were predicted by PSE theory are compared to the DNS results. The comparison shows that the modes are in quantitative agreement. Of the modes that were likely to be excited, all received energy initially. The streamwise vorticity component $(0,2)$ grows rapidly because of the self-interaction of the oblique wave pair. All other modes grow more slowly downstream than the streamwise vortex, and these other modes contain less energy by orders of magnitude. As a result of the rapid growth of the vortex mode $(0,2)$, the oblique waves interact with the vortex, and this leads to an amplified harmonic $(1,3)$. This $(1,3)$ mode gains sufficient energy to overtake the other initially excited modes, but is insufficient to overtake the oblique waves. As shown in Fig. 11.7, the vortex modes self-interact to supply energy to the $(0,4)$ mode that has roughly the same growth rate as the $(0,2)$ mode. Although the computations were discontinued, the disturbances will eventually decay and will not lead to transition because the primary oblique waves decay after they pass the upper branch of the neutral curve, and all other modes are decaying or becoming neutrally stable.

11.6.4 Attachment-Line Flow

On a swept wing, many instability mechanisms exist that can lead to catastrophic breakdown of laminar to turbulent flow. Along the leading edge, Tollmien–Schlichting waves, stationary or traveling crossflow vortices, Taylor–Görtler vortices or combinations of these modes are among the many mechanisms that can lead to such breakdown. In this section we only consider instabilities in the attachment-line region.

Figure 11.8 shows a sketch of the three-dimensional flow field representative of the attachment-line region. The freestream flow approaches the leading

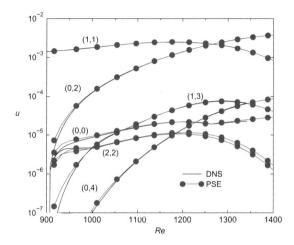

Figure 11.7 Amplitude growth with downstream distance from oblique wave pair with initial amplitudes of $A_{1,1} = 0.001$.

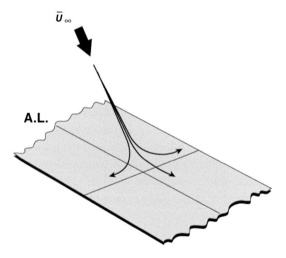

Figure 11.8 Sketch of attachment-line flow.

edge of the wing and is diverted above and below the wing, thus forming an attachment line in the leading edge region. The exact location of the attachment line varies with geometry and angle of attack of the wing. When the wing is swept with respect to the incoming flow, a boundary layer flow forms along the attachment line, flowing from the root of the wing near the fuselage toward the wing tip. The stability or instability of this attachment-line boundary layer flow is the topic problem for this section.

Contamination at the leading edge results from turbulence at a fuselage and wing juncture that travels out over the wing and contaminates otherwise laminar flow on the wing. If the Reynolds number of the attachment-line boundary layer is greater than some critical value, then this contamination inevitably leads to turbulent flow over the complete wing. This phenomenon has been demonstrated by Pfenninger (1965), Maddalon *et al.* (1990) and others. To correct this problem, Gaster (1965c) placed a bump on the leading edge to prevent the turbulent attachment-line boundary layer from sweeping over the entire wing. This bump had to be shaped to create a fresh stagnation point without generating a detrimental adverse pressure gradient. Outboard of the bump, a new laminar boundary layer forms. With this fresh laminar boundary layer, we can study the stability of the flow.

For small-amplitude disturbances introduced into the attachment-line boundary layer flow, experimental results and linear stability theory indicate that these disturbances begin to amplify at a momentum-thickness Reynolds number of approximately 230 to 245.

By neglecting surface curvature, the governing flow simplifies to a similarity solution of the Navier–Stokes equations and is commonly referred to as swept Hiemenz flow. A Cartesian coordinate system $\underline{x} = (x, y, z)$ is used in which x is aligned with the attachment line, y is wall-normal and z corresponds to the direction of flow acceleration away from the attachment line. The fluid comes obliquely down toward the wall and then turns away from the attachment line into the $\pm z$-directions to form a boundary layer. In the x-direction, the flow is uniform. In the absence of sweep, U_0 is equal to 0 and the flow reduces to the two-dimensional stagnation flow first described by Hiemenz (1911). A length scale (factor of the boundary layer thickness) is defined in the (y, z)-plane as $\delta = \sqrt{\nu L / W_0}$, a Reynolds number, $Re = U_0 \delta / \nu = 2.475 Re_\theta$ and a transpiration constant, $\kappa = V_0 \sqrt{L / \nu W_0}$, where $\kappa = 0$ for the zero suction case; U_0, V_0, W_0 are velocity scales; and L is the length scale in the flow acceleration direction z. If the attachment line is assumed to be infinitely long, the velocities become functions of z and y only, and the similarity solution can be found.

The swept Hiemenz formulation was originally described by Hall, Malik & Poll (1984) where a linear stability analysis of the flow was performed. The respective velocities and pressure for swept Hiemenz flow are $\{u, v, w, p\}$, and the governing equations are given by

$$\frac{\partial U}{\partial X} + \frac{\partial V}{\partial Y} + \frac{\partial W}{\partial Z} = 0, \tag{11.51}$$

$$U\frac{\partial U}{\partial X} + V\frac{\partial U}{\partial Y} + W\frac{\partial U}{\partial Z} = -\frac{\partial P}{\partial X} + Re^{-1}\left[\frac{\partial^2 U}{\partial X^2} + \frac{\partial^2 U}{\partial Y^2} + \frac{\partial^2 U}{\partial Z^2}\right], \tag{11.52}$$

$$U\frac{\partial V}{\partial X} + V\frac{\partial V}{\partial Y} + W\frac{\partial V}{\partial Z} = -\frac{\partial P}{\partial Y} + Re^{-1}\left[\frac{\partial^2 V}{\partial X^2} + \frac{\partial^2 V}{\partial Y^2} + \frac{\partial^2 V}{\partial Z^2}\right], \tag{11.53}$$

and

$$U\frac{\partial W}{\partial X} + V\frac{\partial W}{\partial Y} + W\frac{\partial W}{\partial Z} = -\frac{\partial P}{\partial Z} + Re^{-1}\left[\frac{\partial^2 W}{\partial X^2} + \frac{\partial^2 W}{\partial Y^2} + \frac{\partial^2 W}{\partial Z^2}\right], \tag{11.54}$$

where the equations are nondimensionalized with respect to the attachment line velocity U_o, length scale δ and kinematic viscosity v.

A mean or steady solution of the Navier–Stokes equations is sought that obeys the following conditions. At the wall,

$$u = w = 0 \quad v = V_o, \qquad \text{at} \qquad y = 0, \tag{11.55}$$

and sufficiently far away from the wall,

$$u \to U_o, \quad w \to W_o\frac{z}{L} \qquad \text{as} \qquad y \to \infty. \tag{11.56}$$

The velocity field for this similarity solution is

$$\left.\begin{array}{rcl} U(Y) & = & \hat{u}(Y), \\ V(Y) & = & Re^{-1}\hat{v}(Y), \\ W(Y,Z) & = & Re^{-1}Z\hat{w}(Y). \end{array}\right\} \tag{11.57}$$

With the nondimensional velocities (11.57) in the Z momentum equation, (11.54), we have

$$Re^{-2}Z\hat{v}\frac{d\hat{w}}{dY} + R^{-2}Z\hat{w}^2 = -\frac{\partial P}{\partial Z} + Re^{-2}Z\frac{d^2\hat{w}}{dY^2}. \tag{11.58}$$

As $Y \to \infty$, the Z momentum equation, (11.58), reduces to

$$Re^{-2}Z = -\frac{\partial P}{\partial Z}, \tag{11.59}$$

with solution

$$P = P_o - \frac{1}{2}Re^{-2}Z^2, \tag{11.60}$$

where P_o is the constant pressure at the attachment line.

Substitute the velocity from (11.57) and the pressure from (11.60) into the Navier–Stokes equations (11.50)–(11.54). Then, after substituting the continuity equation into the momentum equations and subtracting the Y and Z momentum equations, the following system of ordinary differential equations for $(\hat{u}, \hat{v}, \hat{w})$ results and is

$$\hat{w} + \frac{d\hat{v}}{dY} = 0, \tag{11.61}$$

$$\frac{d^2\hat{u}}{dY^2} - \hat{v}\frac{d\hat{u}}{dY} = 0, \tag{11.62}$$

and

$$\frac{d^3\hat{v}}{dY^3} + \left(\frac{d\hat{v}}{dY}\right)^2 - \hat{v}\frac{d^2\hat{v}}{dY^2} - 1 = 0, \tag{11.63}$$

subject to the boundary conditions given by

$$\frac{d\hat{v}}{dY} = 0, \quad \hat{v} = \kappa, \quad \hat{w} = 0 \quad \text{at} \quad Y = 0, \tag{11.64}$$

and

$$\frac{d\hat{v}}{dY} \to -1, \quad \hat{w} \to 1 \quad \text{as} \quad Y \to \infty, \tag{11.65}$$

where $\kappa = V_0/U_0$ is a parameter of the system. In the absence of sweep, the equations (11.61)–(11.63) reduce to the famous two-dimensional stagnation flow as first described by Hiemenz (1911).

Note that by virtue of the similarity solution, U and V are uniform along the attachment line and W varies linearly with distance from the attachment line. Because of the properties of this base flow, both temporal and spatial DNS approaches should yield equivalent results in the two-dimensional limit for small-amplitude disturbances. However, the temporal DNS assumes that disturbances are growing in time and that there exists a linear transformation from temporal growth to the realistic spatially growing instabilities.

In general, disturbances on and near a three-dimensional attachment-line region are of the three-dimensional nature, requiring solutions of the full three-dimensional Navier–Stokes equations. However, as assumed in the original theoretical study by Hall, Malik & Poll (1984) and confirmed in the DNS computations by Spalart (1989), a single mode in the attachment-line region of swept Hiemenz flow can take the form that had a linear variation of the chordwise velocity component with distance from the attachment line. In the present study, an alternate disturbance form is first used. Namely, the w-velocity component of the disturbance and the transverse shear of the mean flow are negligible; the disturbance becomes truly two-dimensional along the attachment line. This implies that $w = 0$ and $\partial w / \partial Z = 0$ on the attachment line. Although this simplification is not consistent with the equations of motion, it turns out that the neglected terms have little effect on the qualitative behavior of the computed disturbances. This assumption allows us to use a pre-existing DNS solver that has been tested for two-dimensional instabilities and three-dimensional spanwise periodic disturbances in two-dimensional and three-dimensional base flows. This two-dimensional assumption is arguably valid because the flow is overwhelmingly dominated by the flow in the attachment line direction.

Here, an assessment is made with regard to the value of linear stability theory (Orr–Sommerfeld and Squire equations) in attachment-line flow. Note that linear stability theory involves a quasi-parallel flow assumption (i.e., $V = 0$), and that no amplitude information is included in the theory. The simulations are performed on a grid of 661 points ($\simeq$ 60 points per wavelength) along the attachment line and 81 points in the wall-normal direction. The far-field boundary is located at 50δ from the wall, and the computational length along the attachment line is 216.56δ. This attachment line length corresponds to 11 wavelengths for $Re = 570$ and $\omega = 0.1249$. For the time-marching scheme, the disturbance wavelength was divided into 320 time steps per period for small-amplitude disturbances and into 2,560 time steps for large-amplitude disturbances (stability considerations). Disturbances for the first simulations are forced at the computational inflow with an amplitude of $A = 0.001\%$ (i.e., arbitrary small amplitude). Disturbances that evolve in a base flow that complements the quasi-parallel linear stability theory assumptions ($V = 0$) and the full, swept Hiemenz flow are computed with DNS. Figure 11.9 shows the computed disturbance decay rate and the wavelength in the quasi-parallel flow agree exactly with linear stability theory. The disturbance that propagates in the complete swept Hiemenz flow closely retains the wavelength predicted by linear stability theory, but decays at a slower rate than that predicted by linear stability theory.

Hall & Malik (1986) utilized subcritically growing instabilities with a temporal DNS code and therefore the difference between the weakly nonlinear theory and the previous computations should not be attributable to the temporal DNS approximation. Although many previous studies have made use of the temporal approach because of the computational savings over the spatial formulation, the spatial and temporal formulations are only related in the linear limit, with the spatial formulation being more representative of the true physical problem.

Figure 11.10 shows the evolution of the fundamental wave, the mean-flow distortion and the harmonics from a simulation forced at the inflow with a large amplitude of $A = 12\%$ for the Reynolds number $Re = 570$ and frequency

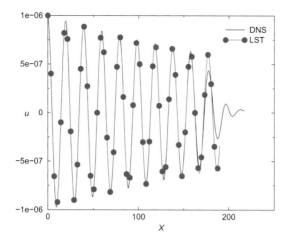

Figure 11.9 DNS and linear stability theory (LST) for parallel attachment-line flow ($Re = 570, \omega = 0.1249$; Samples at $Y = 0.86$).

$\omega = 0.1249$. After a transient region of adjustment, the fundamental wave encounters subcritical growth that is in agreement with the weakly nonlinear theory. Contours of instantaneous streamwise $(U + u)$ and wall-normal $(V + v)$ velocities are shown in Fig. 11.11. Because the disturbance amplitude is sufficiently large, notable distortions in the base flow are observed as a result of the unsteady disturbance forcing.

Finally, the spatial evolution of three-dimensional disturbances is computed by direct numerical simulation that involves the solution to the unsteady nonlinear, three-dimensional Navier–Stokes equations. The simulations are performed on a grid of 661 points ($\simeq 60$ points per wavelength) along the attachment line, 81 points in the wall normal direction, and 25 points in the flow acceleration direction. The far-field boundary is located at 50δ from the wall, the computational length along the attachment line is 216.56δ and the flow acceleration boundaries are located $\pm100\delta$ from the attachment line. For the time-marching scheme, the disturbance wavelength was divided into 320 time steps per period.

To generate three-dimensional disturbances, the flow acceleration length of the harmonic source generator is reduced to enable a more direct transfer of energy to the w velocity component. Disturbances computed in the parameter regime were characterized by a Reynolds number $Re = 570$ and frequency $\omega =$

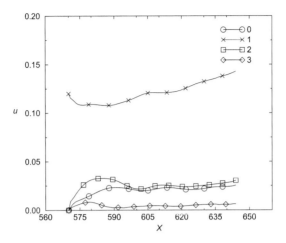

Figure 11.10 DNS and linear stability theory for parallel attachment-line flow
($Re = 570, \omega = 0.1249$).

0.1249. The results of a disturbance generated with a harmonic source located
at $-27.8 < Z < 0.0$ are shown in Fig. 11.12. The top view indicates that the
harmonic source generates a local almost circular pattern that evolves along the
attachment line with spreading both away from and toward the attachment line.
The results imply that a disturbance generated off (but near) the attachment
line can supply energy to the attachment-line region by the spreading of the
wave pattern. In turn, this energy supply may feed an unstable mode on the
attachment line. These results suggest that the flow accelerated shear away
from the attachment line has insufficient strength to deter the spreading of
the disturbance toward the attachment line. More details and results for this
problem can be found in Joslin (1995b, 1997).

11.7 Summary

Here, PSE theory results were evaluated for accuracy in predicting convective
disturbance evolution on a flat plate. PSE theory predictions were compared
with spatial DNS results for two-dimensional Tollmien–Schlichting wave propagation, subharmonic breakdown and oblique wave breakdown.

For two-dimensional Tollmien–Schlichting wave propagation, the modes
predicted by PSE theory were in very good quantitative agreement with the

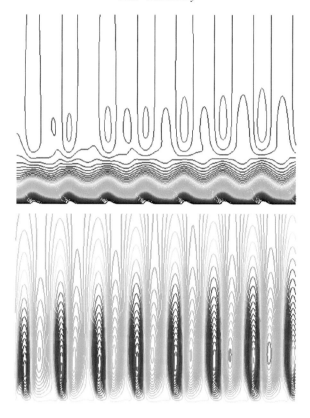

Figure 11.11 Contours of streamwise (top) and wall-normal (bottom) velocities for subcritically growing disturbance in attachment-line boundary layer at $Re = 570$ and $\omega = 0.1249$.

DNS results, except for a small discrepancy in the mean-flow distortion component that was discovered and attributed to far-field boundary condition differences.

For the test case of subharmonic breakdown, the PSE theory results were in very good quantitative agreement with the DNS results for all modes, even the mean-flow distortion component. Also, the present study supports the PSE and DNS comparison made by Herbert (1991) for subharmonic breakdown.

For the complicated test case of oblique-wave breakdown, all modes predicted by PSE theory were shown to be in good quantitative agreement with the DNS results, even for the mean-flow distortion component. Furthermore, these oblique wave pairs were shown to self-interact to excite a streamwise vortex structure, which agrees with the findings of Schmid & Henningson (1992a,b).

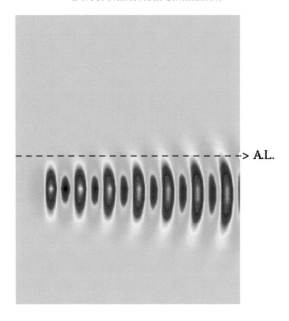

Figure 11.12 Evolution of disturbances in attachment-line boundary layer at $Re =$ 570 and $\omega = 0.1249$, where disturbances are generated with harmonic sources of various lengths. In the figure, A.L. means attachment line.

If the initial wave amplitudes are above a threshold, the interaction of these waves and the vortex can lead to a breakdown that bypasses the secondary instability stage. Irrespective of the initial amplitudes, the streamwise vortex mode becomes the dominant, higher-order mode. This dominance is significant because the presence of small roughness elements may generate oblique-wave packets that can interact and lead to the increased presence of streamwise vorticity.

11.8 Exercises

Exercise 11.1 From equations (11.1)–(11.3), derive the following

(a) the vorticity equation (11.4),
(b) the streamfunction–vorticity equation (11.6),
(c) the velocity–streamfunction equation (11.7).

Exercise 11.2 Beginning with the three-dimensional Navier–Stokes equations in primitive variables, derive disturbances equations in

(a) convective form,
(b) skew symmetric form,
(c) rotational form,
(d) divergent form.

Exercise 11.3   Apply the filter (11.48) to the Navier–Stokes equations to get the LES-filtered equations (11.49) and (11.50).

Exercise 11.4   Develop a core set of subroutine modules that will be used to form a two-dimensional direct numerical simulation code.

(a) Code fourth-order finite difference routines with homogeneous boundary conditions for first and second derivatives in the x-direction. Test this routine by using the function

$$f(x) = A\sin(\alpha x), \quad 0 \le x \le 4\pi.$$

Exercise this routine by varying the amplitude A and wavenumber α. Note that an exact solution exists to compare your numerical solution. These routines will serve to compute directions in the freestream direction.

(b) Code the sixth-order compact finite-difference schemes given by (11.31) and (11.32). Test the routines using the function

$$f(x) = A\sin(\alpha x), \quad 0 \le x \le 4\pi.$$

Exercise this routine by varying the amplitude A and wavenumber α.

(c) Integrate

$$\frac{dT}{dt} = \sin(t)e^{-t}$$

using the Runge–Kutta scheme (11.29)–(11.30). Compute the exact solution and compare to the numerical solution at $t = 1$ for various time steps Δt.

(d) Code Chebyshev collocation routines with homogeneous boundary conditions for first and second derivatives in the y-direction. Test this routine by using the function

$$f(y) = B\cos(n\pi y), \quad -1 \le y \le 1.$$

Exercise this routine by varying the amplitude B. Note that an exact solution exists to compare your numerical solution. These routines will serve to compute derivatives in the wall-normal direction.

(e) Code Fourier transform routines for first and second derivatives in the z-direction. To accomplish this, you must compute the Fourier transform of a function, perform the derivative operation in wavenumber space and inverse Fourier transform back to physical space. Test this routine by using the function

$$f(z) = C + A\sin(\alpha z) + \cos(\beta z), \quad 0 \le z \le 4\pi.$$

Exercise this routine by varying the amplitude A and wavenumber α. Note that an exact solution exists to compare your numerical solution.

Exercise 11.5   Develop a two dimensional Navier–Stokes (DNS) code based on the temporal formulation of Section 11.3. Use either the Adams–Bashforth time-marching scheme or the Runge–Kutta scheme given by (11.29)–(11.30), fourth-order finite differencing for the streamwise direction and Chebyshev collocation for the wall-normal direction.

(a) Use $Re = 688.315$, $\alpha = 0.22$ from the Orr–Sommerfeld equation as initial conditions. Compare your solutions from the DNS code using parallel and nonparallel mean flows with the Orr–Sommerfeld solution.
(b) Show convergence of ω and the maximum u velocity versus time.

Exercise 11.6   Develop a two-dimensional Navier–Stokes (DNS) code based on the spatial formulation of Section 11.4. Use either the Adams–Bashforth time-marching scheme or the Runge–Kutta scheme given by (11.29)–(11.30), fourth-order finite differencing for the streamwise direction and Chebyshev collocation for the wall-normal direction. Repeat the study of Tollmien–Schlichting waves in a Blasius boundary layer as discussed in Section 11.6.

12

Flow Control and Optimization

[In] the realm of science ... what we have achieved will be obsolete in ten, twenty or fifty years. That is the fate, indeed, that is the very meaning of scientific work ... Every scientific "fulfillment" raises new "questions" and cries out to be surpassed rendered obsolete. Everyone who wishes to serve science has to resign himself to this.

– Max Weber (1864–1920)

12.1 Introduction

Earlier chapters have outlined and validated various theoretical and computational methodologies to characterize hydrodynamic instabilities. This chapter serves to cursorily summarize techniques to control flows of interest. In some situations, the instabilities may require suppressive techniques while, in other situations, enhancing the amplification of the disturbance field is desirable. Similarly, enhanced mixing is an application where disturbance amplification may be required to obtain the goal. Small improvements in system performance often lead to beneficial results. For example, Cousteix (1992) noted that 45% percent of the drag for a commercial transport transonic aircraft is due to skin friction drag on the wings, fuselage, fin, etc., and that a 10–15% reduction of the total drag can be expected by maintaining laminar flow over the wings and the fin. Hence, flow control methods that can prevent the onset of turbulence could lead to significant performance benefits to the aircraft industry. For aircraft, as well as many other applications, the flow starts from a smooth laminar state that is inherently unstable and develops instability waves. These instability waves grow exponentially, interact nonlinearly and lead ultimately to fully developed turbulence or flow separation. Therefore, one goal of a good control system is to inhibit, if not eliminate, instabilities

351

that lead to the deviation from laminar to turbulent flow state. Because it is beyond the breadth of this text to cover all possible flow control methodologies, this chapter will primarily highlight passive control techniques, wave-induced forcing, feed-forward and feedback flow control and the optimal flow control approach applied to suppression of boundary layer instabilities that maintains laminar flow. Detailed reviews of available flow control technologies can be found in Gad-el-Hak, Pollard, & Bonnet (1998), Gad-el-Hak (2000), Joslin, Kunz & Stinebring (2000) and Thomas, Choudhari & Joslin (2002).

12.2 Effects of Flexible Boundaries

The literature is abound with techniques for passive flow control. The discovery of these techniques have primarily come from parameter studies using theoretical and computational techniques described in the earlier chapters and through an understanding of the governing flow physics of the application. For the two-dimensional flat-plate boundary layer and flow over two-dimensional wings or engine nacelles, the viscous traveling wave (Tollmien–Schlichting) instability is a dominant mode effecting transition. It is well known that favorable pressure gradients stabilize the Tollmien–Schlichting wave and adverse pressure gradients destabilize the Tollmien–Schlichting wave. Hence, a passive method of flow control would be to effectively make use of the local pressure gradients the disturbance must encounter as it evolves in space. Other techniques may be pseudo active in that, once they are employed, there is no time variance. For example, applying cooling or heating through a surface can stabilize or destabilize a Tollmien–Schlichting wave in air, while the opposite effects are realized in water. In addition, steady suction has been demonstrated through many wind tunnel and flight tests to suppress instabilities, enabling flow to be laminar in regions that would otherwise be turbulent (cf. Joslin, 1998 and Joslin, Kunz & Stinebring, 2000, for an overview of projects that used these flow control strategies). Finally, wall compliance is an additional passive technique that has primarily shown promise for underwater applications (cf. Carpenter, 1990, vis-à-vis compliant walls). With such a technique, the properties of the elastic based wall are optimized to suppress the viscous traveling-wave instability. However, the introduction of wall-induced instability modes is possible, destroying any benefit of using wall compliance to suppress Tollmien–Schlichting waves. This section will review some of the history of the use of flexible or compliant walls and derive the necessary boundary conditions for use with the Orr–Sommerfeld and Squire equations and secondary instability theory that was outlined in Chapter 9.

Research involving flow over flexible walls was started in the late 1950s by Kramer (1957, 1965). Experimentally, Kramer found significant drag reductions using rubber coatings over rigid walls. Investigators in the 1960s focused on the task of experimentally duplicating and theoretically explaining Kramer's results. The majority of these studies failed to produce any comparable results, but the theoretical results laid the foundation for all future studies involving flexible walls. Interest turned toward the use of compliant walls for turbulent drag reduction. NASA (Bushnell, Hefner & Ash, 1977) and the Office of Naval Research (Reischman, 1984) sponsored investigations involving the use of compliant walls for the turbulent problem. Although most of the results from this era were either inconclusive or unsatisfactory, the contributions, together with earlier results, did provide stepping stones to the understanding of the physically complex fluid/wall interaction phenomena.

In the early 1980s, Carpenter & Garrad (1985, 1986) theoretically showed that Kramer type surfaces could lead to potential delays in transition. Further, they indicated deficiencies in previous investigations that may have prevented their achieving results comparable to Kramer's. Later, experiments performed by Willis (1986) and Gaster (1988) have shown favorable results using compliant walls. As outlined in the above-mentioned reviews, a number of investigations of the past thirty years have been conducted involving flexible walls. A main emphasis of these studies was to understand the physical mechanisms involved in the fluid and wall interaction for transitional and turbulent flows.

Most of the studies focused on the two-dimensional instability problem except for Yeo (1986) who showed that a lower critical Reynolds number existed for the isotropic compliant wall for three-dimensional instability waves. Carpenter & Morris (1989) and Joslin, Morris & Carpenter (1991) have shown that three-dimensional Tollmien–Schlichting waves can have greater growth rates over compliant walls than those that are two-dimensional. Even though three-dimensional waves may be dominant, it was demonstrated that transition delays are still obtainable through the use of compliant walls. For this study, they considered a compliant wall model used by Grosskreutz (1975) for his turbulent boundary layer experiments.

The remainder of this next section outlines boundary conditions for the primary instability problem using the Orr–Sommerfeld and Squire equations as well as the secondary instability problem of Grosskreutz (1975) for a wall model. Then, some limited results are presented that demonstrate the suppression of boundary layer transition using this method of flow control.

12.2.1 Primary Wave Model

The derivation of the boundary conditions and results for two- and three-dimensional primary instabilities over compliant and rigid walls have been given by Joslin, Morris & Carpenter (1991). The disturbances are represented as traveling waves that may grow or decay as they propagate. Nonlinear coupling is ignored so that individual components of the frequency spectrum may be studied. Additionally, the quasi-parallel assumption is made.

Consider the incompressible laminar, boundary layer over a smooth flat wall. The Navier–Stokes equations govern the flow and the Blasius profile is used to represent the mean flow. A small amplitude disturbance is introduced into the laminar flow. For this, a normal mode representation is given as

$$\{\tilde{v}, \tilde{\omega}_y\}(x, y, z, t) = \{\hat{v}, \hat{\omega}_y\}(y)e^{i(x\alpha\cos\phi + z\alpha\sin\phi - \omega t)} + c.c., \qquad (12.1)$$

where $\hat{v}$ and $\hat{\omega}_y$ are the complex eigenfunctions of normal velocity and vorticity, respectively. Here, α is the wavenumber, ω is the frequency and ϕ is the wave angle. In general, α and ω are complex making for an ambiguity in the system. For temporal analyses, α is a real specified wavenumber and ω is the complex eigenvalue. For spatial analyses, ω is a real specified frequency and α is the complex eigenvalue. For the compliant wall problem, Joslin, Morris & Carpenter (1991) have shown that the use of equation (12.1) leads to an overestimation of the growth of the wave as it propagates. The wave actually propagates in a nearly streamwise direction that is in the direction of the group velocity rather than normal to the wave fronts. The secondary instabilities were investigated using this simple representation of the primary instabilities. Since the present approach was conservative, it should exemplify the benefits of using compliant walls as a means to obtain transition delays. Also, a major emphasis and motivation of the present study was to determine the behavior or response of the phenomena of secondary instabilities to compliant walls.

If the normal mode relation (12.1) is substituted into the linearized form of the Navier–Stokes equations, the Orr–Sommerfeld (1.17) and Squire (1.38) equations result. The system requires six boundary conditions where the disturbance fluctuations vanish at infinity or

$$\hat{v}(y), \ \hat{v}'(y), \ \hat{\omega}_y(y) \to 0 \qquad \text{as} \qquad y \to \infty. \qquad (12.2)$$

The remaining boundary conditions are determined from the compliant wall model.

The compliant wall model was introduced by Grosskreutz (1975) in his experimental drag reduction studies with turbulent boundary layers. Here it was suggested that the link between streamwise and normal surface displacements would cause a negative production of turbulence near the wall. Although his results for the turbulent flow were disappointing, the surface does react to the fluid fluctuations in transitional flow in such a way as to reduce production of instability growth. Carpenter & Morris (1990) have shown by use of an energy analysis how the many competing energy transfer mechanisms are influenced by the presence of the compliant wall. Of note is the reduced energy production by the Reynolds stress that may cause the reduced growth rates. Further, Joslin, Morris & Carpenter (1991) predicted that transition delays of four to ten times the rigid wall transition Reynolds number were achievable with this coating. As a result, the model has been extended to allow for a secondary instability analysis.

The mechanical model consists of a thin, elastic plate supported by hinged and sprung rigid members inclined to the horizontal and facing upstream at an angle, θ, when in equilibrium. A sketch of the mechanical wall model is shown in Fig. 12.1. The boundary conditions are obtained by enforcing a balance of forces in the streamwise and spanwise directions and the continuity of fluid and wall motion. These are given here in linearized form.

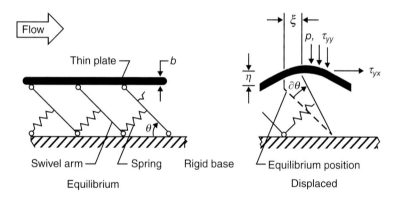

Figure 12.1 Non-isotropic compliant wall (after Grosskreutz, 1975).

For small displacements of an element out of equilibrium, the mechanical surface can be thought to move in a direction perpendicular to the rigid swivel arm. The horizontal and vertical displacements (ξ, η) are linked to the angular displacement $(\delta\theta)$ as

$$\xi = \ell\delta\theta\sin\theta \quad \text{and} \quad \eta = \ell\delta\theta\cos\theta, \tag{12.3}$$

where ℓ is the length of the rigid arm member. Equations of motion for the element in the streamwise and spanwise directions may be obtained by a balance of the forces of the fluid fluctuations acting on the surface and the forces due to the wall motion. These equations are

$$\rho_m b \frac{\partial^2 \eta}{\partial t^2} + \left(B_x \frac{\partial^4 \eta}{\partial x^4} + 2B_{xz} \frac{\partial^4 \eta}{\partial x^2 \partial z^2} + B_z \frac{\partial^4 \eta}{\partial z^4} \right) \cos^2 \theta$$

$$+ K_E \, \eta - E_x b \frac{\partial^2 \xi}{\partial x^2} \sin \theta \cos \theta$$

$$= -(p + \tau_{yy}) \cos^2 \theta + \tau_{yx} \sin \theta \cos \theta, \qquad (12.4)$$

and

$$\rho_m b \frac{\partial^2 (\zeta_z b)}{\partial t^2} + K_S \zeta - E_z b \frac{\partial^2 \zeta}{\partial z^2} = \tau_{yz}, \qquad (12.5)$$

where

$$\{B_x, B_z\} = \frac{\{E_x, E_z\} b^3}{12(1 - v_x v_z)} \quad \text{and} \quad B_{xz} = \sqrt{B_x B_z}.$$

Here, ζ is the spanwise surface displacement, ρ_m and b are the plate density and thickness, (B_x, B_{xz}, B_z) are the flexural rigidities of the plate in the streamwise, transverse and spanwise directions, (E_x, E_z) are the moduli of elasticity of the plate, K_E, K_S are the effective streamwise and spanwise spring-stiffness factors, p is the pressure fluctuation that is obtained from the fluid momentum equations and τ_{yx}, τ_{yy} and τ_{yz} are the streamwise, normal and spanwise viscous shear stress fluctuations in the fluid acting on the wall.

The terms on the left-hand side of equation (12.4) refer to mechanical forces and the terms on the right refer to fluid motion forces due to viscous stress and pressure fluctuations. For the case where the ribs are aligned at $\theta = 0°$, the wall becomes isotropic and reduces to the theoretical model studied by Carpenter & Garrad (1985, 1986). Otherwise the wall is referred to as non-isotropic and the rib angle is determined by θ.

The continuity of fluid and wall motion is given in the streamwise, normal and spanwise directions, respectively, as

$$\frac{\partial \xi}{\partial t} = \tilde{u} + \eta U', \quad \frac{\partial \eta}{\partial t} = \tilde{v} \quad \text{and} \quad \frac{\partial \zeta}{\partial t} = \tilde{w}, \qquad (12.6)$$

where $(\tilde{u}, \tilde{v}, \tilde{w})$ are the disturbance velocity components in the streamwise, normal and spanwise directions, respectively. For the Grosskreutz coating, $K_S \to \infty$ is assumed, which, from equation (12.5), would result in zero effective spanwise surface displacement. From equation (12.6) this implies that $\tilde{w} = 0$. Strictly speaking, if the assumption $K_S \to \infty$ is relaxed, the resulting instabilities have larger growth rates. This suggests that spanwise stiffeners are stabilizing to a disturbed flow. So, with the assumption enforced, a better coating for potential transition delays results. The surface displacement takes the same normal mode form as the primary wave given by equation (12.1). The normal modes are substituted into equations (12.4)–(12.6). These equations can be reduced to three equations in terms of the normal velocity and vorticity by performing operations similar to that of the Orr–Sommerfeld and Squire equations.

12.2.2 Primary Instability Results

The algebraic complexity of the dynamic equations for the secondary disturbance and the compliant wall equations requires that care be taken in applying any numerical technique for solution. Because no theoretical or experimental data are available for the compliant problem, both shooting and spectral approximations are used. Also, since Bertolotti (1985) has shown that for the rigid wall problem with a two-dimensional primary instability, after the transformation from spatial to temporal has been made, the solutions are in good agreement and only the temporal analysis is presented here.

For the spectral method, Chebyshev series are introduced to approximate each mode of the Fourier series. An algebraic transformation is used to change the Chebyshev spectral domain $[-1, 1]$ to the physical domain. Due to the properties of the Chebyshev polynomials, the equations are recast in integral form. Chebyshev polynomials are used to represent the basic flow in the series that are substituted into the integral equations. For the basic flow, 35 polynomials provide sufficient resolution of the eigenfunctions. The series representing the secondary instability requires 40 polynomials for sufficient convergence to the dominant eigenvalue. For the shooting method, beginning with the equations for the compliant wall, integrations of the disturbance equations across the boundary layer are performed using a Runge–Kutta scheme. At the edge of the boundary layer, the numerical solution vectors are matched with the asymptotic solutions. A very accurate initial guess is found to be required for convergence using this method. To demonstrate the accuracy of the numerical techniques, a comparison for the rigid wall case was made with Herbert (1983) for $Re_\delta = 826.36$, $Fr = 83$, $\beta = 0.18$ and $A = 0.02$. Herbert obtained the dominant mode $\sigma = 0.01184$. In good agreement, the present spectral and shooting methods lead to $\sigma = 0.011825$ and $\sigma = 0.011839$, respectively. Additional rigid wall results can be found in Chapter 9.

For all of the results that follow the freestream velocity is 20 m/s, the density is 1.0 kg/m^3 and the kinematic viscosity is 1×10^{-6} m^2/s. The coatings considered consist of both isotropic and anisotropic walls. Both walls were optimized at $Re_{\delta*} = 2{,}240$ for two-dimensional primary instabilities. The isotropic wall has properties $\theta = 0°$, $b = 0.735$mm, $E_x = 1.385$ MN/m^2, $K = 0.354$ GN/m^3 and $\rho_m = 1{,}000$ kg/m^3. The anisotropic wall has properties $\theta =$

$60°$, $b = 0.111$mm, $E_x = 0.509$ MN/m^2, $K = 0.059$ GN/m^3 and $\rho_m = 1,000$ kg/m^3. The equations are nondimensionalized using the freestream velocity U_∞, kinematic viscosity v and an appropriate length scale. Convenient lengths for the boundary layer scale with the x Reynolds number, $Re_x = U_\infty x / v$. These include a thickness, δ, where the Reynolds number is defined $Re_\delta = Re_x^{1/2}$ and a boundary layer displacement thickness, δ^*, where $Re_{\delta^*} = 1.7207 Re_x^{1/2}$.

A Reynolds number of 2,240 was chosen because, for a boundary layer over a rigid wall, the disturbance with the critical frequency (in the e^N sense) reaches its maximum growth rate near this value of Reynolds number. Accordingly, this is a good choice of Reynolds number for optimizing the wall properties. In considering three-dimensional instabilities, the walls optimized for two-dimensional instabilities are used with the addition of isotropic plates. The properties of an isotropic plate are direction independent; that is, $E_x = E_z$. Although complete details of the optimization process and philosophy have been given by Carpenter & Morris (1990), a review is provided here.

With a flexible wall present, other modes of instability arise. And, changes in the compliant wall properties with a stable or marginally stable fluid and wall modes can become unstable and dominant. The present wall properties were varied to achieve an optimal specified condition. The desired condition was to achieve a minimum growth rate for a dominant two-dimensional Tollmien–Schlichting instability while keeping other modes marginally stable. For the secondary analysis, these "optimal" compliant walls led to no additional unstable modes. However, this is not to say that additional growing modes may not appear for different wall properties.

In this section, the concept of stable and unstable regions is considered further. These regions indicate where the instability wave grows or decays. Illustrated in Fig. 12.2 are the neutral curves for the rigid wall, $\theta = 0°$ isotropic wall and the $\theta = 60°$ non-isotropic wall for the two-dimensional Tollmien–Schlichting instability. The $\theta = 60°$ wall has a smaller region of instability located within the rigid wall case. As the Reynolds number increases, the lower branch approaches that of the rigid wall and the upper branch stretches midway between the rigid wall branches. The $\theta = 0°$ wall produces a curve that coincides with the rigid wall curve at high wave frequencies. As the Reynolds number increases, both branches approach the $\theta = 60°$ branches. Nothing is revealed as to the growth rates within the unstable region. Although the region of instability may be smaller for the compliant coatings, the growth rates may very well be greater than the rigid wall growth rates. This is not the case for the coatings under consideration as will be shown in the next section.

Some concern has been expressed with respect to the alignment of the ribs,

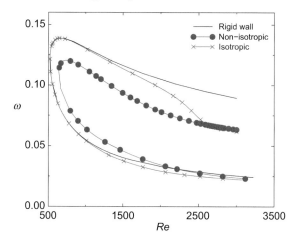

Figure 12.2 Curves of neutral stability over rigid wall (solid curve); $x, \theta = 0°$; and $\bullet, \theta = 60°$ compliant walls for $R_{\delta^*} = 2240$.

or swivel arm. It has been suggested that the same solutions would be expected irrespective of whether the ribs are aligned upstream or downstream. Although Carpenter & Morris (1989) have shown that ribs aligned downstream, or in the direction of the flow, result in higher growth rates than coatings with ribs aligned upstream. We will briefly examine this comparison for the neutral curve. Curves of neutral stability for the rigid wall, $\theta = 60°$ wall and $\theta = -60°$ wall are shown in Fig. 12.3. These coatings result in distinctly different curves where disturbances propagating over the $\theta = -60°$ wall become unstable at lower frequencies and Reynolds numbers than those propagating over the rigid wall.

Figure 12.4 shows the growth rates for the two-dimensional waves for various frequencies for the compliant and rigid walls. For the $\theta = 60°$ wall, the maximum growth rate is about 25% of that for the rigid wall. The width of the unstable region in $\omega - R_{\delta^*}$ space is also reduced considerably for the compliant walls as compared to the rigid surface. Figures 12.5 and 12.6 show the growth rates as functions of frequency for various oblique waves propagating over the same two compliant walls. For both coatings, the maximum growth rates are found for three-dimensional waves traveling at oblique angles of 50–60° to the flow direction. For $\theta = 0°$ an approximately 60% increase in growth rate over the two-dimensional case is found. For the $\theta = 60°$

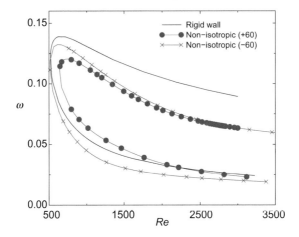

Figure 12.3 Curves of neutral stability over a rigid wall (solid curve); $\bullet, \theta = 60°$; and $x, \theta = -60°$ compliant walls for $R_{\delta^*} = 2240$.

wall the dominance of the three-dimensional waves is considerably reduced but still quite marked. The reduced sensitivity of the anisotropic compliant wall to three-dimensional waves compared to the isotropic case can be attributed to the effects of irreversible energy exchange between the wall and the disturbance due to the work done by the fluctuating shear stress. Carpenter & Morris (1990) showed that this energy exchange has a relatively destabilizing effect on the Tollmien–Schlichting waves that grows as θ increases. This deleterious effect is reduced for oblique waves owing to the reduced magnitude of the fluctuating shear stress in the direction of wave propagation. Hence, the relative improvement is in terms of reductions in the three-dimensional growth rates and range of unstable frequencies for non-isotropic as compared to isotropic compliant walls.

For simplicity we consider the growth of disturbances initiated at the lower branch of the two-dimensional neutral curve for each coating. This is a somewhat more conservative approach compared to the approximate procedure used by Cebeci & Stewartson (1980), who begin their calculations on a sort of three-dimensional neutral curve that they termed the "Zarf." Since the growth rates for both two- and three-dimensional instability waves are small in this region, it is not expected that the predicted transition Reynolds number will be significantly different. The instability is then allowed to seek the angle of wave

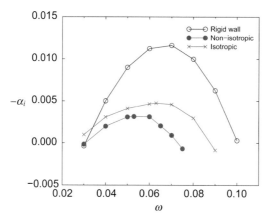

Figure 12.4 Two-dimensional spatial growth rates $-\alpha_i$ as a function of frequency for Tollmien–Schlichting waves over a rigid wall (solid curve); $x, \theta = 0$ wall; and •, $\theta = 60°$ wall at $Re_{\delta^*} = 2{,}240$.

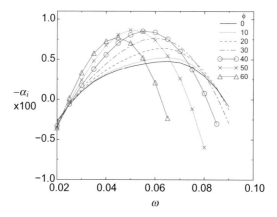

Figure 12.5 Spatial growth rates $-\alpha_i$ as a function of frequency for Tollmien–Schlichting waves over a $\theta = 0°$ wall/isotropic plate at $Re_{\delta^*} = 2{,}240$ for various oblique wave angles.

propagation in which it has a maximum growth rate. The wave is then traced as it convects downstream and the growth rates are used to determine the am-

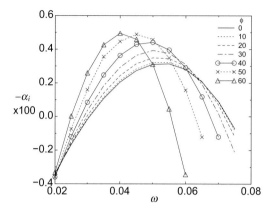

Figure 12.6 Spatial growth rates $-\alpha_i$ as a function of frequency for Tollmien–Schlichting waves over a $\theta = 60°$ wall/isotropic plate at $Re_{\delta^*} = 2{,}240$ for various oblique wave angles.

plification of the wave. The amplification is given by

$$\ln \frac{A}{A_0} = -\int_{(x_0,z_0)}^{(x,z)} \gamma_i(x) d(x,z),$$

where A_0 is the initial amplitude of the disturbance at (x_0, z_0). The e^N method is based on the observation that, when the amplification of the disturbance reaches some value, N, transition occurs (or is imminent). As the waves travel downstream at fixed increments of the streamwise coordinate, x, values of the growth rate and the direction of wave propagation, ϕ, are retained. Although the spanwise incremental step is not known exactly, as it is a continuous function of x, a second-order approximation is made with the known local values of ϕ and x. From this the spanwise increment is obtained and gives a possible error of ± 1 degree in the propagation angle for integrations in the low frequency range.

To confirm these local observations, e^N calculations were performed. Figure 12.7 shows curves of maximum amplification for two-dimensional waves in the frequency range of interest for the compliant and rigid walls. The two-dimensional case agrees with Carpenter & Morris (1990). Some controversy exists as to which value of N is the proper indication of transition. But if we choose a conservative value of $N = 7$ a delay of approximately four to five times the rigid wall transition Reynolds number is realized. However, for higher values of N, the advantages of the wall compliance increase still more.

Figure 12.8 shows similar calculations for the three-dimensional disturbances over the $\theta = 60°$ wall. Instabilities traveling over isotropic and orthotropic plates for the $\theta = 60°$ wall lead to similar maximum amplification curves. This is in agreement with the local calculations given earlier. A decrease from the two-dimensional transition delay occurs, but a transition delay remains compared to the rigid wall results. The same calculations for the $\theta = 0°$ wall illustrated in Fig. 12.9 show a notable difference between the isotropic and orthotropic plate cases. In fact, the results for the isotropic plate case approach the rigid wall results. No transition delay would be expected.

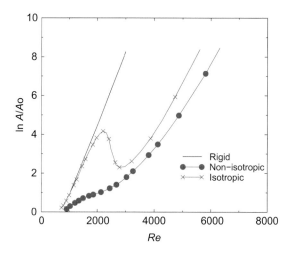

Figure 12.7 Curves of maximum amplification for Tollmien–Schlichting instability waves over a rigid wall (solid); $\theta = 0°$ wall $(-\times-)$; and $\theta = 60°$ $(-\bullet-)$ wall at $Re_{\delta^*} = 2240$.

This concludes the presentation of the primary instability results. It can be concluded that three-dimensional primary instabilities dominate transition over the compliant walls considered and transition delays occur when compared with the rigid wall. In the next section, the effect of compliant walls on secondary instabilities that result from two- and three-dimensional primary waves is discussed.

12.2.3 Secondary Instability Theory

In this section, boundary equations and conditions describing the compliant walls are introduced for secondary instabilities (cf. Chapter 9). The flow is

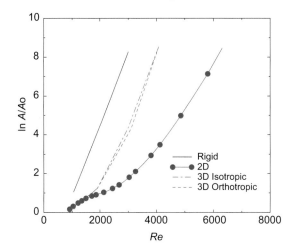

Figure 12.8 Curves of maximum amplification for Tollmien–Schlichting instability waves over a 2D rigid wall; 2D $\theta = 60°$, 3D $\theta = 60°$ orthotropic plate and 3D $\theta = 60°$ isotropic plate.

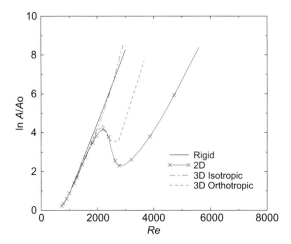

Figure 12.9 Curves of maximum amplification for Tollmien–Schlichting instability waves over a 2D rigid wall; 2D $\theta = 0°$, 3D $\theta = 0°$ orthotropic plate and 3D $\theta = 0°$ isotropic plate.

governed by the Navier–Stokes equations. Instantaneous velocity and pressure components are introduced and given as

$$\begin{aligned}
\underline{v}(\tilde{x},y,\tilde{z},t) &= \underline{v}_2(\tilde{x},y,\tilde{z},t)+B\underline{v}_3(\tilde{x},y,\tilde{z},t), \\
p(\tilde{x},y,\tilde{z},t) &= p_2(\tilde{x},y,\tilde{z},t)+Bp_3(\tilde{x},y,\tilde{z},t),
\end{aligned}\Bigg\} \tag{12.7}$$

where p_3 and $\underline{v}_3 = (u_3,v_3,w_3)$ are the secondary disturbance pressure and velocity in the fixed laboratory reference frame $(\tilde{x},y,\tilde{z})$, and p_2 and $\underline{v}_2 = (u_2,v_2,w_2)$ are the basic pressure and velocity given by

$$\begin{aligned}
\underline{v}_2(\tilde{x},y,\tilde{z},t) &= \{U_0(y),0,0\}+A\{u,v,w\}(\tilde{x},y,\tilde{z},t), \\
p_2(\tilde{x},y,\tilde{z},t) &= Ap(\tilde{x},y,\tilde{z},t).
\end{aligned}\Bigg\} \tag{12.8}$$

The basic flow is given by the Blasius solution and eigenfunctions of the primary wave. Assume locally that the primary wave is periodic in time and periodic in $(\tilde{x},\tilde{z})$ with wavelength $\lambda_r = 2\pi/\alpha_r$ and define a disturbance phase velocity as

$$\underline{c}_r = (c_x = \omega_r/\alpha_r \cos\phi, \; 0, \; c_z = \omega_r/\alpha_r \sin\phi).$$

Then, in a frame moving with the primary wave,

$$\underline{v}(\tilde{x},y,\tilde{z}) = \underline{v}(x,y,z) = \underline{v}(x+\lambda_x,y,z+\lambda_z), \tag{12.9}$$

where (x,z) is the reference frame moving with the wave. With an appropriate normalization of the primary eigenfunctions (u,v,w) the amplitude, A, is directly a measure of the maximum streamwise *rms* fluctuation. This is given by

$$\max_{0<y<\infty} |u(y)|^2 = |u(y_m)|^2 = 1/2. \tag{12.10}$$

As in Chapter 9, the instantaneous velocities and pressure are substituted into the Navier–Stokes equations that have been linearized with respect to the secondary amplitude, B. The disturbance pressure is eliminated, resulting in the vorticity equations and continuity. As with the primary problem, the final secondary disturbance equations take the form of a normal vorticity (9.14) and velocity (9.15).

The compliant wall equations give the remaining boundary conditions in the compliant case. Additionally, the primary amplitude, A, is a parameter in the equations and is assumed to be locally non-varying. As $A \to 0$, the Orr–Sommerfeld and Squire equations result. For the case of interest where $A \neq 0$, the primary eigenfunctions (u,v,w) appear in the equations as coefficients.

The boundary conditions for the secondary disturbance are given as

$$\hat{v}_n, \hat{v}'_n, \hat{\Omega}_n \to 0 \quad \text{as} \quad y \to \infty. \tag{12.11}$$

The analysis for the compliant boundary conditions for secondary instabilities follows the same route as was taken for the primary instabilities, except a number of additional terms arise due to the presence of the primary wave.

The fluid wall motion must be continuous in each direction. In addition, the equations of force (12.4) and (12.5) must balance in the streamwise and spanwise directions in the reference frame moving with the primary wave. Consistent with the fluid equations, the amplitude of the primary wave is assumed to be locally non-varying. In deriving the final form of the wall equations, a significant difference between the primary and secondary form arises from the pressure contribution. The pressure for the secondary disturbance is determined from the momentum equations that are complicated by primary coupling terms.

The continuity of motion between the fluid and solid is given by

$$\frac{\partial \xi_3}{\partial t} = u_3 + \eta_3 U_0' + A\left\{ (\underline{\xi}_1 \cdot \nabla)u_3 + (\underline{\xi}_3 \cdot \nabla)u_1 \right\},   (12.12)$$

$$\frac{\partial \eta_3}{\partial t} = v_3 + A\left\{ (\underline{\xi}_1 \cdot \nabla)v_3 + (\underline{\xi}_3 \cdot \nabla)v_1 \right\},   (12.13)$$

$$\frac{\partial \zeta_3}{\partial t} = w_3 + A\left\{ (\underline{\xi}_1 \cdot \nabla)w_3 + (\underline{\xi}_3 \cdot \nabla)w_1 \right\},   (12.14)$$

where

$$\underline{\xi}_i \cdot \nabla = \xi_i \frac{\partial}{\partial x} + \eta_i \frac{\partial}{\partial y} + \zeta_i \frac{\partial}{\partial z} \qquad \text{for} \qquad i = 1,2,3.$$

Equations (12.12)–(12.14) involve six unknowns for the velocity fluctuations and surface displacement in a highly coupled system. As with the primary boundary conditions, it is possible to derive a set of equations that represent the surface motion in terms of the normal velocity and vorticity only. This is algebraically very tedious. A complete derivation is given by Joslin (1990). Note that if $A = 0$ in the secondary wall equations, the primary wall equations result. This occurs with the fluid equations as well.

The compliant wall dynamic equations for the secondary disturbance are extremely complex and tedious to implement numerically. Hence, we will move on to active control techniques for hydrodynamic instabilities. First, let us summarize the primary and secondary instability results for hydrodynamic instabilities over compliant walls.

The physical nature and makeup of the mechanisms in transition are not altered by the control device (i.e., compliant wall). Rather, only the response of that mechanism is changed. Although three-dimensional primary instabilities theoretically dominate transition with compliant walls, transition delays are found when compared to the rigid wall case. As the primary amplitudes are reduced, the excitement of the secondary instability is delayed. Thus, active or passive devices that suppress primary instability growth should lead to corresponding suppression and delay of succeeding instabilities. The use of passive devices, such as compliant walls, lead to significant reductions in the secondary instability growth rates, and amplification, suppressing the primary growth rates, and subsequent amplification enable delays in the growth of the explosive secondary instability mechanism.

12.3 Wave Induced Forcing

The main deficiency of a passive control technique lies in the fact that the control system has been optimized to operate at a single target design point, whereas it is desirable to have efficient and effective controls over a range of operating conditions. As such, a time varying control system is required. Also, for a given level of control, time varying systems may require much less power input than comparable pseudo time invariant systems. For example, Liepmann & Nosenchuck (1982a) compared the effects of steady and unsteady heating to suppress a Tollmien–Schlichting wave instability, and found that steady heating demanded a 2,000% increase in energy compared with an unsteady "wave cancellation" technique. Hence, unsteady control (i.e., "active flow control") may be more efficient for flow control applications. Wave-induced forcing is one such time varying approach.

For wave-induced forcing, the disturbance frequency, wavelength, phase and amplification rate are all parameters that may be used for control. This information can easily be obtained using two or more wall pressure transducers. By using this disturbance information, a second control wave is forced to either obtain disturbance suppression or enforcing the disturbance amplification. For problems with the goal of instability suppression, the term wave cancellation is commonly used, since the goal of the second forcing wave is to cancel the disturbance present in the flow.

To date, most of the experiments aimed at verifying the wave cancellation concept were conducted on either a flat-plate or an axisymmetric body. Many of these experiments were conducted in water tunnels. Vibrating wires (Milling, 1981), hot strips (Liepmann & Nosenchuck, 1982a,b), suction and blowing (Pupator & Saric, 1989; Ladd, 1990), electromagnetic generators (Thomas, 1983) and adaptive heating element (Ladd & Hendricks, 1988) are some of the methods that were used in experiments to generate the disturbance and control waves. All of these input mechanisms gave the necessary control of the phase and amplitude of the input wave. Among the more successful studies,

Milling (1981) and Thomas (1983) achieved at least an 80% reduction in the amplitude of a two-dimensional disturbance.

Although intuitively obvious, until the work of Bower *et al.* (1987) and Pal, Bower & Meyer (1991), it was not known that perfect cancellation could be obtained within the context of linear theory for which the mean flow is independent of the propagating direction. They used the two-dimensional Orr–Sommerfeld equation to study and control instability wave growth by superposition, and showed that, within the limits of linear stability theory and the parallel flow assumption, both single- and multi-frequency waves can be cancelled. Definitively, Joslin, Erlebacher & Hussaini (1996) performed a numerical experiment that served to unequivocally demonstrate the link between linear superposition and instability suppression. To ensure that linear superposition of individual instabilities was in fact responsible for the results found in previous experiments and computations, they carried out three simulations with (i) only the disturbance, (ii) only the control and (iii) using both disturbance and control, which is the wave cancellation case. By discretely summing the control only and forcing only numerical results, they found that this linear superposed solution is identical to the wave cancellation results. These tests shown in Fig. 12.10 verify the hypothesis that linear superposition is the reason for the previous experimental and computational results. In practice, the disturbance cannot be completely cancelled since some residual disturbance energy will remain in the flow. This residual energy has the potential to amplify and lead to a boundary layer somewhere downstream of the control point. Hence, the wave cancellation flow control strategy would again be required downstream of the initial control point. Incidentally, note that the phase of the residual wave is nearly the same as the original disturbance. This occurs because, at the point of control, the energy in the original disturbance is greater than that of the control wave. A nearly 180° change in phase would occur for the residual wave if the control wave had greater energy than the original disturbance at the point of control.

12.4 Feed-Forward and Feedback Control

The concept of feed-forward or feedback control implies that some measurable quantity in the upstream or downstream location can serve to direct the attributes of the actuator so as to obtain a desired control goal, or objective (see Fig. 12.11). For unsteady flows, Joslin *et al.* (1995) considered feed-forward control for instability suppression. Determining optimal feedback laws is a very difficult proposition, especially in the context of nonlinear problems, so that one usually has to be content with using suboptimal feedback laws. In

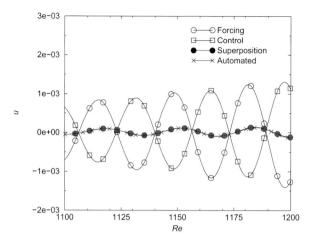

Figure 12.10 Tollmien–Schlichting disturbance amplification using forcing, control, superposition and automated (wave cancellation) methods.

this section the wave cancellation problem is used to discuss feed-forward and feedback control. As discussed by Joslin *et al.* (1995), the computations consist of the integration of the sensors, actuators and controller as shown in Fig. 12.11. The sensors will record the unsteady pressure or shear on the wall; the spectral analyzer (controller) will analyze the sensor data and prescribe a rational output signal; the actuator will use this output signal to control the disturbance growth and stabilize the instabilities within the laminar boundary layer. Although a closed-loop feedback system could be implemented (using an additional sensor downstream of the actuator) to fully automate the control and to lead to nearly exact cancellation of the instability, the feedback will not be introduced here due to the added computational expense of the iterative procedure. The feedback control law would simply compare the measure energy in the residual disturbance and alter the actuation amplitude toward obtaining a near zero residual. This section will describe a simple feed-forward strategy for wave cancellation in order to maintain laminar flow.

Here, the term controller refers to the logic that is used to translate sensor supplied data into a response for the actuator based on some control law. For the present study, a spectral controller is used. Such a controller requires a knowledge of the distribution of energy over frequencies and spatial wavenumbers. For the wave suppression problem, a minimum of two sensors must be used to record either the unsteady pressure or unsteady shear at the wall. By

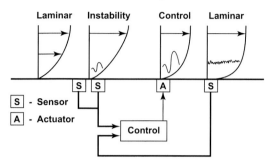

Figure 12.11 Sketch of feed-forward active control.

using fast Fourier transforms, this unsteady data can be transformed as

$$f(\omega) = \int_{-\infty}^{\infty} f(t)e^{-i\omega t}\,dt, \tag{12.15}$$

where $f(t)$ is the signal and ω is the frequency. This transform yields an energy spectrum that indicates which frequencies contain energy and dominate the original signal.

The largest Fourier coefficient indicates which frequency should be used to control the disturbance even though the largest growth rate can be used instead of the largest coefficient. In addition to frequency information, the two sensors provide estimates of both spatial growth rates and phase via the relation

$$\alpha = \frac{1}{A}\frac{dA}{dx}, \tag{12.16}$$

where A is the measured amplitude (complex Fourier coefficients of the dominant frequency mode). The temporal and spatial information are then substituted into the assumed control law or the wall-normal velocity boundary condition. We know that a functional control wave would have nearly the same amplitude but $180°$ out of phase of the original disturbance at the point of control. This information provides a control law for the actuation. Namely, the computational actuator can be described by the following:

$$v_s(x,t) = v_w\left[p_w^1 e^{i(\omega+\phi_t)t+\alpha x_s} + c.c.\right]. \tag{12.17}$$

Here, p_w^1 is the complex pressure (or shear) for the dominant frequency mode (or largest growth rate mode) at the first sensor; ω is the dominant mode determined from equation (12.15); ϕ_t is the phase shift parameter, t is the time, α is the growth rate and wavenumber information calculated from equation (12.16) and x_s is the distance between the first sensor and the actuator. Because the sensor information can be used only to approximate the actuator amplitude and temporal phase, v_w and ϕ_t are parameters that must be optimized to obtain exact wave cancellation. This may be accomplished through a gradient descent algorithm and no attempt is made to demonstrate exact wave cancellation.

To demonstrate the effectiveness of feed-forward control for wave cancellation, some sample results are presented from Joslin *et al.* (1995). For the computations, the Reynolds number is $Re = 900$ (based on displacement thickness) and the disturbance frequency is $Fr = \omega/Re \times 10^6 = 86$, which is reminiscent of an unstable mode. The disturbance forcing slot has a length $5.13\delta_0^*$ and is centered $23.10\delta_0^*$ downstream of the computational inflow boundary. The first sensor is located $57.88\delta_0^*$ downstream of the inflow, and the second sensor is located $2.33\delta_0^*$ downstream of the first sensor. The actuator has a slot length $4.67\delta_0^*$ and is located $77.94\delta_0^*$ downstream of the inflow boundary. These separation distances were chosen arbitrarily for this demonstration. Ideally, the forcing, sensors and actuator should have a minimal separation distance to improve the accuracy of the sensor information provided to the actuator.

A small-amplitude disturbance ($v_f = 0.01\%$) is forced and controlled via the feed-forward control law (12.17) without feedback. Figure 12.12 shows the Tollmien–Schlichting wave amplitudes with downstream distance for the present spectrally controlled results compared with the control case ($v_w = 0.9v_f$; $\phi_t = 1.2\pi/\omega$) of Joslin, Erlebacher & Hussaini (1996) and the uncontrolled wave. The present results demonstrate that a measure of wave cancellation can be obtained from the feed-forward system alone. Feedback is, however, necessary to optimize the control amplitude and phase for exact cancellation of the disturbance.

12.5 Optimal Control Theory

The optimal solution to suppress a single instability wave in a flat-plate boundary layer is well understood and therefore would serve as a good test problem here to demonstrate the significant advantages of optimal control theory. Taken from the research of Joslin *et al.* (1997), a self-contained, automated methodology is presented for active flow control. This methodology couples the time-dependent Navier–Stokes system with an adjoint Navier–Stokes system and optimality conditions from which optimal states, i.e., unsteady flow fields and controls (e.g., actuators), may be determined. For wave cancellation, the objective of the control approach is to match the stress vector along a portion of the boundary to the desired steady laminar boundary layer value. Control is effected through the injection or suction of fluid through a single orifice on the boundary. The system determines whether injection or suction is warranted

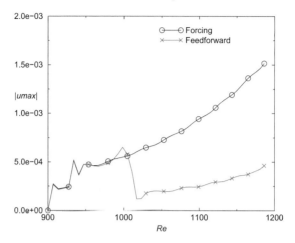

Figure 12.12 Feed-forward control of Tollmien–Schlichting waves in flat-plate boundary layer.

and at what point in time actuation is effected. The results for this sample test problem will demonstrate that instability suppression can be achieved without any a priori knowledge of the disturbance field, which is significant because other control techniques have required some knowledge of the flow unsteadiness such as frequencies, energy content, etc.

The goal of optimal control theory is to minimize or maximize an objective function in a robust manner. When the flow is time dependent, and is a strong function of initial conditions, it becomes difficult to establish the precise controls that will achieve the desired effect. While still in the linear regime, wave cancellation, as discussed above in this section, works well using a feedback mechanism. In practice, there are many waves that can interact nonlinearly in ways not always known in advance. Rather than try to cancel the incoming waves, one seeks appropriate controls in other ways. One means of achieving this, without an extensive search over the space of possible controls, is to postulate a family of desired controls. For example, an arbitrary time-dependent amplitude and a specified spatial distribution to find an objective function (i.e., stress over a region of the plate). Then, through a formal minimization process, one derives a set of differential equations and their adjoints, whose solutions produce the optimal actuator profile among the specified set. While the solution to this set of equations cannot be accomplished in real time, the results can be applied using standard passive or active control mechanisms. The advantage of this approach is that entire collections of controls can be studied simultaneously rather than one at a time. Optimal control techniques will not provide the real time control where there is ultimate interest but, by systematically computing the best control within specified tolerances and with a given objective function, it will be possible to develop strategies (active or passive) to control a wide variety of disturbances. For example, to effectively control boundary layer transition due to the interaction of a crossflow vortex and a Tollmien–Schlichting wave using periodic heating and cooling, optimal control would allow (1) a determination of the best objective function to use for a given type of control (some are better than others), and (2) provide insight into the relationship between the time dependence of the control and the input waves. This insight could then be built into a neural network, or other type of self-learning system, to allow effective control over a wide range of input parameters.

Optimal control methodologies have been recently applied to a variety of problems involving drag reduction, flow and temperature matching, etc. to provide more sophisticated flow control strategies in engineering applications. Computational fluid dynamics (CFD) algorithms have reached a sufficiently high level of maturity, generality and efficiency so that it is now feasible to implement sophisticated flow optimization methods that lead to a large number of coupled partial differential equations. Optimal control theory is quite mathematical, and its formal nature is amenable to the derivation of mathematical theorems related to existence of solutions and well posedness of the problem. Only partial results of this type are possible in three dimensions since, in this case, the Navier–Stokes equations themselves do not enjoy a full theoretical foundation; in two dimensions, a complete theory is available. Two very nice surveys of the mathematical theories of optimal flow control are Gunzburger (1995) and Borggaard *et al.* (1995). A mathematical study of a simplified problem related to the one considered in this paper can be found in Fursikov, Gunzburger & Hou (1996).

12.5.1 Optimization Methodology

In the present setting, an objective or cost functional is defined that measures the difference between the measured stresses, and the desired laminar values along a limited section of the bounding wall and over a specified length of time. One may interpret the objective functional as a sensor, i.e., the objective functional senses how far the flow stresses along the wall are from the corresponding desired values. To control the flow, time-dependent injection and suction are imposed along a small orifice in the bounding wall. Although the spatial dependence of the suction profile is specified (for simplicity), the optimal control methodology determines the time variation of this profile. However, unlike feedback control methodologies wherein the sensed data determines the control through a specified feedback law or controller, here the time dependence of the control is the natural result of the minimization of the objective functional. However, in the optimal control setting, the sensor is actually an objective functional and the controller is a coupled system of partial differential equations that determine the control that does the best job of minimizing the objective functional.

12.5.2 The State Equations

Let Ω denote the flow domain that is the semi-infinite channel or boundary layer $[x \geq 0, 0 \leq y \leq h]$, where h is the location of the upper wall for the channel or the truncated freestream distance for the boundary layer. Let Γ denote its boundary and let $(0, T)$ be the time interval of interest. The inflow part of the boundary $[x = 0, 0 \leq y \leq h]$ is denoted by Γ_i. The part of the boundary on which control is applied (i.e., along which the suction and blowing actuator is placed) by Γ_a and is assumed to be a finite connected part of the lower boundary (or wall) $[x \geq 0, y = 0]$. Solid walls are denoted by Γ_w; for the channel flow, Γ_w is the lower boundary $[x \geq 0, y = 0]$ with Γ_a excluded and the upper boundary $[x \geq 0, y = h]$; for the boundary layer flow, Γ_w is only the lower boundary with Γ_a excluded. For the boundary layer case, the upper boundary $[x \geq 0, y = h]$, which is not part of Γ_w, is denoted by Γ_e. Controls are only activated over the given time interval $T_0 < t < T_1$, where $0 \leq T_0 < T_1 \leq T$.

The flow field is described by the velocity vector (u, v) and the scalar pressure p and is obtained by solving the following momentum and mass conservation equations (11.9)–(11.11) subject to the initial and boundary conditions:

$$(u, v)\Big|_{t=0} = (u_0, v_0) \quad \text{in} \quad \Omega, \tag{12.18}$$

$$(u, v)\Big|_{\Gamma_a} = \begin{cases} (g_1, g_2) & \text{in} \quad (T_0, T_1), \\ (0, 0) & \text{in} \quad (0, T_0) \quad \text{and} \quad (T_1, T), \end{cases} \tag{12.19}$$

$$(u, v)\Big|_{\Gamma_i} = (u_i, v_i) \quad \text{and} \quad (u, v)\Big|_{\Gamma_w} = (0, 0) \quad \text{in} \quad (0, T), \tag{12.20}$$

and

$$(u, v, p) \to \text{base flow}, \quad \frac{\partial u}{\partial x}, \frac{\partial v}{\partial x} \to 0 \quad \text{as} \quad x \to \infty. \tag{12.21}$$

Here, the initial velocity vector $[u_0(x, y), v_0(x, y)]$ and the inflow velocity vector $[u_i(t, y), v_i(t, y)]$ are assumed given and the base flow is assumed to be

Poiseuille flow for the channel case, or the Blasius flow for the boundary layer case. This system holds for both the channel and Blasius flow cases. In the latter case, the upper boundary is not part of Γ_w and the additional boundary conditions

$$u\Big|_{\Gamma_e} = U_\infty \quad \text{and} \quad p - 2v\frac{\partial v}{\partial y}\Big|_{\Gamma_e} = P_\infty \quad \text{in } (0,T) \tag{12.22}$$

are imposed, where U_∞ and P_∞ denote the freestream flow speed and pressure, respectively.

The control functions $g_1(t,x)$ and $g_2(t,x)$, which give the rate at which fluid is injected or sucked tangentially and perpendicularly, respectively, through Γ_a are to be determined as part of the optimization process. In order to make sure that the control remains bounded at T_0, it is required that

$$g_1\Big|_{t=T_0} = g_{10}(x) \quad \text{and} \quad g_2\Big|_{t=T_0} = g_{20}(x) \quad \text{on } \Gamma_a, \tag{12.23}$$

where $g_{10}(x)$ and $g_{20}(x)$ are specified functions defined on Γ_a. Commonly, one chooses $g_{10}(x) = g_{20}(x) = 0$.

12.5.3 The Objective Functional and the Optimization Problem

Assume that Γ_s is a finite, connected part of the lower boundary $[x \geq 0, y = 0]$, which is disjoint from Γ_a and that (T_a, T_b) is a time interval such that $0 \leq T_a < T_b \leq T$. Then, consider the functional

$$\begin{aligned}
\mathscr{J}(u,v,p,g_1,g_2) = & \frac{\alpha_1}{2}\int_{T_a}^{T_b}\int_{\Gamma_s}|\tau_1 - \tau_a|^2\,d\Gamma dt \\
& + \frac{\alpha_2}{2}\int_{T_a}^{T_b}\int_{\Gamma_s}|\tau_2 - \tau_b|^2\,d\Gamma dt \\
& + \frac{\beta_1}{2}\int_{T_0}^{T_1}\int_{\Gamma_a}\left(\left|\frac{\partial g_1}{\partial t}\right|^2 + |g_1|^2\right)d\Gamma dt \\
& + \frac{\beta_2}{2}\int_{T_0}^{T_1}\int_{\Gamma_a}\left(\left|\frac{\partial g_2}{\partial t}\right|^2 + |g_2|^2\right)d\Gamma dt,
\end{aligned} \tag{12.24}$$

where g_1 and g_2 denote the controls and $\tau_a(t,x)$ and $\tau_b(t,x)$ are given functions defined on $(T_a, T_b) \times \Gamma_s$. Note that since Γ_s is part of the lower boundary of the channel or boundary layer wall, $\tau_1 = v\partial u/\partial y$ and $\tau_2 = -p + 2v\partial v/\partial y$ are the shear and normal stresses, respectively, exerted by the fluid on the bounding wall along Γ_s and thus τ_a and τ_b may be interpreted as given shear and normal stresses, respectively. Then, the boundary segment Γ_s can be thought of as a sensor that measures the stresses on the wall. Thus, in (12.24), Γ_s is the part of the boundary Γ along which one wishes to match the shear and normal stresses to the given functions τ_a and τ_b, respectively, and (T_a, T_b) is the time interval over which this matching is to take place. Other than notational, there are no difficulties introduced if one wishes to match each component of the stress vector over a different boundary segment or over a different time interval.

The third and fourth terms in (12.24) are used to limit the size of the control. Indeed, no bounds are a priori placed on g_1 or g_2, and their magnitudes are limited by adding a penalty to the stress matching functional defined by the first two terms in (12.24). The particular form that these penalty terms take, i.e, the third and fourth terms in (12.24), is motivated by the necessity to limit not only the size of the controls g_1 and g_2, but also to limit oscillations. The constants α_1, α_2, β_1 and β_2 can be used to adjust the relative importance of the terms appearing in the functional (12.24).

The (constrained) optimization problem is given as follows:

> *Find u, v, p, g_1, and g_2 such that the functional $\mathcal{J}(u, v, p, g_1, g_2)$*
> *given in (12.24) is minimized subject to the requirement that*
> *(11.9)–(11.11) and (12.18)–(12.21) and (12.24) are satisfied*
> *and, for the boundary layer flow case, (12.22) is also satisfied.*

12.5.4 The Adjoint System

The method of Lagrange multipliers is formally used to enforce the constraints (11.9)–(11.11) and (12.19). To this end, the Lagrangian functional

$$\mathcal{L}(u, v, p, g_1, g_2, \hat{u}, \hat{v}, \hat{p}, s_1, s_2)$$

$$= \frac{\alpha_1}{2} \int_{T_a}^{T_b} \int_{\Gamma_s} |\tau_1 - \tau_a|^2 \, d\Gamma dt + \frac{\alpha_2}{2} \int_{T_a}^{T_b} \int_{\Gamma_s} |\tau_2 - \tau_b|^2 \, d\Gamma dt$$

$$+ \frac{\beta_1}{2} \int_{T_0}^{T_1} \int_{\Gamma_a} \left(\left| \frac{\partial g_1}{\partial t} \right|^2 + |g_1|^2 \right) d\Gamma dt + \frac{\beta_2}{2} \int_{T_0}^{T_1} \int_{\Gamma_a} \left(\left| \frac{\partial g_2}{\partial t} \right|^2 + |g_2|^2 \right) d\Gamma dt$$

$$- \int_0^T \int_\Omega \hat{u} \left[\frac{\partial u}{\partial t} + u \frac{\partial u}{\partial x} + v \frac{\partial u}{\partial y} + \frac{\partial p}{\partial x} - 2\nu \frac{\partial^2 u}{\partial x^2} - \nu \frac{\partial}{\partial y} \left(\frac{\partial u}{\partial y} + \frac{\partial v}{\partial x} \right) \right] d\Omega dt$$

$$- \int_0^T \int_\Omega \hat{v} \left[\frac{\partial v}{\partial t} + u \frac{\partial v}{\partial x} + v \frac{\partial v}{\partial y} + \frac{\partial p}{\partial y} - \nu \frac{\partial}{\partial x} \left(\frac{\partial u}{\partial y} + \frac{\partial v}{\partial x} \right) - 2\nu \frac{\partial^2 v}{\partial y^2} \right] d\Omega dt$$

$$- \int_0^T \int_\Omega \hat{p} \left(\frac{\partial u}{\partial x} + \frac{\partial v}{\partial y} \right) d\Omega dt$$

$$-\int_{T_0}^{T_1}\int_{\Gamma_a} s_1(u-g_1)d\Gamma dt - \int_0^{T_0}\int_{\Gamma_a} s_1 u\, d\Gamma dt - \int_{T_1}^{T}\int_{\Gamma_a} s_1 u\, d\Gamma dt$$

$$-\int_{T_0}^{T_1}\int_{\Gamma_a} s_2(v-g_2)d\Gamma dt - \int_0^{T_0}\int_{\Gamma_a} s_2 v\, d\Gamma dt - \int_{T_1}^{T}\int_{\Gamma_a} s_2 v\, d\Gamma dt \quad (12.25)$$

is introduced. In (12.25), $\hat{u}$ and $\hat{v}$ are Lagrange multipliers that are used to enforce the x and y components of the momentum equation (11.9) and (11.10), respectively, $\hat{p}$ is a Lagrange multiplier that is used to enforce the continuity equation (11.11) and s_1 and s_2 are Lagrange multipliers that are used to enforce the x and y components of the boundary condition (12.19), respectively. Note that Lagrange multipliers have not been introduced to enforce the constraints (12.18), (12.20), (12.22) and (12.24), so that these conditions must be required of all candidate functions u, v, p, g_1 and g_2.

Through the introduction of Lagrange multipliers, the constrained optimization problem is converted into the unconstrained problem:

Find $u, v, p, g_1, g_2, \hat{u}, \hat{v}, \hat{p}, s_1$ and s_2 satisfying (12.18), (12.20), (12.21) *and* (12.23), *such that the Lagrangian functional $\mathscr{L}(u, v, p, g_1, g_2, \hat{u}, \hat{v}, \hat{p}, s_1, s_2)$ given by* (12.25) *is rendered stationary.*

In this problem, each argument of the Lagrangian functional is considered to be an independent variable so that each may be varied independently.

The first-order necessary condition that stationary points must satisfy is that the first variation of the Lagrangian with respect to each of its arguments vanishes at those points. One easily sees that the vanishing of the first variations with respect to the Lagrange multipliers recovers the constraint equations (11.9)–(11.11) and (12.19). Specifically,

$$\frac{\delta\mathscr{L}}{\delta\hat{u}}, \frac{\delta\mathscr{L}}{\delta\hat{v}} = 0 \implies \quad x\text{- and } y\text{-momentum equations (11.9) and (11.10)},$$

$$\frac{\delta\mathscr{L}}{\delta\hat{p}} = 0 \implies \quad \text{continuity equation (11.11)},$$

$$\frac{\delta\mathscr{L}}{\delta s_1}, \frac{\delta\mathscr{L}}{\delta s_2} = 0 \implies \quad x \text{ and } y \text{ components of (12.19)},$$

where $\delta\mathscr{L}/\delta\hat{u}$ denotes the first variation of $\mathscr{L}$ with respect to $\hat{u}$, etc.

Next, set the first variations of the Lagrangian with respect to the state variables u, v and p equal to zero. These result in the *adjoint* or *co-state equations*. Note that for the channel flow, candidate solutions must satisfy (12.18),

(12.20), (12.21) and (12.23), and thus

$$
\left.
\begin{aligned}
&\delta u|_{t=0} = \delta v|_{t=0} = 0 \quad \text{on} \quad \Omega\delta g_2|_{t=T_0} = 0 \quad \text{on} \quad \Gamma_a, \\
&\delta u|_{\Gamma_i} = \delta v|_{\Gamma_i} = 0; \quad \delta u|_{\Gamma_w} = \delta v|_{\Gamma_w} = 0 \quad \text{for} \quad (0,T), \\
&\delta p, \delta u, \delta v, \frac{\partial \delta u}{\partial x}, \frac{\partial \delta v}{\partial x} \to 0 \text{ as } x \to \infty \text{ for } (0,T).
\end{aligned}
\right\}
\tag{12.26}
$$

Consider $\delta \mathscr{L}/\delta p = 0$ and

$$
\alpha_2 \int_{T_a}^{T_b} \int_{\Gamma_s} \delta p\,(\tau_2 - \tau_b)\,d\Gamma dt + \int_0^T \int_\Omega \left(\hat{u}\frac{\partial \delta p}{\partial x} + \hat{v}\frac{\partial \delta p}{\partial y} \right) d\Omega dt = 0 \tag{12.27}
$$

for arbitrary variations δp in the pressure. After applying Gauss' theorem to the above, we get

$$
\alpha_2 \int_{T_a}^{T_b} \int_{\Gamma_s} \delta p\,(\tau_2 - \tau_b)\,d\Gamma dt - \int_0^T \int_\Omega \delta p \left(\frac{\partial \hat{u}}{\partial x} + \frac{\partial \hat{v}}{\partial y} \right) d\Omega dt
$$

$$
+ \int_0^T \int_\Gamma \delta p(\hat{u}n_1 + \hat{v}n_2)\,d\Gamma dt = 0, \tag{12.28}
$$

where n_1 and n_2 denote the x and y components of the outward normal to Ω, respectively, along Γ by choosing variations δp that vanish on the boundary Γ but which are arbitrary in the interior Ω of the flow domain. Then,

$$
\frac{\partial \hat{u}}{\partial x} + \frac{\partial \hat{v}}{\partial y} = 0 \quad \text{on} \quad (0,T) \times \Omega. \tag{12.29}
$$

Now, choosing variations δp that are arbitrary along the boundary Γ, gives

$$
\hat{u}n_1 + \hat{v}n_2 =
\begin{cases}
0 \quad \text{on } (0,T) \times \Gamma\backslash\Gamma_s, (0,T_a) \times \Gamma_s, (T_b,T) \times \Gamma_s, \\
-\alpha_2 \left(-p + 2v\frac{\partial v}{\partial y} - \tau_b \right) \quad \text{on } (T_a,T_b) \times \Gamma_s,
\end{cases}
\tag{12.30}
$$

where $\Gamma\backslash\Gamma_s$ denotes the boundary Γ with Γ_s deleted. Note that in the above derivation of (12.29) and (12.30), as in those found later in the section, the last relation in (12.26) implies that the boundary integrals at infinity make no contribution.

Next, consider $\delta\mathscr{L}/\delta v = 0$, where the conditions (12.26) have been used to eliminate boundary integrals along Γ_i, Γ_w and as $x \to \infty$, an integral over Ω at $t = 0$. First, variations δv that vanish at $t = 0$, $t = T$ and in a neighborhood of Γ are chosen, but which are otherwise arbitrary. Such a choice implies that all boundary integrals in (12.27) vanish, allowing for

$$
-\frac{\partial \hat{v}}{\partial t} + \hat{u}\frac{\partial u}{\partial y} + \hat{v}\frac{\partial v}{\partial y} - u\frac{\partial \hat{v}}{\partial x} - v\frac{\partial \hat{v}}{\partial y} - \frac{\partial \hat{p}}{\partial y}
$$

$$
-v\frac{\partial}{\partial x}\left(\frac{\partial \hat{u}}{\partial y} + \frac{\partial \hat{v}}{\partial x} \right) - v\frac{\partial}{\partial y}\left(2\frac{\partial \hat{v}}{\partial y} \right) = 0 \quad \text{in } (0,T) \times \Omega, \tag{12.31}
$$

where equation (11.11) is used to effect a simplification. Next, variations that vanish in a neighborhood of Γ, but which are otherwise arbitrary, are chosen to obtain

$$\hat{v}|_{t=T} = 0 \quad \text{in} \quad \Omega. \tag{12.32}$$

Now, along Γ, δv and $\partial \delta v / \partial n$ may be independently selected, provided that (12.26) is satisfied. Also, $\partial / \partial n$ denotes the derivative in the direction of the outward normal to Ω along Γ. If $\delta v = 0$ and $\partial \delta v / \partial n$ varies arbitrarily along Γ, then

$$\hat{v} = \begin{cases} 0 \quad \text{on} \ (0,T) \times \Gamma \backslash \Gamma_s, (0,T_a) \times \Gamma_s, (T_b,T) \times \Gamma_s, \\ \alpha_2 \left(-p + 2v \frac{\partial v}{\partial y} - \tau_b \right) \quad \text{on} \quad (T_a,T_b) \times \Gamma_s. \end{cases} \tag{12.33}$$

To see this, note that along the inflow, Γ_i, $n_2 = 0$ and $\partial / \partial n = -\partial / \partial x$ while, along the top and bottom boundaries $n_1 = 0$, $\partial / \partial n = \pm \partial / \partial y$, respectively, and, since $\delta v = 0$, $\partial \delta v / \partial x = 0$. Now (12.30) and (12.32) agree on the boundary segments where they simultaneously apply. Finally, δv is arbitrarily chosen along Γ_a to obtain

$$s_2 = -\hat{p} n_2 - \hat{v}(u n_1 + v n_2) - v \left(\frac{\partial \hat{u}}{\partial y} + \frac{\partial \hat{v}}{\partial x} \right) n_1 - 2v \frac{\partial \hat{v}}{\partial y} n_2 \ \text{on} \ (0,T) \times \Gamma_a. \tag{12.34}$$

Next, consider $\delta \mathcal{L} / \delta u = 0$. Applying to the resulting equation the same process that led to (12.31)–(12.34) yields

$$-\frac{\partial \hat{u}}{\partial t} + \hat{u} \frac{\partial u}{\partial x} + \hat{v} \frac{\partial v}{\partial x} - u \frac{\partial \hat{u}}{\partial x} - v \frac{\partial \hat{u}}{\partial y} - \frac{\partial \hat{p}}{\partial x}$$
$$- v \frac{\partial}{\partial x} \left(2 \frac{\partial \hat{u}}{\partial x} \right) - v \frac{\partial}{\partial y} \left(\frac{\partial \hat{u}}{\partial y} + \frac{\partial \hat{v}}{\partial x} \right) = 0 \quad \text{in} \ (0,T) \times \Omega, \tag{12.35}$$

$$\hat{u}|_{t=T} = 0 \quad \text{in} \quad \Omega, \tag{12.36}$$

$$\hat{u} = \begin{cases} 0 \quad \text{on} \ (0,T) \times \Gamma \backslash \Gamma_s, (0,T_a) \times \Gamma_s, (T_b,T) \times \Gamma_s, \\ \alpha_1 \left(v \frac{\partial u}{\partial y} - \tau_a \right) \quad \text{on} \ (T_a,T_b) \times \Gamma_s \end{cases} \tag{12.37}$$

and

$$s_1 = -\hat{p} n_1 - \hat{u}(u n_1 + v n_2) - 2v \frac{\partial \hat{u}}{\partial x} n_1 - v \left(\frac{\partial \hat{u}}{\partial y} + \frac{\partial \hat{v}}{\partial x} \right) n_2 \ \text{on} \ (0,T) \times \Gamma_a. \tag{12.38}$$

In deriving (12.37) we have used the assumption that Γ_s is part of the lower boundary of the channel so that along Γ_s we have that $n_2 = -1$. Again, there is no conflict between (12.28) and (12.37) along boundary segments on which both apply.

12.5.5 The Optimality Conditions

The only first-order necessary conditions left to consider are $\delta\mathcal{L}/\delta g_1 = 0$ and $\delta\mathcal{L}/\delta g_2 = 0$. These conditions are usually called the "optimality conditions." Now, since all candidate functions g_1 and g_2 must satisfy (12.23), it follows that $\delta g_1 = 0$ and $\delta g_2 = 0$ at $t = T_0$. Then, take $\delta\mathcal{L}/\delta g_2 = 0$ and apply Gauss' theorem to remove all derivatives from the variation δg_2. Choose variations δg_2 that vanish at $t = T_1$, but which are otherwise arbitrary and, using (12.34),

$$-\frac{\partial^2 g_2}{\partial t^2} + g_2 = -\frac{1}{\beta_2}\left(\hat{p} + 2v\frac{\partial\hat{v}}{\partial y}\right) \quad \text{on} \quad (T_0, T_1) \times \Gamma_a \qquad (12.39)$$

results, where (12.33) and the assumption that Γ_a is part of the lower boundary so that, along Γ_a, $n_1 = 0$ and $n_2 = -1$, have been used. Now, choosing variations that are arbitrary at $t = T_1$ allows that $\partial g_2/\partial t = 0$ along Γ_a at $t = T_1$ so that, invoking (12.34), $g_2(t,x)$ satisfies

$$g_2|_{t=T_0} = g_{20}(x) \quad \text{and} \quad \frac{\partial g_2}{\partial t}\Big|_{t=T_1} = 0 \quad \text{on } \Gamma_a. \qquad (12.40)$$

Note that, given $\hat{p}$ and $\hat{v}$, (12.39) and (12.40) constitute, at each point x on Γ_a, a two point boundary value problem in time over the interval (T_0, T_1).

In a similar manner, setting $\delta\mathcal{L}/\delta g_1 = 0$ leads to

$$-\frac{\partial^2 g_1}{\partial t^2} + g_1 = -\frac{1}{\beta_1}\left(v\frac{\partial\hat{u}}{\partial y}\right) \quad \text{on} \quad (T_0, T_1) \times \Gamma_a, \qquad (12.41)$$

$$g_1|_{t=T_0} = g_{10}(x) \quad \text{and} \quad \frac{\partial g_1}{\partial t}\Big|_{t=T_1} = 0 \quad \text{on } \Gamma_a. \qquad (12.42)$$

12.5.6 Finite Computational Domains

Since we are still only considering the channel flow case, for the computations the semi-infinite domain Ω is replaced by a finite domain Ω_C, defined by the introduction of the outflow boundary Γ_o given by $[x = L, 0 \le y \le h]$. Thus, we have that Ω_C is the rectangle $[0 \le x \le L, 0 \le y \le h]$. The outflow does not require the imposition of boundary conditions along the outflow boundary Γ_o because a buffer zone (Streett & Macaraeg, 1989) is attached to the end of the physical computational domain, where the governing equations are parabolized in this buffer region.

A similar treatment of the adjoint variables should have required consideration of an infinite domain $[-\infty < x < \infty, 0 < y < h]$. If this had been done, the
boundary conditions (12.34) and (12.38) would not have been obtained along
the inflow Γ_i. In fact, the inflow boundary Γ_i for the state equation is the outflow boundary for the adjoint equations, and, conversely, the outflow boundary
Γ_o for the state equation is the inflow boundary for the adjoint equations. This
is easily seen by comparing the leading inertial terms of the state equations
with t increasing and the adjoint equations with t decreasing. Now, on both Γ_i
and Γ_o we have that $u > 0$ and $v \approx 0$ which is why Γ_i is an inflow boundary
and Γ_o is an outflow boundary for the state. On the other hand, the fact that t is
decreasing in the adjoint equations implies that now Γ_i is an outflow boundary
and Γ_o is an inflow boundary for those equations.

Thus, to be consistent with the treatment of the state equations, the adjoint
outflow Γ_i should be treated in a manner similar to the earlier treatment of
the state outflow Γ_o. This treatment of the adjoint outflow does not require
the imposition of any boundary conditions for the adjoint variables along Γ_i.
Finally, since Γ_o is an inflow boundary for the adjoint equations, one has that

$$\hat{u} = 0 \quad \text{and} \quad \hat{v} = 0 \qquad \text{on} \qquad (0, T) \times \Gamma_o. \tag{12.43}$$

12.5.7 *The Optimality System for Channel Flow*

We now have the full "optimality system" for channel flow whose solutions
determine the optimal states, controls and adjoint states. These are:

> *State equations:* (11.9)–(11.11), (12.18)–(12.21)
> *Co-state equations:* (12.29)–(12.33), (12.35)–(12.37)
> *Optimality equations:* (12.39) to (12.42)

Since (12.34) and (12.38) merely serve to determine the uninteresting Lagrange multipliers s_2 and s_1, they can be ignored.

The state equations are driven by the given initial velocity (u_0, v_0), the given
inflow velocity (u_i, v_i) and the controls (g_1, g_2). Indeed, the purpose of this
study is to determine g_1 and g_2 that optimally counteracts instabilities created
upstream of Γ_a. The adjoint equations are homogeneous except for the boundary condition along Γ_s, the part of the boundary along which we are trying to
match the stresses. The data in that boundary condition are exactly the discrep-
ancy between the desired stresses τ_a and τ_b and the stresses $\tau_1 = v \partial u / \partial y$ and
$\tau_2 = -p + 2v \partial u / \partial y$ along Γ_s, weighted by the factors α_1 and α_2. The equations for the controls are driven by the negative of the adjoint stresses along
Γ_a, the part of the boundary along which we apply the control, weighted by

the factors $1/\beta_1$ and $1/\beta_2$. Of course this division into equations for the state, the adjoint state and the control is really obscured by the fact that equations are all intimately coupled.

12.5.8 The Optimality System for Boundary Layer Flow

If one follows a similar process to that used for the channel flow, one may derive an optimality system for the boundary layer. The only difference is that in the latter case Γ_w denotes only the lower boundary with Γ_a excluded, and that the additional boundary condition (12.22) along the upper boundary Γ_e must be taken into account.

With the new interpretation for Γ_w, one can still define the Lagrangian functional (12.25) and use the constraints (12.26) on allowable variations. However, due to (12.22), allowable variations are further constrained by

$$\delta u\Big|_{\Gamma_e} = \left(\delta p - 2v\frac{\partial \delta v}{\partial y}\right)\Big|_{\Gamma_e} = 0 \quad \text{for } (0,T), \tag{12.44}$$

which implies that, along Γ_e, one may not independently choose the variations in δp and $\partial \delta v/\partial y$. By simultaneously considering variations in p, v and $\partial v/\partial y$ along Γ_e, one can show that

$$\hat{u} = 0 \quad \text{on} \quad (0,T) \times \Gamma_e. \tag{12.45}$$

Then, letting δv be arbitrary along Γ_e,

$$\hat{p} + 2v\frac{\partial \hat{v}}{\partial y} + v\hat{v} = 0 \quad \text{on} \quad (0,T) \times \Gamma_e. \tag{12.46}$$

The resulting system for the boundary layer now includes (12.45) and (12.46) in addition to the channel flow system.

12.5.9 Numerical Experiments

The optimal control methodology (Joslin *et al.*, 1997) was developed for the fully nonlinear Navier–Stokes system, and thus is applicable to the case of nonlinear and three-dimensional flow control. Here, we simply demonstrate the methodology for the wave cancellation problem because the optimal control is known a priori to be wave superposition and a single instability wave is evolving in a flat-plate boundary layer.

The formidable coupled system is solved in an iterative manner. First, the Navier–Stokes equations are solved for the state variables, i.e., the velocity field (u,v) and pressure p with control information (i.e., no control $g_1 = g_2 = 0$

for first iteration). Then co-state equations are solved for the adjoint or co-state variables $(\hat{u}, \hat{v})$ and $\hat{p}$. Then, using these adjoint variables, the controls g_1 and g_2 are then found by solving the optimality equations. The procedure is repeated until satisfactory convergence is achieved.

The nonlinear unsteady Navier–Stokes equations and linear adjoint Navier–Stokes equations are solved by direct numerical simulation (DNS) of disturbances that evolve spatially within the boundary layer. The spatial DNS (Joslin, Streett & Chang, 1993; Joslin & Streett, 1994) approach involves spectral and high-order finite difference methods and a three-stage Runge–Kutta method (Williamson, 1980) for time advancement. The influence matrix technique is employed to solve the resulting pressure equation (Streett & Hussaini, 1991). Disturbances are forced into the boundary layer by unsteady suction and blowing through a slot in the wall. The buffer domain technique (Streett & Macaraeg, 1989) is used for the outflow boundary treatment.

In the present study, only normal injection or suction control is allowed, so that we set $g_1 = 0$ in (12.19), $\beta_1 = 0$ in the functional (12.24) and ignore (12.41) and (12.42). Also, we only match the normal stress along Γ_s so that we choose $\alpha_1 = 0$ in the functional (12.24) and in (12.37). The Reynolds number based on the inflow displacement thickness (δ_0^*) is $Re = 900$ and the nondimensional frequency for the forced disturbance is $F = \omega/Re \times 10^6 = 86$. The forcing amplitude is $v_f = 0.1\%$. The disturbance forcing slot Γ_f, the control or actuator orifice Γ_a and the matching or sensor segment Γ_s have equal length $4.48\delta_0^*$. The forcing is centered downstream at $389.62\delta_0^*$. The Reynolds number based on the displacement thickness at that location is $R = 1,018.99$. The actuator is centered at $403.62\delta_0^*$ $(Re = 1037.13)$ and the sensor is centered at $417.62\delta_0^*$ $(Re = 1054.97)$. These separation distances were arbitrarily chosen for this demonstration. In practice, the control and matching segments should have a minimal separation distance so that the pair can be packaged as a single unit, or bundle, for distributed application of many bundles.

All simulations allow the flow field to develop for one period, i.e., from $t = 0 \rightarrow T_a = T_p$ before control is initiated. In the first series of simulations, the interval during which control is applied is arbitrarily chosen to be $T_a \rightarrow T_b = 2T_p$. Based on $\alpha_1 = \beta_1 = 0$, $\alpha_2 = 1$ and $\beta_2 = 10$, the convergence history for the wall-normal velocity and measured normal shear τ_2 are shown in Fig. 12.13. The velocities are obtained at a fixed distance from the wall corresponding to $1.18\delta_0^*$ and at the fixed time T_b. Convergence is obtained with four iterations. The results demonstrate that a measure of wave cancellation can be obtained from the DNS control theory system. The wall-normal amplitude of the modified wave at $Re = 1092.5$ is 40% of the uncontrolled wave. The control without optimizing the choice of α_1, α_2, β_1 and β_2 has led to a 60%

Flow Control and Optimization

decrease in the amplitude of the traveling wave. Clearly, Fig. 12.13 shows that
a net reduction of the disturbance energy is obtained by energy input due to
the control. This results in a delay of transition by way of a suppression of the
instability evolution.

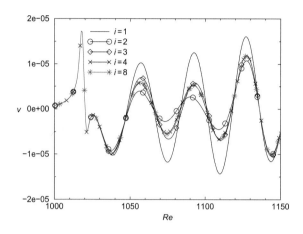

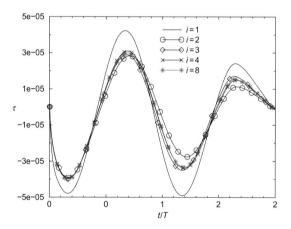

Figure 12.13 Convergence of disturbance wall-normal velocity with downstream
distance (top) and measured shear stress with discrete time (bottom) for control
of Tollmien–Schlichting waves in flat-plate boundary layer.

From the wave cancellation study of Joslin, Erlebacher & Hussaini (1996),
the relationship between amplitude of the actuator (v_a) with resulting instabil-

Table 12.1 *Normal stress for two values of* β_2.

β_2	normal stress
10	9.369×10^{-6}
11	8.814×10^{-6}

ity was similar to the channel flow wave cancellation study in Biringen (1984). The trend indicates that, beginning with a small actuation amplitude, as the actuation level is increased, the amount of wave cancellation by energy extraction from the disturbance increases. At some optimal actuation, nearly exact wave cancellation is achieved for the instability wave. As the actuation amplitude further increases the resulting instability amplitude increases. This was clearly explained in Joslin *et al.* (1996) to occur because, in the wave superposition process, the actuator wave becomes dominant over the forced wave. At this point, the resulting instability undergoes a phase shift corresponding to the phase of the wave generated by the actuator. The relationship is encouraging for the DNS optimal control theory approach and suggests that a gradient descent type algorithm might further enhance the wave suppression capability of the present approach. Namely, an approach for the optimal selection of α_1, α_2, β_1 and β_2 might lead to a more useful theoretical and computational tool for flow control.

To simply demonstrate this concept, Lagrange interpolation (or perhaps extrapolation) is introduced for β_1 and β_2 based on imposed values for α_1 and α_2:

$$\beta_{1,2}^{n+1} = \frac{\beta_{1,2}^{n}(\tau_{1,2}^{*} - \tau_{1,2}^{n-1}) - \beta_{1,2}^{n-1}(\tau_{1,2}^{*} - \tau_{1,2}^{n})}{(\tau_{1,2}^{n} - \tau_{1,2}^{n-1})}, \qquad (12.47)$$

where $\tau_{1,2}^{*}$ are some desired values of the stress components and $\tau_{1,2}^{n}$ are the stress components based on the choice $\beta_{1,2}^{n}$. Although τ_1^{*} and τ_2^{*} may be equivalent to the target values τ_a and τ_b in the functional (12.25), this may lead to significant over/under shoots for the iteration process. Instead, τ_1^{*} and τ_2^{*} is the incremental decrease, or target value, for interpolation to more desirable β_1 and β_2 values. To illustrate this process, the $\beta_2 = 10$ and $\beta_2 = 11$ control results are obtained with the iteration procedure. The measures of normal stress are somewhat arbitrarily obtained at some time as measured by the sensor or matching segment Γ_s. The values of the normal stress are given in the Table 12.1. These values are used for a desired normal stress τ_2^{*}, which in this case is 65% of the $\beta_2 = 11$ results.

Using the results for $\beta_2 = 10$ and $\beta_2 = 11$ in (12.47) yields the value $\beta_2 =$

16.5, which is used in a simulation to obtain a greater degree of instability suppression. The WC[1] results and the enhanced optimal control solution are shown in Fig. 12.14. This interpolation approach based on relationship of Fig. 12.14 indicates that optimizing β_2 has led to results very close to WC. The solutions differ somewhat near $t = T_a$ and $t = T_b$ because of the conditions (12.40) and (12.42) that serve to control the levels of g_1 and g_2. For all practical purposes, the solutions obtained with the present DNS control theory methodology provide the desired flow control features without prior knowledge of the forced instability.

The adjoint system requires that the velocity field (u, v) obtained from the Navier–Stokes equations be known for all time. For the iteration sequence and a modestly course grid, 82 M-bytes of disk (or runtime) space are required to store the velocities at all time steps and for all grid points. For $T_a \rightarrow T_b = 3T_p$, 246 M-bytes are necessary for the computation. Clearly for three-dimensional problems the control scheme becomes prohibitively expensive. Therefore, a secondary goal of this study is to determine if this limitation can be eliminated.

Because the characteristics of the actuator (g_1 and g_2) and resulting solutions are comparable to WC, some focus should be placed on eliminating the enormous memory requirements discussed earlier in this section. This limitation can easily be removed if the flow control problem involves small amplitude unsteadiness (or instabilities). The time-dependent coefficients of the adjoint system (12.31) and (12.35) reduce to the steady state solution and no additional memory is required over the Navier–Stokes system in terms of coefficients. This has been verified by a comparison of a simulation with steady coefficients compared with the C2[2] control case. As expected, the results for both cases are identical. Additionally, if the instabilities have small amplitudes, then a linear Navier–Stokes solver can be used instead of the full nonlinear solver that was used in the present study. This linear system would be very useful for the design of flow control systems. However, if the instabilities in the flow have sufficient amplitude to interact nonlinearly, then some measure of unsteady coefficient behavior is likely required and then, depending on the amplitudes, the coefficients saved at every time step may be replaced with storing coefficients every ten or more time steps, thereby reducing the memory requirements by an order of magnitude. This hypothesis will require further study.

[1] Wave Cancellation.
[2] See Chapter 10 for definition of the C2 criterion.

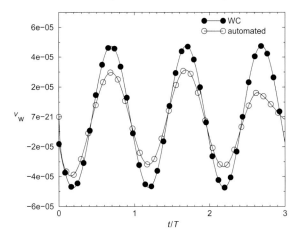

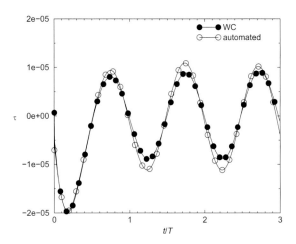

Figure 12.14 Actuator response (top) and sensor measured shear stress (bottom) for the control of Tollmien–Schlichting waves in a flat-plate boundary layer.

12.5.10 Summary

The coupled Navier–Stokes equations, adjoint Navier–Stokes and equations for optimality were solved and validated for the flow control problem of instability wave suppression in a flat-plate boundary layer. By solving the system in the previous section, optimal controls were determined that met the objective of minimizing the perturbation normal stress along a portion of the bounding wall. As a result, the optimal control was found to be an effective means for suppressing two-dimensional, unstable Tollmien–Schlichting traveling waves. The results indicate that the DNS control theory solution is comparable to the wave cancellation result but, unlike the latter, requires no a priori knowledge of the instability characteristics.

12.6 Exercises

Begin with the baseline solutions that you obtained from DNS and linear stability theory for Reynolds number based on displacement thickness ($R = 900$) and nondimensional frequency ($F = 86 = \omega/Re \times 10^6$) and the profiles from the Orr–Sommerfeld equation and Blasius basic flows for the initial conditions.

Exercise 12.1 Introduce an oscillatory suction and blowing condition downstream of the wave-forcing location (control actuator). Do a parameter analysis on the amplitude and frequency of the actuator holding the forcing conditions fixed. Discuss your results. Did you observe any wave cancellation or suppression downstream of the actuator? Why or why not?

Exercise 12.2 Write a simple routine to represent the feed-forward strategy discussion in Section 12.4. Implement this strategy as discussion in the section using two grid points upstream of the actuator but downstream of the forcing wave generator. With two sensors, is it easy or difficult to cancel the wave? Why or why not?

Exercise 12.3 Discuss the process you might use to introduce optimal flow control theory into your DNS code. Would you see these changes to the code as being extremely difficult, somewhat difficult, or easy to implement? Explain the rationale for your answer.

13

Investigating Hydrodynamic Instabilities with Experiments

Science begets knowledge; opinion, ignorance.

– Hippocrates (460–370 B.C.)

This text has covered some historical and more advanced theoretical and computational techniques to predict the onset of transitional flows with linear methods, the amplification and interaction of these linear modes in the non-linear regime and the matching of these predictions with empirical models. Furthermore, some methods of control have been developed and discussed in the previous chapter on flow control. Here, we address issues associated with investigating hydrodynamic instabilities using experimental techniques. These issues include the experimental facility, model configuration and instrumentation, all of which impact the understanding of hydrodynamic instabilities.

Because the authors have primary expertise in theory and computation, we readily acknowledge the topics in this chapter are based on literature from leading scientists and engineers in the field of transitional flows. This chapter serves as an introduction to the experimental process. The content of this chapter is primarily based on the review by Saric (1994b) and a text by Smol'yakov & Tkachenko (1983).

13.1 Experimental Facility

Since the theoretical and computational modeling of a hydrodynamic instability process is the goal, two key aspects of the flow must be carefully documented in the experiment before studying the instabilities. First, the physical properties of the flow environment must be understood within the experimental facility. The makeup of the facility dictates the background (or freestream) disturbances and the spatial-temporal characteristics of the flow environment.

The incoming freestream environment should be understood and characterized before commencing with a discussion of the use of artificial disturbances, which are typically the manner in which hydrodynamic instabilities are investigated. This freestream environment is dictated by the facility. Here, we will restrict our discussion to wind tunnel facilities as opposed to free jet or water tunnel facilities. Typically, we design the experimental environment to mimic the environment that the application would encounter. For an aircraft in cruise flight this would be still air (low freestream turbulence and acoustic levels). For a turbine blade, this would be higher freestream levels associated with internal engine flows. As such, the wind tunnel must be constructed to achieve certain environmental goals to mimic the application. Most wind tunnels built to date have been designed for steady force balance measurements, and typically have large freestream turbulence levels that make them inappropriate for use in hydrodynamic stability investigations. Such high turbulence levels overwhelm the potential existence and characterization of infinitesimal instability modes, making the background noise levels in the instrumentation far above the instability signal.

One of the first, and now classical, successful experimental investigations of hydrodynamic instabilities was conducted by Schubauer & Skramstad (1947) in a low turbulence tunnel ("Dryden tunnel") at the National Bureau of Standards. Additionally, Liepmann (1943) investigated hydrodynamic instabilities on curved walls. To minimize the turbulence levels, we now know that the diffuser design plays a significant role in the resulting flow characteristics. By introducing bends and a diverging diffuser, the absence of sudden changes in the flow is ensured. Also, turbulence is damped by way of fine anti-turbulence screens as far upstream of the core measurement region as possible. For supersonic low disturbance wind tunnels, the turbulent boundary layer upstream of the chock location is removed by suction. The successful design of a low disturbance supersonic facility is extremely challenging because of the dominance of acoustic disturbances in the facility. The facility noise, such as vibration or the motor, must be suppressed to avoid contamination of the natural hydrodynamic modes with acoustically induced modes. Whereas the facility vibration can be inhibited with a mechanical vibration absorption means, the motor acoustics should be cancelled with mufflers. Finally, the air must be free from debris that could either stick to a model and act as a roughness element or impact or damage the instrumentation that is typically very small and delicate. This can be accomplished using dust filters at the air intake point and upstream of the anti-turbulence screens.

The velocity fluctuations and turbulence levels should be documented in the freestream. Spatial correlations should be undertaken to decouple the turbulence and any existing acoustics fields. These measures will indicate whether the tunnel is a low turbulence or quiet facility. Although gaining an understanding of the facility attributes is essential to contributing to the study of hydrodynamic instability, this step in the experimental process is often purposely not undertaken for two reasons. First, there can be considerable cost, in terms of funds and people resources, to an organization to perform this every time changes are made to the facility. Many contracts or grants will not cover the cost of such tests. The second rational that may deter an organization from performing such a facility analysis resides with the meaning of the results. If the tunnel has extremely high disturbance levels, future business opportunities may be quenched because of the public knowledge of the tunnel deficiencies. So the topic of facility flow quality becomes a topic of debate. However, the characterization of the flow quality in the facility is key to understanding any hydrodynamic instability investigation.

13.2 Model Configuration

Whereas the first key aspect of the flow was governed by the facility, the second salient aspect of the flow involves the installation of the model configuration in the facility and resulting basic state characteristics. For example, the leading edge of a flat-plate model will have a nonzero pressure gradient. Downstream of the leading edge the measurements can indicate that the desired zero pressure gradient field is present or a misalignment would yield an adverse or favorable pressure gradient. For a proper aligned, zero-pressure gradient flow, the now Blasius boundary layer will have an effective virtual leading edge, which is different from the model's leading edge. If the measurements of instability modes in the flow do not account for this virtual leading edge, Reynolds number errors as high as 10–15% may result when comparing with theory. To ascertain a leading-edge correction, one should measure the mean boundary layer and calculate the displacement thickness. From our understanding of the Blasius similarity scaling, the connection between the displacement thickness, streamwise location and Reynolds number fall directly from the scaling.

In addition to the virtual leading-edge correction that must be understood, small pressure gradients as small as fractions of a percent can significantly alter the stability or instability of traveling waves. To reduce this uncertainty and better understand the true characteristics of the basic flow, the shape factor should be measured using the boundary layer profiles at different stations in the downstream direction. Such minute pressure gradient features would become evident with changes in the shape factor. Further, any deviations in the spanwise direction should be documented because spanwise nonuniformities can induce secondary instability modes.

13.3 Inducing Hydrodynamics Instabilities

As discussed in the previous sections, the facility and model directly impact the characteristics of experimentally observed hydrodynamic flow instabilities. Whereas turbulent flows are a robust and chaotic environment, the laminar counterflow is extremely sensitive to disturbances and, at the right flow conditions (e.g., Reynolds number), readily admit hydrodynamic instabilities. These instabilities can be induced by the natural tunnel environment or via more controlled artificial disturbance generators.

13.3.1 Natural Disturbances

Although introducing artificial disturbances can be extremely beneficial to study numerous physical phenomena associated with the transition process, understanding the natural ingestion of disturbances has in recent years become a major research topic area. Under natural transition, the freestream turbulence, vorticity or acoustics can interact with the attributes of the model to introduce energy in the wavelength and frequency range relevant to the most unstable modes. This process known as receptivity has been discussed in Chapter 10. However, trying to understand what is measured downstream to the cause and effect attribute at disturbance inception is difficult and assumes that the direct receptivity mechanism can be inferred from downstream measurements.

13.3.2 Artificial Disturbances

The process of hydrodynamic instability inception, amplification and breakdown has been studied for over a century. While numerous techniques are available to study the later stages of this flow phenomena, the inception portion of this problem involves fluctuations too small to measure. Hence, comparisons between theory, computation and experiments must recognize this deficiency in the comparisons. To minimize the unknowns in the upstream freestream environment and to control the experiment, artificial disturbances are introduced into the flow. As far back as the famous experiment by Schubauer & Skramstad (1947), artificial disturbances were introduced into a boundary by using a vibrating ribbon. In the presence of a stationary magnet, alternating current through the ribbon leads to a Lorentz force. This method leads to fluctuations with a prescribed dominant frequency and wavelength. A sufficiently long ribbon must be used or end effects from the ribbon can contaminate the flow. Even with a long ribbon, the end effects spread inwardly downstream at an angle of approximately 12°. So, there is a cone of effectiveness that is somewhat similar to side-wall model and side-wall end effects. Such end effects can alter the disturbance evolution process. The ribbon does not introduce a single mode but rather a disturbance whose dominant mode is the Tollmien–Schlichting wave for a flat-plate boundary layer mean flow. This means a relaxation distance must be maintained until the more stable modes decay. This relaxation distance may be as much as ten boundary layer thicknesses. The ribbon is an intrinsic device, and so its presence may induce a wake that effects the basic flow state. Such an alteration of the basic state may induce otherwise less dominant instabilities. Furthermore, the disturbances induced by the vibrating ribbon may interact with other random disturbances already present in the flow and may potentially alter the amplification process as well as the nonlinear interactions in the nonlinear regime of the flow. The study of the nonlinear interaction of waves is complicated by the limitations of the disturbance generator. A large-amplitude vibrating ribbon at a given frequency cannot introduce only these distinct and desired modes, but rather the ribbon will introduce fundamental, harmonics and differences in the desired modes. Consequently the desired nonlinear instability study may be contaminated by the presence of additional modes that are related to the desired fundamental mode. Additionally, roughness elements can be placed at the branch I neutral point to maximize the receptivity in a flat-plate boundary layer flow (King & Breuer, 2001). So, similar to issues of natural disturbance induction, the artificial disturbance generation can lead to complications and requires care in the experimental study.

13.4 Measurement Instrumentation

In this section, qualitative visual and quantitative measurement techniques are outlined for the study of hydrodynamic instabilities. Widely used techniques to visualize the instabilities include liquid crystals, smoke wire and tracer techniques. Thermo-anemometry (hot wires and hot films) are discussed as quantitative measuring techniques.

13.4.1 Liquid Crystals

Liquid crystals can be applied to a model using an air brush and should be applied evenly over the model. The approach is useful to measure abrupt changes in the surface shear stress properties by distinct color changes that can be recorded with a camera. Such abrupt changes of surface shear stress occur in flows that have separated or the onset of transition and are most relevant to our discussion. This robust technique is useful in the later stages of transition, and such information is a valuable aid to the placement of quantitative information and to visualize potential three-dimensionality in the flow transition process. The use of liquid crystals could induce additional instabilities within the flow due to the potentially non-smooth application on a model. The flow could take this non-smooth surface to be a rough surface, which, in turn, contributes to the receptivity and amplification of infinitesimal modes. Also, chloroform is the solvent used to remove the remaining liquid crystals from the model and is somewhat cumbersome to use in a closed laboratory environment.

13.4.2 Smoke Wires

The second flow visualization technique summarized here is the smoke wire. The smoke wire has a diameter typically ranging from 50 to 80 μm. A computer is used to initiate a set voltage with a time delayed shutter release. The connected wire that has a coating of oil is heated via the voltage and generates a burst of smoke streaks. This smoke then travels downstream with the flow and is distorted with the flow. Problems can arise with the smoke wire similar to any intrusive measurement technique. Any upstream mode basing by the smoke wire will feel the interference of the wire. The traveling wave would involve a step like change in its amplitude resulting from the wire. Saric (1994b) also carefully notes that the quantitative measurements of a wave pattern should be acquired with the flow visualization technique in the flow. Furthermore, measurements within 15 diameters of the smoke wire should be interpreted with caution because of the wire induced affects on the flow.

13.4.3 Bubbles and Dyes

Finally, tracer techniques that have been successfully used to visually study transitional flows include hydrogen bubbles in water and dyes. The benefits and difficulties of these techniques are similar to the smoke wire technique, with the addition of possible buoyancy effects for bubbles.

13.4.4 Thermo-Anemometry

Thermo-anemometry consists of hot wire and hot film techniques. Shown in Fig. 13.1 are schematics of various thermo-anemometry concepts. For quantification of the fluctuations in the flow, the hot wire anemometer has been one of the most widely used techniques to date. A diameter for a hot wire is typically $3–5\mu$m and is therefore extremely delicate. The straight wire and slant wire pair can be used to accurately measure the streamwise and spanwise velocity fluctuations. For thin boundary layers, the wall-normal cannot be measured due to the span of the wire. Because this technique relies on the correlation between temperature and velocity on a wheat stone bridge, the temperatures during calibration and testing should be within a few degrees to avoid errors in the measurements. Similar to the visualization techniques, the hot wires or traverse mechanism may cause blockage and interfere with the flow field. The essential element of the hot wire is the miniature metal element that is heated by an electrical current. The metals most typically used are tungsten, platinum or platinum iridium. In a flow on situation, the cooling of the element or heat transfer to the fluid increase with increasing flow velocity. Hence, recording the cooling process with a bridge circuit leads to a relationship between the electrical resistance and the flow velocity. This relationship requires a calibration with a know velocity field. The sensitivity of the hot wire enables quantitative measures down to a few percent of the mean flow velocity.

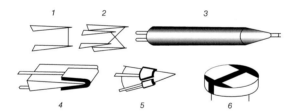

Figure 13.1 Hot-wire anemometers: 1-single wire probe; 2-two wire probe (X-probe); 3-general appearance of probe and its body; 4-wedge shaped film probe; 5-conical film probe; 6-thin film anemometer probe (flush mounted). (Reprinted from Springer Nature: Smol'yakov & Tkachenko, 1983.)

A second technique to measure flow instabilities in a laminar flow (as well as turbulent flows) is the hot film. Here, the film is attached to the surface of the model and is used to measure the spectra and shear stress from the oncoming flow. Multiple films can provide phase and group velocity directions. This robust technique is often used in flight experiments primarily for shear stress measurements, ascertaining whether the flow is laminar, intermittent or turbulent. See Fischer & Anders (1999) for a description of hot film usage in a supersonic laminar flow control flight experiment. The metals used for hot films are typically platinum and nickel.

For a better understanding of the mechanics of hot films, the discussion of Hosder & Simpson (2001) is summarized to describe how an experimental measurement is turned from a voltage signal into a shear stress. The hot film sensors heat the near wall region of the fluid by forced convection. The heat transfer gives a measure of the shear because of the similarity between the gradient transport of heat and momentum. The time averaged voltage (v) and shear stress (τ_w) are connected through a constant temperature anemometer by King's Law or

$$\frac{v^2}{(T_w - T_\infty)} = A + B(\tau_w)^{1/3}, \tag{13.1}$$

where T_w is the sensor temperature and T_∞ is the freestream or tunnel temperature. The constants A and B are found by a linear regression through the calibration procedure. Hence, from this relationship, one can easily see that large changes in the fluid temperature during an investigation can lead to erroneous results or cause the experimentalist to recalibrate the sensor.

Recently, advanced measurement techniques, such as laser Doppler velocimetry and particle image velocimetry, have successfully been used to measure two- and three-dimensional fluctuations in turbulent flows. Such techniques have not been demonstrated for transition flows because of insufficient frequency response of the systems.

13.5 Signal Analysis

The devices described in the previous section for obtaining quantitative information on hydrodynamic instabilities involve the transformation of a physical quantity into electrical current or voltage fluctuations. The electrical signals must then be processed to obtain velocities. This processing can be accomplished with analog or digital electronics. For analog systems, the continuous electrical signal is transformed in a similar fashion as an operator is applied to a mathematical function. An oscillatory electrical signal that is tuned to a specific frequency will transmit Fourier components of only that frequency, just as a Fourier transform. For a digital system, the continuous signal is encoded into a series of discrete levels. The more frequent the encoding, the higher the quantization frequency and the more detailed the correspondence between the discrete and continuous values. The quantization frequency is then extremely important and is usually referred to as the analog to digital conversion. The coded signals are then fed to computer memory for immediate or later analysis by a program. For a given quantization frequency, this post-analysis program dictates resulting statistics of the measurements. The discrete approach is advantageous because it does not depend on the origin of the data; however, highly fluctuating flow fields require large quantization frequencies and hence a large amount of stored data. So the analysis in a post experimental program will lead to the final spatial and temporal modal information for the disturbances. As such, we will not elaborate on various analysis approaches. However, one can easily begin a Fourier transform analysis routine to obtain dominant modal information; this analyzer is typically referred to as a spectral analyzer.

13.6 Summary

Many issues associated with experimentally measuring hydrodynamic instabilities has been discussed in this chapter. Since this text is primarily associated with theoretical and computational issues with hydrodynamic instabilities, this chapter is meant to be a cursory look at experimentation. The topic is extremely challenging and requires an understanding of the instability processes before moving to the laboratory.

Appendix A
Mathematical Formulas

$$\sin(0) = \sin(\pi) = 0, \quad \sin(\pi/2) = 1, \quad \sin(3\pi/2) = -1 \tag{A.1}$$

$$\cos(0) = 1, \quad \cos(\pi/2) = \cos(3\pi/2) = 0, \quad \cos(\pi) = -1 \tag{A.2}$$

$$e^{iu} = \sin(u) + i\cos(u)) \tag{A.3}$$

$$\sin(u+v) = \sin(u)\cos(v) + \cos(u)\sin(v) \tag{A.4}$$

$$\sin(u-v) = \sin(u)\cos(v) - \cos(u)\sin(v) \tag{A.5}$$

$$\cos(u+v) = \cos(u)\cos(v) - \sin(u)\sin(v) \tag{A.6}$$

$$\cos(u-v) = \cos(u)\cos(v) + \sin(u)\sin(v) \tag{A.7}$$

$$\sinh(u) = \frac{1}{2}\left[e^u - e^{-u}\right], \qquad \cosh(u) = \frac{1}{2}\left[e^u + e^{-u}\right] \tag{A.8}$$

$$\tanh(u) = \frac{\sinh(u)}{\cosh(u)} = \frac{(e^u - e^{-u})}{(e^u + e^{-u})} = \frac{1}{\coth(u)} \tag{A.9}$$

$$\sinh(0) = \sinh(i\pi) = 0 \tag{A.10}$$

$$\cosh(0) = 1, \qquad \cosh(i\pi) = -1 \tag{A.11}$$

$$\text{sech}^2(u) = 1 - \tanh^2(u) \tag{A.12}$$

$$\sinh(u+v) = \sinh(u)\cosh(v) + \cosh(u)\sinh(v) \tag{A.13}$$

$$\sinh(u+iv) = \sinh(u)\cos(v) + \cosh(u)\sin(v) \tag{A.14}$$

$$d[\sin(u)]/dx = \cos(u)\, du/dx \tag{A.15}$$

$$d[\cos(u)]/dx = -\sin(u)\, du/dx \tag{A.16}$$

$$d[e^u]/dx = e^u\, du/dx \tag{A.17}$$

$$d[\sinh(u)]/dx = \cosh(u)\, du/dx \tag{A.18}$$

$$d[\cosh(u)]/dx = \sinh(u)\, du/dx \tag{A.19}$$

$$d[\tanh(u)]/dx = \text{sech}^2(u)\, du/dx \tag{A.20}$$

$$d[\text{sech}(u)]/dx = -\tanh(u)\text{sech}(u)\, du/dx \tag{A.21}$$

Mathematical Formulas

Quadratic equation: $\qquad ax^2 + bx + c = 0$ $\hfill$ (A.22)

Solution to quadratic equation: $\qquad x = \dfrac{[-b \pm \sqrt{b^2 - 4ac}]}{2a}$ $\hfill$ (A.23)

Integration by parts: $\quad \displaystyle\int u(x)v'(x)dx = u(x)v(x) - \int u'(x)v(x)dx$ $\quad$ (A.24)

Appendix B
Numerical Methods

B.1 Riccati Transformation

There are basically two different methods or approaches for determining the eigenvalues to Rayleigh's equation, given a particular mean profile. The first method is called a "global method," in which Rayleigh's equation is discretized on a uniform mesh having **M** grid points, and all the eigenvalues of the resulting **MxM** matrix are found by an appropriate eigenvalue package. Care must be exercised when employing this method since the number of eigenvalues change as **M** is changed; i.e., as the mesh is refined, the number of eigenvalues increase. One must be able to distinguish between physical eigenvalues and unphysical ones associated with mesh refinement.

The second method is a local method, in that Rayleigh's equation is solved by some numerical procedure and each eigenvalue is determined in succession. Since this method is most often quoted in the literature we present it here. One popular numerical procedure for solving Rayleigh's equation first involves a Riccati transformation, or

$$G = \frac{\phi'}{\phi}, \tag{B.1}$$

that reduces the second-order linear equation to the following first-order, non-linear equation:

$$G' + G^2 = \alpha^2 + \frac{U''}{U - c}. \tag{B.2}$$

For unbounded flows, where $\phi \approx e^{\mp \alpha y}$ for $y \to \pm\infty$, the appropriate boundary conditions are

$$G(\infty) = -\alpha, \qquad G(-\infty) = \alpha. \tag{B.3}$$

Alternatively, one can first appeal to the pressure disturbance equation (2.33)

401

and employ the Riccati transformation

$$G = \frac{p'}{\alpha p}. \tag{B.4}$$

By substituting (B.4) into (2.33) the first-order nonlinear equation

$$G' + \alpha G^2 - \frac{2U'}{U - c} G - \alpha = 0, \tag{B.5}$$

with boundary conditions

$$G(\infty) = -1, \qquad G(-\infty) = +1, \tag{B.6}$$

is found. The advantage of using (B.5) over (B.2) is that the boundary conditions (B.6) are no longer functions of α, which can cause numerical problems when the wavenumber becomes small. The results presented in Section 2.6 used (B.5) and (B.6) to compute the eigenvalues.

There are several advantages to solving the Riccati equation rather than Rayleigh's equation directly. These include:

- Rayleigh's equation is homogeneous and so has the trivial solution in its solution space. Unless one is close to the correct eigenvalue, the solution may converge only to the trivial solution. The Riccati equation is nonhomogeneous and hence does not have the trivial solution in its solution space.
- Rayleigh's equation has exponentially growing solutions as $y \to \pm\infty$, and so care must be exercised to eliminate these nonphysical solutions. The Riccati equation has finite conditions and so the problem of eliminating the growing solutions is avoided.

The numerical procedure for the determination of the eigenvalues is as follows:

Step 1: Fix a value of the wavenumber α and make an initial estimate on the phase speed c.

Step 2: Integrate the Riccati equation from L to 0 and from $-L$ to 0 (for the unbounded case, L is large and typically taken in the range of 6 to 8).

Step 3: Compute $[G]|_{y=0}$, the jump in G at the origin.

Step 4: If $[G] < tol$, where tol is a small number, then stop. The value of c is the required eigenvalue. If $[G] \geq tol$, compute an updated value of c by some numerical root finding procedure. Go to Step 2 and repeat until convergence is reached.

For the integrator, a simple method is to use a fourth-order, variable step Runge–Kutta method. For updating the eigenvalue, we typically use Muller's method, which is essentially a Secant method for complex eigenvalues. To start the procedure in Step 1, we must have a good estimate for c. This value is usually obtained by perturbing by a small distance from the neutral value. Recall that the neutral eigenvalue is given by $c_N = U(y_s)$, where y_s is the inflection point where $U''(y_s) = 0$. The value of α_N, and hence ω_N, can be found by integrating the Ricatti equation numerically. However, a problem exists when integrating along the real axis since $U - c_N = 0$ at y_s. To avoid this, we analytically continue the path of integration below the real y-axis (Lessen, 1950). That is, we let $y = y_r + iy_i$, and fix $y_i < 0$. However, care must be taken so that y_i does not pass through a branch point or singularity of the mean profile when analytically continued into the complex plane. For example, consider the profile $U = \tanh(y)$. Extending into the complex plane we get

$$U = \tanh(y_r + iy_i), \tag{B.7}$$

which has a singularity at $(0, -\pi/2)$ in the lower half-plane.

Another numerical procedure for solving Rayleigh's equation involves the Compound Matrix method. This method has been shown to be essentially the same as the Ricatti method, but applied to Rayleigh's equation directly. Although it also offers the advantage of eliminating the exponentially growing solutions, we shall not discuss this method here but postpone the discussion until Chapter 3, where viscous disturbances are addressed.

B.2 Compound Matrix Method

Although many numerical methods exist that can be used to solve the Orr–Sommerfeld equation, e.g., initial-value, finite difference, Galerkin, spectral, etc., we shall present here only the Compound Matrix method because of its relative ease of implementation. Basically, one only needs to be familiar with Runge–Kutta methods to be successful in solving the Orr–Sommerfeld equation. Other methods require much more knowledge. We do not, however, pretend that the Compound Matrix method is as efficient as others; some methods work better depending on the circumstances. The use of compound matrices to solve the Orr–Sommerfeld equation was first presented by Ng & Reid (1979, 1979), and the presentation here draws from their work (the method is also presented in Drazin & Reid , 1984; see also Davey, 1980).

Consider the general fourth-order differential equation

$$\phi^{iv} - a_1\phi''' - a_2\phi'' - a_3\phi' - a_4\phi = 0, \quad y_1 \leq y \leq y_2, \tag{B.8}$$

where primes denote differentiation with respect to y, and $a_1 - a_4$ are coefficients, which may be functions of the independent variable. This equation must be solved subject to appropriate boundary conditions at y_1 and y_2. For the particular case of the Orr–Sommerfeld equation,

$$\left.\begin{array}{ll} a_1 = 0, & a_2 = 2\alpha^2 + i\alpha Re(U - c), \\ a_3 = 0, & a_4 = -\alpha^4 - i\alpha Re\left[\alpha^2(U - c) + U''\right], \end{array}\right\} \qquad (B.9)$$

where $U(y)$ is the mean profile, α and Re are real parameters and c is the complex eigenvalue.

Let ϕ_1 and ϕ_2 be any two solutions that satisfy the boundary conditions at y_2. We now consider the matrix

$$\Phi = \begin{bmatrix} \phi_1 & \phi_2 \\ \phi_1' & \phi_2' \\ \phi_1'' & \phi_2'' \\ \phi_1''' & \phi_2''' \end{bmatrix} \qquad (B.10)$$

The 2×2 minors of Φ, in lexical order, are

$$\left.\begin{array}{ll} Y_1 = \phi_1\phi_2' - \phi_1'\phi_2, & Y_4 = \phi_1'\phi_2'' - \phi_1''\phi_2', \\ Y_2 = \phi_1\phi_2'' - \phi_1''\phi_2, & Y_5 = \phi_1'\phi_2''' - \phi_1'''\phi_2', \\ Y_3 = \phi_1\phi_2''' - \phi_1'''\phi_2, & Y_6 = \phi_1''\phi_2''' - \phi_1'''\phi_2'', \end{array}\right\} \qquad (B.11)$$

and they satisfy the quadratic identity

$$Y_1Y_6 - Y_2Y_5 + Y_3Y_4 = 0, \qquad (B.12)$$

which is a useful check on the accuracy of the numerical integration. Differential equations for Y_1 to Y_6 may be found by differentiating (B.11) and using (B.8) to eliminate ϕ_1, ϕ_2, yielding the linear system

$$\left.\begin{array}{rcl} Y_1' & = & Y_2, \\ Y_2' & = & Y_3 + Y_4, \\ Y_3' & = & a_3Y_1 + a_2Y_2 + a_1Y_3 + Y_5, \\ Y_4' & = & Y_5, \\ Y_5' & = & -a_4Y_1 + a_2Y_4 + a_1Y_5 + Y_6, \\ Y_6' & = & -a_4Y_2 - a_3Y_4 + a_1Y_6. \end{array}\right\} \qquad (B.13)$$

In general, the system is always integrated from y_2 down to y_1. The boundary conditions at y_2 depend on the mean profile. For channel flows we take $y_2 = 1$, the centerline of the channel. For symmetric disturbances we impose the condition $\phi'(1) = \phi'''(1) = 0$. This leads to the conditions $Y_2(1) = 1$ with all other quantities set to zero. The proper boundary conditions for the Blasius boundary layer is somewhat more complicated. We begin by noting that as $y \to \infty$, $U \to 1$, $U'' \to 0$ and

$$\phi_1 \to e^{-\alpha y}, \qquad \phi_2 \to e^{-py}, \tag{B.14}$$

where

$$p = \sqrt{\alpha^2 + i\alpha Re(1-c)}, \quad Re(p) > 0. \tag{B.15}$$

Substitution into (B.11) yields the asymptotic boundary conditions

$$\left. \begin{array}{ll} Y_1 = 1, & Y_4 = \alpha p, \\ Y_2 = -(\alpha + p), & Y_5 = -\alpha p(\alpha + p), \\ Y_3 = \alpha^2 + \alpha p + p^2, & Y_6 = \alpha^2 p^2, \end{array} \right\} \tag{B.16}$$

where the exponential factors have been eliminated and normalized Y_1 for convenience. Typically y_2 ranges from 5 to 10, depending on the value of the Reynolds number.

When y_1 corresponds to a wall (as it does for channel flows and semi-infinite flows), the proper boundary condition is $\phi(0) = \phi'(0) = 0$. Examining (B.11) yields the condition $Y_1(0) = 0$, which determines the eigenvalue c; setting $D = Y_1(0)$ is then the discriminate. This condition is usually normalized by the largest growth rate of the variables $Y_i(0)$ to make the actual quantity finite when c is not an eigenvalue.

The numerical solution now proceeds as follows. Given a particular mean profile, the values of α and Re are fixed and an initial value for the eigenvalue c is selected. Integrate the system (B.13) from y_2 to $y_1 = 0$ using a standard fourth-order Runge–Kutta scheme (although variable step schemes with error control would also work), and determine the value of the discriminate $D = Y_1(0)$. If the discriminate is not zero, an iterative procedure is used to update the value of c until $|D|$ is less than some prescribed tolerance; the value of c is then the required eigenvalue. The iterative method that is usually employed for finding complex eigenvalues is Muller's method (see, e.g., Press, Teukolsky, Vetterling & Flannery, 1992). This procedure determines one eigenvalue at a time, and usually requires a sufficiently close initial guess to determine the eigenvalue. Alternatively, one can loop over some region in the complex c-plane, determine the value of D at each (c_r, c_i) point, plot the contours of D_r and D_i and write a short post-processing code which determines the intersection points. These then provide initial guesses that can be substituted back into the iterative method to determine the eigenvalues with more precision ("polishing"). The entire process can be written within a single code, if desired.

Once an eigenvalue has been found, the Compound Matrix method can be used to determine the corresponding eigenfunction. Having found c and Y_i, it can be reasoned that there must exist constants λ_1 and λ_2 such that

$$\phi = \lambda_1 \phi_1 + \lambda_2 \phi_2. \tag{B.17}$$

By differentiating the above relation three times, and eliminating the constants λ_1 and λ_2, the following four relations are obtained

$$\left.\begin{aligned}
Y_1 \phi'' - Y_2 \phi' + Y_4 \phi &= 0, \\
Y_1 \phi''' - Y_3 \phi' + Y_5 \phi &= 0, \\
Y_2 \phi''' - Y_3 \phi'' + Y_6 \phi &= 0, \\
Y_4 \phi''' - Y_5 \phi'' + Y_6 \phi' &= 0.
\end{aligned}\right\} \tag{B.18}$$

The eigenfunction ϕ is then determined by integrating any one of the four relations from $y = 0$ to $y = y_2$. Since $Y_1(0) = 0$, the first two relations cannot be integrated starting at $y = 0$ (Ng & Reid (1979) showed that this is only a minor problem, and one can start at one grid point away from the origin). By trial and error, Davey (1984) noted that the third relation gave slightly more accurate results than the fourth for the specific case of Blasius boundary layer flow, with the normalization $\phi''(0) = (1 - i)Re^{1/2}$. The solutions Y_i need to be stored in tables, and then are used as reference tables as part of the integration scheme for ϕ.

B.3 Chebyshev Series Formulas

The Chebyshev polynomials, $T_n(x)$, are defined on the interval $x \in [-1, +1]$ and are derived from and related to the cosine function by

$$T_n(\cos \theta) = \cos n\theta \tag{B.19}$$

with the initial few polynomials appearing as

$$\left.\begin{aligned}
T_0(x) &= 1, \\
T_1(x) &= x, \\
T_2(x) &= 2x^2 - 1, \\
T_3(x) &= 4x^3 - 3x, \quad \text{etc.}
\end{aligned}\right\} \tag{B.20}$$

The following trigonometric identity can be obtained.

$$\cos(n+1)\theta = 2\cos\theta \cdot \cos n\theta - \cos(n-1)\theta. \tag{B.21}$$

This results in a Chebyshev recurrence formula for higher-order polynomials.

$$T_{n+1}(x) = 2xT_n(x) - T_{n-1}(x). \tag{B.22}$$

The product formula is

$$T_n(x)T_m(x) = \frac{1}{2}\left[T_{n+m}(x) + T_{|n-m|}(x)\right] \tag{B.23}$$

and the indefinite integral relation is

$$\int T_n(x)dx = \begin{cases} T_1(x) & n = 0, \\ \frac{1}{4}\left(T_0(x) + T_2(x)\right) & n = 1, \\ \frac{1}{2}\left[\frac{T_{n+1}(x)}{n+1} - \frac{T_{n-1}(x)}{n-1}\right] & n \geq 2. \end{cases} \tag{B.24}$$

The series boundary conditions for a polynomial of order n are

$$T_n(\pm 1) = (\pm 1)^n \tag{B.25}$$

and the differential relation for Chebyshev polynomials at the boundaries is

$$\frac{d^p}{dx^p}T_n(\pm 1) = (\pm 1)^{n+p}\prod_{k=0}^{p-1}(n^2 - k^2)/(2k+1). \tag{B.26}$$

Another efficient relation useful when performing the summation of a Chebyshev series to determine a functional value of x is given by

$$f(x) = \sum_{n=o}^{N}{}' a_n T_n(x) = \frac{1}{2}\left[b_0(x) - b_2(x)\right], \tag{B.27}$$

where the prime signifies that the leading term is to be halved. The recurrence system required to evaluate (B.27) is

$$\left.\begin{aligned} b_n(x) &= 2xb_{n+1}(x) - b_{n+2}(x) + a_n, \\ b_{N+1}(x) &= b_{N+2}(x) = 0. \end{aligned}\right\} \tag{B.28}$$

A Chebyshev formula useful in approximating a known function in a Chebyshev series can be defined as

$$\Phi(x) = \sum_{n=0}^{N}{}' \phi_n T_n(x), \tag{B.29}$$

where $\Phi(x)$ is a known function. The coefficients, ϕ_n, are given by

$$\phi_n = \frac{2}{N}\sum_{k=0}^{N}{}'' \Phi(x_k)T_n(x_k) \tag{B.30}$$

with

$$x_k = \cos\frac{k\pi}{N} \quad \text{for } k = 0,1,2,...,N. \tag{B.31}$$

The double prime on the summation signifies that the leading and trailing coefficients are to be halved. This approximation of a known function is required
for the Blasius profile and the primary eigenfunctions. As an example, the Blasius profile will be represented by the series after listing the remaining general
formulae.

The final Chebyshev property that will be given prior to listing practical
integral formulae is the approximation of the differential of a known function
in Chebyshev series. The derivative is given by

$$\phi'(x) = \sum_{n=0}^{\infty} b_n T_n(x), \tag{B.32}$$

where

$$b_n = \frac{2}{c_n} \sum_{\substack{p=n+1 \\ p+n \text{ odd}}}^{\infty} p a_p \tag{B.33}$$

and

$$c_n = \begin{cases} 2 & n = 0, \\ 1 & n > 0. \end{cases} \tag{B.34}$$

The coefficients, a_n, are obtained from the series approximation to the known
function, $\phi(x)$.

To obtain the solution of a differential equation by a Chebyshev series approximation, it is convenient, although not necessary, to convert the differential
equation to an integral form. As such, a function is represented by the following finite, Chebyshev series:

$$\phi(x) = \sum_{n=0}^{N} {}' a_n T_n(x). \tag{B.35}$$

By applying the integral relation (B.24) appropriately and repeatedly, the following relations are obtained.

$$\int \phi(x)dx = \sum_{n=0}^{N+1} {}' b_n T_n(x) \tag{B.36}$$

where

$$b_n = \frac{1}{2n}(a_{n-1} - a_{n+1}) \quad \text{for } n \geq 1, \tag{B.37}$$

$$\iint \phi(x)dx^2 = \sum_{n=0}^{N+2}{}' b_n T_n(x), \tag{B.38}$$

$$b_n = \left[\frac{a_{n-2}}{4n(n-1)} - \frac{a_n}{2(n^2-1)} + \frac{a_{n+2}}{4n(n+1)}\right] \quad \text{for } n \geq 2, \tag{B.39}$$

$$\iiint \phi(x)dx^3 = \sum_{n=0}^{N+3}{}' b_n T_n(x), \tag{B.40}$$

$$b_n = \frac{a_{n-3}}{8n(n-1)(n-2)} - \frac{3a_{n-1}}{8n(n-2)(n+1)} + \frac{3a_{n+1}}{8n(n-1)(n+2)}$$
$$- \frac{a_{n+3}}{8n(n+1)(n+2)} \quad \text{for } n \geq 3, \tag{B.41}$$

$$\iiiint \phi(x)dx^4 = \sum_{n=0}^{N+4}{}' b_n T_n(x), \tag{B.42}$$

and

$$b_n = \frac{a_{n-4}}{16n(n-1)(n-2)(n-3)} - \frac{a_{n-2}}{4n(n^2-1)(n-3)} + \frac{3a_n}{8(n^2-1)(n^2-4)}$$
$$- \frac{a_{n+2}}{4n(n^2-1)(n+3)} + \frac{a_{n+4}}{16n(n+1)(n+2)(n+3)} \quad \text{for } n \geq 4. \tag{B.43}$$

When the coefficients in the differential equations are nonconstant, the Chebyshev product formula (B.23) is needed. Introducing a function, $u(x)$, representing the nonconstant coefficient, the following is obtained:

$$u(x)\phi(x) = \sum_{n=0}^{\infty}{}' d_n T_n(x) \tag{B.44}$$

with

$$u(x) = \sum_{n=0}^{\infty}{}' u_n T_n(x) \tag{B.45}$$

and

$$d_n = \frac{1}{2}u_n a_0 + \frac{1}{2}\sum_{m=1}^{N}(u_{|m-n|} + u_{m+n})a_m \quad \text{for } n \geq 0. \tag{B.46}$$

Integrations are performed in a straightforward manner using the integral relation (B.24). The following integral relations prove useful for the problems presented here:

$$\int u(x)\phi(x)dx = \sum_{n=0}^{N+1}{}' d_n T_n(x), \tag{B.47}$$

where

$$d_n = \frac{1}{4n}(u_{n-1} - u_{n+1})a_0 + \frac{1}{4n}\sum_{m=1}^{N}(u_{|m-n+1|} - u_{|m-n-1|}$$

$$+u_{m+n-1} - u_{m+n+1})a_m \quad \text{for } n \geq 1, \tag{B.48}$$

$$\iint u(x)\phi(x)dx^2 = \sum_{n=0}^{N+2}{}' d_n T_n(x), \tag{B.49}$$

$$d_n = \left[\frac{u_{n-2}}{8n(n-1)} - \frac{u_n}{4(n^2-1)} + \frac{u_{n+2}}{8n(n+1)}\right]a_0$$

$$+\sum_{m=1}^{N}\left[\frac{u_{|m-n+2|} + u_{m+n-2}}{8n(n-1)} - \frac{u_{m+n} + u_{|m-n|}}{4(n^2-1)}\right.$$

$$\left.+\frac{u_{|m-n-2|} + u_{m+n+2}}{8n(n+1)}\right]a_m \quad \text{for} \quad n \geq 2, \tag{B.50}$$

$$\iiint u(x)\phi(x)dx^3 = \sum_{n=0}^{N+3}{}' d_n T_n(x), \tag{B.51}$$

$$d_n = \left[\frac{u_{n-3}}{16n(n-1)(n-2)} - \frac{3u_{n-1}}{16n(n+1)(n-2)}\right.$$

$$+\frac{3u_{n+1}}{16n(n-1)(n+2)} - \frac{u_{n+3}}{16n(n+1)(n+2)}\right]a_0$$

$$+\sum_{m=1}^{N}\left[\frac{u_{|m-n+3|} + u_{m+n-3}}{16n(n-1)(n-2)} - \frac{3(u_{|m-n+1|} + u_{m+n-1})}{16n(n+1)(n-2)}\right.$$

$$\left.+\frac{3(u_{|m-n-1|} + u_{m+n+1})}{16n(n-1)(n+2)} - \frac{u_{|m-n-3|} + u_{m+n+3}}{16n(n+1)(n+2)}\right]a_m \quad \text{for } n \geq 3, \tag{B.52}$$

$$\iiiint u(x)\phi(x)dx^4 = \sum_{n=0}^{N+4}{}' d_n T_n(x), \tag{B.53}$$

and

$$d_n = \left[\frac{u_{n-4}}{32n(n-1)(n-2)(n-3)} - \frac{u_{n-2}}{8n(n^2-1)(n-3)} + \frac{3u_n}{16(n^2-1)(n^2-4)}\right.$$

$$\left.-\frac{u_{n+2}}{8n(n^2-1)(n+3)} + \frac{u_{n+4}}{32n(n+1)(n+2)(n+3)}\right]a_0$$

$$+ \sum_{m=1}^{N} \left[\frac{u_{|m-n+4|} + u_{m+n-4}}{32n(n-1)(n-2)(n-3)} - \frac{u_{|m-n+2|} + u_{m+n-2}}{8n(n^2-1)(n-3)} \right.$$

$$+ \frac{3(u_{|m-n|} + u_{m+n})}{16(n^2-1)(n^2-4)} - \frac{u_{|m-n-2|} + u_{m+n+2}}{8n(n^2-1)(n-3)}$$

$$\left. + \frac{u_{|m-n-4|} + u_{m+n+4}}{32n(n+1)(n+2)(n+3)} \right] a_m \quad \text{for } n \geq 4. \qquad \text{(B.54)}$$

These relations replace the appropriate terms in an integral equation in order to obtain a solution. The integral formulae require the order of the Chebyshev terms to begin with the order of the integral equation. The proof of this will not be given here, but can be found in Gottlieb & Orszag (1977).

B.4 Routines

In this section we present several numerical methods that may be useful to solve some of the exercises in this book.

The following is a fifth-order Runge–Kutta method given by Luther (1966). Luther refers to this as a Newton–Cotes type, and is given by

$$y_{n+1} = y_n + \{7k_1 + 7k_3 + 32k_4 + 12k_5 + 32k_6\}/90, \qquad \text{(B.55)}$$

where

$$\begin{aligned}
k_1 &= hf(x_n, y_n), \\
k_2 &= hf(x_n + h, y_n + k_1), \\
k_3 &= hf(x_n + h, y_n + \{k_1 + k_2\}/2), \\
k_4 &= hf(x_n + h/4, y_n + \{14k_1 + 5k_2 - 3k_3\}/64), \\
k_5 &= hf(x_n + h/2, y_n + \{-12k_1 - 12k_2 + 8k_3 + 64k_4\}/96), \\
k_6 &= hf(x_n + 3h/4, y_n + \{-9k_2 + 5k_3 + 16k_4 + 36k_5\}/64),
\end{aligned} \right\} \tag{B.56}$$

and h is the step size.

The following are used for the shooting approach to find the zero of a function with Newton and three-point inverse Lagrange interpolations as listed by Burden & Faires (1985) and the False Position method listed by Gear (1978). These are:

Newton:

$$\alpha_{i+1} = \alpha_i - (\alpha_i - \alpha_{i-1}) \frac{\Delta_i}{\Delta_i - \Delta_{i-1}}; \qquad \text{(B.57)}$$

False Position:

$$\alpha_{i+1} = \frac{\Delta_i \alpha_{i-1} - \Delta_{i-1}\alpha_i}{\Delta_i - \Delta_{i-1}};$$ (B.58)

Inverse Lagrange:

$$\alpha_{i+1} = \frac{\alpha_i \Delta_{i-1}\Delta_{i-2}}{(\Delta_i - \Delta_{i-1})(\Delta_i - \Delta_{i-2})} + \frac{\alpha_{i-1}\Delta_i\Delta_{i-2}}{(\Delta_{i-1} - \Delta_i)(\Delta_{i-1} - \Delta_{i-2})}$$

$$+ \frac{\alpha_{i-2}\Delta_i\Delta_{i-1}}{(\Delta_{i-2} - \Delta_i)(\Delta_{i-2} - \Delta_{i-1})}.$$ (B.59)

Here, α_{i+1} is the eigenvalue for the next iteration $(i+1)$ and Δ is the matrix determinant of the numerical vector asymptotic matching, which goes to zero as α converges to the proper eigenvalue.

In some circumstances a cubic spline can be useful. This is determined from the following

$$\begin{pmatrix} x_1^3 & x_1^2 & x_1 \\ x_2^3 & x_2^2 & x_2 \\ x_3^3 & x_3^2 & x_3 \end{pmatrix} \begin{pmatrix} A \\ B \\ C \end{pmatrix} = \begin{pmatrix} p_1 \\ p_2 \\ p_3 \end{pmatrix},$$ (B.60)

where x_i are the locations for the spline and p_i are the function values. These equations are solved simultaneously to obtain the coefficients

$$C = \frac{p_3 x_1^2 - p_1 x_3^2}{x_3 x_1^2 - x_3^2 x_1},$$ (B.61)

$$B = \frac{p_2 x_1^3 - p_1 x_2^3 - C(x_2 x_1^3 - x_1 x_2^3)}{x_1^2 x_2^2 (x_1 - x_2)},$$ (B.62)

and

$$A = \frac{p_1 - C x_1 - B x_1^2}{x_1^3}.$$ (B.63)

Additionally, Simpson's rules are given by

$$\int_{x_0}^{x_2} f(x)dx = \frac{h}{3}[f(x_0) + 4f(x_1) + f(x_2)] - \frac{h^5}{90}f^{(4)}(\zeta),$$ (B.64)

where $x_0 < \zeta < x_2$, and

$$\int_{x_0}^{x_4} f(x)dx = \frac{2h}{45}\big[7f(x_0) + 32f(x_1) + 12f(x_2) + 32f(x_3)$$

$$+ 7f(x_4)\big] - \frac{8h^7}{954}f^{(6)}(\zeta),$$ (B.65)

where $x_0 < \zeta < x_4$.

Appendix C
Solutions to Exercises

Solutions to selected questions are provided here.

C.1 Solutions for Chapter 2

Exercises 2.1–2.7 are straightforward derivations. The student will gain a deep understanding of the mathematics and assumptions by performing the derivations themselves.

Exercise 2.8 Consider the two eigenvalue problems

$$y'' - \alpha^2 y = 0, \quad y(0) = 0, \quad y(h) = 0, \tag{C.1}$$

and

$$y'' - \alpha^2 y = 0, \quad y(0) = 0, \quad y(\infty) = \text{bounded}. \tag{C.2}$$

Determine the eigenvalues and show that the eigenvalues are different for bounded and unbounded regions. What happens in the limit $h \to \infty$ in the first case? Do the eigenvalues approach those of the second? Explain.

This example illustrates that care must be exercised when approximating unbounded regions by bounded domains for numerical considerations.

Solution
Eigenvalue Problem 1:
First, assume α is real and positive. The solution to equation (C.1) takes the form

$$y(x) = Ae^{mx}. \tag{C.3}$$

Substitute (C.3) into (C.1) to give

$$A(m^2 - \alpha^2)e^{mx} = 0.$$

Either $A = 0$ or $m^2 - \alpha^2 = 0$, i.e., $m = \pm\alpha$. The general solution becomes

$$y(x) = Ae^{\alpha x} + Be^{-\alpha x}. \tag{C.4}$$

Equation (C.4) is subject to the boundary conditions

$$y(0) = 0 \quad \text{and} \quad y(h) = 0. \tag{C.5}$$

With these boundary conditions, the following two equations result for the two unknowns (A, B).

$$y(0) = A + B = 0, \tag{C.6}$$

$$y(h) = Ae^{\alpha h} + Be^{-\alpha h} = 0. \tag{C.7}$$

From the first boundary condition, $A = -B$. Substituting this into (C.7) leads to

$$B\left(e^{\alpha h} - e^{-\alpha h}\right) = 0. \tag{C.8}$$

Then either

$$B = 0 \quad \text{or} \quad \left(e^{\alpha h} - e^{-\alpha h}\right) = 0.$$

From this, $\alpha = 0$. Thus the trivial solution is the only solution if α is real.

Now, assume that α may be complex. Using an identity from Appendix A, boundary condition (C.5), and let $\alpha = \alpha_r + i\alpha_i$ be complex, then

$$\sinh(\alpha_r h + i\alpha_i h) = 0. \tag{C.9}$$

Note that identities from Appendix A are often used in deriving the solutions that follow. As such, (C.9) becomes

$$\sinh(\alpha_r + i\alpha_i) = \sinh(\alpha_r h)\cos(\alpha_i h) + i\cosh(\alpha_r h)\sin(\alpha_i h).$$

So

$$\sinh(\alpha_r h)\cos(\alpha_i h) = 0 \quad \text{and} \quad \cosh(\alpha_r h)\sin(\alpha_i h) = 0. \tag{C.10}$$

If $\alpha_r = 0$, and using identities (A.10) and (A.11), then $\sin(\alpha_i h) = 0$. This implies that

$$\alpha_i = \frac{n\pi}{h}, n = 0, \pm 1, \pm 2\ldots,$$

or,

$$\alpha = i\alpha_i = \frac{n\pi}{h}, n = 0, \pm 1, \pm 2\ldots \tag{C.11}$$

Therefore the eigenvalue solution to *eigenvalue problem* 1 is an infinite number of discrete eigenvalues given by:

$$y(x) = \sin\left(\frac{n\pi x}{h}\right), n = 0, \pm 1, \pm 2 \dots \tag{C.12}$$

Eigenvalue Problem 2:

As before, first assume α is real and positive. The general solution to equation (C.2) is (C.4), which is subject to the boundary conditions

$$y(0) = 0 \quad \text{and} \quad y(\infty) = \text{bounded.} \tag{C.13}$$

With these boundary conditions, bounded implies $A = 0$. Then $y(0) = B = 0$. This implies that only the trivial solution exists if α is real. As before assume that α is complex, or $\alpha = i\alpha_i$. Equation (C.2) becomes

$$y'' - \alpha_i^2 y = 0. \tag{C.14}$$

The general solution takes the form of

$$y = A\cos(\alpha_i x) + B\sin(\alpha_i x) \tag{C.15}$$

and is subject to the boundary conditions (C.13). The first condition yields

$$y(0) = A = 0. \tag{C.16}$$

As $x \to \infty$, the solution remains bounded. The solution with $B = 1$ is

$$y = \sin(\alpha_i x). \tag{C.17}$$

Since α_i can be any real number, we find the solution to be a continuum. Therefore the eigenvalues and corresponding eigenfunctions for the two problems are different.

Bounded domain: $\quad 0 \le x \le h, \quad \alpha = \frac{in\pi}{h}, \quad n = 0, \pm 1, \pm 2, \dots$
Unbounded domain: $0 \le x \le \infty, \quad \alpha = i\alpha_i, \quad 0 \le \alpha_i \le \infty$

In the first case, as $h \to \infty$, we see that the discrete set of eigenvalues collapses onto a single eigenvalue. This example demonstrates that if one is trying to find the eigenvalues numerically to a problem defined on a semi-infinite domain, truncating the domain to a finite domain can lead to incorrect results. *One must be careful in solving eigenvalue problems numerically on semi-infinite domains.*

Exercise 2.9 Consider the simple pendulum as shown in Fig. 2.12. The differential equation governing the motion of the pendulum is given by

$$mL^2\frac{d\theta^2}{dt^2} + k\frac{d\theta}{dt} + mgL\sin\theta = 0, \tag{C.18}$$

where the first term is the acceleration term, the second term is the viscous damping term and the last term is the restoring moment.

(a) Nondimensionalize the equation and show that only one parameter exists that governs the behavior of the system.

(b) Compute the two equilibrium states.

(c) Determine the stability of each of the two states.

Solution

(a) Let the independent variable be: $t = A\tau$. Substitute into (C.18) to find

$$\frac{mL^2}{A^2}\frac{d^2\theta}{d\tau^2} + \frac{k}{A}\frac{d\theta}{d\tau} + mgL\sin\theta = 0. \tag{C.19}$$

To remove the coefficient from the first term, multiply each term of (C.19) by A^2/mL^2 to give

$$\frac{d^2\theta}{d\tau^2} + \frac{Ak}{mL^2}\frac{d\theta}{d\tau} + \frac{A^2g}{L}\sin\theta = 0. \tag{C.20}$$

Therefore, to remove the coefficient dependence of the second term, choose

$$A = \frac{mL^2}{k}.$$

Note: alternatively, the factor in the third term can be removed by defining $A = \sqrt{L/g}$. Substitute the definition of A into equation (C.20)

$$\frac{d^2\theta}{d\tau^2} + \frac{d\theta}{d\tau} + \frac{m^2L^4}{k^2}\frac{g}{L}\sin\theta = 0. \tag{C.21}$$

The only remaining parameter in the equation is

$$\alpha = \frac{m^2L^3g}{k^2} \quad \text{with} \quad \alpha > 0. \tag{C.22}$$

The one-parameter equation now becomes

$$\frac{d^2\theta}{d\tau^2} + \frac{d\theta}{d\tau} + \alpha\sin\theta = 0. \tag{C.23}$$

(b) For an equilibrium state, we set $d/d\tau = 0$; therefore, $\sin(\theta) = 0$, or, $\theta = 0, \pi$.

(c) For stability, a perturbation θ' to θ is introduced at the equilibrium states. At $\theta = 0$; we have

$$\theta = 0 + \theta'. \tag{C.24}$$

Then

$$\sin\theta = \sin(0 + \theta') = \sin(\theta').$$

Since θ' is a small perturbation, the first term in a Taylor series expansion gives

$$\sin\theta \approx \theta'. \tag{C.25}$$

Substitute (C.25) into (C.23) and subtract the base equation to get

$$\frac{d^2\theta'}{d\tau^2} + \frac{d\theta'}{d\tau} + \alpha\theta' = 0. \tag{C.26}$$

Let the perturbation take the form $\theta' = e^{mt}$ and substitute into (C.26).

$$m^2 + m + \alpha = 0. \tag{C.27}$$

Solving the quadratic equation

$$m = \frac{-1 \pm \sqrt{1 - 4\alpha}}{2}. \tag{C.28}$$

Then, the perturbation analysis yields

$$\theta' = e^{-t/2} e^{\pm 1/2\sqrt{1 - 4\alpha}}. \tag{C.29}$$

Since the solution decays in time for any α (the reader should show this by considering the two cases $0 < \alpha < 1/4$ and $\alpha > 1/4$), the solution is *stable*.

Repeating the analysis for the second equilibrium state, begin with

$$\theta = \pi + \theta'. \tag{C.30}$$

Assume equilibrium and substitute (C.30) to give

$$\sin\theta = \sin(\pi + \theta') \approx -\theta'. \tag{C.31}$$

Then

$$\frac{d^2\theta'}{d\tau^2} + \frac{d\theta'}{d\tau} - \alpha\theta' = 0. \tag{C.32}$$

As before, let $\theta' = e^{mt}$ and substitute into (C.32) to give

$$m^2 + m - \alpha = 0. \tag{C.33}$$

Solving the quadratic equation

$$m = \frac{-1 \pm \sqrt{1 + 4\alpha}}{2} = -\frac{1}{2} + \frac{\sqrt{1 + 4\alpha}}{2}, \quad -\frac{1}{2} - \frac{\sqrt{1 - 4\alpha}}{2}. \tag{C.34}$$

Note that the first solution is positive for any $\alpha > 0$, and hence will grow in time. The second solution is negative for any $\alpha > 0$, and hence will decay in time. The net effect is that the full solution grows in time, and therefore the solution is *unstable*.

Exercise 2.10 Compute the temporal inviscid stability characteristics of the following piecewise linear profiles: (a) Top hat jet, (b) Triangular jet and (c) Linear shear layer profile.

(a) Top hat (or rectangular) jet

$$U(y) = \begin{cases} 0 & L < y < \infty, \\ U_c & -L < y < L, \\ 0 & -\infty < y < -L. \end{cases} \qquad \text{(C.35)}$$

Sketch the velocity profile. Note that there are two modes, one that is even about the $y = 0$ axis (i.e., $\hat{v}'(0) = 0$), and one that is odd (i.e., $\hat{v}(0) = 0$). Determine the eigenrelation for each mode.

Solution

The sketch of the velocity profile is shown in Fig. C.1. Since $\hat{v}$ is continuous at

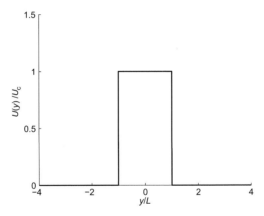

Figure C.1 Top hat (or rectangular) jet.

the origin, it may make the calculations easier if you can show

$$\hat{v} = \begin{cases} e^{-\alpha y} & L < y < \infty, \\ A\cosh(\alpha y) + B\sinh(\alpha y) & 0 < y < L. \end{cases} \qquad \text{(C.36)}$$

Rayleigh's equation (2.29) for $U'' = 0$ is

$$\hat{v}'' - \alpha^2 \hat{v} = 0. \qquad \text{(C.37)}$$

The solution to this equation is (C.36). The jump conditions (2.58) and (2.65) in the text at $y = L$ are

$$\|\rho\{(\alpha U - \omega)\hat{v}' - \alpha U'\hat{v}\}\| = 0, \qquad \text{(C.38)}$$

$$\left\| \frac{\hat{v}}{\alpha U - \omega} \right\| = 0. \tag{C.39}$$

Case I. For even modes $\hat{v}'(0) = 0$, then the derivative of (C.36) becomes

$$\hat{v}' = A\alpha \sinh(\alpha y) + B\alpha \cosh(\alpha y)$$

or

$$\hat{v}'(0) = A\alpha \sinh(0) + B\alpha \cosh(0) = 0.$$

From (A.10) and (A.11), we have $B = 0$. The solution becomes

$$\hat{v} = A\cosh(\alpha y). \tag{C.40}$$

Equations (C.38) with $U' = 0$ at $y = L$ reduces to

$$\left\| (\alpha U - \omega)\hat{v}' \right\| = 0. \tag{C.41}$$

Applying the matching conditions, we have

$$-\omega(-\alpha)e^{-\alpha L} = (\alpha U_c - \omega)\alpha A \sinh(\alpha L),$$

$$\omega e^{-\alpha L} = (\alpha U_c - \omega)A \sinh(\alpha L). \tag{C.42}$$

Matching condition (C.39) is

$$\frac{e^{-\alpha L}}{-\omega} = \frac{A\cosh(\alpha L)}{\alpha U_c - \omega} \tag{C.43}$$

or

$$A = -\left[\frac{\alpha U_c - \omega}{\omega} \right] \frac{e^{-\alpha L}}{\cosh(\alpha L)}. \tag{C.44}$$

Substitute (C.44) into (C.42) to give

$$\omega e^{-\alpha L} = (\alpha U_c - \omega)\left[\frac{\alpha U_c - \omega}{\omega} \right] \frac{e^{-\alpha L}}{\sinh(\alpha L)} \cosh(\alpha L) \tag{C.45}$$

or

$$\omega^2 = -(\alpha U_c - \omega)^2 \tanh(\alpha L). \tag{C.46}$$

Finally, the eigenrelation solution is

$$\omega^2 + (\alpha U_c - \omega)^2 \tanh(\alpha L) = 0. \tag{C.47}$$

Case II for odd *modes* $\hat{v}(0) = 0$. From (A.10), $\sinh(0) = 0$, then $A = 0$ in (C.36). Therefore

$$\hat{v} = B\sinh(\alpha y). \tag{C.48}$$

As before this is subject to (2.58) and (2.65), or

$$\|(\alpha U - \omega)\hat{v}'\| = 0, \tag{C.49}$$

$$\left\|\frac{\hat{v}}{\alpha U - \omega}\right\| = 0. \tag{C.50}$$

Applying the interface condition (C.49) gives

$$-\omega(-\alpha)e^{-\alpha L} = (\alpha U_c - \omega)\alpha B\cosh(\alpha L),$$
$$\omega e^{-\alpha L} = (\alpha U_c - \omega)B\cosh(\alpha L). \tag{C.51}$$

Using the interface condition (C.50),

$$\frac{e^{-\alpha L}}{-\omega} = \frac{B\sinh(\alpha L)}{\alpha U_c - \omega}. \tag{C.52}$$

Therefore the constant B is given by

$$B = -\left[\frac{\alpha U_c - \omega}{\omega}\right]\frac{e^{-\alpha L}}{\sinh(\alpha L)}. \tag{C.53}$$

Substitute (C.53) into (C.51) to yield

$$\omega e^{-\alpha L} = -(\alpha U_c - \omega)\left[\frac{\alpha U_c - \omega}{\omega}\right]\frac{e^{-\alpha L}}{\sinh(\alpha L)}\cosh(\alpha L). \tag{C.54}$$

Finally, the eigenrelation is

$$\omega^2 + (\alpha U_c - \omega)^2 \coth(\alpha L) = 0. \tag{C.55}$$

(b) Triangular jet

$$U(y) = \begin{cases} 0 & 1 < y < \infty, \\ 1-y & 0 < y < 1, \\ 1+y & -1 < y < 0, \\ 0 & -\infty < y < -1. \end{cases} \tag{C.56}$$

Sketch the velocity profile. Again there are two modes, one that is even ($\hat{v}'(0) = 0$), and one that is odd ($\hat{v}(0) = 0$). Determine the eigenrelation for each mode, and plot the growth rate curve in the (α, ω_i)-plane.

The sketch of the velocity profile is shown in Fig. C.2. As before, Rayleigh's equation (2.29) with $U'' = 0$ is

$$\hat{v}'' - \alpha^2 \hat{v} = 0. \tag{C.57}$$

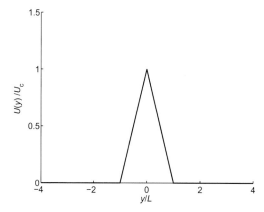

Figure C.2 Triangular jet.

For $0 < y < \infty$, the general solution is

$$\hat{v} = \begin{cases} e^{-\alpha(y-1)} & y > 1, \\ A\cosh\left[\alpha(1-y)\right] + B\sinh\left[\alpha(1-y)\right] & 0 < y < 1. \end{cases} \quad \text{(C.58)}$$

The solutions for $-\infty < y < 0$ can be constructed using either even or odd symmetry conditions. The interface condition (C.38) is repeated

$$\left\|(\alpha U - \omega)\hat{v}' - \alpha U'\hat{v}\right\| = 0. \quad \text{(C.59)}$$

Substituting the solution (C.58) into (C.59) gives

$$(\alpha U - \omega)e^{-\alpha(y-1)} - \alpha U'e^{-\alpha(1-y)}$$
$$= -(\alpha U - \omega)\left[\alpha A\sinh\{\alpha(1-y)\} + \alpha B\cosh\{\alpha(1-y)\}\right]$$
$$- \alpha U'\left[A\cosh\{\alpha(1-y)\} + B\sinh\{\alpha(1-y)\}\right].$$

At $y = 1$, the interface condition becomes

$$(\alpha U - \omega) - U' = (\alpha U - \omega)B - U'A.$$

Simplify using $U(1) = 0$, $U'_U(1) = 0$ and $U'_L(1) = 1$ to find

$$\omega = \omega B + A. \quad \text{(C.60)}$$

The interface condition (C.39) with (C.58) gives

$$\frac{e^{-\alpha(y-1)}}{\alpha U - \omega} = \frac{A\cosh\left[\alpha(1-y)\right] + B\sinh\left[\alpha(1-y)\right]}{\alpha U - \omega}.$$

Again, with $U(1) = 0$

$$\frac{1}{\omega} = \frac{A}{\omega}.$$

Therefore $A = 1$. From (C.60)

$$\omega = \omega B + 1$$

or

$$B = \frac{\omega - 1}{\omega}. \tag{C.61}$$

These coefficients can then be substituted into (C.58) to complete the solution for $0 < y < \infty$:

$$\hat{v} = \begin{cases} e^{-\alpha(y-1)} & y > 1, \\ \cosh[\alpha(1-y)] + \frac{\omega-1}{\omega}\sinh[\alpha(1-y)] & 0 < y < 1. \end{cases} \tag{C.62}$$

Case I: odd mode (varicose mode). If $\hat{v}$ is an odd function, then $\hat{v}(y) = -\hat{v}(y)$ and the solutions are

$$\hat{v} = \begin{cases} e^{-\alpha(y-1)} & y > 1, \\ \cosh[\alpha(1-y)] + B\sinh[\alpha(1-y)] & 0 < y < 1, \\ -\cosh[\alpha(1+y)] - B\sinh[\alpha(1+y)] & -1 < y < 0, \\ -e^{\alpha(y+1)} & y < -1. \end{cases} \tag{C.63}$$

Note that the first interface condition (C.38) is satisfied automatically by choosing the proper odd extension. The interface condition (C.39) at $y = 0$ with $U(0) = 1$ is

$$\frac{\cosh\alpha + B\sinh\alpha}{\alpha - \omega} = \frac{-\cosh\alpha - B\sinh\alpha}{\alpha - \omega}$$

or

$$2B\sinh\alpha = -\cosh\alpha.$$

Therefore using (A.9)

$$B = -\coth\alpha. \tag{C.64}$$

From (C.61), the eigenrelation is

$$\omega - 1 = \omega(-\coth\alpha)$$

or

$$\omega(1 + \coth\alpha) = 1. \tag{C.65}$$

Using (A.8) and (A.9)

$$1 + \coth \alpha = \frac{e^\alpha - e^{-\alpha}}{e^\alpha - e^{-\alpha}} + \frac{e^\alpha + e^{-\alpha}}{e^\alpha - e^{-\alpha}} = \frac{2e^\alpha}{e^\alpha - e^{-\alpha}}. \tag{C.66}$$

Divide the numerator and denominator of (C.66) by e^α and substitute into (C.65) to give

$$\omega \left(\frac{2}{1 - e^{-2\alpha}} \right) = 1$$

or

$$\omega = \frac{1}{2} \left(1 - e^{-2\alpha} \right). \tag{C.67}$$

Since ω is always real when α is real, the flow is always *stable* for this mode.

Case II: Even mode (Sinuous mode). If $\hat{v}$ is an even function, then $\hat{v}(y) = \hat{v}(-y)$. The solution is

$$\hat{v} = \begin{cases} e^{-\alpha(y-1)} & y > 1, \\ \cosh[\alpha(1-y)] + B \sinh[\alpha(1-y)] & 0 < y < 1, \\ \cosh[\alpha(1+y)] + B \sinh[\alpha(1+y)] & -1 < y < 0, \\ -e^{\alpha(y+1)} & y < -1. \end{cases} \tag{C.68}$$

The interface condition (C.38) at $y = 0$ with $U(0) = 1$, $U'(0^-) = -1$ and $U'(0^+) = 1$, is

$$\frac{\cosh \alpha + B \sinh \alpha}{\alpha - \omega} = \frac{-\cosh \alpha - B \sinh \alpha}{\alpha - \omega}, \tag{C.69}$$

which is satisfied automatically. Note the even mode may imply that $\hat{v}'(0) = 0$; however, the jump condition does not permit this since U' is discontinuous at $y = 0$. Next, the interface condition (C.39) is used at $y = 0$ to give

$$-\alpha(\alpha - \omega)(\sinh \alpha + B \cosh \alpha) + \alpha(\cosh \alpha + B \sinh \alpha)$$

$$= \alpha(\alpha - \omega)(\sinh \alpha + B \cosh \alpha) - \alpha(\cosh \alpha + B \sinh \alpha).$$

Solving for B:

$$2\alpha(\alpha - \omega)(\sinh \alpha + B \cosh \alpha) = 2\alpha(\cosh \alpha + B \sinh \alpha).$$

Rearranging to solve for

$$B[(\alpha - \omega) \cosh \alpha - \sinh \alpha] = \cosh \alpha - (\alpha - \omega) \sinh \alpha.$$

Therefore

$$B = \frac{\cosh \alpha - (\alpha - \omega) \sinh \alpha}{(\alpha - \omega) \cosh \alpha - \sinh \alpha}. \tag{C.70}$$

Substitute (C.70) into (C.61) to give the eigenrelation

$$(\omega - 1)\left[(\alpha - \omega)\cosh\alpha - \sinh\alpha\right] = \omega\left[\cosh\alpha - (\alpha - \omega)\sinh\alpha\right].$$

Dividing through by e^α yields

$$(\omega - 1)\left[(\alpha - \omega)(1 + e^{2\alpha}) - 1 + e^{-2\alpha}\right] - \omega\left[1 + e^{-2\alpha} - (\alpha - \omega)(1 - e^{-2\alpha})\right]$$
$$= 0$$

or

$$(\omega - 1)\left[\alpha + \alpha e^{-2\alpha} - \omega - \omega e^{-2\alpha} - 1 + e^{-2\alpha}\right]$$
$$- \omega - \omega e^{-2\alpha} + (\alpha\omega - \omega^2)(1 - e^{-2\alpha}) = 0.$$

Further simplifying, we see that the eigenrelation is

$$2\omega^2 + (1 - 2\alpha - e^{-2\alpha})\omega - \left[(1 - \alpha) - (1 + \alpha)e^{-2\alpha}\right] = 0. \qquad \text{(C.71)}$$

For an unstable mode, ω must be complex. The solution to the quadratic equation gives

$$\omega = \frac{1}{4}\left[-(1 - 2\alpha - e^{-2\alpha})\right.$$
$$\left.\pm\sqrt{(1 - 2\alpha - e^{-2\alpha})^2 + 8\{(1 - \alpha) - (1 + \alpha)e^{-2\alpha}\}}\right]. \qquad \text{(C.72)}$$

When the discriminate is negative, ω is complex, yielding an unstable solution. Thus, for an unstable mode

$$(1 - 2\alpha - e^{-2\alpha})^2 + 8\{(1 - \alpha) - (1 + \alpha)e^{-2\alpha}\} \le 0. \qquad \text{(C.73)}$$

The neutral mode is $\alpha_n \approx 1.833$, $\omega_n \approx 0.6729$ and $c_n \approx 0.367$. Modes are *unstable* for $0 < \alpha < \alpha_n$; the eigenrelation yields maximum values at $\alpha \approx 1.222$ and $\omega \approx (0.3827, 0.2469)$. The eigenrelation is plotted in Fig. C.3.

(c) Linear shear layer profile

$$U(y) = \begin{cases} 1 & 1 < y < \infty, \\ \frac{1}{2}[1 + \beta_U + (1 - \beta_U)y] & -1 < y < 1, \\ \beta_U & -\infty < y < -1. \end{cases} \qquad \text{(C.74)}$$

Sketch the velocity profiles. Here, $\beta_U = U_{-\infty}/U_{+\infty}$ is the ratio of freestream velocities. Sketch the velocity profile for several values of the parameter β_U. Determine the eigenrelation, and show that it reduces to that found by Rayleigh when $\beta_U = -1$. Plot the growth rate curve in the (α, ω_i)-plane for $\beta_U = 0.5, 0, -0.5, -1$. The sketch of the velocity profile is shown in Fig. C.4. For conve-

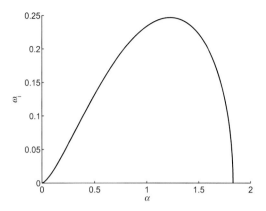

Figure C.3 Plot of ω_i as a function of α for triangular jet.

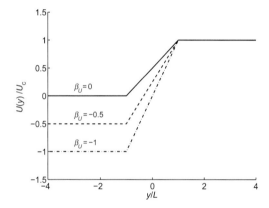

Figure C.4 Linear shear layer profile for $\beta_U = 0$ (solid), $\beta_U = -0.5$ (dash) and $\beta_U = -1$ (dot-dash).

nience, let $b = (1 - \beta_U)/2$, then the velocity profile is given by

$$U(y) = \begin{cases} 1, \\ 1 + b(y - 1), \\ \beta_U, \end{cases} \qquad U'(y) = \begin{cases} 0 & y > 1, \\ b & -1 < y < 1, \\ 0 & y < -1. \end{cases} \qquad (C.75)$$

The general solution is given by

$$\hat{v} = \begin{cases} Ae^{-\alpha(y-1)}, \\ Be^{-\alpha(y-1)} + Ce^{\alpha(y+1)}, \\ e^{\alpha(y+1)}. \end{cases} \qquad (C.76)$$

At the interface $y = 1$, we have $\|\hat{v}\| = 0$, or from (C.76)

$$A = B + Ce^{2\alpha}. \tag{C.77}$$

Substituting (C.76) into interface condition (C.38) at $y = 1$ with $U_U = U_L = 1$, $U_U' = 0$, and $U_L' = b$, we have

$$(\alpha - \omega)(-\alpha)A = (\alpha - \omega)(-\alpha)(B - Ce^{2\alpha}) - \alpha b(B + Ce^{2\alpha}).$$

Using (C.77) and rearranging

$$(\alpha - \omega - b)A = (\alpha - \omega)(B - Ce^{2\alpha}). \tag{C.78}$$

Eliminate A by substituting (C.77) into (C.78) and solve for

$$B = \frac{C}{b}e^{2\alpha}(2\alpha - 2\omega - b). \tag{C.79}$$

At the interface, $y = -1$, $\|\hat{v}\| = 0$, giving

$$1 = Be^{2\alpha} + C. \tag{C.80}$$

Using interface condition (C.38) at $y = -1$ with $U_U = 1 - 2b$, $U_L = \beta_U$, $U_U' = b$ and $U_L' = 0$ to give

$$[\alpha(1 - 2b) - \omega]\left[-\alpha Be^{2\alpha} + \alpha C\right] - \alpha b\left[Be^{2\alpha} + C\right] = [\alpha\beta_U - \omega]\alpha. \tag{C.81}$$

Divide (C.81) by α and rearrange the terms to give

$$(\alpha - 2b\alpha - \omega)(C - Be^{2\alpha}) - b(Be^{2\alpha} + C) = \alpha\beta_U - \omega.$$

Substitute $b = (1 - \beta_U)/2$:

$$(\alpha\beta_U - \omega) = (\alpha\beta_U - \omega)(C - Be^{2\alpha}) - b(Be^{2\alpha} + C). \tag{C.82}$$

Eliminating C by substituting (C.80) into (C.79) yields

$$B = \frac{1}{b}(1 - Be^{2\alpha})(2\alpha - 2\omega - b)e^{2\alpha}$$

or

$$bB + Be^{4\alpha}(2\alpha - 2\omega - b) = (2\alpha - 2\omega - b)e^{2\alpha}.$$

Rearranging, we get

$$B\left[b + e^{4\alpha}(2\alpha - 2\omega - b)\right] = (2\alpha - 2\omega - b)e^{2\alpha}. \tag{C.83}$$

Now substitute (C.80) into (C.82), giving

$$\alpha\beta_U - \omega = (\alpha\beta_U - \omega)(1 - 2Be^{2\alpha}) - b(Be^{2\alpha} + 1 - Be^{2\alpha})$$

or

$$(\alpha\beta_U - \omega)2Be^{2\alpha} = -b.$$

Then

$$B = \frac{-b}{2e^{2\alpha}(\alpha\beta_U - \omega)}. \tag{C.84}$$

This provides two equations (C.83) and (C.84) for one unknown, B. Eliminating the unknown by substituting (C.83) into (C.84) yields

$$\frac{e^{2\alpha}(2\alpha - 2\omega - b)}{b + e^{4\alpha}(2\alpha - 2\omega - b)} = \frac{-b}{2e^{2\alpha}(\alpha\beta_U - \omega)}.$$

Cross-multiply to give

$$2e^{4\alpha}(\alpha\beta_U - \omega)(2\alpha - 2\omega - b) = -b^2 - be^{4\alpha}(2\alpha - 2\omega - b).$$

Rearranging, we find

$$4\alpha^2\beta_U e^{4\alpha} - 4\alpha\omega e^{4\alpha} - 4\alpha\beta_U \omega e^{4\alpha} + 4\omega^2 e^{4\alpha} - 2\alpha b\beta_U e^{4\alpha}$$

$$+2b\omega e^{4\alpha} - b^2(e^{4\alpha} - 1) + 2\alpha b e^{4\alpha} - 2b\omega e^{4\alpha} = 0.$$

Divide by $e^{4\alpha}$:

$$4\omega^2 - 4\alpha\omega(1 + \beta_U) + 4\alpha^2\beta_U + \frac{(1-\beta_U)^2}{4}(4\alpha - 1) + \frac{(1-\beta_U)^2}{4}e^{-4\alpha}$$

$$= 0$$

or

$$\omega = \frac{1 + \beta_U}{2}\alpha \pm \frac{1 - \beta_U}{4}\sqrt{(1 - 2\alpha)^2 - e^{-4\alpha}}. \tag{C.85}$$

The mode is neutral if $\Delta = (1 - 2\alpha)^2 - e^{-4\alpha} = 0$ and unstable if $\Delta < 0$. The neutral mode is given by:

$$\alpha_n \approx 0.6392, c_n = \frac{1 + \beta_U}{2}, \omega_n = \frac{1 + \beta_U}{2}\alpha_n.$$

The eigenrelation is plotted in Fig. C.5.

Exercise 2.11 Note that the equations of motion (2.1) to (2.3) for an inviscid flow, with gravity acting in the $-\hat{j}$-direction, can be written as

$$\nabla \cdot \underline{u} = 0, \tag{C.86}$$

$$\frac{\partial \underline{u}}{\partial t} + (\nabla \times \underline{u}) \times \underline{u}, = -\nabla\left(\frac{p}{\rho} + \frac{1}{2}\underline{u} \cdot \underline{u} + gy\right), \tag{C.87}$$

where $\underline{u} = (u, v)$.

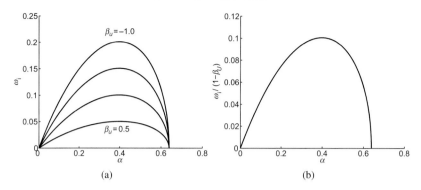

Figure C.5 Growth rates as a function of wavenumber for linear shear layer pro-file. In (b), the growth rates are normalized by $1 - \beta_U$.

(a) Assume the flow is irrotational. Show for this case that there exists a po-tential function $\phi(x,y,t)$, defined by $\underline{u} = \nabla\phi$, which satisfies the equations

$$\nabla^2\phi = 0, \tag{C.88}$$

$$\frac{\partial \nabla\phi}{\partial t} + \nabla\left(\frac{p}{\rho} + \frac{1}{2}|\nabla\phi|^2 + gy\right) = 0. \tag{C.89}$$

(b) Integrate the second equation to get

$$\frac{\partial\phi}{\partial t} + \frac{p}{\rho} + \frac{1}{2}|\nabla\phi|^2 + gy = c, \tag{C.90}$$

where c is a constant of integration. This last equation is Bernoulli's equa-tion for unsteady incompressible flows.

Solution
(a–b) The flow is irrotational, therefore $\nabla \times \underline{u} = 0$. Let the velocity potential be defined such that $\underline{u} = \nabla\phi$. Then $\nabla \times \underline{u} = \nabla \times \nabla\phi = 0$ and $\nabla \cdot \underline{u} = \nabla \cdot \nabla\phi = \nabla^2\phi = 0$. Equation (C.87) becomes

$$\frac{\partial \nabla\phi}{\partial t} + \nabla\left(\frac{p}{\rho} + \frac{1}{2}|\nabla\phi|^2 + gy\right) = 0. \tag{C.91}$$

Integrating equation (C.91), we find

$$\frac{\partial\phi}{\partial t} + \frac{p}{\rho} + \frac{1}{2}|\nabla\phi|^2 + gy = c. \tag{C.92}$$

This is Bernoulli's equation and is interpreted as a constant along streamlines.

(c) Using the above equations, we now wish to investigate the temporal stability characteristics for the following (irrotational) mean flow profile

$$U(y) = \begin{cases} U_1 & 0 < y < \infty, \\ U_2 & -\infty < y < 0. \end{cases} \tag{C.93}$$

Assuming a normal mode solution, show that the eigenrelation is given by

$$\omega = \frac{\alpha(\rho_1 U_1 + \rho_2 U_2)}{\rho_1 + \rho_2} \pm i \sqrt{\frac{\alpha^2 \rho_1 \rho_2 (U_2 - U_1)^2}{(\rho_1 + \rho_2)^2} - \frac{\alpha g(\rho_2 - \rho_1)}{\rho_1 + \rho_2}}. \tag{C.94}$$

Hint: Write ϕ in terms of normal modes, and then solve the Laplacian equation to determine the perturbation flow. Next, use Bernoulli's equation to get a jump condition that involves gravity.

First, we find the mean flow solution. Since $\underline{u} = \nabla\phi$, then $u = \phi_x$ and $v = \phi_y$. Then the mean potential is

$$\phi_{mean} = \begin{cases} U_1 x & 0 < y < \infty, \\ U_2 x & -\infty < y < 0. \end{cases} \tag{C.95}$$

Substituting the velocity profile (C.95) into the Bernoulli's equation (C.92) gives

$$p_1 + \frac{\rho_1}{2} U_1^2 + gy\rho_1 = c_1 \rho_1 \quad \text{for} \quad 0 < y < \infty, \tag{C.96}$$

$$p_2 + \frac{\rho_2}{2} U_2^2 + gy\rho_2 = c_2 \rho_2 \quad \text{for} \quad -\infty < y < 0. \tag{C.97}$$

Since $\|p\| = p_2 - p_1 = 0$ at $y = 0$, then Bernoulli's equation at $y = 0$ requires

$$\rho_1 \left(c_1 - \frac{1}{2} U_1^2 \right) = \rho_2 \left(c_2 - \frac{1}{2} U_2^2 \right). \tag{C.98}$$

For the disturbances, write

$$\phi = \phi_{mean}(x) + \hat{\phi}(y) e^{i(\alpha x - \omega t)},$$

$$p = p_{mean}(x) + \hat{p}(y) e^{i(\alpha x - \omega t)}. \tag{C.99}$$

For the location of the free surface

$$y = 0 + a e^{i(\alpha x - \omega t)}. \tag{C.100}$$

Substituting the disturbance form (C.99) into the Laplace equation

$$-\alpha^2 \hat{\phi}(y) + \frac{d^2 \hat{\phi}(y)}{dy^2} = 0. \tag{C.101}$$

Since decaying solutions are desired, the solutions take the form of

$$\hat{\phi}(y) = \begin{cases} Ae^{-\alpha y} & 0 < y < \infty, \\ Be^{\alpha y} & -\infty < y < 0. \end{cases} \tag{C.102}$$

With the conditions for the free surface, the displacement of the interface condition (2.58) is given by

$$\left\| (\alpha U - \omega)\hat{v}' \right\| = 0. \tag{C.103}$$

Since $U' = 0$, recall (2.64), namely

$$\hat{v}(y) = \begin{cases} i(\alpha U_1 - \omega)a, & y = 0^+, \\ i(\alpha U_2 - \omega)a, & y = 0^-. \end{cases} \tag{C.104}$$

Since $\hat{v} = d\hat{\phi}/dy$ at $y = 0$, using (C.102) evaluated at $y = 0$ and (C.104),

$$\begin{aligned} -\alpha A = i(\alpha U_1 - \omega)a, \quad y = 0^+, \\ \alpha B = i(\alpha U_2 - \omega)a, \quad y = 0^-. \end{aligned} \tag{C.105}$$

Given the two equations for a and equate to get

$$\frac{-\alpha A}{i(\alpha U_1 - \omega)} = \frac{\alpha B}{i(\alpha U_2 - \omega)} = a. \tag{C.106}$$

The linearized version of Bernoulli's equation is found by substituting (C.99) and (C.100) into (C.92)

$$\begin{aligned} -i\omega\rho_1\hat{\phi} + \hat{p}_1 + i\alpha\rho_1 U_1\hat{\phi} + g\rho_1 a &= 0, \\ -i\omega\rho_2\hat{\phi} + \hat{p}_2 + i\alpha\rho_2 U_2\hat{\phi} + g\rho_2 a &= 0. \end{aligned} \tag{C.107}$$

Since $\hat{p}_1 = \hat{p}_2$ at $y = 0$, use (C.102) and (C.107)

$$\rho_1\left[i(\alpha U_1 - \omega)A + ga\right] = \rho_2\left[i(\alpha U_2 - \omega)B + ga\right]. \tag{C.108}$$

Use (C.106) to eliminate A and a to get an equation for B:

$$\rho_1\left[-i(\alpha U_1 - \omega)^2\frac{B}{\alpha U_2 - \omega} + \frac{\alpha g B}{i(\alpha U_2 - \omega)}\right]$$

$$= \rho_2\left[i(\alpha U_2 - \omega)B + \frac{\alpha g B}{i(\alpha U_2 - \omega)}\right]. \tag{C.109}$$

Since $B = 0$ is not desired

$$\rho_1\left[(\alpha U_1 - \omega)^2 + \alpha g\right] = \rho_2\left[-(\alpha U_2 - \omega)^2 + \alpha g\right]. \tag{C.110}$$

Solving for ω:

$$\rho_1\left[\alpha^2 U_1^2 - 2\alpha U_1\omega + \omega^2 + \alpha g\right] = \rho_2\left[-\alpha^2 U_2^2 + 2\alpha U_2\omega - \omega^2 + \alpha g\right]$$

or

$$(\rho_1 + \rho_2)\omega^2 - 2\alpha(\rho_1 U_1 + \rho_2 U_2)\omega$$
$$+\alpha^2(\rho_1 U_1^2 + \rho_2 U_2^2) + \alpha g(\rho_1 - \rho_2) = 0. \qquad \text{(C.111)}$$

Using the quadratic equation yields

$$\omega = \frac{2\alpha(\rho_1 U_1 + \rho_2 U_2) \pm \sqrt{d}}{2(\rho_1 + \rho_2)}, \qquad \text{(C.112)}$$

where

$$d = -4\alpha^2 \rho_1 \rho_2 (U_1^2 - 2U_1 U_2 + U_2^2) - 4\alpha g(\rho_1 - \rho_2)(\rho_1 + \rho_2).$$

Therefore (C.112) becomes

$$\omega = \frac{\alpha(\rho_1 U_1 + \rho_2 U_2)}{\rho_1 + \rho_2} \pm i \sqrt{\frac{\alpha^2 \rho_1 \rho_2 (U_1 - U_2)^2}{(\rho_1 + \rho_2)^2} - \frac{\alpha g(\rho_2 - \rho_1)}{\rho_1 + \rho_2}}, \qquad \text{(C.113)}$$

which is the desired result.

(d) For internal gravity waves, set $U_1 = U_2 = 0$. Give the eigenrelation, and state the conditions for stability.

From (C.113), we set $U_1 = U_2 = 0$ and find

$$\omega = \pm i \sqrt{\frac{\alpha g(\rho_1 - \rho_2)}{\rho_1 + \rho_2}}. \qquad \text{(C.114)}$$

The flow is unstable if $\omega_i > 0$. Thus, the flow is:

- Unstable if $\rho_1 > \rho_2$, i.e., heavier fluid on top;
- Stable if $\rho_2 > \rho_1$, i.e., heavier fluid on bottom.

(e) For surface gravity waves, set $\rho_1 = 0$ and $U_1 = U_2 = 0$. Give the eigenrelation, and state the conditions for stability.

From (C.114), we have

$$\omega = \pm i \sqrt{\frac{-\alpha g \rho_2}{\rho_2}} = \pm\sqrt{\alpha g}. \qquad \text{(C.115)}$$

The flow is stable.

(f) If surface tension is present then the jump in pressure across the interface is no longer zero, but must be modified to

$$\|p\| = T \frac{\partial^2 f}{\partial x^2}, \qquad \text{(C.116)}$$

where T is the coefficient for surface tension and $F = y - f(x,t)$ gives the location of the surface. Show that the eigenrelation can be written as

$$\omega = \frac{\alpha(\rho_1 U_1 + \rho_2 U_2)}{\rho_1 + \rho_2}$$

$$\pm i \sqrt{\frac{\alpha^2 \rho_1 \rho_2 (U_2 - U_1)^2}{(\rho_1 + \rho_2)^2} - \frac{\alpha^2}{\rho_1 + \rho_2} \left[\frac{g(\rho_2 - \rho_1)}{\alpha} + \alpha T \right]}. \quad \text{(C.117)}$$

What role does surface tension play in regards to the stability of the flow?

If surface tension is present, then the jump in the normal stress is proportional to the curvature

$$\|p\| = T \frac{\partial^2 f}{\partial x^2} \quad \text{(C.118)}$$

where $F = y - f(x,t)$. Now write

$$f = a e^{i(\alpha x - \omega t)}.$$

Then

$$\frac{\partial^2 f}{\partial x^2} = -\alpha^2 a e^{i(\alpha x - \omega t)}. \quad \text{(C.119)}$$

Thus (C.118) becomes

$$\hat{p}_1 - \hat{p}_2 = -T \alpha^2 a. \quad \text{(C.120)}$$

With this, Bernoulli's equation becomes

$$\rho_1 \left[i(\alpha U_1 - \omega)\hat{\phi}_1 + ga \right] = \rho_2 \left[i(\alpha U_2 - \omega)\hat{\phi}_2 + ga \right] + T\alpha^2 a. \quad \text{(C.121)}$$

The other jump condition (C.103) remains the same, and gives the relationship (C.106). Eliminating A and B as before yields the equation

$$(\rho_1 + \rho_2)\omega^2 - 2\alpha(\rho_1 U_1 + \rho_2 U_2)\omega + \alpha^2(\rho_1 U_1^2 + \rho_2 U_2^2)$$
$$+ \alpha g(\rho_1 - \rho_2) - T\alpha^3 = 0.$$

The eigenrelation is

$$\omega = \frac{\alpha(\rho_1 U_1 + \rho_2 U_2)}{\rho_1 + \rho_2}$$

$$\pm i \sqrt{\frac{\alpha^2 \rho_1 \rho_2 (U_1 - U_2)^2}{(\rho_1 + \rho_2)^2} - \frac{\alpha^2}{(\rho_1 + \rho_2)} \left[\frac{g(\rho_2 - \rho_1)}{\alpha} + T\alpha \right]}. \quad \text{(C.122)}$$

As $T \to \infty$, $\sqrt{}$ becomes imaginary, and thus the flow is stable; i.e., surface tension is a stabilizing effect. The same is true for g. As $g \to \infty$, the discriminant becomes negative so that the square root becomes imaginary, and thus the flow is stable; i.e., gravity is a stabilizing effect.

Exercise 2.12 Prove Result 2.7. Use the following to aid in the construction of the proof.

(a) Let $G(y) = \phi(y)/\sqrt{U-c}$, and assume $c_i > 0$.

(b) Show that G satisfies

$$\left[(U-c)G'\right]' - \left[\frac{1}{2}U'' + \alpha^2(U-c) + \frac{1}{4}\frac{(U')^2}{U-c}\right]G = 0. \qquad \text{(C.123)}$$

(c) Show

$$-\int_a^b \left[|G'|^2 + \alpha^2|G|^2\right]dy + \frac{1}{4}\int_a^b \frac{U'^2}{|U-c|^2}|G|^2dy = 0. \qquad \text{(C.124)}$$

(d) Note that

$$|U-c|^2 = (U-c_r)^2 + c_i^2 \geq c_i^2 \qquad \text{(C.125)}$$

and thus

$$|U-c|^{-2} \leq c_i^{-2}. \qquad \text{(C.126)}$$

Solution
(a–b) Let $G(y) = \phi(y)/\sqrt{U-c}$, and assume $c_i > 0$, then

$$\phi(y) = \sqrt{U-c}\, G(y). \qquad \text{(C.127)}$$

Begin with the Rayleigh equation (2.92)

$$(U-c)(\phi'' - \alpha^2\phi) - U''\phi = 0. \qquad \text{(C.128)}$$

Derivatives of ϕ using (C.127) are given by

$$\phi' = \sqrt{U-c}\, G' + \frac{U'G}{2\sqrt{U-c}},$$

$$\phi'' = \sqrt{U-c}\, G'' + \frac{U'G'}{\sqrt{U-c}} + \frac{U''G}{2\sqrt{U-c}} - \frac{(U')^2G}{4(U-c)^{3/2}}. \qquad \text{(C.129)}$$

Substitute (C.127) and (C.129) into (C.128) to yield

$$(U-c)\left[\sqrt{U-c}\, G'' + \frac{U'G'}{\sqrt{U-c}} + \frac{U''G}{2\sqrt{U-c}} - \frac{(U')^2G}{4(U-c)^{3/2}}\right.$$

$$\left. -\alpha^2\sqrt{U-c}G\right] - U''\sqrt{U-c}\, G = 0. \qquad \text{(C.130)}$$

Dividing (C.130) by $\sqrt{U-c}$ gives

$$(U-c)G'' + U'G' + \frac{1}{2}U''G - \frac{(U')^2}{4(U-c)}G - \alpha^2(U-c)G - U''G = 0. \quad \text{(C.131)}$$

Simplifying (C.131) yields

$$\left[(U-c)G'\right]' - \left[\frac{U''}{2} + \alpha^2(U-c) + \frac{(U')^2}{4(U-c)}\right]G = 0. \quad \text{(C.132)}$$

This is the desired result.

(c) Multiply (C.132) by G^*, which is the complex conjugate of G, integrating over $y \in [a,b]$ and noting that $G^*G = |G|^2$ gives

$$\int_a^b G^* \left[(U-c)G'\right]' dy - \int_a^b \left[\frac{U''}{2} + \alpha^2(U-c) + \frac{(U')^2}{4(U-c)}\right]|G|^2 dy = 0. \quad \text{(C.133)}$$

Integrate by parts. The first term yields

$$\int_a^b G^*[(U-c)G']'dy = G^*(U-c)G' \Big|_a^b - \int_a^b (U-c)(G^*)'G'dy. \quad \text{(C.134)}$$

Since G, and hence G^*, vanishes at a,b, we have

$$\int_a^b G^* \left[(U-c)G'\right]' dy = - \int_a^b (U-c)|G'|^2 dy, \quad \text{(C.135)}$$

so that (C.133) becomes

$$\int_a^b (U-c)|G'|^2 dy + \int_a^b \left[\frac{U''}{2} + \alpha^2(U-c) + \frac{(U')^2}{4(U-c)}\right]|G|^2 dy = 0. \quad \text{(C.136)}$$

Note that

$$\frac{1}{U-c} = \frac{1}{U-c}\frac{U-c^*}{U-c*} = \frac{U-c^*}{|U-c|^2}. \quad \text{(C.137)}$$

Now write $c = c_r + ic_i$ and $c^* = c_r - ic_i$, yielding

$$\int_a^b (U - c_r - ic_i)|G'|^2 dy + \int_a^b \left[\frac{U''}{2} + \alpha^2(U - c_r - ic_i)\right.$$

$$\left. + \frac{(U')^2}{4|U-c|^2}(U - c_r + ic_i)\right]|G|^2 dy = 0. \quad \text{(C.138)}$$

Assuming $c_i > 0$, then set the imaginary part to zero and divide (C.138) by $(U - c_r)$, yielding

$$\int_a^b \left[|G'|^2 + \alpha^2|G|^2\right] dy - \int_a^b \frac{(U')^2}{4|U-c|^2}|G|^2 dy = 0, \quad \text{(C.139)}$$

which is the desired result.

(d) Note that $|U - c|^2 = (U - c_r)^2 + c_i^2 \geq c_i^2$ and thus

$$\frac{1}{|U - c|^2} \leq \frac{1}{c_i^2}. \tag{C.140}$$

From the expression derived above in (c), which is (C.139), for the left-hand side, rewrite

$$\int_a^b \alpha^2 |G|^2 dy \leq \int_a^b \left[|G'|^2 + \alpha^2 G^2 \right] dy. \tag{C.141}$$

Since $|G'|^2$ and $\alpha^2 |G|^2$ are both positive. Then

$$\int_a^b \alpha^2 |G|^2 dy \leq \int_a^b \left[|G'|^2 + \alpha^2 G^2 \right] dy = \int_a^b \frac{(U')^2}{4|U - c|^2} |G|^2 dy$$

or

$$\int_a^b \alpha^2 |G|^2 dy \leq \int_a^b \frac{(U')^2}{4|U - c|^2} |G|^2 dy$$

or

$$\int_a^b \alpha^2 |G|^2 dy \leq \int_a^b \frac{(U')^2}{4c_i^2} |G|^2 dy$$

or

$$\alpha^2 \int_a^b |G|^2 dy \leq \frac{1}{4c_i^2} \max \left[(U')^2 \right] \int_a^b |G|^2 dy. \tag{C.142}$$

Therefore,

$$\alpha^2 \leq \frac{1}{4c_i^2} \max \left[(U')^2 \right]. \tag{C.143}$$

Taking the square root of both sides,

$$\alpha c_i \leq \frac{1}{2} \max |U'|. \tag{C.144}$$

The maximum is taken over the open interval $[a,b]$. This proves Result 2.7.

Exercise 2.13 Why might using the Ricatti transformation $G = \phi'/\alpha\phi$ instead of $G = \phi'/\phi$ not be a good idea? See Appendix B for a discussion of the Ricatti transformation for solving Rayleigh's equation.

Solution

As $\alpha \to 0$, $G = \phi'/\alpha\phi$ is not well behaved, whereas $G = \phi'/\phi$ is well behaved; see Appendix B.

Exercise 2.14 A numerical method for solving Rayleigh's equation is presented in Appendix B. Compute the temporal inviscid stability characteristics of the following continuous profiles. In each case plot the mean profile, and give the location of the inflection point and the corresponding neutral phase speed.

Solution

(a) *Error function profile.* The solution is presented in Fig. C.6, where we plot the error function,

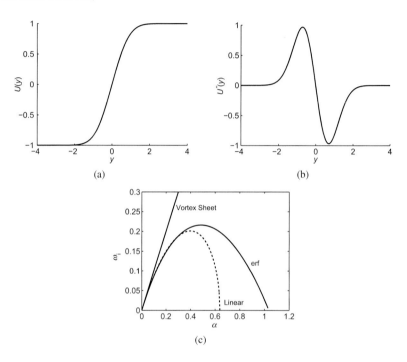

(a)   (b)

(c)

Figure C.6 Plot of (a) error function, (b) second derivative of error function, and (c) instability spectrum ω_i as a function of α.

$$U(y) = \mathrm{erf}(y) \equiv \frac{2}{\sqrt{\pi}} \int_0^y e^{-u^2} \, du,$$

the second derivative, and the instability spectrum ω_i as a function of α. Note that the second derivative gives the inflection point at $y = 0$.

(b) *Hyperbolic tangent profile.* The solution is presented in Fig. C.7, where we plot the hyperbolic tangent profile

$$U(y) = \frac{1}{2}\left[1 + \beta_U + (1 - \beta_U)\tanh y\right],$$

with $\beta_U = 0.5, 0, -0.2$ and the instability spectrum ω_i as a function of α.

Note that the second derivative gives the inflection point at $y = 0$. Here, $\beta_U = U_{-\infty}/U_{+\infty}$ is the ratio of freestream velocities.

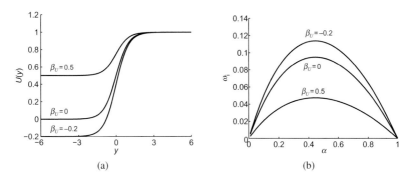

Figure C.7 Hyperbolic tangent profile and instability spectrum.

(c) *Laminar mixing layer.* The solution is presented in Fig. C.8, where we plot the laminar mixing layer

$$2f''' + ff'' = 0, \quad f'(-\infty) = \beta_U, \quad f(0) = 0, \quad f'(+\infty) = 1$$

and the instability spectrum ω_i as a function of α, with $\beta_U = 0.5$ and 0.

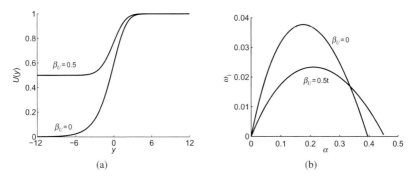

Figure C.8 Laminar mixing layer profile and instability spectrum.

(d) *Symmetric jet.* The solution is presented in Fig. C.9, where we plot the symmetric jet

$$U(y) = \operatorname{sech}^2(y)$$

and the instability spectrum ω_i as a function of α. We note that two neutral modes exist with

$$\phi_I = \operatorname{sech}^2(y), \quad \alpha_N = 2, \quad c_N = 2/3,$$

$$\phi_{II} = \text{sech}\,(y)\,\tanh(y), \quad \alpha_N = 1, \quad c_N = 2/3.$$

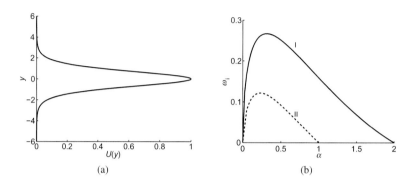

Figure C.9 Symmetric jet profile and instability spectrum.

(e) *Symmetric wake.* The solution is presented in Fig. C.10, where we plot the symmetric wake

$$U(y) = 1 - Q\text{sech}^2 y \tag{C.145}$$

and the instability spectrum ω_i as a function of α, with $Q = 0.3, 0.6, 0.9$. Here, Q is a measure of the wake deficit. Note that two modes exist, and plots of the growth rate curves for each are given.

(f) *Gaussian jet.* The solution is presented in Fig. C.11, where we plot the Gaussian jet

$$U(y) = e^{-y^2 \ln 2}$$

and the instability spectrum ω_i as a function of α.

(g) *Combination shear plus jet.* The solution is presented in Fig. C.12, where we plot the combined shear and jet profile

$$U(y) = \frac{1}{2}\left[1 + \beta_U + (1 - \beta_U)\tanh y\right] - Qe^{-y^2 \ln 2}$$

and the instability spectrum ω_i as a function of α, with $\beta_U = 0.5$ and $Q = 0.4, 0.8$.

(h) *Asymptotic suction boundary layer profile.* The asymptotic suction profile

$$U(y) = 1 - e^{-y}, \qquad 0 < y < \infty$$

is stable to inviscid disturbances since there is no inflection point in the flow domain!

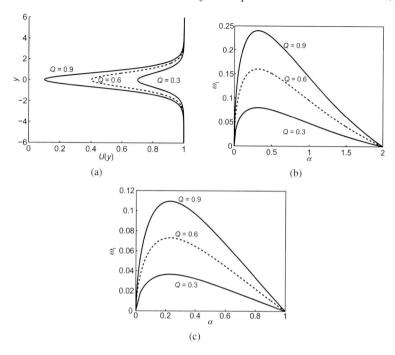

Figure C.10 (a) Symmetric wake profile and instability spectrum for (b) symmetric mode and (c) asymmetric mode.

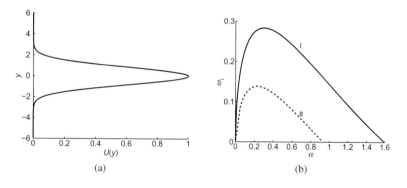

Figure C.11 Gaussian jet profile and instability spectrum.

(i) *Falkner–Skan flow.* The solution is presented in Fig. C.13, where we plot the Falkner–Skan profile

$$2f''' + ff'' + \beta(1 - f'^2) = 0, \qquad f(0) = f'(0) = 0, \qquad f'(\infty) = 1,$$

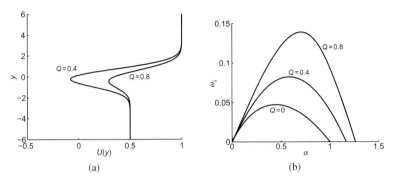

Figure C.12 Shear+jet profile and instability spectrum.

the second derivative and the instability spectrum ω_i as a function of α. Note that there is an inflection point ($U''(y) = 0$) inside the flow domain for $\beta < 0$. Also note from the stability diagram that as β is reduced towards zero, the unstable branch will vanish, in agreement with the fact that the Blasius boundary layer is stable to inviscid disturbances since there is no inflection point inside the flow domain.

Exercise 2.15 Derive the corresponding Rayleigh equation for an incompressible, density stratified flow

$$\hat{v}'' + (\rho'/\rho)\hat{v}' - \left[\alpha^2 + \frac{U'' + \rho'U'/\rho}{U - c}\right]\hat{v} = 0. \tag{C.146}$$

Hint: Assume that the mean density is a function of y, but ignore density fluctuations when deriving Rayleigh's equation. The same equation results for compressible flows in the limit of small Mach number, as will be shown later in Chapter 5.

Assume the mean flow profile

$$U(y) = \frac{1}{2}\left[1 + \beta_U + (1 - \beta_U)\tanh y\right], \tag{C.147}$$

$$\rho(y) = \frac{1}{2}\left[1 + \beta_\rho + (1 - \beta_\rho)\tanh y\right]. \tag{C.148}$$

Solve the temporal stability problem for $\beta_U = 0.5$ and $\beta_\rho = 0.5, 1.0, 1.5$ and graph the eigenrelation in the (α, ω_i)-plane.

Solution

The appropriate equations for a 2-D inviscid, incompressible, density-stratified flow are

$$u_x + v_y = 0, \tag{C.149}$$

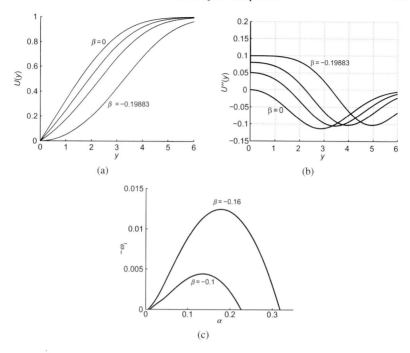

Figure C.13 Plot of (a) Falkner–Skan profile, (b) second derivative and (c) instability spectrum. For the stability diagram we used $Re = 6 \times 10^4$ as an approximation for the inviscid results.

$$\rho\left[u_t + uu_x + vu_y\right] + p_x = 0, \tag{C.150}$$

$$\rho\left[v_t + uv_x + vv_x\right] + p_y = 0. \tag{C.151}$$

The mean plus disturbances have the following normal mode form

$$\rho = \rho(y), \tag{C.152}$$

$$u = U(y) + \hat{u}(y)e^{i(\alpha x - \omega t)}, \tag{C.153}$$

$$v = \hat{v}(y)e^{i(\alpha x - \omega t)}, \tag{C.154}$$

$$p = P + \hat{p}(y)e^{i(\alpha x - \omega t)}; \tag{C.155}$$

we have assumed in the above that there are no fluctuations in ρ. Substituting
(C.152)–(C.155) into (C.149)–(C.151) leads to

$$i\alpha\hat{u} + \hat{v}' = 0, \tag{C.156}$$

$$i\alpha\rho(U - c)\hat{u} + \rho U'\hat{v} + i\alpha\hat{p} = 0, \tag{C.157}$$

$$i\alpha\rho(U-c)\hat{v}+\hat{p}'=0. \tag{C.158}$$

Substituting (C.156) into equation (C.157) yields

$$(U-c)\rho\hat{v}'-\rho U'\hat{v}-i\alpha\hat{p}=0, \tag{C.159}$$

$$i\alpha\rho(U-c)\hat{v}+\hat{p}'=0. \tag{C.160}$$

Take the derivative of (C.159) with respect to y, multiply equation (C.160) by $i\alpha$ and add the two equations to give

$$\rho(U-c)\hat{v}''+(U-c)\rho'\hat{v}'-U'\rho'\hat{v}-\rho U''\hat{v}-\rho\alpha^2(U-c)\hat{v}=0. \tag{C.161}$$

Dividing (C.161) by $\rho(U-c)$, we find

$$\hat{v}''+\left(\frac{\rho'}{\rho}\right)\hat{v}'-\left[\alpha^2+\frac{U''+\rho'U'/\rho}{U-c}\right]\hat{v}=0. \tag{C.162}$$

This is the desired result.

Comment: This equation can also be rewritten as follows. Multiply equation (C.162) by ρ and simplify:

$$(\rho\hat{v}')'-\left[\rho\alpha^2+\frac{(\rho U')'}{U-c}\right]\hat{v}=0. \tag{C.163}$$

If $\rho=1/T$, then this equation is identical to (5.28) in the limit of $M\to 0$.

C.2 Solutions for Chapter 3

Exercise 3.2 The asymptotic suction profile is an exact solution of the Navier–Stokes equations under the assumptions

$$U(\infty)=U_0, \quad U(0)=0, \quad V(0)=-V_s,$$

where U_0 is the freestream crossflow velocity, and V_s is the blowing ($V_s<0$) or suction ($V_s>0$) parameter.

(a) Assuming the mean profile is a function of y only, deduce the nondimensional solution

$$U=1-e^{-y}, \quad V=-Re^{-1}, \quad P=\text{constant},$$

where the reference length is $L=v/V_s$, the reference velocity is U_0 and the Reynolds number is defined as $Re=U_0L/v=U_0/V_s$.

(b) Using the Orr–Sommerfeld equation (2.31), compute the temporal stability characteristics. In particular, show that the critical Reynolds number is $Re_{crit} = 47,047$, with $\alpha_{crit} = 0.1630$ and $c_{crit} = 0.1559$ (see Hughes & Reid, 1965a,b; Drazin & Reid, 1984). Note that e^{-y} has a rather slow decay rate as $y \to \infty$, and so typically one must choose $y_2 = 16$ or larger. Note that $V_{s,crit} = U_0 Re_{crit}^{-1} = 2.13 \times 10^{-5} U_0$, and so only a small fraction of the suction parameter is needed to stabilize the flow when compared to the Blasius boundary layer flow.

(c) Derive, from first principles, the modified Orr–Sommerfeld equation

$$(U - c)(D^2 - \alpha^2)\phi - U''\phi = (i\alpha Re)^{-1} \left[(D^2 - \alpha^2)^2 + (D^2 - \alpha^2)D \right] \phi,$$

for the asymptotic suction profile (Hughes & Reid, 1965a,b). Here, $D = d/dy$ and prime denotes differentiation with respect to y.

(d) For the modified Orr–Sommerfeld equation, compute the continuous spectrum and sketch it in the complex c-plane.

(e) For the modified Orr–Sommerfeld equation, compute the temporal stability characteristics. Show graphically that there are a finite number of stable modes, and that the stable modes pop off the continuum as the Reynolds number increases (alternatively, show that the eigenvalues merge towards the continuum as the Reynolds number decreases, starting at a sufficiently large Reynolds number so as to have more than one mode). Finally, show that the critical Reynolds number is $Re_{crit} = 54,370$ with $\alpha_{crit} = 0.1555$ and $c_{crit} = 0.150$ (Hocking, 1975).

(f) Comparing the critical Reynolds number obtained in parts (b) and (e), discuss the reason for the differences. What lesson should be learned here?

Solution
(a) Substitute the mean profile into the nondimensional equations (2.6) and (2.8). The desired results follows.

(b) Modify your Orr–Sommerfeld solver for the asymptotic suction profile, and verify that $Re_{crit} = 47,047$.

(c) The nondimensional equations of motion are

$$u_x + v_y = 0, \tag{C.164}$$

$$u_t + uu_x + vu_y + p_x = Re^{-1}\nabla^2 u, \tag{C.165}$$

$$v_t + uv_x + vv_y + p_y = Re^{-1}\nabla^2 v. \tag{C.166}$$

Write the base flow and perturbations using normal mode approach as

$$
\begin{aligned}
u &= U(y) + \hat{u}(y)e^{i(\alpha x - \omega t)}, \\
v &= V + \hat{v}(y)e^{i(\alpha x - \omega t)}, \\
p &= P + \hat{p}(y)e^{i(\alpha x - \omega t)},
\end{aligned}
\tag{C.167}
$$

where $U(y) = 1 - e^{-y}$, $V = -Re^{-1}$. Substitute (C.167) into (C.164)–(C.166) yields

$$
i\alpha\hat{u} + \hat{v}' = 0,
\tag{C.168}
$$

$$
i\alpha(U - c)\hat{u} + V\hat{u}' + U'\hat{v} + i\alpha\hat{p} = Re^{-1}(D^2 - \alpha^2)\hat{u},
\tag{C.169}
$$

$$
i\alpha(U - c)\hat{v} + V\hat{v}' + \hat{p}' = Re^{-1}(D^2 - \alpha^2)\hat{v},
\tag{C.170}
$$

where $c = \omega/\alpha$ and $D = ()' = d/dy$. Multiply (C.170) by $i\alpha$, differentiate equation (C.169) with respect to y and subtract them to eliminate $\hat{p}$. Substitute (C.168) to eliminate $\hat{u}$ giving

$$
\begin{aligned}
i\alpha(U - c)\hat{v}'' \quad + \quad & V\hat{v}''' - i\alpha U''\hat{v} + \alpha^2\left[Re^{-1}(D^2 - \alpha^2)\hat{v}\right. \\
 - \quad & \left. i\alpha(U - c)\hat{v} - V\hat{v}'\right] = Re^{-1}D^2(D^2 - \alpha^2)\hat{v}.
\end{aligned}
\tag{C.171}
$$

Note $V = -Re^{-1}$. Divide (C.171) by $i\alpha$, then we have

$$
(U - c)(D^2 - \alpha^2)\hat{v} - U''\hat{v} = (i\alpha Re)^{-1}\left[(D^2 - \alpha^2)^2 + D(D^2 - \alpha^2)\right]\hat{v},
\tag{C.172}
$$

which is the desired result.

(d) For the continuous spectrum, let $y \to \infty$, then $U \to 1$, $U'' \to 0$ and assume as

$$
\hat{v} \sim e^{\pm iky} \quad \text{as} \quad y \to \infty,
\tag{C.173}
$$

with $k > 0$ real and positive. Substitute (C.173) into (C.172) as $y \to \infty$

$$
(1 - c)(-k^2 - \alpha^2) = (i\alpha Re)^{-1}\left[(k^2 + \alpha^2)^2 \pm ik(-k^2 - \alpha^2)\right]
$$

or,

$$
(k^2 + \alpha^2)\left[1 - c + (i\alpha Re)^{-1}(k^2 + \alpha^2 \pm ik)\right] = 0
$$

or, since $(k^2 + \alpha^2) \neq 0$

$$
(1 - c) + (i\alpha Re)^{-1}(k^2 + \alpha^2 \pm ik) = 0.
$$

Solve for c

$$
c = 1 - i(\alpha Re)^{-1}(k^2 + \alpha^2 \pm ik).
\tag{C.174}
$$

Two branches yield

$$c_+ = 1 - i(\alpha Re)^{-1}(k^2 + \alpha^2 + ik),$$
$$c_- = 1 - i(\alpha Re)^{-1}(k^2 + \alpha^2 - ik),$$

or,

$$c_+ = 1 + k(\alpha Re)^{-1} - i(\alpha Re)^{-1}(k^2 + \alpha^2),$$
$$c_- = 1 - k(\alpha Re)^{-1} - i(\alpha Re)^{-1}(k^2 + \alpha^2). \tag{C.175}$$

Note that as $k \to 0$, $c_+ = c_- = 1 - i\alpha Re^{-1}$. The continuum for c_+ is irrelevant since it yields phase speeds greater than 1: i.e., speeds are greater than the base flow. Figure C.14 plots the c_- spectrum.

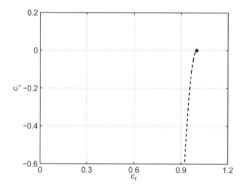

Figure C.14 Continuous spectrum plotted in the complex phase space. Here, $\alpha = 0.1$, $Re = 10^3$.

(f) For the standard Orr–Sommerfeld equation, we get $Re_{\text{crit}} = 47,047$, and for the modified Orr–Sommerfeld equation, we get $Re_{\text{crit}} = 54,370$, which has a broader stable region than that corresponding to the standard Orr–Sommerfeld equation; i.e., the standard Orr–Sommerfeld equation underpredicts the correct stability region. Therefore, it is important to always use the proper stability equation, consistent with the mean flow, to determine the stability characteristics.

Exercise 3.3 Consider the Falkner–Skan family (3.9). Using asymptotic expansions, show that in the limit $\beta \to \infty$, the solution of (3.9) is given by

$$U = 3 \tanh^2 \left[\frac{z}{2} + \tanh^{-1} \sqrt{2/3} \right] - 2,$$

where $\eta = z/\sqrt{\beta}$ is the scaled coordinate. Using this profile, compute the temporal stability characteristics using the Orr–Sommerfeld equation (2.31).

In particular, show that the critical Reynolds number is $Re_{\delta,\text{crit}} = 21,675$, with $\alpha_{\delta,\text{crit}} = 0.1738$ and $c_{\text{crit}} = 0.1841$.

Solution

Equation (3.9) is given by

$$2f''' + ff'' + \beta(1 - f'^2) = 0. \tag{C.176}$$

Let $\eta = z/\sqrt{\beta}$, then

$$\frac{\partial}{\partial \eta} = \frac{\partial}{\partial z}\frac{dz}{d\eta} = \sqrt{\beta}\frac{\partial}{\partial z},$$

and so the equation, in terms of z, can be written as

$$2\beta^{3/2}f''' + \beta ff'' + \beta(1 - \beta f'^2) = 0, \tag{C.177}$$

where now $()' = d/dz$. Dividing (C.177) by β, we have

$$2\sqrt{\beta}f''' + ff'' + 1 - \beta f'^2 = 0. \tag{C.178}$$

As $\beta \to \infty$, expand f as

$$f = \frac{1}{\sqrt{\beta}}f_0 + \frac{1}{\beta}f_1 + \cdots. \tag{C.179}$$

To leading order

$$2f_0''' + 1 - f_0'^2 = 0. \tag{C.180}$$

Assume

$$f_0' = A + B\tanh^2(az + b). \tag{C.181}$$

Then

$$f_0'' = 2Ba\tanh(az + b)\text{sech}^2(az + b), \tag{C.182}$$

$$f_0''' = 2Ba^2\text{sech}^4(az + b) - 4Ba^2\tanh^2(az + b)\text{sech}^2(az + b). \tag{C.183}$$

Substitute (C.181) and (C.183) into (C.180); we find

$$4Ba^2\text{sech}^4(az + b) - 8Ba^2\tanh^2(az + b)\text{sech}^2(az + b) + 1$$
$$- A^2 - 2AB\tanh^2(az + b) - B^2\tanh^4(az + b) = 0. \tag{C.184}$$

Now use the identity $\text{sech}^2(x) = 1 - \tanh^2(x)$ and substitute into (C.184) to find

$$4Ba^2\left[1 - 2\tanh^2(az + b) + \tanh^4(az + b)\right]$$
$$- 8Ba^2\tanh^2(az + b)\left[1 - \tanh^2(az + b)\right] + 1 - A^2$$

$$- 2AB \tanh^2(az+b) - B^2 \tanh^4(az+b) = 0. \tag{C.185}$$

Equate like terms to get

$$O(1)4Ba^2 + 1 - A^2 = 0,$$

$$O(\tanh^2)16Ba^2 + 2AB = 0,$$

$$O(\tanh^4)12Ba^2 - B^2 = 0. \tag{C.186}$$

We now have three equations and three unknowns. Use the first equation to find B:

$$B = \frac{A^2 - 1}{4a^2}. \tag{C.187}$$

Divide the second equation by $-2B$ to find A:

$$A = -8a^2. \tag{C.188}$$

Divide the third equation by B, giving

$$B = 12a^2. \tag{C.189}$$

Substitute (C.188) and (C.189) into (C.187) and combine like terms to yield

$$16a^4 = 1 \quad \text{or} \quad a = \frac{1}{2}. \tag{C.190}$$

Then from (C.188) and (C.189), we find

$$A = -2 \quad \text{and} \quad B = 3. \tag{C.191}$$

Therefore from (C.181), we have

$$f'_o = 3 \tanh^2\left[\frac{z}{2} + b\right] - 2. \tag{C.192}$$

To satisfy the boundary condition at $z = 0$, we have

$$f'_o = 0 = 3 \tanh^2[b] - 2, \tag{C.193}$$

$$b = \tanh^{-1}\left[\sqrt{\frac{2}{3}}\right]. \tag{C.194}$$

Therefore, we find

$$f'_o = U = 3 \tanh^2\left[\frac{z}{2} + \tanh^{-1}\sqrt{2/3}\right] - 2. \tag{C.195}$$

This is the desired result.

Exercise 3.5 Consider the constant mean profile

$$U = U_1, \quad V = 0, \quad P = \text{constant}, \quad 0 < y < \infty,$$

which allows slip along a bounding plate at $y = 0$. Although the velocity does not vanish at the plate, assume that the disturbance velocity does.

(a) Show that the general solution to the Orr–Sommerfeld equation (2.31) can be written as

$$\phi = Ae^{-\alpha y} + Be^{+\alpha y} + Ce^{-py} + De^{+py},$$

where $p^2 = \alpha^2 + i\alpha Re(U_1 - c)$.

(b) If $\text{Re}(p) > 0$, show that the only solution that satisfies the boundary conditions $\phi(0) = \phi'(0) = 0$ is the trivial solution.

(c) Show that only a continuum exists if $p = ik, 0 < k < \infty$ is purely imaginary, and deduce the solution

$$\phi_k = A\left[e^{-\alpha y} - \cos(ky) + \alpha k^{-1}\sin(ky)\right],$$

with eigenvalue

$$\omega_k = \alpha U_1 - i(Re)^{-1}(\alpha^2 + k^2).$$

Solution

(a) The Orr–Sommerfeld equation (2.31), with $U = U_1$ is given by

$$(U_1 - c)(\phi'' - \alpha^2\phi) = (i\alpha Re)^{-1}(\phi^{iv} - 2\alpha^2\phi'' + \alpha^4\phi). \qquad \text{(C.196)}$$

Since this is a linear ordinary differential equation with constant coefficients, the general solution is of the form

$$\phi = e^{my}. \qquad \text{(C.197)}$$

Substituting (C.197) into (C.196) yields

$$(U_1 - c)(m^2 - \alpha^2) = (i\alpha Re)^{-1}(m^2 - \alpha^2)^2, \qquad \text{(C.198)}$$

or,

$$(m^2 - \alpha^2)\left[U_1 - c - (i\alpha Re)^{-1}(m^2 - \alpha^2)\right] = 0. \qquad \text{(C.199)}$$

The solutions are

$$m = \pm\alpha, \pm p,$$

where

$$p^2 = \alpha^2 + i\alpha Re(U_1 - c). \qquad \text{(C.200)}$$

Therefore the general solution is

$$\phi = Ae^{-\alpha y} + Be^{\alpha y} + Ce^{-py} + De^{py}. \qquad \text{(C.201)}$$

This is the desired result.

(b) If $\text{Re}(p) > 0$, then the general solution that is bounded as $y \to \infty$ is

$$\phi = Ae^{-\alpha y} + Ce^{-py} \tag{C.202}$$

If we try to satisfy the boundary conditions $\phi(0) = \phi'(0) = 0$, then we find

$$\phi(0) = A + C = 0,$$

$$\phi'(0) = -\alpha A - pC = 0. \tag{C.203}$$

Therefore we have

$$A = -C \quad \text{and} \quad C(\alpha - p) = 0. \tag{C.204}$$

Since $\alpha \neq p$, we have $C = 0$. Therefore the only solution is the trivial solution.

(c) Suppose $p = ik, 0 < k < \infty$. Then the general solution that decays as $y \to \infty$ is given by

$$\phi = Ae^{-\alpha y} + Ce^{-iky} + De^{iky}, \tag{C.205}$$

which can also be written as

$$\phi = Ae^{-\alpha y} + c\cos(ky) + d\sin(ky). \tag{C.206}$$

Satisfying the boundary conditions yields

$$\phi(0) = A + c = 0 \quad \text{hence} \quad c = -A, \tag{C.207}$$

$$\phi'(0) = -\alpha A + dk = 0 \quad \text{hence} \quad d = \frac{\alpha}{k}A. \tag{C.208}$$

Hence, the solution is given by

$$\phi = A\left[e^{-\alpha y} - \cos(ky) + \frac{\alpha}{k}\sin(ky)\right]. \tag{C.209}$$

The eigenvalue is found from the definition of p, i.e., $p = ik$

$$\sqrt{\alpha^2 + i\alpha \text{Re}(U_1 - c)} = ik, \tag{C.210}$$

or

$$\alpha^2 + i\alpha \text{Re}(U_1 - c) = -k^2. \tag{C.211}$$

Let $\omega = \alpha c$, and solve for ω:

$$\omega = \alpha U_1 - i\text{Re}^{-1}(\alpha^2 + k^2). \tag{C.212}$$

This is the desired result.

C.3 Solutions for Chapter 4

Exercise 4.2 For the Gaussian wake profile

$$U(y) = 1 - Qe^{-y^2 \ln 2},$$

compute the inviscid spatial stability characteristics for various values of the wake deficit parameter Q. Show that the flow becomes absolutely unstable when $Q \geq 0.943$ (see Hultgren & Aggarwal, 1987).

Solution

The inflection point, found by setting $U''(y_c) = 0$, gives for the neutral phase speed $c_n = 1 - Q\exp(-1/2)$. The solution is presented in Fig. C.15, where we plot the instability spectrum $-\alpha_i$ as a function of ω. Only the symmetric mode is shown. Note that a cusp is beginning to form at $Q = 0.94$, indicating the onset of absolute instability.

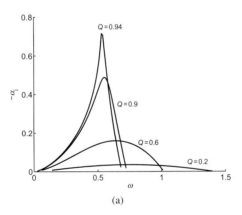

(a)

Figure C.15 Plot of instability spectrum $-\alpha_i$ as a function of ω.

Exercise 4.3 Recall the laminar mixing layer of Section 2.6, given by

$$2f''' + ff'' = 0,$$

where

$$f'(-\infty) = \beta_U, \qquad f(0) = 0, \qquad f'(+\infty) = 1.$$

The profile was shown in Fig. 2.10. Compute the spatial growth rate curves for $\beta_U = 0$ and 0.5, and compare them to that obtained using the hyperbolic tangent profile (4.36).

Solution

The inflection point, found by setting $U''(y_c) = 0$, gives for the neutral phase speed $c_n \approx 0.587$ for $\beta_U = 0$ and $c_n \approx 0.766$ for $\beta_U = 0.5$. The solution is presented in Fig. C.16, where we plot the instability spectrum $-\alpha_i$ as a function of ω. The instability spectrum for the hyperbolic tangent is also plotted.

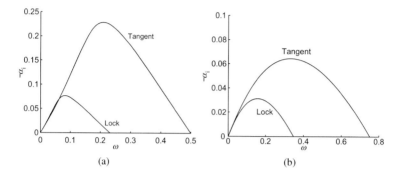

(a)   (b)

Figure C.16 Plot of instability spectrum $-\alpha_i$ as a function of ω and for (a) $\beta_U = 0$ and (b) $\beta_U = 0.5$.

Exercise 4.4 For viscous disturbances, modify your numerical code built in Chapter 3 for the Orr–Sommerfeld equation to allow for spatial stability calculations. Be sure to use double precision. Use the code to solve the following problems.

(a) Compute the spatial growth rate for the Falkner–Skan profile of (3.9). Take $\beta = 10, 1, 0.5, -0.1$ and -0.19 with $Re = 1,000$. For one of the values of β, compute the temporal growth rate and use Gaster's transformation to compare to the curve obtained using the spatial code.
(b) Compute the continuum and the spatial growth rate for the asymptotic suction profile of Exercise 3.2 using the modified Orr–Sommerfeld equation.

Solution
(a) If you first set $\beta = 0$ for Blasius flow, you should be able to obtain Fig. 4.15 using your spatial stability code, and then use Gaster's transformation (4.38) to obtain the dashed curve.

(b) The modified Orr–Sommerfeld equation is given by

$$(U - c)(D^2 - \alpha^2)\phi - U''\phi = (i\alpha Re)^{-1} \left[(D^2 - \alpha^2)^2 + (D^2 - \alpha^2)D \right] \phi.$$
$$(C.213)$$

Note that the temporal continuous spectrum (α real, ω complex) was the subject of Exercise 3.2. Here, we want to compute the spatial continuous spectrum (α complex, ω real). Assuming, $\phi \sim e^{iky}$, and substituting this into equation

(C.213), we find

$$\underbrace{(k^2+\alpha^2)}_{(1)}\underbrace{\left[1-c+(i\alpha Re)^{-1}(k^2+\alpha^2-ik)\right]}_{(2)}=0. \tag{C.214}$$

There are four spatial branches determined by setting (1) and (2) equal to zero. For the first two branches, we have

$$k^2+\alpha^2=0 \quad \text{or} \quad \alpha=\pm ik. \tag{C.215}$$

Recall that $c=\omega/\alpha$. For the remaining branches, we have

$$i\alpha Re - i\omega Re + k^2+\alpha^2 - ik = 0. \tag{C.216}$$

Grouping like terms of α, we find

$$\alpha^2+i\alpha Re+(k^2-i\omega Re-ik)=0. \tag{C.217}$$

Solving the quadratic equation (C.217), we find

$$\alpha=\frac{iRe}{2}\left[-1\pm\sqrt{1+4Re^{-2}(k^2-iRe\omega-ik)}\right]. \tag{C.218}$$

These four branches, (C.215) and (C.218), are the desired result.

C.4 Solutions for Chapter 5

Exercise 5.1 Consider the linear stability system (5.22)–(5.26):

$$i(U-c)r+ip f+(\rho\phi)'=0, \tag{C.219}$$

$$\rho\left[i(U-c)f+U'\phi\right]+\frac{i\Pi}{\gamma M^2}=0, \tag{C.220}$$

$$i\alpha^2\rho(U-c)\phi+\frac{\Pi'}{\gamma M^2}=0, \tag{C.221}$$

$$\rho\left[i(U-c)\theta+T'\phi\right]-i\frac{\gamma-1}{\gamma}(U-c)\Pi=0, \tag{C.222}$$

and

$$\Pi=\rho\theta+rT. \tag{C.223}$$

(a) Show that the system can be reduced to the two first-order equations

$$\alpha^2\rho(U-c)\phi=\frac{i\Pi'}{\gamma M^2}, \tag{C.224}$$

and

$$(U - c)\phi' - U'\phi = \frac{i\Pi}{\gamma M^2}\xi, \qquad \text{(C.225)}$$

where

$$\xi = T - M^2(U - c)^2.$$

(b) Combine the two equations (C.224) and (C.225) to get the second-order equation (5.27) for pressure:

$$\Pi'' + \left[\frac{T'}{T} - \frac{2U'}{U - c}\right]\Pi' - \frac{\alpha^2}{T}\left[T - M^2(U - c)^2\right]\Pi = 0, \qquad \text{(C.226)}$$

or (5.28) for ϕ:

$$\left[\frac{\phi'}{\xi}\right]' - \left[\frac{\alpha^2}{T} + \frac{1}{(U - c)}\left(\frac{U'}{\xi}\right)'\right]\phi = 0. \qquad \text{(C.227)}$$

(c) Following Gropengiesser (1969), introduce the transformation

$$\chi = \frac{i\Pi}{\gamma M^2 \phi} \qquad \text{(C.228)}$$

and derive the following first-order nonlinear differential equation

$$\chi' = \rho\alpha^2(U - c) - \chi\left[\frac{\chi\xi + U'}{U - c}\right] \qquad \text{(C.229)}$$

where ξ is defined in (5.29). State the proper boundary conditions in the freestream as $y = \pm\infty$.

Solution
Recall from equation (5.21), the velocity, pressure, temperature and density as a wave disturbance, takes the form

$$(u, v, p, T, \rho) = (U, 0, 1, T, \rho)(y) + (f, \alpha\phi, \Pi, \theta, r)(y)e^{i(\alpha x - \omega t)}.$$

Although it is not necessary to decompose in order to obtain the solutions, it is important to remember that the solutions fold back into the physics of wave instabilities.

(a) Equation (C.221) can be written to give

$$\alpha^2\rho(U - c)\phi = \frac{i\Pi'}{\gamma M^2}, \qquad \text{(C.230)}$$

which is the first desired result.

To get the second result, we start with (C.219) and (C.220) to eliminate f,

(C.223) to eliminate r, and then (C.222) to eliminate θ as follows. First, rearrange (C.219) to get

$$i\rho f = -i(U-c)r - (\rho\phi)'. \tag{C.231}$$

Substituting this into (C.220) leads to

$$-(U-c)\left[i(U-c)r + (\rho\phi)'\right] + \rho U'\phi + \frac{i\Pi}{\gamma M^2} = 0. \tag{C.232}$$

From (C.223) we have

$$r = \frac{\Pi - \rho\theta}{T}. \tag{C.233}$$

Substitute into (C.232) to give

$$-\frac{i(U-c)^2}{T}\left[\Pi - \rho\theta\right] - (U-c)(\rho\phi)' + \rho U'\phi + \frac{i\Pi}{\gamma M^2} = 0. \tag{C.234}$$

Substituting (C.222) into (C.234) yields

$$\frac{(U-c)}{T}\left[i\frac{\gamma-1}{\gamma}(U-c)\Pi - \rho T'\phi\right] - (U-c)(\rho\phi)' + \rho U'\phi$$

$$+i\Pi\left[\frac{1}{\gamma M^2} - \frac{(U-c)^2}{T}\right] = 0, \tag{C.235}$$

or

$$-(U-c)(\rho\phi' + \rho'\phi) + \rho U'\phi - \rho(U-c)\frac{T'}{T}\phi$$

$$+i\Pi\left[\frac{1}{\gamma M^2} + \frac{\gamma-1}{\gamma}\frac{(U-c)^2}{T} - \frac{(U-c)^2}{T}\right] = 0. \tag{C.236}$$

Let $\rho T = 1$ from the mean flow, then

$$\rho = \frac{1}{T} \quad \text{and} \quad \rho' = -\frac{T'}{T^2}.$$

Equation (C.236) becomes

$$-(U-c)\frac{\phi'}{T} + (U-c)\frac{T'}{T^2}\phi + \frac{U'}{T}\phi - (U-c)\frac{T'}{T^2}\phi$$

$$+i\Pi\left[\frac{1}{\gamma M^2} - \frac{(U-c)^2}{\gamma T}\right] = 0. \tag{C.237}$$

Multiply equation (C.237) by T to find

$$(U-c)\phi' - U'\phi = \frac{i\pi}{\gamma M^2}\xi, \tag{C.238}$$

where $\xi = T - M^2(U-c)^2$, which is the desired result.

(b) Rearrange (C.224) to give

$$\phi = \frac{i\Pi'}{\gamma M^2}\frac{T}{\alpha^2(U-c)} \tag{C.239}$$

and take the derivative:

$$\phi' = \frac{i\Pi''}{\gamma M^2}\frac{T}{\alpha^2(U-c)} + \frac{i\Pi'}{\gamma M^2}\frac{T'}{\alpha^2(U-c)} - \frac{i\Pi'}{\gamma M^2}\frac{TU'}{\alpha^2(U-c)^2}, \tag{C.240}$$

where $\phi' = d\phi/dy$. Substitute (C.239) and (C.240) into (C.225) and multiply by $\alpha^2\gamma M^2/iT$, to get

$$\Pi'' + \left[\frac{T'}{T} - \frac{2U'}{U-c}\right]\Pi' - \frac{\alpha^2}{T}\left[T - M^2(U-c)^2\right]\Pi = 0, \tag{C.241}$$

which is the desired result (C.226). Then from (C.225), we divide by ξ, taking a derivative, yielding

$$\frac{i\Pi'}{\gamma M^2} = (U-c)\left(\frac{\phi'}{\xi}\right)' + \left(\frac{U'}{\xi}\right)\phi' - \left(\frac{U'}{\xi}\right)'\phi - \left(\frac{U'}{\xi}\right)\phi'. \tag{C.242}$$

Substitute (C.224) into (C.242) and divide by $(U-c)$ to obtain

$$\left(\frac{\phi'}{\xi}\right)' - \left[\frac{\alpha^2}{T} + \frac{1}{U-c}\left(\frac{U'}{\xi}\right)'\right]\phi = 0, \tag{C.243}$$

which is the desired result (C.227).

(c) Let

$$\chi = \frac{i\Pi}{\gamma M^2\phi} \quad \text{or} \quad \phi\chi = \frac{i\Pi}{\gamma M^2}. \tag{C.244}$$

Taking the derivative:

$$\phi'\chi + \phi\chi' = \frac{i\Pi'}{\gamma M^2}. \tag{C.245}$$

Now use (C.224) and (C.225) to eliminate ϕ' and Π'. From (C.224), we find

$$\frac{i\Pi'}{\gamma M^2} = \frac{\alpha^2}{T}(U-c)\phi, \tag{C.246}$$

and from (C.225) we find

$$\phi' = \frac{U'}{U-c}\phi + \frac{i\Pi}{\gamma M^2}\frac{\xi}{U-c}. \tag{C.247}$$

Substitute (C.246) and (C.247) into (C.245) to obtain

$$\chi\left[\frac{U'}{U-c}\phi + \frac{i\Pi}{\gamma M^2}\frac{\xi}{U-c}\right] + \phi\chi' = \frac{\alpha^2}{T}(U-c)\phi. \tag{C.248}$$

Divide (C.248) by ϕ and use the definition of χ in equation (C.244):

$$\chi\left[\frac{U'}{U-c}+\chi\frac{\xi}{U-c}\right]+\chi'=\frac{\alpha^2(U-c)}{T}, \tag{C.249}$$

or

$$\chi'=\frac{\alpha^2}{T}(U-c)-\chi\left[\frac{\chi\xi+U'}{U-c}\right], \tag{C.250}$$

which is the desired result. For the boundary conditions, we first look at $y\to\infty$. From (C.226), we have

$$\Pi''-\alpha^2\left[1-M^2(1-c)^2\right]\Pi=0. \tag{C.251}$$

Let

$$\Pi=Ae^{-\Omega_+y}. \tag{C.252}$$

Substituting (C.252) into (C.251), we find

$$\Omega_+^2=\alpha^2\left[1-M^2(1-c)^2\right]. \tag{C.253}$$

Then evaluating (C.224) at $y\to\infty$, we have

$$\alpha^2(1-c)\phi=\frac{-iA}{\gamma M^2}\Omega_+e^{-\Omega_+y}, \tag{C.254}$$

so that

$$\chi=-\frac{\alpha^2(1-c)}{\Omega_+}\quad\text{as}\quad y\to\infty. \tag{C.255}$$

Similarly, as $y\to-\infty$,

$$\Pi=Be^{\Omega_-y}, \tag{C.256}$$

where

$$\Omega_-^2=\frac{\alpha^2}{\beta_T}\left[\beta_T-M^2(\beta_U-c)^2\right]. \tag{C.257}$$

Then from (C.224) as $y\to-\infty$, we have

$$\frac{\alpha^2(\beta_U-c)}{\beta_T}\phi=\frac{iB\Omega_-e^{\Omega_-y}}{\gamma M^2} \tag{C.258}$$

and therefore we get

$$\chi=\frac{\alpha^2(\beta_U-c)}{\beta_T\Omega_-}\quad\text{as}\quad y\to-\infty. \tag{C.259}$$

Equations (C.255) and (C.259) are the proper boundary conditions for χ.

Exercise 5.3 Use the following steps outlined to prove Results 5.2–5.4.

(a) Show that the compressible Rayleigh equation (5.28) can be written as

$$\left(\frac{\Omega^2 \psi'}{\xi}\right)' - \frac{\alpha^2 \Omega^2}{T}\psi = 0, \tag{C.260}$$

where

$$\Omega = U - c, \qquad \xi = T - M^2\Omega^2, \qquad \psi = \phi/(U-c).$$

(b) To prove Results 5.2 and 5.3, multiply the above equation by ψ^*, the complex conjugate of ψ, integrate over the region $[a,b]$ and set the real and imaginary parts to zero to get the desired results. It is helpful when proving Result 5.3 to use the relation

$$(U - U_{\min})(U - U_{\max}) \le 0,$$

which is valid for monotone profiles.

(c) To prove Result 5.4 show that the above equation can be written as

$$\left[\frac{\Omega\chi'}{\xi}\right]' - \chi\left[\frac{1}{2}\left(\frac{U'}{\xi}\right)' + \frac{(U')^2}{4\xi\Omega} + \alpha^2\Omega\right] = 0 \tag{C.261}$$

where

$$\chi = \Omega^{1/2}\psi.$$

Multiply by χ^*, integrate over the region $[a,b]$ and examine the imaginary part.

Solution

(a) The compressible Rayleigh equation (5.28) is

$$\left[\frac{\phi'}{\xi}\right]' - \left[\frac{\alpha^2}{T} + \frac{1}{U-c}\left(\frac{U'}{\xi}\right)'\right]\phi = 0. \tag{C.262}$$

Let $\Omega = U - c$, then $\xi = T - M^2(U-c)^2 = T - M^2\Omega^2$. Define $\psi = \phi/(U-c)$, so that $\Omega\psi = \phi$. Taking the derivative,

$$\Omega'\psi + \Omega\psi' = \phi'.$$

With $\Omega' = U'$, substituting these relations into (C.262), we find

$$\left[\frac{U'\psi + \Omega\psi'}{\xi}\right]' - \left[\frac{\alpha^2}{T} + \frac{1}{\Omega}\left(\frac{U'}{\xi}\right)'\right]\Omega\psi = 0. \tag{C.263}$$

Multiply (C.263) by Ω:

$$\Omega\left(\frac{U'\psi}{\xi}\right)' + \Omega\left(\frac{\Omega\psi'}{\xi}\right)' - \left[\frac{\alpha^2\Omega^2}{T} + \Omega\left(\frac{U'}{\xi}\right)'\right]\psi = 0. \tag{C.264}$$

Expanding, we have

$$\Omega \left(\frac{U'}{\xi}\right)' \psi + \Omega \left(\frac{U'}{\xi}\right) \psi' \Omega \left(\frac{\Omega' \psi'}{\xi}\right) + \Omega^2 \left(\frac{\psi'}{\xi}\right)'$$
$$- \left(\frac{\alpha^2}{T}\right) \Omega^2 \psi - \Omega \left(\frac{U'}{\xi}\right)' \psi = 0. \tag{C.265}$$

Simplifying,

$$2\Omega\Omega' \left(\frac{\psi'}{\xi}\right) + \Omega^2 \left(\frac{\psi'}{\xi}\right)' - \frac{\alpha^2}{T}\Omega^2 \psi = 0, \tag{C.266}$$

or, combining the first two terms,

$$\left(\frac{\Omega^2 \psi'}{\xi}\right)' - \frac{\alpha^2 \Omega^2}{T} \psi = 0. \tag{C.267}$$

This is the desired result.

(b) Multiply (C.267) by ψ^*, where $*$ denotes complex conjugates for this exercise and integrate:

$$\int_a^b \left(\frac{\Omega^2 \psi'}{\xi}\right)' \psi^* dy - \alpha^2 \int_a^b \frac{\Omega^2}{T} \psi\psi^* dy = 0. \tag{C.268}$$

Integrate the first term by parts to yield

$$\frac{\Omega^2 \psi'}{\xi} \psi^* \Big|_a^b - \int_a^b \frac{\Omega^2 \psi'}{\xi} (\psi^*)' dy - \alpha^2 \int_a^b \frac{\Omega^2}{T} |\psi|^2 dy = 0. \tag{C.269}$$

Now $\psi = 0$ at $y = a, b$, yields $\psi^* = 0$ at $y = a, b$, so

$$\int_a^b \left[\frac{\Omega^2}{\xi} |\psi'|^2 + \alpha^2 \frac{\Omega^2}{T} |\psi|^2\right] dy = 0 \tag{C.270}$$

or

$$\int_a^b \left[\frac{\Omega^2 \xi^*}{|\xi|^2} |\psi'|^2 + \frac{\alpha^2 \Omega^2}{T} |\psi|^2\right] dy = 0. \tag{C.271}$$

Now

$$\Omega^2 = (U - c)^2 = (U - c_r - ic_i)^2, \tag{C.272}$$

which equals

$$\Omega^2 = (U - c_r)^2 - c_i^2 - 2ic_i(U - c_r). \tag{C.273}$$

Recall

$$\xi^* = T - M^2(\Omega^*)^2. \tag{C.274}$$

Substituting (C.273) gives

$$\xi^* = T - M^2\left[(U - c_r)^2 - c_i^2\right] - 2iM^2 c_i(U - c_r). \tag{C.275}$$

Looking at the imaginary part,

$$\mathrm{Im}[\Omega^2 \xi^*] = -2c_i(U - c_r)T. \tag{C.276}$$

Thus

$$\int_a^b \left[\frac{-2c_i(U - c_r)T}{|\xi|^2}|\psi'|^2 - \frac{2c_i(U - c_r)\alpha^2}{T}|\psi|^2\right] dy = 0 \tag{C.277}$$

or

$$-2c_i \int_a^b \left[\frac{T|\psi'|^2}{|\xi|^2} + \frac{\alpha^2}{T}|\psi|^2\right](U - c_r)dy = 0. \tag{C.278}$$

Thus if $c_i \neq 0$, then $(U - c_r)$ must change sign somewhere in $[a, b]$, i.e., $c_r \in [U_{\min}, U_{\max}]$. This proves Result 5.2.

We now look at the real part:

$$\int_a^b \left[(U - c_r)^2 - c_i^2\right]\left[T - M^2\left\{(U - c_r)^2 - c_i^2\right\}\right]\frac{|\psi'|^2}{|\xi|^2}dy$$

$$-\int_a^b \left[4M^2 c_i^2(U - c_r)^2\right]\frac{|\psi'|^2}{|\xi|^2}dy$$

$$+\int_a^b \frac{\alpha^2}{T}\left[(U - c_r)^2 - c_i^2\right]|\psi|^2 dy = 0, \tag{C.279}$$

or

$$\int_a^b \left[(U - c_r)^2 - c_i^2\right]\left[\frac{T|\psi'|^2}{|\xi|^2} + \frac{\alpha^2}{T}|\psi|^2\right] dy$$

$$= \int_a^b \left[M^2\left\{(U - c_r)^2 - c_i^2\right\}^2\right.$$

$$\left. + 4M^2 c_i^2(U - c_r)^2\right]\frac{|\psi'|^2}{|\xi|^2}dy = 0. \tag{C.280}$$

Let

$$Q^2 = \frac{T|\psi'|^2}{|\xi|^2} + \frac{\alpha^2}{T}|\psi|^2,$$

and note that the integral on the right-hand side is always positive. Then we have

$$\int_a^b \left[(U - c_r)^2 - c_i^2\right]Q^2 dy \geq 0. \tag{C.281}$$

If we assume the mean profile is monotonic in $U(y)$, then it is true that the following holds

$$(U - U_{\min})(U - U_{\max}) \le 0. \tag{C.282}$$

Multiplying (C.282) by Q^2 and integrating, we find

$$\int_a^b (U - U_{\min})(U - U_{\max})Q^2 dy \le 0,$$

or

$$\int_a^b \left[U^2 - U(U_{\min} + U_{\max}) + U_{\min}U_{\max} \right] Q^2 dy \le 0. \tag{C.283}$$

Now recall the following from the imaginary part (C.278),

$$-2c_i \int_a^b Q^2(U - c_r) dy = 0. \tag{C.284}$$

If $c_i \ne 0$, then we can rewrite (C.284) as

$$\int_a^b U Q^2 dy = \int_a^b c_r Q^2 dy. \tag{C.285}$$

The inequality (C.283) thus becomes

$$\int_a^b \left[U^2 - c_r(U_{\min} + U_{\max}) + U_{\min}U_{\max} \right] Q^2 dy \le 0. \tag{C.286}$$

Now (C.281) can be written as

$$\int_a^b (U - c_r)^2 Q^2 dy \ge \int_a^b c_i^2 Q^2 dy, \tag{C.287}$$

or

$$\int_a^b U^2 Q^2 dy \ge \int_a^b [2U c_r - c_r^2 + c_i^2] Q^2 dy. \tag{C.288}$$

Use (C.288) in (C.286) to eliminate the U^2 term, paying close attention to the inequality signs:

$$\int_a^b \left[2U c_r - c_r^2 + c_i^2 - c_r(U_{\min} + U_{\max}) + U_{\min}U_{\max} \right] Q^2 dy \le 0. \tag{C.289}$$

But again recall (C.285) and so we can eliminate U in the inequality (C.289) to get

$$\left[2c_r^2 - c_r^2 + c_i^2 - c_r(U_{\min} + U_{\max}) + U_{\min}U_{\max} \right] \int_a^b Q^2 dy \le 0. \tag{C.290}$$

Since the integral is positive, we have

$$c_r^2 - c_r(U_{\min} + U_{\max}) + \left[\frac{U_{\min} + U_{\max}}{2} \right]^2$$

$$- \left[\frac{U_{\min} + U_{\max}}{2} \right]^2 + c_i^2 + U_{\min}U_{\max} \le 0. \tag{C.291}$$

Note that we added and subtracted a term. Simplifying, we get

$$\left[c_r - \frac{U_{min} + U_{max}}{2}\right]^2 + c_i^2 \leq \left[\frac{U_{min} + U_{max}}{2}\right]^2. \tag{C.292}$$

The equality is a circle centered at

$$(c_r, c_i) = \left[\frac{U_{min} + U_{max}}{2}, 0\right] \tag{C.293}$$

with radius

$$\left[\frac{U_{min} + U_{max}}{2}\right]. \tag{C.294}$$

Since $c_i > 0$ for an unstable temporal mode, we have a semicircle. Thus, the inequality is Howard's semicircle theorem, which is Result 5.3 for compressible inviscid flow.

(c) Note: This problem might prove too difficult for students with weak mathematical skills. According to Chimonas (1970), we begin with the equation

$$\left[\frac{\Omega^2 \psi'}{\xi}\right]' - \frac{\alpha^2 \Omega^2}{T} \psi = 0, \tag{C.295}$$

where $\Omega = U - c$, $\xi = T - M^2\Omega^2$ and $\psi = \phi/(U-c)$. Now let $\chi = \Omega^{1/2}\psi$. Then a new equation for χ can be derived, yielding

$$\left[\frac{\Omega\chi'}{\xi}\right]' - \chi\left[\frac{1}{2}\left(\frac{U'}{\xi}\right)' + \frac{(U')^2}{4\xi\Omega} + \frac{\alpha^2\Omega}{T}.\right] = 0. \tag{C.296}$$

Multiply (C.296) by χ^* and integrate, and note that

$$\int_a^b \left[\frac{\Omega\chi'}{\xi}\right]' \chi^* dy = \left.\frac{\Omega\chi'\chi^*}{\xi}\right|_a^b - \int_a^b \frac{\Omega}{\xi}|\chi'|^2 dy. \tag{C.297}$$

We see that (C.296) simplifies to

$$\int_a^b \frac{\Omega|\chi'|^2}{\xi} dy + \int_a^b \left[\frac{1}{2}\left(\frac{U'}{\xi}\right)' + \frac{(U')^2}{4\xi\Omega} + \frac{\alpha^2\Omega}{T}\right]|\chi|^2 dy = 0. \tag{C.298}$$

Look at the imaginary part of each term. Define

$$I_1 = \int_a^b \frac{\Omega|\chi'|^2}{\xi} dy = \int_a^b \frac{\Omega\xi^*|\chi'|^2}{|\xi|^2} dy, \tag{C.299}$$

$$\text{Im}(I_1) = -c_i \int_a^b \frac{T + M^2[(U - c_r)^2 + c_i^2]}{|\xi|^2}|\chi'| dy, \tag{C.300}$$

$$I_2 = \int_a^b \frac{1}{2} \left(\frac{U'}{\xi}\right)' |\chi|^2 dy = \frac{1}{2} \left(\frac{U'}{\xi}\right) |\chi|^2 \Big|_a^b - \int_a^b \frac{1}{2} \left(\frac{U'}{\xi}\right) \frac{d}{dy}(\chi\chi^*) dy$$

$$= -\frac{1}{2} \int_a^b \frac{U'\xi^*}{|\xi|^2} \frac{d}{dy}(\chi\chi^*) dy, \tag{C.301}$$

$$\text{Im}(I_2) = c_i \int_a^b \frac{M^2 U'(U - c_r)}{|\xi|^2} \frac{d}{dy}(\chi\chi^*) dy, \tag{C.302}$$

$$I_3 = \int_a^b \frac{(U')^2}{4\xi\Omega} |\chi|^2 dy = \int_a^b \frac{(U')^2 \xi^* \Omega^*}{4|\xi|^2 |\Omega|^2} |\chi|^2 dy, \tag{C.303}$$

$$\text{Im}(I_3) = \frac{c_i}{4} \int_a^b (U')^2 \frac{T + M^2 c_i^2 - 3M^2 (U - c_r)^2}{|\xi|^2 |\Omega|^2} |\chi|^2 dy, \tag{C.304}$$

$$I_4 = \int_a^b \alpha^2 \frac{\Omega}{T} |\chi|^2 dy, \tag{C.305}$$

$$\text{Im}(I_4) = -c_i \int_a^b \frac{\alpha^2}{T} |\chi|^2 dy. \tag{C.306}$$

Thus, (C.298) can be written as

$$I_1 + I_2 + I_3 + I_4 = 0. \tag{C.307}$$

Setting the imaginary part to zero, we get

$$-c_i \int_a^b \frac{\hat{I}_1 + \hat{I}_2 + \hat{I}_3}{|\xi|^2} dy - c_i \int_a^b \frac{\alpha^2}{T} |\chi|^2 dy = 0. \tag{C.308}$$

For $c_i > 0$, we can reduce as follows. First, define

$$A^2 = T + M^2 \left[(U - c_r)^2 + c_i^2\right] \geq 0.$$

Let $\hat{I} = \sum_{i=1}^3 \hat{I}_i$, then

$$\hat{I} = A^2 |\chi'|^2 - M^2 U'(U - c_r) \frac{d}{dy}(\chi\chi^*) - \frac{1}{4} \frac{(U')^2}{|\Omega|^2} \left[T + M^2 c_i^2\right.$$

$$\left. -3M^2 (U - c_r)^2\right] |\chi|^2 = A^2 \left[|\chi'| - \frac{MU'}{2\sqrt{T}} |\chi|\right]^2 + \frac{A^2 MU'}{\sqrt{T}} |\chi||\chi'|$$

$$-\frac{A^2}{T} \left(\frac{MU'}{2}\right)^2 |\chi|^2 - M^2 U'(U - c_r) \frac{d}{dy}(\chi\chi^*)$$

$$-\frac{1}{4} \frac{(U')^2}{|\Omega|^2} \left[T + M^2 c_i^2 - 3M^2 (U - c_r)^2\right] \chi|^2. \tag{C.309}$$

In the last expression we have added and subtracted terms. Now,

$$T + M^2 c_i^2 - 3M^2(U - c_r)^2 + M^2|\Omega|^2 - M^2|\Omega|^2$$
$$= T + 2M^2 c_i^2 - 2M^2(U - c_r)^2 - M^2|\Omega|^2, \tag{C.310}$$

since $|\Omega|^2 = \Omega\Omega^* = (U - c_r)^2 + c_i^2$. Define the following for convenience:

$$P_1^2 = \left[|\chi'| - \frac{MU'}{2\sqrt{T}}|\chi| \right]^2, \tag{C.311}$$

$$P_2^2 = \frac{A^2 MU'}{\sqrt{T}}|\chi||\chi'| - M^2 U'(U - c_r)\frac{d}{dy}(\chi\chi^*) \tag{C.312}$$

$$= \left\{ T + M^2\left[(U - c_r)^2 + c_i^2\right] \right\} \frac{MU'}{\sqrt{T}}|\chi||\chi'| - M^2 U'(U - c_r)\frac{d}{dy}(\chi\chi^*)$$

$$= \sqrt{T}U'\left\{ \left[1 + \left(\frac{M|\Omega|}{\sqrt{T}}\right)^2 \right] M|\chi||\chi'| - \frac{M^2}{\sqrt{T}}(U - c_r)\frac{d}{dy}(|\chi||\chi'|) \right\}.$$

Then

$$\hat{I} = A^2 P_1^2 + P_2^2 - A^2\left(\frac{MU'}{2}\right)^2 \frac{|\chi|^2}{T} + \frac{(U')^2 M^2}{4}|\chi|^2$$

$$- \frac{1}{4}\frac{(U')^2}{|\Omega|^2}\left\{ T + 2M^2\left[c_i^2 - (U - c_r)^2 \right] \right\}|\chi|^2. \tag{C.313}$$

Now, after some algebra, we have

$$\hat{I} = A^2 P_1^2 + P_2^2 - \frac{|\xi|^2}{|\Omega|^2}\frac{(U')^2}{4T}|\chi|^2. \tag{C.314}$$

As a side note, we will show that $P_2^2 \geq 0$. From complex variables, we have the inequality

$$Z + Z^* \leq 2|Z|. \tag{C.315}$$

Thus,

$$\frac{d}{dy}(\chi\chi^*) = \chi(\chi^*)' + \chi'\chi^* = \chi(\chi^*)' + \left[\chi(\chi^*)'\right]^* \leq 2|\chi||\chi'| \tag{C.316}$$

so that

$$-\frac{1}{2}\frac{d}{dy}(\chi\chi^*) \geq -|\chi||\chi'|.$$

Therefore

$$P_2^2 \geq \sqrt{T}U' \left\{ \left[1 + \left(\frac{M|\Omega|}{\sqrt{T}} \right)^2 \right] M - \frac{2M^2(U - c_r)}{\sqrt{T}} \right\} |\chi||\chi'|. \qquad \text{(C.317)}$$

Now

$$1 + \left(\frac{M|\Omega|}{\sqrt{T}} \right)^2 \geq \frac{2M|\Omega|}{\sqrt{T}} = \frac{2M\sqrt{(U - c_r)^2 + c_i^2}}{\sqrt{T}} \sqrt{T} \geq \frac{2M(U - c_r)}{\sqrt{T}}. \qquad \text{(C.318)}$$

Thus,

$$P_2^2 > \sqrt{T}U' \left[\frac{2M(U - c_r)}{\sqrt{T}} M - \frac{2M^2(U - c_r)}{\sqrt{T}} \right] |\chi||\chi'| = 0, \qquad \text{(C.319)}$$

i.e., $P_2^2 \geq 0$ is a positive quantity.

Putting it all together, we first have from (C.308), with $c_i > 0$

$$\int_a^b \frac{\hat{I}}{\cdot} |\xi|^2 dy = - \int_a^b \frac{\alpha^2}{T} |\chi|^2 dy. \qquad \text{(C.320)}$$

Or, using (C.314) with $P = A^2 P_1^2 + P_2^2$ for convenience, we rewrite (C.320) as

$$\int_a^b \frac{P^2}{|\xi|^2} dy = \int_a^b \frac{(U')^2}{4|\Omega|^2 T} |\chi|^2 dy - \int_a^b \frac{\alpha^2}{T} |\chi|^2 dy. \qquad \text{(C.321)}$$

Since $P^2 \geq 0$, we have the inequality

$$\int_a^b \frac{(U')^2}{4|\Omega|^2 T} |\chi|^2 dy \geq \int_a^b \frac{\alpha^2}{T} |\chi|^2 dy. \qquad \text{(C.322)}$$

But recall that

$$|\Omega|^2 = (U - c_r)^2 + c_i^2 \geq c_i^2.$$

Thus

$$\frac{1}{|\Omega|^2} \leq \frac{1}{c_i^2}$$

$$\int_a^b \frac{(U')^2}{4c_i^2 T} |\chi|^2 dy \geq \int_a^b \frac{(U')^2}{4|\Omega|^2 T} |\chi|^2 dy \geq \int_a^b \frac{\alpha^2}{T} |\chi|^2 dy, \qquad \text{(C.323)}$$

or

$$\left[\frac{(U'_{\max})^2}{4c_i^2} - \alpha^2 \right] \int_a^b \frac{|\chi|^2}{T} dy \geq 0. \qquad \text{(C.324)}$$

Since the integral is not zero, we have

$$\frac{(U'_{max})^2}{4c_i^2} \geq \alpha^2. \tag{C.325}$$

Since $\omega_i = \alpha c_i$ for temporal stability, we have

$$\omega_i \leq \frac{|U'_{max}|}{2}, \tag{C.326}$$

which proves Result 5.4. Note that if we are dealing with η instead of y, then the upper bound on the growth rate would be

$$\omega_i \leq \frac{|U'_{max}|}{2T_{min}}. \tag{C.327}$$

Exercise 5.4 Rewrite the compressible Rayleigh equation (5.28) in terms of the similarity variable η.

Solution

Begin with equation (5.28),

$$\left(\frac{\phi'}{\xi}\right)' - \left[\frac{\alpha^2}{T} + \frac{1}{U-c}\left(\frac{U'}{\xi}\right)'\right]\phi = 0. \tag{C.328}$$

Let $\eta = \int_0^y \rho\,dy$; then

$$\frac{d}{dy} = \frac{d}{d\eta}\frac{d\eta}{dy} = \rho\frac{d}{d\eta}. \tag{C.329}$$

So (C.328) becomes

$$\rho\left(\frac{\rho\phi'}{\xi}\right)' - \left[\frac{\alpha^2}{T} + \frac{\rho}{U-c}\left(\frac{\rho U'}{\xi}\right)'\right]\phi = 0, \tag{C.330}$$

where prime now means differentiation with respect to the similarity variable η. Also, note that $\rho T = 1$ from the mean flow. So we just have

$$\left(\frac{\phi'}{T\xi}\right)' - \left[\alpha^2 + \frac{1}{U-c}\left(\frac{U'}{T\xi}\right)'\right]\phi = 0. \tag{C.331}$$

Exercise 5.5 Compute the elements of the matrices **B** and **C** of equation (5.103).

Solution

The instability equations in matrix form are given by equation (5.103) or

$$[\mathbf{A}D^2 + \mathbf{B}D + \mathbf{C}] = \Phi, \tag{C.332}$$

where $\Phi = (f, \phi, \Pi, \theta)$ are the two-dimensional velocities, pressure and temperature perturbations. Note that we will need the following relationships

$$\rho = \frac{1}{T}, \quad s = \theta\frac{d\mu}{dT}, \quad \mu' = T'\frac{d\mu}{dT}, \tag{C.333}$$

where again the mean density and temperature are (ρ, T) and the viscosity and perturbation viscosity are (μ, s). The $()'$ are now derivatives with respect to η.

Rewrite (5.98) and multiply by $\alpha Re/\mu$ and substitute (C.333) as necessary.

$$f'' + \frac{T'}{\mu}\frac{d\mu}{dT}f' - \left[\frac{4}{3}\alpha^2 + \frac{i\alpha Re}{T\mu}(U-c)\right]f + \frac{i\alpha^2}{3}\phi'$$

$$+ \left[i\alpha^2\frac{T'}{\mu}\frac{d\mu}{dT} - \frac{\alpha ReU'}{T\mu}\right]\phi - \frac{i\alpha Re}{\mu\gamma M^2}\Pi$$

$$+ \frac{U'}{\mu}\frac{d\mu}{dT}\theta' + \left[\frac{U''}{\mu}\frac{d\mu}{dT} + \frac{U'T'}{\mu}\frac{d^2\mu}{dT^2}\right]\theta = 0. \qquad (C.334)$$

The matrix coefficients for the first row of (C.332) from (C.333) are

$$B_{1,1} = \frac{T'}{\mu}\frac{d\mu}{dT} \qquad B_{1,2} = \frac{i\alpha^2}{3} \qquad B_{1,3} = 0 \qquad B_{1,4} = \frac{U'}{\mu}\frac{d\mu}{dT}$$

$$C_{1,1} = -\left[\frac{4}{3}\alpha^2 + \frac{i\alpha Re(U-c)}{\mu T}\right] \qquad C_{1,2} = i\alpha^2\frac{T'}{\mu}\frac{d\mu}{dT} - \frac{\alpha ReU'}{\mu T}$$

$$C_{1,3} = -\frac{i\alpha Re}{\gamma M^2\mu} \qquad C_{1,4} = \frac{U''}{\mu}\frac{d\mu}{dT} + \frac{T'U'}{\mu}\frac{d^2\mu}{dT^2}.$$

Rewrite equation (5.99) and multiply by $3\alpha Re/4\mu$ and substitute (C.333) as necessary to give

$$\phi'' + \frac{i}{4}f' - \frac{T'}{2\mu}\frac{d\mu}{dT}if + \frac{T'}{\mu}\frac{d\mu}{dT}\phi' - \frac{3Re}{4\alpha\mu\gamma M^2}\Pi'$$

$$- \left[\frac{3\alpha^2}{4} + \frac{3i\alpha Re}{4T\mu}(U-c)\right]\phi + \frac{3iU'}{4\mu}\frac{d\mu}{dT}\theta = 0, \qquad (C.335)$$

yielding the coefficients

$$B_{2,1} = \frac{i}{4} \qquad B_{2,2} = \frac{T'}{\mu}\frac{d\mu}{dT} \qquad B_{2,3} = \frac{-3Re}{4\alpha\gamma M^2\mu} \qquad B_{2,4} = 0$$

$$C_{2,1} = -\frac{i}{2}\frac{T'}{\mu}\frac{d\mu}{dT} \qquad C_{2,2} = -\left[\frac{3\alpha^2}{4} + \frac{3i\alpha Re(U-c)}{4\mu T}\right]$$

$$C_{2,3} = 0 \qquad C_{2,4} = \frac{3iU'}{4\mu}\frac{d\mu}{dT}.$$

Rewrite equation (5.101) for r

$$r = \frac{\Pi - \rho\theta}{T}.$$

Substitute into (5.97) and divide by ρ to give

$$if + \phi' - \frac{T'}{T}\phi + i(U-c)\Pi - \frac{i(U-c)}{T}\theta = 0, \qquad \text{(C.336)}$$

yielding the coefficients

$$B_{3,1} = B_{3,3} = B_{3,4} = 0 \qquad B_{3,2} = 1$$

$$C_{3,1} = i \qquad C_{3,2} = -\frac{T'}{T} \qquad C_{3,3} = i(U-c) \qquad C_{3,4} = -\frac{i(U-c)}{T}.$$

Rewrite equation (5.100) and multiply by $\alpha RePr/\mu\gamma$ and substitute (C.333) as necessary to give

$$\theta'' + \frac{2T'}{\mu}\frac{d\mu}{dT}\theta' + \left[-i\frac{\alpha RePr}{\mu T\gamma}(U-c) - \alpha^2 + \frac{(T')^2}{\mu}\frac{d^2\mu}{dT^2} + \frac{T''}{\mu}\frac{d\mu}{dT} \right.$$

$$\left. +(\gamma-1)M^2 Pr\frac{(U')^2}{\mu}\frac{d\mu}{dT} \right]\theta + 2(\gamma-1)M^2 PrU'f'$$

$$- \frac{(\gamma-1)i\alpha RePr}{\gamma\mu}f - \frac{(\gamma-1)\alpha RePr}{\gamma\mu}\phi'$$

$$+ \left[2i\alpha^2(\gamma-1)M^2 PrU' - \frac{\alpha RePrT'}{\gamma\mu T} \right]\phi = 0, \qquad \text{(C.337)}$$

yielding the coefficients

$$B_{4,1} = 2(\gamma-1)M^2 PrU', \quad B_{4,2} = -\frac{(\gamma-1)\alpha RePr}{\gamma\mu},$$

$$B_{4,3} = 0, \qquad B_{4,4} = \frac{2T'}{\mu}\frac{d\mu}{dT},$$

$$C_{4,1} = -\frac{(\gamma-1)i\alpha RePr}{\gamma\mu},$$

$$C_{4,2} = 2i\alpha^2(\gamma-1)M^2 PrU' - \frac{\alpha RePrT'}{\gamma\mu T},$$

$$C_{4,3} = 0,$$

$$C_{4,4} = -i\frac{\alpha RePr}{\mu\gamma T}(U-c) - \alpha^2 + \frac{(T')^2}{\mu}\frac{d^2\mu}{dT^2} + \frac{T''}{\mu}\frac{d\mu}{dT}$$

$$+(\gamma-1)M^2 Pr\frac{(U')^2}{\mu}\frac{d\mu}{dT}.$$

Exercise 5.6 Compute the elements $a_{i,j}$ of equation (5.107).

Solution
From equation (5.107) we have

$$f' = a_{1,1}f + a_{1,2}f' + a_{1,3}\phi + a_{1,4}\frac{\Pi}{\gamma M^2} + a_{1,5}\theta + a_{1,6}\theta', \qquad \text{(C.338)}$$

$$f'' = a_{2,1}f + a_{2,2}f' + a_{2,3}\phi + a_{2,4}\frac{\Pi}{\gamma M^2} + a_{2,5}\theta + a_{2,6}\theta', \qquad \text{(C.339)}$$

$$\phi' = a_{3,1}f + a_{3,2}f' + a_{3,3}\phi + a_{3,4}\frac{\Pi}{\gamma M^2} + a_{3,5}\theta + a_{3,6}\theta', \qquad \text{(C.340)}$$

$$\left(\frac{\Pi}{\gamma M^2}\right)' = a_{4,1}f + a_{4,2}f' + a_{4,3}\phi + a_{4,4}\frac{\Pi}{\gamma M^2} + a_{4,5}\theta + a_{4,6}\theta', \qquad \text{(C.341)}$$

$$\theta' = a_{5,1}f + a_{5,2}f' + a_{5,3}\phi + a_{5,4}\frac{\Pi}{\gamma M^2} + a_{5,5}\theta + a_{5,6}\theta', \qquad \text{(C.342)}$$

$$\theta'' = a_{6,1}f + a_{6,2}f' + a_{6,3}\phi + a_{6,4}\frac{\Pi}{\gamma M^2} + a_{6,5}\theta + a_{6,6}\theta'. \qquad \text{(C.343)}$$

For (C.338), we have

$$a_{1,2} = 1 \quad a_{1,1} = a_{1,3} = a_{1,4} = a_{1,5} = a_{1,6} = 0.$$

From (C.334) we find the coefficients for (C.339).

$$a_{2,1} = \alpha^2 + \frac{i\alpha Re(U-c)}{\mu T}, \qquad a_{2,2} = -\frac{T'}{\mu}\frac{d\mu}{dT},$$

$$a_{2,3} = -i\alpha^2 \frac{T'}{\mu}\frac{d\mu}{dT} + \frac{\alpha Re U'}{\mu T} - \frac{i\alpha^2 T'}{3T}, \qquad a_{2,4} = \frac{i\alpha Re}{\mu} - \frac{\gamma M^2 \alpha^2(U-c)}{3},$$

$$a_{2,5} = \frac{\alpha^2(U-c)}{3T} - \left[\frac{U''}{\mu}\frac{d\mu}{dT} + \frac{T'U'}{\mu}\frac{d^2\mu}{dT^2}\right], \qquad a_{2,6} = -\frac{U'}{\mu}\frac{d\mu}{dT}.$$

In deriving these coefficients we made use of (C.336).
From (C.336) we find the coefficients for (C.340).

$$a_{3,1} = -i, \quad a_{3,2} = a_{3,6} = 0, \quad a_{3,3} = \frac{T'}{T},$$

$$a_{3,4} = -i\gamma M^2(U-c), \quad a_{3,5} = \frac{i(U-c)}{T}.$$

From (C.335), and using (C.336) to eliminate ϕ'', we find the coefficients for (C.341):

$$a_{4,1} = -\frac{i}{L}\left[\frac{2T'}{\mu}\frac{d\mu}{dT} + \frac{4}{3}\frac{T'}{T}\right], \qquad a_{4,2} = -\frac{i}{L}, \qquad a_{4,6} = \frac{4i(U-c)}{3LT},$$

$$a_{4,3} = \frac{1}{L}\left[-\alpha^2 + \frac{4(T')^2}{3\mu T}\frac{d\mu}{dT} + \frac{4T''}{3T} - \frac{i\alpha Re(U-c)}{\mu T}\right],$$

$$a_{4,4} = -\frac{4i\gamma M^2}{3L}\left[\frac{(U-c)T'}{\mu}\frac{d\mu}{dT} + U' + (U-c)\frac{T'}{T}\right],$$

$$a_{4,5} = \frac{i}{L}\left[\frac{U'}{\mu}\frac{d\mu}{dT} + \frac{4(U-c)T'}{3\mu T}\frac{d\mu}{dT} + \frac{4U'}{3T}\right],$$

where

$$L = \frac{Re}{\alpha\mu} + i\frac{4}{3}\gamma M^2(U-c).$$

For (C.342), we have

$$a_{5,6} = 1, \qquad a_{5,1} = a_{5,2} = a_{5,3} = a_{5,4} = a_{5,5} = 0.$$

From (C.337) we find the coefficients for (C.343) after using (C.336) to eliminate f,

$$a_{6,1} = 0, \qquad a_{6,2} = -2(\gamma-1)M^2 PrU',$$

$$a_{6,3} = \frac{\alpha RePr}{\mu}\frac{T'}{T} - 2i\alpha^2(\gamma-1)M^2 PrU',$$

$$a_{6,4} = -\frac{i\alpha RePr}{\mu}(\gamma-1)M^2(U-c),$$

$$a_{6,5} = \frac{i\alpha RePr}{\mu T}(U-c) + \alpha^2 - \frac{(T')^2}{\mu}\frac{d^2\mu}{dT^2} - \frac{T''}{\mu}\frac{d\mu}{dT} - (\gamma-1)M^2 Pr\frac{(U')^2}{\mu}\frac{d\mu}{dT},$$

$$a_{6,6} = -\frac{2T'}{\mu}\frac{d\mu}{dT}.$$

C.5 Solutions for Chapter 6

Exercise 6.1 Derive Rayleigh's stability condition (6.36) in the inviscid limit. You may assume axisymmetric disturbances, but do not invoke the small gap approximation.

Solution
The dimensional linearized equations of motion (6.13)–(6.16) in cylindrical polar coordinates for an inviscid fluid are given by

$$\frac{\partial \tilde{u}_r}{\partial r} + \frac{\tilde{u}_r}{r} + \frac{1}{r}\frac{\partial \tilde{v}_\theta}{\partial \theta} + \frac{\partial \tilde{w}}{\partial z} = 0,$$

$$\frac{\partial \tilde{u}_r}{\partial t} + \Omega \frac{\partial \tilde{u}_r}{\partial \theta} - 2\Omega \tilde{v}_\theta = -\frac{1}{\rho} \frac{\partial \tilde{p}}{\partial r},$$

$$\frac{\partial \tilde{v}_\theta}{\partial t} + \Omega \frac{\partial \tilde{v}_\theta}{\partial \theta} + \left(\frac{dV}{dr} + \Omega \right) \tilde{u}_r = -\frac{1}{\rho r} \frac{\partial \tilde{p}}{\partial \theta},$$

$$\frac{\partial \tilde{w}}{\partial t} + \Omega \frac{\partial \tilde{w}}{\partial \theta} = -\frac{1}{\rho} \frac{\partial \tilde{p}}{\partial z}, \tag{C.344}$$

where the basic steady solution is given by

$$u_r = w = 0, \quad v_\theta = V(r), \quad p = P(r) \quad \text{and} \quad \Omega(r) = V(r)/r.$$

The no-slip boundary conditions for the inner and outer cylinder yield

$$\tilde{u}_r = \tilde{v}_\theta = \tilde{w} = 0 \quad \text{at} \quad r = R_1 \quad \text{and} \quad r = R_2.$$

For axisymmetric disturbances ($\partial/\partial \theta = 0$), we find from equations (C.344) that the equations reduce to

$$\frac{\partial \tilde{u}_r}{\partial r} + \frac{\tilde{u}_r}{r} + \frac{\partial \tilde{w}}{\partial z} = 0,$$

$$\frac{\partial \tilde{u}_r}{\partial t} - 2\Omega \tilde{v}_\theta = -\frac{1}{\rho} \frac{\partial \tilde{p}}{\partial r},$$

$$\frac{\partial \tilde{v}_\theta}{\partial t} + \left(\frac{dV}{dr} + \Omega \right) \tilde{u}_r = 0,$$

$$\frac{\partial \tilde{w}}{\partial t} = -\frac{1}{\rho} \frac{\partial \tilde{p}}{\partial z}. \tag{C.345}$$

Eliminating $\tilde{v}_\theta, \tilde{w}, \tilde{p}$, we find

$$\frac{\partial^2}{\partial t^2} \left(\frac{\partial^2}{\partial r^2} + \frac{1}{r} \frac{\partial}{\partial r} - \frac{1}{r^2} + \frac{\partial^2}{\partial z^2} \right) \tilde{u}_r + \Phi \frac{\partial^2 \tilde{u}_r}{\partial z^2} = 0, \tag{C.346}$$

where $\Phi = 2\Omega(dV/dr + \Omega)$. Let the normal modes be

$$\tilde{u}_r(r, z, t) = \hat{u}(r) e^{\sigma t + i\lambda z}.$$

Substitute the normal mode into (C.346), we find

$$\left(\frac{d^2}{dr^2} + \frac{1}{r} \frac{d}{dr} - \frac{1}{r^2} - \lambda^2 \right) \hat{u} - \frac{\lambda^2}{\sigma^2} \Phi \hat{u} = 0. \tag{C.347}$$

This second-order linear ordinary differential equation is of the Sturm–Liouville type, where the eigenvalues λ^2/σ^2 are all negative if $\Phi > 0$ positive if $\Phi < 0$ and negative and positive if Φ changes sign. If Φ is negative over some region for a given basic flow, the flow may be unstable according to Rayleigh's criterion; i.e., the gradient of the mean vorticity must change signs within the bounds of the flow.

An alternative way of establishing stability is to first note that equation (C.347) can be rewritten as

$$\frac{d}{dr}\left(\frac{1}{r}\frac{d}{dr}(r\hat{u})\right) - \lambda^2\hat{u} = \frac{\lambda^2}{\sigma^2}\Phi\hat{u}. \tag{C.348}$$

Now multiply by $r\hat{u}^*$ and integrate by parts

$$r\hat{u}^*\frac{1}{r}\frac{d}{dr}(r\hat{u})\Big|_{R_1}^{R_2} - \int_{R_1}^{R_2}\frac{1}{r}\left|\frac{d}{dr}(r\hat{u})\right|^2 dr - \lambda^2\int_{R_1}^{R_2} r|\hat{u}|^2 dr = \frac{\lambda^2}{\sigma^2}\int_{R_1}^{R_2} r\Phi|\hat{u}|^2 dr. \tag{C.349}$$

The first term is zero due to the boundary conditions. The equation then becomes

$$-I_1 - \lambda^2 I_2 = \frac{\lambda^2}{\sigma^2}\int_{R_1}^{R_2} r\Phi|\hat{u}|^2 dr. \tag{C.350}$$

Since the integrals I_1 and I_2 are strictly positive over the domain $[R_1, R_2]$, the left-hand side must be negative. Now if $\Phi > 0$ throughout the domain, the integral on the right-hand side is also positive, and therefore the characteristic values λ^2/σ^2 are all negative. Setting $\lambda^2/\sigma^2 = -s^2$, where s is a positive real number, we see that $\sigma = i\lambda/s$. This implies that the disturbances do not grow in time; i.e., the linear disturbances are stable. On the other hand, if $\Phi < 0$ in some region, then we can have $\lambda^2/\sigma^2 > 0$; i.e., $\sigma^2 > 0$ and $\sigma_r > 0$, which implies instability.

Note that

$$r^3\Phi = \frac{d}{dr}(rV)^2 = \frac{d}{dr}(r^2\Omega)^2, \tag{C.351}$$

so the stability criteria for rotating cylinders requires

$$\frac{d(r^2\Omega)^2}{dr} > 0 \tag{C.352}$$

throughout the domain for stable solutions. This is Rayleigh's stability criteria (6.36).

We can examine the condition $\Phi > 0$ for the Taylor problem. Upon using $V(r) = Ar + B/r$, we see that

$$\Phi = 2\frac{V}{r}\left(\frac{dV}{dr} + \frac{V}{r}\right)$$

$$= 4AB\left(\frac{A}{B} + \frac{1}{r^2}\right)$$

$$= \frac{4R_1^2 R_2^2 (\Omega_2 R_2^2 - \Omega_1 R_1^2)(\Omega_1 - \Omega_2)}{(R_2^2 - R_1^2)^2} \left[\frac{\Omega_2 R_2^2 - \Omega_1 R_1^2}{R_1^2 R_2^2 (\Omega_1 - \Omega_2)} + \frac{1}{r^2} \right]. \quad \text{(C.353)}$$

Now for neutral stability we have $\Phi = 0$, which implies the condition

$$\frac{\Omega_2}{\Omega_1} = \frac{R_1^2}{R_2^2}. \quad \text{(C.354)}$$

This line in the (Ω_2, Ω_1)-plane is sometimes referred to as the Rayleigh line. Its plot is shown in Fig. C.17. To determine the regions of instability or stability, we only need consider the following limits:

1. $\Omega_2 = 0$, $\Omega_1 \neq 0$. In this case we get

$$\Phi = \frac{4R_1^4 \Omega_1^2}{(R_2^2 - R_1^2)^2} \left[1 - \frac{R_2^2}{r^2} \right], \quad \text{(C.355)}$$

   which is negative for $R_1 < r < R_2$. This corresponds to an unstable solution.
2. $\Omega_1 = 0$, $\Omega_2 \neq 0$. In this case we get

$$\Phi = \frac{4R_2^4 \Omega_2^2}{(R_2^2 - R_1^2)^2} \left[1 - \frac{R_1^2}{r^2} \right], \quad \text{(C.356)}$$

   which is positive for $R_1 < r < R_2$. This corresponds to a stable solution.

The regions of stability are shown in Fig. C.17.

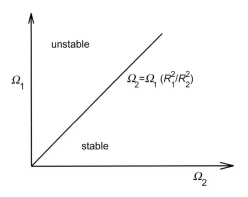

Figure C.17 Regions of stability for inviscid disturbances. The Rayleigh line is given by $\Omega_2 = \Omega_1 (R_1^2/R_2^2)$.

Exercise 6.2 Using asymptotic analysis appropriate for the small gap approximation, derive equations (6.20)–(6.21) from (6.17)–(6.18).

Solution

The dimensional linearized equations of motion (6.17-6.18) in cylindrical polar coordinates for a viscous fluid are given by

$$\left[\frac{\partial}{\partial t} - \nu\left(\nabla^2 - \frac{1}{r^2}\right)\right]\left[\nabla^2\tilde{u}_r - \frac{\tilde{u}_r}{r^2}\right] = 2\frac{V}{r}\frac{\partial^2\tilde{v}_\theta}{\partial z^2},$$

$$\left[\frac{\partial}{\partial t} - \nu\left(\nabla^2 - \frac{1}{r^2}\right)\right]\tilde{v}_\theta = -\left(\frac{dV}{dr} + \frac{V}{r}\right)\tilde{u}_r, \tag{C.357}$$

where

$$\nabla^2 = \frac{\partial^2}{\partial r^2} + \frac{1}{r}\frac{\partial}{\partial r} + \frac{1}{r^2}\frac{\partial^2}{\partial\theta^2} + \frac{\partial^2}{\partial z^2}. \tag{C.358}$$

Assume axisymmetric solutions, and let the normal modes be given by (6.19), i.e.,

$$\tilde{u}_r(r,z,t) = \hat{u}(r)e^{\sigma t + i\lambda z}, \quad \tilde{v}_\theta(r,z,t) = \hat{v}(r)e^{\sigma t + i\lambda z}. \tag{C.359}$$

Substituting into (C.357)

$$\left[\nu\left(\frac{d^2}{dr^2} + \frac{1}{r}\frac{d}{dr} - \frac{1}{r^2} - \lambda^2\right) - \sigma\right]\left[\frac{d^2}{dr^2} + \frac{1}{r}\frac{d}{dr} - \frac{1}{r^2} - \lambda^2\right]\hat{u} = 2\lambda^2\frac{V}{r}\hat{v},$$

$$\left[\nu\left(\frac{d^2}{dr^2} + \frac{1}{r}\frac{d}{dr} - \frac{1}{r^2} - \lambda^2\right) - \sigma\right]\hat{v} = \left(\frac{dV}{dr} + \frac{V}{r}\right)\hat{u}. \tag{C.360}$$

Now for the small gap approximation, we write

$$\delta = R_2 - R_1 \leq \frac{1}{2}(R_1 + R_2), \tag{C.361}$$

and we introduce the scaled coordinate

$$r = R_1 + (R_2 - R_1)\xi$$

$$= R_1 + \delta\xi. \tag{C.362}$$

Note that at $r = R_1$, $\xi = 0$ and, at $r = R_2$, $\xi = 1$. We now make the small gap approximation and carry out an asymptotic expansion in the limit $\delta \ll 1$. The derivatives transform to

$$\frac{d}{dr} = \frac{d}{d\xi}\frac{d\xi}{dr} = \frac{1}{\delta}\frac{d}{d\xi},$$

$$\frac{d^2}{dr^2} = \frac{1}{\delta^2}\frac{d^2}{d\xi^2},$$

$$\frac{1}{r}\frac{d}{dr} = \frac{1}{R_1 + \delta\xi}\frac{1}{\delta}\frac{d}{d\xi} \approx \frac{1}{R_1}\frac{1}{\delta}\frac{d}{d\xi},$$

$$\frac{1}{r^2} = \frac{1}{(R_1 + \delta\xi)^2} \approx \frac{1}{R_1^2},$$ (C.363)

where we have kept only the leading-order terms in the last two expressions. Substituting into (C.360), and multiplying the first equation for u by δ^4 and the second equation for v by δ^2, we get

$$\left[\frac{d^2}{d\xi^2} + \frac{\delta}{R_1}\frac{d}{d\xi} - \frac{\delta^2}{R_1^2} - \delta^2\lambda^2 - \frac{\delta^2\sigma}{\nu}\right]$$

$$\left[\frac{d^2}{d\xi^2} + \frac{\delta}{R_1}\frac{d}{d\xi} - \frac{\delta^2}{R_1^2} - \delta^2\lambda^2\right]\hat{u} = \frac{2\delta^4\lambda^2}{\nu}\left(\frac{V}{r}\right)\hat{v},$$

$$\left[\frac{d^2}{d\xi^2} + \frac{\delta}{R_1}\frac{d}{d\xi} - \frac{\delta^2}{R_1^2} - \delta^2\lambda^2 - \frac{\delta^2\sigma}{\nu}\right]\hat{v} = \frac{\delta^2}{\nu}\left(\frac{dV}{dr} + \frac{V}{r}\right)\hat{u}. \quad \text{(C.364)}$$

In order to keep the wavenumber λ and frequency σ in the equations, we introduce the scalings

$$\hat{\lambda} = \delta\lambda, \quad \hat{\sigma} = \frac{\delta^2\sigma}{\nu}.$$ (C.365)

Note that we also scaled the frequency by ν; if we were to consider inviscid disturbances we would set $\nu = 0$ in (C.360) and scale the frequency by δ^2. Dropping all $O(\delta)$ terms and higher, we get

$$\left[\frac{d^2}{d\xi^2} - \hat{\lambda}^2 - \hat{\sigma}\right]\left[\frac{d^2}{d\xi^2} - \hat{\lambda}^2\right]\hat{u} = \frac{2\delta^2\hat{\lambda}^2}{\nu}\left(\frac{V}{r}\right)\hat{v},$$

$$\left[\frac{d^2}{d\xi^2} - \hat{\lambda}^2 - \hat{\sigma}\right]\hat{v} = \frac{\delta^2}{\nu}\left(\frac{dV}{dr} + \frac{V}{r}\right)\hat{u}.$$ (C.366)

These equations are identical to that of (6.21)–(6.22), suitably interpreted.

Now, recall for the Taylor–Couette problem we have for the mean flow

$$V = Ar + \frac{B}{r},$$ (C.367)

and note that

$$\frac{dV}{dr} + \frac{V}{r} = 2A.$$ (C.368)

The constants A and B can be expanded as

$$A = \frac{\Omega_2 R_2^2 - \Omega_1 R_1^2}{R_2^2 - R_1^2} \approx -\frac{R_2(\Omega_1 - \Omega_2)}{2\delta} - \frac{1}{4}(\Omega_1 - \Omega_2) + \Omega_1,$$ (C.369)

$$B = \frac{R_1^2 R_2^2(\Omega_1 - \Omega_2)}{R_2^2 - R_1^2} \approx \frac{R_2^3(\Omega_1 - \Omega_2)}{2\delta} + \frac{R_2^2}{4}(\Omega_1 - \Omega_2) - R_2^2(\Omega_1 - \Omega_2),$$

(C.370)

where we kept the first two terms in the expansion. Also note that we replaced R_1 with $R_1 = R_2 - \delta$ in the expansion so that all terms now involve R_2 only. In addition we expand

$$\frac{1}{r^2} = \frac{1}{(R_2 + (\xi - 1)\delta)^2} \approx \frac{1}{R_2^2}\left[1 - \frac{2(\xi - 1)\delta}{R_2}\right] \tag{C.371}$$

so that the expression V/r on the right-hand side of C.366(a) becomes

$$\frac{V}{r} = A + \frac{B}{r^2} \approx \Omega_1[1 - (1 - \mu)\xi], \quad \text{where } \mu = \Omega_2/\Omega_1 \tag{C.372}$$

to leading order in δ. (The reader should go through the algebra to verify this.) The equations (C.366) becomes

$$\left[\frac{d^2}{d\xi^2} - \hat{\lambda}^2 - \hat{\sigma}\right]\left[\frac{d^2}{d\xi^2} - \hat{\lambda}^2\right]\hat{u} = \frac{2\delta^2\hat{\lambda}^2}{v}\Omega_1[1 - (1 - \mu)\xi]\hat{v},$$

$$\left[\frac{d^2}{d\xi^2} - \hat{\lambda}^2 - \hat{\sigma}\right]\hat{v} = \frac{2A\delta^2}{v}\hat{u}. \tag{C.373}$$

It is convenient to make the transformation

$$\hat{u} \to \frac{2\delta^2\hat{\lambda}^2\Omega_1}{v}\hat{u} \tag{C.374}$$

and the final set of equations becomes

$$\left[\frac{d^2}{d\xi^2} - \hat{\lambda}^2 - \hat{\sigma}\right]\left[\frac{d^2}{d\xi^2} - \hat{\lambda}^2\right]\hat{u} = [1 - (1 - \mu)\xi]\hat{v},$$

$$\left[\frac{d^2}{d\xi^2} - \hat{\lambda}^2 - \hat{\sigma}\right]\hat{v} = -Ta\hat{\lambda}^2\hat{u}, \tag{C.375}$$

where Ta is the Taylor number, defined as

$$Ta = -\frac{4A\Omega_1}{v^2}\delta^4. \tag{C.376}$$

Alternatively, one could have used the transformation

$$\frac{2A\delta^2}{v}\hat{u} \to \hat{u}, \tag{C.377}$$

to get

$$\left[\frac{d^2}{d\xi^2} - \hat{\lambda}^2 - \hat{\sigma}\right]\left[\frac{d^2}{d\xi^2} - \hat{\lambda}^2\right]\hat{u} = -Ta\hat{\lambda}^2[1 - (1 - \mu)\xi]\hat{v},$$

$$\left[\frac{d^2}{d\xi^2} - \hat{\lambda}^2 - \hat{\sigma}\right]\hat{v} = \hat{u}. \tag{C.378}$$

Either form is appropriate.

Exercise 6.3 Compute the inviscid solutions for the Taylor problem in the small gap approximations and for $\Omega_1 = \Omega_2$. Show that the solutions are stable.

Solution
We start with the dimensional system (C.360) and set $v = 0$, to get a single equation for $\hat{u}$

$$\left[\frac{d^2}{dr^2} + \frac{1}{r}\frac{d}{dr} - \frac{1}{r^2} - \lambda^2\right]\hat{u} = \frac{4A\lambda^2}{\sigma^2}\left(\frac{V}{r}\right)\hat{u}. \qquad (C.379)$$

We now make the small gap approximation (see Exercise 6.2), and note that for $\Omega_1 = \Omega_2$, $A = \Omega_1$ and $V/r \approx \Omega_1$ to leading order, and we get

$$\left[\frac{d^2}{d\xi^2} - \lambda^2\right]\hat{u} = \frac{4\Omega_1^2\lambda^2}{\sigma^2}\hat{u}, \qquad (C.380)$$

subject to the boundary conditions $u = 0$ at $\xi = 0$ and $\xi = 1$. The solutions are given by

$$\hat{u} = C\sin(n\pi\xi), \qquad (C.381)$$

where C and n are integers. Substitution yields for the eigenvalue

$$\sigma^2 = \frac{-4\Omega_1^2}{1 + n^2\pi^2/\hat{\lambda}^2} \qquad (C.382)$$

or

$$\sigma = \frac{2i\Omega_1}{\sqrt{1 + n^2\pi^2/\hat{\lambda}^2}}. \qquad (C.383)$$

Note that the solution is stable.

Exercise 6.5 Determine the governing equations for the perturbations for the rotating disk and indicate the terms that are due to viscosity, rotation and streamline curvature.

Solution
We begin with the Navier–Stokes equations for an incompressible fluid, augmented on the right-hand side by the Coriolis term

$$-\frac{2}{r}\underline{\Omega} \times \underline{u} \equiv \frac{2}{r}(\Omega^*v, -\Omega^*u, 0), \qquad (C.384)$$

where we have assumed the angular frequency is a constant and acts in the z-direction. The augmented dimensional equations are

$$\frac{1}{r}\frac{\partial(ru)}{\partial r} + \frac{1}{r}\frac{\partial v}{\partial\theta} + \frac{\partial w}{\partial z} = 0,$$

$$\frac{\partial u}{\partial t}+u\frac{\partial u}{\partial r}+\frac{v}{r}\frac{\partial u}{\partial \theta}+w\frac{\partial u}{\partial z}-\frac{v^2}{r}-\frac{2\Omega^* v}{r}=-\frac{1}{\rho}\frac{\partial p}{\partial r}+\nu\left(\nabla^2 u-\frac{u}{r^2}-\frac{2}{r^2}\frac{\partial v}{\partial \theta}\right),$$

$$\frac{\partial v}{\partial t}+u\frac{\partial v}{\partial r}+\frac{v}{r}\frac{\partial v}{\partial \theta}+w\frac{\partial v}{\partial z}+\frac{uv}{r}+\frac{2\Omega^* u}{r}=-\frac{1}{r\rho}\frac{\partial p}{\partial \theta}+\nu\left(\nabla^2 v-\frac{v}{r^2}+\frac{2}{r^2}\frac{\partial u}{\partial \theta}\right),$$

$$\frac{\partial w}{\partial t}+u\frac{\partial w}{\partial r}+\frac{v}{r}\frac{\partial w}{\partial \theta}+w\frac{\partial w}{\partial z}=-\frac{1}{\rho}\frac{\partial p}{\partial z}+\nu\nabla^2 w,$$

where

$$\nabla^2=\frac{\partial^2}{\partial r^2}+\frac{1}{r}\frac{\partial}{\partial r}+\frac{1}{r^2}\frac{\partial^2}{\partial \theta^2}+\frac{\partial^2}{\partial z^2}.$$

To nondimensionalize the equations, we use L^*, $r_a^*\Omega^*$, $\rho^* r_a^{*2}\Omega^{*2}$ to define the length, velocity and pressure scales, respectively. Here Ω^* is the constant angular frequency and $L^*=\sqrt{\nu^*/\Omega^*}$. The Reynolds number is then defined as $Re=r_a^*\Omega^* L^*/\nu^*$. The nondimensional equations are given by

$$\frac{1}{r}\frac{\partial(ru)}{\partial r}+\frac{1}{r}\frac{\partial v}{\partial \theta}+\frac{\partial w}{\partial z}=0,$$

$$\frac{\partial u}{\partial t}+u\frac{\partial u}{\partial r}+\frac{v}{r}\frac{\partial u}{\partial \theta}+w\frac{\partial u}{\partial z}-\frac{v^2}{r}-\frac{2v}{r}=-\frac{\partial p}{\partial r}+Re^{-1}\left(\nabla^2 u-\frac{u}{r^2}-\frac{2}{r^2}\frac{\partial v}{\partial \theta}\right),$$

$$\frac{\partial v}{\partial t}+u\frac{\partial v}{\partial r}+\frac{v}{r}\frac{\partial v}{\partial \theta}+w\frac{\partial v}{\partial z}+\frac{uv}{r}+\frac{2u}{r}=-\frac{1}{r}\frac{\partial p}{\partial \theta}+Re^{-1}\left(\nabla^2 v-\frac{v}{r^2}+\frac{2}{r^2}\frac{\partial u}{\partial \theta}\right),$$

$$\frac{\partial w}{\partial t}+u\frac{\partial w}{\partial r}+\frac{v}{r}\frac{\partial w}{\partial \theta}+w\frac{\partial w}{\partial z}=-\frac{\partial p}{\partial z}+Re^{-1}\nabla^2 w.$$

To derive the linear stability equations we write

$$u=\frac{r}{Re}U(z)+\hat{u}(z)e^{i(\alpha r+\beta\theta-\omega t)},$$

$$v=\frac{r}{Re}V(z)+\hat{v}(z)e^{i(\alpha r+\beta\theta-\omega t)},$$

$$w=\frac{1}{Re}W(z)+\hat{w}(z)e^{i(\alpha r+\beta\theta-\omega t)},$$

$$p=\frac{1}{Re^2}P(z)+\hat{p}(z)e^{i(\alpha r+\beta\theta-\omega t)}. \tag{C.385}$$

Lingwood (1995) notes that in order to make the linearized perturbation equations separable in r, θ and t, it is necessary to ignore variations in Reynolds number with radius. This involves replacing r by Re in the linearized equations, and is usually referred to as the parallel flow approximation. Substituting into the nondimensional equations we get, after some rearrangement,

$$\frac{d\hat{w}}{dz} = -i[(\alpha - i/Re)\hat{u} + \overline{\beta}\hat{v}], \tag{C.386}$$

$$\frac{1}{Re}\frac{d^2\hat{u}}{dz^2} = \frac{1}{Re}(\alpha^2 + \overline{\beta}^2)\hat{u} + i\alpha\hat{p} + i(\alpha U + \overline{\beta} V - \omega)\hat{u}$$
$$+ \frac{U}{Re}\hat{u} + \frac{W}{Re}\frac{d\hat{u}}{dz} + \frac{dU}{dz}\hat{w} - \frac{2V}{Re}\hat{v} - \frac{2\hat{v}}{Re}, \tag{C.387}$$

$$\frac{1}{Re}\frac{d^2\hat{v}}{dz^2} = \frac{1}{Re}(\alpha^2 + \overline{\beta}^2)\hat{v} + i\overline{\beta}\hat{p} + i(\alpha U + \overline{\beta} V - \omega)\hat{v}$$
$$+ \frac{U}{Re}\hat{v} + \frac{W}{Re}\frac{d\hat{v}}{dz} + \frac{dV}{dz}\hat{w} + \frac{2V}{Re}\hat{u} + \frac{2\hat{u}}{Re}, \tag{C.388}$$

$$\frac{d\hat{p}}{dz} = -i(\alpha U + \overline{\beta} V - \omega)\hat{w} - \frac{W}{Re}\frac{d\hat{w}}{dz}$$
$$- \frac{1}{Re}\frac{dW}{dz}\hat{w} + \frac{1}{Re}\left[\frac{d^2\hat{w}}{dz^2} - (\alpha^2 + \overline{\beta}^2)\hat{w}\right], \tag{C.389}$$

where $\overline{\beta} = \beta/Re$. Following Lingwood (1995), we can write the system as six first-order equations by defining

$$z_1 = (\alpha - i/Re)\hat{u} + \overline{\beta}\hat{v}, \ z_2 = (\alpha - i/Re)\frac{d\hat{u}}{dz} + \overline{\beta}\frac{d\hat{v}}{dz},$$

$$z_3 = \hat{w}, \ z_4 = \hat{p},$$

$$z_5 = (\alpha - i/Re)\hat{v} - \overline{\beta}\hat{u}, \ z_6 = (\alpha - i/Re)\frac{d\hat{v}}{dz} - \overline{\beta}\frac{d\hat{u}}{dz}. \tag{C.390}$$

The form of z_1 is motivated by the continuity equation (C.386). It is now an easy matter to derive the following first-order equations

$$\frac{dz_1}{dz} = z_2, \tag{C.391}$$

$$\frac{1}{Re}\frac{dz_2}{dz} = \frac{1}{Re}\left[(\alpha^2 + \overline{\beta}^2) + iRe(\alpha U + \overline{\beta} V - \omega) + U\right]z_1 + \frac{W z_2}{Re} \tag{C.392}$$
$$+ \left[(\alpha - i/Re)\frac{dU}{dz} + \overline{\beta}\frac{dV}{dz}\right]z_3 + i\left[\alpha^2 + \overline{\beta}^2 - \frac{i\alpha}{Re}\right]z_4 - \frac{2(1+V)z_5}{Re},$$

$$\frac{dz_3}{dz} = -iz_1, \tag{C.393}$$

$$\frac{dz_4}{dz} = \frac{iWz_1}{Re} - \frac{iz_2}{Re} - \frac{1}{Re}\left[(\alpha^2 + \overline{\beta}^2) + iRe(\alpha U + \overline{\beta}V - \omega) + \frac{dW}{dz}\right]z_3,$$

(C.394)

$$\frac{dz_5}{dz} = z_6,$$

(C.395)

$$\frac{1}{Re}\frac{dz_6}{dz} = \frac{2(1+V)z_1}{Re} + \left[(\alpha - i/Re)\frac{dV}{dz} - \overline{\beta}\frac{dU}{dz}\right]z_3 + \frac{\overline{\beta}z_4}{Re}$$

$$+ \frac{1}{Re}\left[(\alpha^2 + \overline{\beta}^2) + iRe(\alpha U + \overline{\beta}V - \omega) + U\right]z_5 + \frac{Wz_6}{Re}.$$

(C.396)

Note that to derive equation (C.392), we multiply (C.387) by $(\alpha - i/Re)$, then add and subtract $(\overline{\beta}/Re)d^2\hat{v}/dz^2$ and use (C.388) to eliminate the term $-(\overline{\beta}/Re)d^2\hat{v}/dz^2$. Similarly, to derive (C.396), we multiply (C.388) by $(\alpha - i/Re)$, then add and subtract $(\overline{\beta}/Re)d^2\hat{u}/dz^2$ and use (C.387) to eliminate $(\overline{\beta}/Re)d^2\hat{u}/dz^2$. Finally, (C.394) is found by substituting (C.386) to eliminate derivatives in $\hat{w}$. These equations are identical to those of Lingwood (1995).

Exercise 6.6 Determine the linear equations for the trailing vortex. Then examine the energy equation and that for vorticity.

Solution
We follow Khorrami (1991) and write for the mean flow

$$U = U(r), \quad V = \frac{q}{r}\left(1 - e^{-r^2}\right), \quad W = W_\infty + e^{-r^2}.$$

(C.397)

Now let

$$u = U(r) + iF(r)e^{i(\alpha z + n\theta - \omega t)}, \qquad v = V(r) + G(r)e^{i(\alpha z + n\theta - \omega t)},$$

$$w = W(r) + H(r)e^{i(\alpha z + n\theta - \omega t)}, \qquad p = P_0 + P(r)e^{i(\alpha z + n\theta - \omega t)},$$

(C.398)

where n is the azimuthal wavenumber, being zero or an integer; α the axial wavenumber; and ω the frequency. For temporal stability, α is real and ω is complex. Substituting into (6.8)–(6.11) yields

$$\frac{dF}{dr} + \frac{F}{r} + \frac{nG}{r} + \alpha H = 0,$$

$$\frac{-i}{Re}\frac{d^2F}{dr^2} + i\left[U - \frac{1}{rRe}\right]\frac{dF}{dr}$$

$$+ \left[\omega + i\frac{dU}{dr} - \frac{nV}{r} - \alpha W + \frac{i}{Re}\left(\frac{n^2+1}{r^2} + \alpha^2\right)\right]F$$

$$+ \left[\frac{i2n}{r^2Re} - \frac{2V}{r}\right]G + \frac{dP}{dr} = 0,$$

$$\frac{-1}{Re}\frac{d^2G}{dr^2} + \left[U - \frac{1}{rRe}\right]\frac{dG}{dr}$$

$$+ \left[-i\omega + \frac{inV}{r} + i\alpha W + \frac{U}{r} + \frac{1}{Re}\left(\frac{n^2+1}{r^2} + \alpha^2\right)\right]G$$

$$+ \left[i\frac{dV}{dr} + \frac{iV}{r} + \frac{2n}{r^2 Re}\right]F + \frac{inP}{r} = 0,$$

$$\frac{-1}{Re}\frac{d^2H}{dr^2} + \left[U - \frac{1}{rRe}\right]\frac{dH}{dr} + \left[-i\omega + \frac{inV}{r} + i\alpha W + \frac{1}{Re}\left(\frac{n^2}{r^2} + \alpha^2\right)\right]H$$

$$+ i\frac{dW}{dr}F + i\alpha P = 0.$$

These equations must satisfy six boundary conditions. These are

$$F(\infty) = G(\infty) = H(\infty) = 0, \tag{C.399}$$

and at the axis of the vortex, $r = 0$, we have

if $n = 0$ $F(0) = G(0) = 0$, $H'(0) = 0$,

if $n = \pm 1$ $F(0) \pm G(0) = 0$, $F'(0) = 0$, $H(0) = 0$,   (C.400)

if $|n| > 1$ $F(0) = G(0) = H(0) = 0$.

Exercise 6.7 Determine the solution for the perturbations in pipe flow in the inviscid limit.

Solution
We start with equation (6.48). First note that for Poiseuille flow $U(r) = 1 - r^2$, so that

$$\left(\frac{U'}{r}\right)' = 0. \tag{C.401}$$

Then the inviscid equation becomes

$$(U - c)(D^2 - \alpha^2)\hat{\psi} = 0, \tag{C.402}$$

where $\omega = \alpha c$, with c the complex wave speed. Note that $(U - c) = 0$ defines the critical layer so that in the limit $\alpha Re \to +\infty$ we have $c = 1$ to leading order in an asymptotic expansion in αRe. The inviscid solutions are

$$\hat{\psi}_1 = rI_1(\alpha r), \hat{\psi}_2 = rK_1(\alpha r), \tag{C.403}$$

where I_1 and K_1 are the modified Bessel functions. Since K_1 becomes unbounded as $r \to 0$, the inviscid solution is just

$$\psi = ArI_1(\alpha r). \tag{C.404}$$

Exercise 6.8 For the round jet answer the following.

(a) Starting with equation (6.58), derive Rayleigh's inflection point theorem.

(b) Starting with equation (6.58), and introducing the transformation $g(r) = r\hat{v}/(U - c)$, derive Howard's semicircle theorem.

(c) Derive the linearized equations (6.60)–(6.63).

Solution

(a) To derive Rayleigh's inflection point theorem, we start with equation (6.58). We first write $\omega = \alpha c$, divide by $(U - c)$ and then multiply by $r\hat{v}^*$, to get

$$\frac{d}{dr}\left[\frac{r(r\hat{v})'}{\alpha^2 r^2 + n^2}\right] r\hat{v}^* - r|\hat{v}|^2 - \frac{d}{dr}\left[\frac{rU'}{\alpha^2 r^2 + n^2}\right]\frac{|r\hat{v}|^2}{|U - c|^2}(U - c^*) = 0. \quad \text{(C.405)}$$

Now take the complex conjugate of the whole equation

$$\frac{d}{dr}\left[\frac{r(r\hat{v}^*)'}{\alpha^2 r^2 + n^2}\right] r\hat{v} - r|\hat{v}|^2 - \frac{d}{dr}\left[\frac{rU'}{\alpha^2 r^2 + n^2}\right]\frac{|r\hat{v}|^2}{|U - c|^2}(U - c) = 0. \quad \text{(C.406)}$$

Subtracting gives

$$\frac{d}{dr}\left[\frac{r(r\hat{v})'}{\alpha^2 r^2 + n^2}\right] r\hat{v}^* - \frac{d}{dr}\left[\frac{r(r\hat{v}^*)'}{\alpha^2 r^2 + n^2}\right] r\hat{v}$$

$$-2ic_i\frac{d}{dr}\left[\frac{rU'}{\alpha^2 r^2 + n^2}\right]\frac{|r\hat{v}|^2}{|U - c|^2} = 0. \quad \text{(C.407)}$$

We now integrate over $(0, \infty)$, integrate by parts the first two terms of (C.407), and note that the result is zero (the reader should show this). This all yields

$$-2ic_i\frac{d}{dr}\left[\frac{rU'}{\alpha^2 r^2 + n^2}\right]\frac{|r\hat{v}|^2}{|U - c|^2} = 0. \quad \text{(C.408)}$$

Thus, for instability, $c_i \neq 0$, which implies

$$\frac{d}{dr}\left[\frac{rU'}{\alpha^2 r^2 + n^2}\right] \quad \text{(C.409)}$$

and it must change sign on the interval $(0, \infty)$. This is Rayleigh's inflection point theorem for round jets.

(b) We start by introducing the transformation

$$g(r) = \frac{r\hat{v}}{U - c} \quad \text{(C.410)}$$

so that equation (6.58) becomes

$$\frac{d}{dr}\left[\frac{r}{\alpha^2 r^2 + n^2}(U - c)^2 g'\right] - \frac{1}{r}(U - c)^2 g = 0. \quad \text{(C.411)}$$

We now multiply by $g*$ and integrate by parts to get

$$\int_0^\infty (U-c)^2 \Phi dr = 0, \tag{C.412}$$

where

$$\Phi = \frac{r}{\alpha^2 r^2 + n^2}|g'|^2 + \frac{1}{r}|g|^2, \tag{C.413}$$

which is a real positive function over $(0, \infty)$. We now separate (C.412) into real and imaginary parts. We first note that

$$(U-c)^2 = U^2 + c_r^2 - c_i^2 - 2Uc_r - 2ic_i(U - c_r). \tag{C.414}$$

For $c_i \neq 0$, the imaginary part of (C.412) can be written as

$$\int_0^\infty (U - c_r)\Phi dr = 0. \tag{C.415}$$

Note that $U - c_r$ must change sign; i.e., $U_{\min} < c_r < U_{\max}$. The real part of (C.412) becomes

$$\int_0^\infty U^2 \Phi dr = \int_0^\infty (2c_r U - c_r^2 + c_i^2)\Phi dr$$

$$= (c_r^2 + c_i^2)\int_0^\infty \Phi dr. \tag{C.416}$$

Now suppose $U_{\min} \leq U(r) \leq U_{\max}$; then

$$(U - U_{\min})(U - U_{\max}) \leq 0, \tag{C.417}$$

since the first term is strictly positive while the second term is strictly negative. We now multiple the inequality by Φ and integrate to get

$$0 \geq \int_0^\infty (U - U_{\min})(U - U_{\max})\Phi dr$$

$$= \int_0^\infty (U^2 - U(U_{\min} + U_{\max}) + U_{\min}U_{\max})\Phi dr$$

$$= (c_r^2 + c_i^2 - c_r(U_{\min} + U_{\max}) + U_{\min}U_{\max})\int_0^\infty \Phi dr$$

$$= \left[\left(c_r - \frac{1}{2}(U_{\min} + U_{\max})\right)^2 + c_i^2 - \frac{1}{4}(U_{\min} - U_{\max})^2\right]\int_0^\infty \Phi dr, \tag{C.418}$$

where we used the real and imaginary parts to eliminate U from consideration.

The last result implies

$$\left(c_r - \frac{U_{\min} + U_{\max}}{2}\right)^2 + c_i^2 \le \frac{1}{4}(U_{\min} - U_{\max})^2 . \tag{C.419}$$

Therefore, the complex velocity must lie within the semicircle, thus proving Howard's semicircle theorem.

(c) To derive the linearized equations (6.60)–(6.63), we start with the inviscid form of equations (6.40)–(6.43)

$$\left.\begin{array}{rcl} \dfrac{1}{r}\dfrac{\partial(r\tilde{v})}{\partial r} + \dfrac{1}{r}\dfrac{\partial \tilde{w}}{\partial \theta} + \dfrac{\partial \tilde{u}}{\partial x} &=& 0, \\[2mm] \dfrac{\partial \tilde{v}}{\partial t} + U\dfrac{\partial \tilde{v}}{\partial x} &=& -\dfrac{\partial \tilde{p}}{\partial r}, \\[2mm] \dfrac{\partial \tilde{w}}{\partial t} + U\dfrac{\partial \tilde{w}}{\partial x} &=& -\dfrac{1}{r}\dfrac{\partial \tilde{p}}{\partial \theta}, \\[2mm] \dfrac{\partial \tilde{u}}{\partial t} + U\dfrac{\partial \tilde{u}}{\partial x} + \tilde{v}\dfrac{dU}{dr} &=& -\dfrac{\partial \tilde{p}}{\partial x}. \end{array}\right\} \tag{C.420}$$

We let $\alpha\xi = \alpha x + n\theta$, so that

$$\frac{\partial}{\partial x} = \frac{\partial}{\partial \xi}, \qquad \frac{\partial}{\partial \theta} = \frac{n}{\alpha}\frac{\partial}{\partial \xi}. \tag{C.421}$$

The equations transform to

$$\left.\begin{array}{rcl} \dfrac{1}{r}\dfrac{\partial(r\tilde{v})}{\partial r} + \dfrac{n}{\alpha r}\dfrac{\partial \tilde{w}}{\partial \xi} + \dfrac{\partial \tilde{u}}{\partial \xi} &=& 0, \\[2mm] \dfrac{\partial \tilde{v}}{\partial t} + U\dfrac{\partial \tilde{v}}{\partial \xi} &=& -\dfrac{\partial \tilde{p}}{\partial r}, \\[2mm] \dfrac{\partial \tilde{w}}{\partial t} + U\dfrac{\partial \tilde{w}}{\partial \xi} &=& -\dfrac{n}{\alpha r}\dfrac{\partial \tilde{p}}{\partial \xi}, \\[2mm] \dfrac{\partial \tilde{u}}{\partial t} + U\dfrac{\partial \tilde{u}}{\partial \xi} + \tilde{v}\dfrac{dU}{dr} &=& -\dfrac{\partial \tilde{p}}{\partial \xi}. \end{array}\right\} \tag{C.422}$$

Note that the equations are now two-dimensional. We decompose into normal modes

$$(\tilde{u}, \tilde{v}, \tilde{w}, \tilde{p})(r, \xi, t) = (\hat{u}, \hat{v}, \hat{w}, \hat{p})(r)e^{i\alpha\xi - i\omega t} \tag{C.423}$$

to get

$$\left.\begin{array}{rcl} \dfrac{1}{r}\dfrac{d(r\hat{v})}{dr} + i\left(\alpha\hat{u} + \dfrac{n}{r}\hat{w}\right) &=& 0, \\[2mm] i\alpha(U - c)\hat{v} &=& -\hat{p}', \\[2mm] i\alpha(U - c)\hat{w} &=& -\dfrac{in}{r}\hat{p}, \\[2mm] i\alpha(U - c)\hat{u} + \hat{v}U' &=& -i\alpha\hat{p}. \end{array}\right\} \tag{C.424}$$

We now define the following

$$
\begin{aligned}
\overline{\alpha u} &= \alpha \hat{u} + \frac{n\hat{w}}{r}, \\
\overline{p/\alpha} &= \hat{p}/\alpha, \\
\overline{v} &= \hat{v}, \\
\overline{\alpha w} &= \alpha \hat{w} - \frac{n\hat{u}}{r},
\end{aligned}
\right\}
\tag{C.425}
$$

where

$$
\overline{\alpha} = \sqrt{\alpha^2 + n^2/r^2}.
\tag{C.426}
$$

Note that the first term is motivated by equation (C.424a). Substitution yields the desired result, namely equations (6.60)–(6.63).

C.6 Solutions for Chapter 7

Exercise 7.2 Investigate the stability of the flow represented in Fig. 7.13.

Solution
We follow Chandrasekhar (1981). We consider two-dimensional disturbances so that $k_1 = k$. The solution to Rayleigh's equation (2.32) with $U'' = 0$, in terms of the perturbation variable ϕ, is given by

$$
\phi = \begin{cases}
Ae^{-kz} & z > +1, \\
A_0 e^{-kz} - B_0 e^{kz} & -1 < z < +1, \\
-Be^{kz} & z < -1.
\end{cases}
\tag{C.427}
$$

The generalized jump conditions at $z = \pm 1$ are

$$
\begin{aligned}
\left[\frac{\phi'}{kU - \omega}\right] &= 0, \\
[\rho(kU - \omega)\phi'' - \rho kU'\phi'] &= (gk^2[\rho] - k^4 T)\frac{\phi'}{kU - \omega},
\end{aligned}
\right\}
\tag{C.428}
$$

where the right-hand side of (C.428b) is to be evaluated at the interface, and $[\ ]$ denotes the jump across the interface. In what follows we ignore surface tension effects. Upon using (C.428a), and noting that U is continuous at the interfaces, we get

$$
\begin{aligned}
Ae^{-k} &= A_0 e^{-k} + B_0 e^k, \\
Be^{-k} &= A_0 e^k + B_0 e^{-k}.
\end{aligned}
\right\}
\tag{C.429}
$$

Now using the second interface condition (C.428b), and making use of (C.429), we get

$$(1-\varepsilon)(k-\omega)^2\left(A_0e^{-k}+B_0e^k\right)-(k-\omega)^2\left(A_0e^{-k}-B_0e^k\right)$$
$$-(k-\omega)\left(A_0e^{-k}+B_0e^k\right)=gk\varepsilon\left(A_0e^{-k}+B_0e^k\right),\qquad\text{(C.430)}$$
$$(1+\varepsilon)(k+\omega)^2\left(A_0e^k+B_0e^{-k}\right)=-(k+\omega)^2\left(A_0e^k-B_0e^{-k}\right)$$
$$+(k+\omega)\left(A_0e^k+B_0e^{-k}\right)+gk\varepsilon\left(A_0e^k+B_0e^{-k}\right).\qquad\text{(C.431)}$$

These equations can be rewritten as

$$-\frac{A_0}{B_0}e^{-2k}=\frac{\varepsilon[(k-\omega)^2+gk]+(k-\omega)-2(k-\omega)^2}{\varepsilon[(k-\omega)^2+gk]+(k-\omega)},\qquad\text{(C.432)}$$

$$-\frac{B_0}{A_0}e^{-2k}=\frac{\varepsilon[(k+\omega)^2-gk]-(k+\omega)+2(k+\omega)^2}{\varepsilon[(k+\omega)^2-gk]-(k+\omega)}\qquad\text{(C.433)}$$

or, equivalently,

$$-\frac{A_0}{B_0}e^{-2k}=1-\frac{2(k-\omega)^2}{\varepsilon[(k-\omega)^2+gk]+(k-\omega)},\qquad\text{(C.434)}$$

$$-\frac{B_0}{A_0}e^{-2k}=1+\frac{2(k+\omega)^2}{\varepsilon[(k+\omega)^2-gk]-(k+\omega)}.\qquad\text{(C.435)}$$

Now define $J=\varepsilon g$ as the Richardson number for this problem. In dimensional units,

$$J=\varepsilon g=-\frac{g^*}{\rho_0}\frac{[\rho]/2d}{(dU/dz)^2},\qquad\text{(C.436)}$$

where $[\rho]$ is the jump across the transition layer, dU/dz is the local shear in the transition layer, g^* is the dimensional gravity parameter and $2d$ is the width of the transition layer.

Now define $v=-\omega/k$ and $\kappa=2k$, then the expressions (C.434) and (C.435) become

$$\left.\begin{aligned}-\frac{A_0}{B_0}e^{-\kappa}&=1-\frac{\kappa(v+1)^2}{J+1+v+\frac{1}{2}\varepsilon\kappa(v+1)^2},\\[2mm]-\frac{B_0}{A_0}e^{-\kappa}&=1-\frac{\kappa(v-1)^2}{J-v+1-\frac{1}{2}\varepsilon\kappa(v-1)^2}.\end{aligned}\right\}\qquad\text{(C.437)}$$

Eliminating A_0/B_0 we get

$$e^{-2\kappa}=\left[1-\frac{\kappa(v+1)^2}{J+1+v+\frac{1}{2}\varepsilon\kappa(v+1)^2}\right]\left[1-\frac{\kappa(v-1)^2}{J-v+1-\frac{1}{2}\varepsilon\kappa(v-1)^2}\right].$$
$$\text{(C.438)}$$

This is the dispersion relation that determines $\omega = \omega(k, g, \varepsilon)$, and the stability characteristics can be determined numerically. However, we can make some analytical progress if we assume uniform density, so that we take the limit $\varepsilon \to 0$ while keeping $J = O(1)$. Then the dispersion relation reduces to

$$\left[(J+1)^2 - v^2\right] e^{-2\kappa} = [J + v + 1 - \kappa(v+1)^2][J - v + 1 - \kappa(v-1)^2]. \tag{C.439}$$

This is a quadratic equation in v^2, namely,

$$\kappa^2 v^4 - (2J\kappa + 2\kappa^2 - 2\kappa + 1 - e^{-2\kappa})v^2$$
$$+ [(J+1)(1 - e^{-\kappa}) - \kappa][(J+1)(1 + e^{-\kappa}) - \kappa] = 0. \tag{C.440}$$

It can be shown (Chandrasekhar, 1981) that for instability we must have

$$\frac{\kappa}{1 + e^{-\kappa}} < J + 1 < \frac{\kappa}{1 - e^{-\kappa}}. \tag{C.441}$$

The regions of instability are plotted in Fig. C.18.

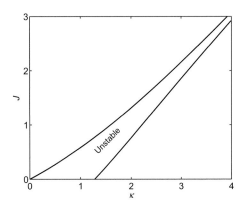

Figure C.18 Plot of instability region; after Chandrasekhar (1981).

Exercise 7.3 Derive equation (7.14).

Solution
We start with the linearized disturbance equations for mass and momentum (7.9)–(7.12)

$$\tilde{u}_x + \tilde{v}_y + \tilde{w}_z = 0, \tag{C.442}$$

$$\tilde{u}_t + U\tilde{u}_x + U'\tilde{w} = -\tilde{p}_x/\rho_0, \tag{C.443}$$

$$\tilde{v}_t + U\tilde{v}_x = -\tilde{p}_y/\rho_0, \tag{C.444}$$

$$\tilde{w}_t + U\tilde{w}_x = -\tilde{p}_z/\rho_0 - g\tilde{\rho}/\rho_0. \tag{C.445}$$

If you take the divergence of the momentum equations, you find

$$\tilde{u}_{xt} + U\tilde{u}_{xx} + U'\tilde{w}_x = -\tilde{p}_{xx}/\rho_0, \tag{C.446}$$

$$\tilde{v}_{yt} + U\tilde{v}_{xy} = -\tilde{p}_{yy}/\rho_0, \tag{C.447}$$

$$\tilde{w}_{zt} + U\tilde{w}_{xz} + U'\tilde{w}_x = -\tilde{p}_{zz}/\rho_0 - g\tilde{\rho}_z/\rho_0. \tag{C.448}$$

Now add the three equations and make use of the continuity equation to find the reduced equation

$$[\tilde{p}_{xx} + \tilde{p}_{yy} + \tilde{p}_{zz}]/\rho_0 = -2U'\tilde{w}_x - g\tilde{\rho}_z/\rho_0, \tag{C.449}$$

which is the desired solution, equation (7.14).

Exercise 7.4 Derive equation (7.15).

Solution

Begin by operating on (C.445) with the three-dimensional Laplacian ∇^2, to obtain

$$\nabla^2 \tilde{w}_t + U\nabla^2 \tilde{w}_x + 2U'\tilde{w}_{xz} + U''\tilde{w}_x = -\nabla^2 \tilde{p}_z/\rho_0 - g\nabla^2 \tilde{\rho}/\rho_0. \tag{C.450}$$

Now substitute (7.14) into (C.450) to eliminate the pressure

$$\nabla^2 \tilde{w}_t + U\nabla^2 \tilde{w}_x + U''\tilde{w}_x = 2U''\tilde{w}_x + g\tilde{\rho}_{zz}/\rho_0 - g\nabla^2 \tilde{\rho}/\rho_0. \tag{C.451}$$

Now operate on (C.451) using $\partial/\partial t + U\partial/\partial x$ to find

$$\left(\frac{\partial}{\partial t} + U\frac{\partial}{\partial x}\right)^2 \nabla^2 \tilde{w} - U''\left(\frac{\partial}{\partial t} + U\frac{\partial}{\partial x}\right)\tilde{w}_x$$

$$+ \frac{g}{\rho_0}\left(\frac{\partial}{\partial t} + U\frac{\partial}{\partial x}\right)(\tilde{\rho}_{xx} + \tilde{\rho}_{yy}) = 0. \tag{C.452}$$

Now substitute (7.13) into (C.452) to eliminate the density, but first recognize that (7.13) can be rearranged to give

$$\left(\frac{\partial}{\partial t} + U\frac{\partial}{\partial x}\right)\tilde{\rho} = -\bar{\rho}'\tilde{w}$$

so that

$$\left(\frac{\partial}{\partial t} + U\frac{\partial}{\partial x}\right)(\tilde{\rho}_{xx} + \tilde{\rho}_{yy}) = -\bar{\rho}'(\tilde{w}_{xx} + \tilde{w}_{yy}).$$

After substituting, (C.452) becomes

$$\left(\frac{\partial}{\partial t} + U\frac{\partial}{\partial x}\right)^2 \nabla^2 \tilde{w} - U''\left(\frac{\partial}{\partial t} + U\frac{\partial}{\partial x}\right)\tilde{w}_x - \frac{g\bar{\rho}'}{\rho_0}(\tilde{w}_{xx} + \tilde{w}_{yy}) = 0. \tag{C.453}$$

If we now introduce the Brunt–Väisälä frequency

$$N^2 = N^2(z) = -\frac{g}{\rho_0}\bar{\rho}',$$

then (C.453) becomes the desired result, equation (7.15).

Exercise 7.5 Derive the perturbation energy equation (7.38).

Solution
Begin with the perturbation equations (7.10)–(7.12). We multiply (7.10) by $\tilde{u}$, (7.11) by $\tilde{v}$, (7.12) by $\tilde{w}$, keeping only second-order terms in the perturbation quantities, then adding, we get

$$\frac{\rho_0}{2}\frac{\partial}{\partial t}(\tilde{u}^2 + \tilde{v}^2 + \tilde{w}^2) + \rho_0 U(\tilde{u}\tilde{u}_x + \tilde{v}\tilde{v}_x + \tilde{w}\tilde{w}_x) + \rho_0 U'\tilde{u}\tilde{w}$$
$$= -(\tilde{u}\tilde{p}_x + \tilde{v}\tilde{p}_y + \tilde{w}\tilde{p}_z) - g\tilde{w}\tilde{\rho}. \tag{C.454}$$

We can now use (7.9) to rewrite the pressure term, and since U is a function of z it can be brought under the x-derivative sign, to get

$$\frac{\rho_0}{2}\frac{\partial}{\partial t}(\tilde{u}^2 + \tilde{v}^2 + \tilde{w}^2) = -\frac{\rho_0}{2}\frac{\partial}{\partial x}\left(U(\tilde{u}^2 + \tilde{v}^2 + \tilde{w}^2)\right)$$
$$-\rho_0 U'\tilde{u}\tilde{w} - \nabla\cdot(\tilde{u}\tilde{p}) - g\tilde{w}\tilde{\rho}. \tag{C.455}$$

We now integrate over the volume, and note that the first and third terms on the right-hand side of (C.455) can be replaced by surface integrals using the Reynolds transport theorem (aka, the divergence theorem), which then vanish due to fact that the perturbation velocities are zero at the boundaries. The end result is the perturbation energy equation

$$\frac{dE}{dt} = \frac{\partial}{\partial t}\iiint \frac{\rho_0}{2}(\tilde{u}^2 + \tilde{v}^2 + \tilde{w}^2)dV$$
$$= \iiint \frac{\rho_0}{2}\frac{\partial}{\partial t}(\tilde{u}^2 + \tilde{v}^2 + \tilde{w}^2)dV$$
$$= -\int \rho_0\tilde{u}\tilde{w}U'dz - \int g\tilde{w}\tilde{\rho}dz, \tag{C.456}$$

which is the desired result, equation (7.38).

Exercise 7.6 State the linearized equations that result in (7.73).

Solution
Ignoring w, the governing equations that result in (7.73) are

$$\left.\begin{array}{rcl} u_x + v_y &=& 0, \\ u_t + Uu_x + U'v - fv &=& -p_x/\rho, \\ v_t + Uv_x + fu &=& -p_y/\rho, \end{array}\right\} \tag{C.457}$$

where $U(y)$ is the mean flow and $f = f_0 + \beta y$ is the β-plane approximation. The reader should verify that (7.73) results from the above equations.

Exercise 7.7 Derive the coupled pair of instability equations (7.91)–(7.92), beginning with the disturbance form of the Navier–Stokes equations (7.86)–(7.89).

Solution

Begin with disturbance equations (7.86)–(7.89)

$$\tilde{u}_x + \tilde{v}_y + \tilde{w}_z = 0, \tag{C.458}$$

$$\tilde{u}_t + U\tilde{u}_x + V\tilde{u}_y + U'\tilde{w} - f\tilde{v} = -\tilde{p}_x/\rho_0 + \nu\nabla^2\tilde{u}, \tag{C.459}$$

$$\tilde{v}_t + U\tilde{v}_x + V\tilde{v}_y + V'\tilde{w} + f\tilde{u} = -\tilde{p}_y/\rho_0 + \nu\nabla^2\tilde{v}, \tag{C.460}$$

$$\tilde{w}_t + U\tilde{w}_x + V\tilde{w}_y = -\tilde{p}_z/\rho_0 + \nu\nabla^2\tilde{w}. \tag{C.461}$$

Take the divergence of the momentum equations, add the resulting three equations and make use of the continuity equation to find reduced equation (similar to Exercise 7.7)

$$\nabla^2\tilde{p}/\rho_0 = -2U'\tilde{w}_x - 2V'\tilde{w}_y + f\tilde{v}_x - f\tilde{u}_y. \tag{C.462}$$

Now, operating on (C.461) with the three-dimensional Laplacian ∇^2, we obtain

$$\left(\frac{\partial}{\partial t} + U\frac{\partial}{\partial x} + V\frac{\partial}{\partial y}\right)\nabla^2\tilde{w} + U''\tilde{w}_x + 2U'\tilde{w}_{xz} + V''\tilde{w}_y + 2V'\tilde{w}_{yz}$$
$$= -\nabla^2\tilde{p}_z/\rho_0 + \nu\nabla^4\tilde{w}. \tag{C.463}$$

We now take the z-derivative of (C.462) and substitute into (C.463) to eliminate the pressure, getting

$$\left(\frac{\partial}{\partial t} + U\frac{\partial}{\partial x} + V\frac{\partial}{\partial y}\right)\nabla^2\tilde{w}$$
$$= U''\tilde{w}_x + V''\tilde{w}_y - f\tilde{v}_{xz} + f\tilde{u}_{yz} + \nu\nabla^4\tilde{w}. \tag{C.464}$$

To derive an equation for $\tilde{v}$, we take the y-derivative of (C.459), take the x-derivative of (C.460), and subtract, yielding

$$\left(\frac{\partial}{\partial t} + U\frac{\partial}{\partial x} + V\frac{\partial}{\partial y}\right)(\tilde{u}_y - \tilde{v}_x) + U'\tilde{w}_y - V'\tilde{w}_x - f(\tilde{v}_y + \tilde{u}_x)$$
$$= \nu\nabla^2(\tilde{u}_y - \tilde{v}_x). \tag{C.465}$$

Modal solutions combined with Fourier decomposition can be used as before to reduce this system to one of ordinary differential equations. Assume that all variables are a sum of modes written as

$$\{\tilde{u}, \tilde{v}, \tilde{w}, \tilde{p}\} = \{\hat{u}, \hat{v}, \hat{w}, \hat{p}\}(z) e^{i(\alpha_1 x + \alpha_2 y - \omega t)}. \tag{C.466}$$

Substituting (C.465) into (C.464) and (C.465), and using (C.458), gives

$$(\alpha \overline{U} - \omega)\Delta\hat{w} - \alpha \overline{U}''\hat{w} + iv\Delta^2\hat{w} = f(\alpha\overline{v})', \tag{C.467}$$

$$(\alpha \overline{U} - \omega)(\alpha\overline{v}) + iv\Delta(\alpha\overline{v}) = i\alpha\overline{V}'\hat{w} + f\hat{w}', \tag{C.468}$$

where

$$\Delta = \frac{d^2}{dz} - \alpha^2, \qquad \alpha^2 = \alpha_1^2 + \alpha_2^2$$

and full use of the Squire transformation has been used, namely

$$\alpha\overline{U} = \alpha_1 U + \alpha_2 V, \qquad \alpha\overline{u} = \alpha_1\hat{u} + \alpha_2\hat{v}, \tag{C.469}$$

$$\alpha\overline{V} = \alpha_2 U - \alpha_1 V, \qquad \alpha\overline{v} = \alpha_2\hat{u} - \alpha_1\hat{v}, \tag{C.470}$$

in order to put the equations in this form. Equations (C.467) and (C.468) are the desired result.

C.7 Solutions for Chapter 8

Exercise 8.1 Derive the disturbance equations equivalent to (8.25) and (8.26) when the mean profile U is a continuous function of y. Assume incompressibility, three-dimensional disturbances and include viscous effects. In addition, derive the corresponding vorticity components.

Solution
We begin with the disturbance equations in nondimensional form

$$\left.\begin{array}{rcl} \tilde{u}_x + \tilde{v}_y + \tilde{w}_z &=& 0, \\ \tilde{u}_t + U\tilde{u}_x + U'\tilde{v} + \tilde{p}_x &=& Re^{-1}\nabla^2\tilde{u}, \\ \tilde{v}_t + U\tilde{v}_x + \tilde{p}_y &=& Re^{-1}\nabla^2\tilde{v}, \\ \tilde{w}_t + U\tilde{w}_x + \tilde{p}_z &=& Re^{-1}\nabla^2\tilde{w}, \end{array}\right\} \tag{C.471}$$

where the subscripts denote partial derivatives and the prime denotes ordinary derivative with respect to y. The pressure equation can be found by taking the divergence of the momentum equations, resulting in

$$\nabla^2\tilde{p} = -2U'\tilde{v}_x. \tag{C.472}$$

An equation for $\tilde{v}$ can also be found by taking the Laplacian ∇^2 of the y-momentum equation

$$\left(\frac{\partial}{\partial t}+U\frac{\partial}{\partial x}\right)\nabla^2\tilde{v}+2U'\tilde{v}_{xy}+U''\tilde{v}_x+\nabla^2\tilde{p}_y=Re^{-1}\nabla^4\tilde{v}. \qquad (C.473)$$

Now substituting in for pressure, we get

$$\left(\frac{\partial}{\partial t}+U\frac{\partial}{\partial x}\right)\nabla^2\tilde{v}=U''\tilde{v}_x+Re^{-1}\nabla^4\tilde{v}. \qquad (C.474)$$

An equation for the vorticity component, $\tilde{\omega}_y=\tilde{u}_z-\tilde{w}_x$, is found by taking the z-derivative of the x-momentum equation, x-derivative of the z-momentum and subtracting to eliminate pressure, resulting in

$$\left(\frac{\partial}{\partial t}+U\frac{\partial}{\partial x}\right)\tilde{\omega}_y+U'\tilde{v}_z=Re^{-1}\nabla^2\tilde{\omega}_y. \qquad (C.475)$$

We now apply the moving coordinate transformation

$$T=t,\quad \xi=x-U(y)t,\quad \eta=y,\quad \zeta=z,$$

so that

$$\left.\begin{array}{rcl} \dfrac{\partial}{\partial t} &=& \dfrac{\partial}{\partial T}-U\dfrac{\partial}{\partial\xi},\quad \dfrac{\partial}{\partial x}=\dfrac{\partial}{\partial\xi},\quad \dfrac{\partial}{\partial z}=\dfrac{\partial}{\partial\zeta},\\[2mm] \dfrac{\partial}{\partial y} &=& \dfrac{\partial}{\partial\eta}-U'T\dfrac{\partial}{\partial\xi},\\[2mm] \dfrac{\partial^2}{\partial y^2} &=& \dfrac{\partial^2}{\partial\eta^2}-2U'T\dfrac{\partial^2}{\partial\xi\partial\eta}+(U')^2T^2\dfrac{\partial^2}{\partial\xi^2}-U''T\dfrac{\partial}{\partial\xi}, \end{array}\right\} \qquad (C.476)$$

and thus

$$\nabla^2=\frac{\partial^2}{\partial\eta^2}-2U'T\frac{\partial^2}{\partial\xi\partial\eta}+(U')^2T^2\frac{\partial^2}{\partial\xi^2}-U''T\frac{\partial}{\partial\xi}+\frac{\partial^2}{\partial\xi^2}+\frac{\partial^2}{\partial\zeta^2}. \quad (C.477)$$

The transformed equations become

$$\left.\begin{array}{rcl} \dfrac{\partial}{\partial T}\nabla^2\tilde{v} &=& U''\dfrac{\partial\tilde{v}}{\partial\xi}+Re^{-1}\nabla^4\tilde{v},\\[2mm] \dfrac{\partial}{\partial T}\tilde{\omega}_y+U'\dfrac{\partial\tilde{v}}{\partial\zeta} &=& Re^{-1}\nabla^2\tilde{\omega}_y, \end{array}\right\} \qquad (C.478)$$

where now prime denotes an ordinary derivative with respect to η.

We now define the Fourier transform as

$$\check{v}(\alpha,\eta,\gamma,T)=\int_{-\infty}^{\infty}\int_{-\infty}^{\infty}\tilde{v}(\xi,\eta,\zeta,T)e^{i(\alpha\xi+\gamma\zeta)}d\xi d\zeta, \qquad (C.479)$$

where α and γ are taken to be real and are the nondimensional wavenumbers in the ξ- and ζ-directions, respectively. Then, say, the ξ-derivative transforms to

$$\int_a^b \int_a^b \frac{\partial \tilde{v}}{\partial \xi} e^{i(\alpha\xi+\gamma\zeta)} d\xi d\zeta = \int_a^b \left\{ \tilde{v} e^{i\alpha\xi} \Big|_a^b - \int_a^b i\alpha\tilde{v} e^{i\alpha\xi} d\xi \right\} e^{i\gamma\zeta} d\zeta,$$

(C.480)

where we have used integration by parts. Taking the limits as $a \to -\infty$ and $b \to \infty$ yields

$$\int_{-\infty}^{\infty} \int_{-\infty}^{\infty} \frac{\partial \tilde{v}}{\partial \xi} e^{i(\alpha\xi+\gamma\zeta)} d\xi d\zeta = -i\alpha \int_{-\infty}^{\infty} \int_{-\infty}^{\infty} \tilde{v} e^{i(\alpha\xi+\gamma\zeta)} d\xi d\zeta = -i\alpha\check{v}.$$

(C.481)

Similarly, the ζ-derivative transforms according to

$$\int_{-\infty}^{\infty} \int_{-\infty}^{\infty} \frac{\partial \tilde{v}}{\partial \zeta} e^{i(\alpha\xi+\gamma\zeta)} d\xi d\zeta = -i\gamma\check{v}.$$

(C.482)

By employing integration by parts twice the second derivatives transforms according to

$$\left. \begin{array}{rcl} \int_{-\infty}^{\infty} \int_{-\infty}^{\infty} \frac{\partial^2 \tilde{v}}{\partial \xi^2} e^{i(\alpha\xi+\gamma\zeta)} d\xi d\zeta & = & -\alpha^2 \check{v}, \\ \int_{-\infty}^{\infty} \int_{-\infty}^{\infty} \frac{\partial^2 \tilde{v}}{\partial \zeta^2} e^{i(\alpha\xi+\gamma\zeta)} d\xi d\zeta & = & -\gamma^2 \check{v}. \end{array} \right\}$$

(C.483)

The attentive student will note that the Fourier transformation of the first derivatives gives a negative sign by virtue of employing integration by parts, whereas using normal modes one would get a positive sign for the first derivative when using $e^{i(\alpha\xi+\gamma\zeta)}$. The Fourier transformation and normal mode approach give the same sign for the second derivatives. The two methods (Fourier transformation, normal modes) are essentially equivalent.

Returning to the governing equations, the Fourier transformation yields

$$\left. \begin{array}{rcl} \frac{\partial}{\partial T} \Delta\check{v} & = & -i\alpha U''\check{v} + Re^{-1}\Delta^2\check{v}, \\ \frac{\partial}{\partial T}\check{\omega}_y - i\gamma U'\check{v} & = & Re^{-1}\Delta\check{\omega}_y, \end{array} \right\}$$

(C.484)

where

$$\Delta = \frac{\partial^2}{\partial \eta^2} + 2i\alpha U'T\frac{\partial}{\partial \eta} - \left(\tilde{\alpha}^2 - i\alpha U''T + \alpha^2 (U')^2 T^2 \right),$$

(C.485)

where $\tilde{\alpha}^2 = \alpha^2 + \gamma^2$. Equation (C.484) is now the equivalent versions of (8.25) and (8.26) when the mean flow is continuous and viscous effects are included. This is the desired result.

The perturbation vorticity components are defined as

$$
\left.\begin{aligned}
\tilde{\omega}_x &= \tilde{w}_y - \tilde{v}_z, \\
\tilde{\omega}_y &= \tilde{u}_z - \tilde{w}_x, \\
\tilde{\omega}_z &= \tilde{v}_x - \tilde{u}_y.
\end{aligned}\right\}
\tag{C.486}
$$

Applying the coordinate transformation

$$
\left.\begin{aligned}
\tilde{\omega}_x &= \tilde{w}_\eta - U'T\tilde{w}_\xi - \tilde{v}_\zeta, \\
\tilde{\omega}_y &= \tilde{u}_\zeta - \tilde{w}_\xi, \\
\tilde{\omega}_z &= \tilde{v}_\xi - \tilde{u}_\eta + U'T\tilde{u}_\xi.
\end{aligned}\right\}
\tag{C.487}
$$

Taking the Fourier transformation

$$
\left.\begin{aligned}
\check{\omega}_x &= \check{w}_\eta + i\alpha U'T\check{w} + i\gamma\check{v}, \\
\check{\omega}_y &= -i\gamma\check{u} + i\alpha\check{w}, \\
\check{\omega}_z &= -i\alpha\check{v} - \check{u}_\eta - i\alpha U'T\check{u}.
\end{aligned}\right\}
\tag{C.488}
$$

If we introduce the Squire transformation $\tilde{\alpha}\overline{w} = -\gamma\check{u} + \alpha\check{w}$, then one immediately sees that

$$
\check{\omega}_y = i\tilde{\alpha}\overline{w}.
\tag{C.489}
$$

Then the vorticity equation (C.484b) can be recast in terms of $\overline{w}$,

$$
\frac{\partial\overline{w}}{\partial T} = U'\sin\phi\,\check{v} + Re^{-1}\Delta\hat{w},
\tag{C.490}
$$

where $\sin\phi = \gamma/\tilde{\alpha}$. This equation is identical to (8.26) when the mean flow here is replaced by $U = \sigma y$ and viscous effects are ignored.

By appealing to polar coordinates, the vorticity components are

$$
\left.\begin{aligned}
\tilde{\omega}_{\tilde{\alpha}} &= \cos\phi\,\tilde{\omega}_x + \sin\phi\,\tilde{\omega}_z, \\
\tilde{\omega}_\phi &= \sin\phi\,\tilde{\omega}_x - \cos\phi\,\tilde{\omega}_z,
\end{aligned}\right\}
\tag{C.491}
$$

where $\sin\phi = \gamma/\tilde{\alpha}$ and $\cos\phi = \alpha/\tilde{\alpha}$. Applying the coordinate transformation, and then taking the Fourier transformation, we get

$$
\left.\begin{aligned}
\check{\omega}_{\tilde{\alpha}} &= \left(\frac{\partial}{\partial\eta} + i\alpha U'T\right)\overline{w}, \\
\check{\omega}_\phi &= -\frac{i}{\tilde{\alpha}}\Delta\check{v},
\end{aligned}\right\}
\tag{C.492}
$$

which are equations (8.27). This is the desired result.

Exercise 8.2 Consider the baroclinic plane Couette flow of Section 7.4. Redo the problem using the normal mode approach, and compare the eigensolution to that found using the moving coordinate transformation.

Solution

The general solution to Rayleigh's equation (2.29) is given by

$$\hat{v} = Ae^{-\alpha y} + Be^{\alpha y}. \tag{C.493}$$

For baroclinic flow, we require the pressure to vanish at $y = \pm H$, where the pressure disturbance is given by (2.57):

$$-\frac{i\alpha^2 \hat{p}}{\rho} = (\omega - \alpha U)\hat{v}' + \alpha U'\hat{v}. \tag{C.494}$$

Applying the condition at $y = H$ with $U = \sigma H$, $U' = \sigma$, and $\hat{p} = 0$,

$$(\omega - \alpha\sigma H)\left[-Ae^{-\alpha H} + Be^{\alpha H}\right] + \sigma\left[Ae^{-\alpha H} + Be^{\alpha H}\right] = 0. \tag{C.495}$$

Applying the condition at $y = -H$ with $U = -\sigma H$, $U' = \sigma$ and $\hat{p} = 0$, we find

$$(\omega + \alpha\sigma H)\left[-Ae^{\alpha H} + Be^{-\alpha H}\right] + \sigma\left[Ae^{\alpha H} + Be^{-\alpha H}\right] = 0. \tag{C.496}$$

We have two equations for the two unknowns A and B. Solving, we get the eigenrelation

$$\left[\omega^2 - \alpha^2\sigma^2 H^2 - 2\alpha\sigma^2 H - \sigma^2\right]e^{-2\alpha H}$$
$$= \left[\omega^2 - \alpha^2\sigma^2 H^2 + 2\alpha\sigma^2 H - \sigma^2\right]e^{2\alpha H}, \tag{C.497}$$

or, solving for ω,

$$\omega^2 = \alpha^2\sigma^2 H^2 + \sigma^2 - 2\alpha\sigma^2 H\left(\frac{e^{-2\alpha H} + e^{2\alpha H}}{e^{2\alpha H} - e^{-2\alpha H}}\right), \tag{C.498}$$

i.e.,

$$\omega = \pm\frac{\sigma}{2}\sqrt{4 - 4q\coth q + q^2}, \tag{C.499}$$

where $q = 2\alpha H$. Since the normal modes are proportional to $e^{-i\omega t}$, we see that the solution is unstable if the discriminant of the square root is negative. Note that this is exactly the eigenrelation (7.84) when using the moving coordinate transformation, except that the solutions for the moving coordinate transformation are proportional to $e^{\omega T}$.

Exercise 8.3 Consider the inviscid boundary layer flow given by

$$U(y) = \begin{cases} U_0 & y \geq H, \\ \sigma y & 0 < y < H. \end{cases}$$

Compute the solutions using the normal mode approach and the moving coordinate approach, and compare the two.

Solution

We first solve using normal modes. The general solution to Rayleigh's equation (2.29) is given by

$$\hat{v} = \begin{cases} Ae^{-\alpha(y-H)} & y > H, \\ Be^{-\alpha(y-H)} + Ce^{\alpha y} & 0 < y < H. \end{cases} \tag{C.500}$$

Now recall the jump conditions from Chapter 2,

$$\left\| (\alpha U - \omega)\hat{v}' - \alpha U' \hat{v} \right\| = 0, \qquad \|\hat{v}\| = 0. \tag{C.501}$$

At $y = H^+$, $U = \sigma H$, $U' = 0$, and at $y = H^-$, $U = \sigma H$, $U' = \sigma$, so that the two jump conditions give

$$(\alpha\sigma H - \omega)A = (\alpha\sigma H - \omega)\left[B - Ce^{\alpha H}\right] + \sigma\left[B + Ce^{\alpha H}\right], \tag{C.502}$$

$$A = B + Ce^{\alpha H}. \tag{C.503}$$

At $y = 0$, the disturbance velocity vanishes, $\hat{v} = 0$, which gives

$$Be^{\alpha H} + C = 0. \tag{C.504}$$

We have three equations for three unknowns. Solving we get

$$(\alpha\sigma H - \omega)(1 - e^{2\alpha H}) = (\alpha\sigma H - \omega + \sigma) + (\alpha\sigma H - \omega - \sigma)e^{2\alpha H} \tag{C.505}$$

or, solving for ω,

$$\omega = \frac{\sigma}{2}\left[2\alpha H - 1 + e^{-2\alpha H}\right]. \tag{C.506}$$

Since the normal modes are proportional to $e^{-i\omega t}$, we see that the disturbances are neutrally stable, and only oscillate in time. The fact that the disturbance is neutrally stable is expected, since viscosity is needed to make the boundary layer unstable.

We now solve using the moving coordinate transformation. Recall that the general form of the moving coordinate transformation from Exercise 8.1 is given (in two dimensions) by

$$\frac{\partial}{\partial T}\Delta\check{v} = -i\alpha U''\check{v} + Re^{-1}\Delta^2\check{v}, \tag{C.507}$$

where

$$\Delta = \frac{\partial^2}{\partial\eta^2} + 2i\alpha U'T\frac{\partial}{\partial\eta} - \left(\alpha^2 - i\alpha U''T + \alpha^2(U')^2T^2\right). \tag{C.508}$$

In the freestream we have $U' = U'' = 0$, so the inviscid equation is given by

$$\frac{\partial}{\partial T}\Delta\check{v} = 0, \quad \Delta = \frac{\partial^2}{\partial\eta^2} - \alpha^2. \tag{C.509}$$

The homogeneous solution is given by

$$\check{v}_H = C(T)e^{-\alpha(\eta-H)}. \tag{C.510}$$

In the wall layer we have $U' = \sigma$, $U'' = 0$, so that the equation is

$$\frac{\partial}{\partial T}\Delta\check{v} = 0, \quad \Delta = \frac{\partial^2}{\partial\eta^2} + 2i\alpha\sigma T\frac{\partial}{\partial\eta} - \alpha^2(1 + \sigma^2 T^2). \tag{C.511}$$

The homogeneous solution is given by

$$\check{v}_H = A(T)e^{-\alpha\eta(1+i\sigma T)} + B(T)e^{\alpha\eta(1-i\sigma T)}. \tag{C.512}$$

Therefore, the general solution to the boundary layer problem is given by

$$\check{v} = \begin{cases} C(T)e^{-\alpha(\eta-H)} + F_1(\eta,T), & y > H, \\ A(T)e^{-\alpha\eta(1+i\sigma T)} + B(T)e^{\alpha\eta(1-i\sigma T)} + F_2(\eta,T), & 0 < y < H, \end{cases} \tag{C.513}$$

where the functions F_1 and F_2 are the rotational components of $\check{v}$ and their form depends on the initial vorticity specified at time $T = 0$; i.e., they are the particular solutions to the inhomogeneous equation (C.509) and (C.512). Also, note that we require F_1 to be bounded as $\eta \to \infty$. In addition, $\check{v}$ must vanish at $\eta = 0$ and at $T = 0$ so that $A(0) = B(0) = F_2(0,0) = 0$. Here, we shall not specify a particular form for F_1 and F_2.

Since we have 3 unknowns A, B and C, we need three boundary conditions. The first requires $\check{v} = 0$ at $\eta = 0$, which gives

$$A(T) + B(T) + F_2(0,T) = 0. \tag{C.514}$$

The other two conditions are applied at $y = H$. The first of these is that $\check{v}$ must be continuous, yielding

$$Ae^{-\alpha H(1+i\sigma T)} + Be^{\alpha H(1-i\sigma T)} + F_2(H,T) = C + F_1(H,T). \tag{C.515}$$

The second requires that the pressure be continuous at $y = H$. Using (7.81), this implies

$$\frac{\partial^2\check{v}_1}{\partial T\partial\eta} + i\sigma_1 T\alpha\frac{\partial\check{v}_1}{\partial T} + 2i\sigma_1\alpha\check{v}_1 = \frac{\partial^2\check{v}_2}{\partial T\partial\eta} + i\sigma_2 T\alpha\frac{\partial\check{v}_2}{\partial T} + 2i\sigma_2\alpha\check{v}_2, \tag{C.516}$$

where here the subscript 1 means values for $y = H^+$ and the subscript 2 means values for $y = H^-$. Applying the boundary condition we get

$$-\alpha \overset{\circ}{C} - \alpha^2 \check{p}_{1,R} = \left\{ -\alpha \overset{\circ}{A} e^{-\alpha H} + i\alpha \sigma (1 + \alpha H) A e^{-\alpha H} \right.$$

$$\left. + \alpha \overset{\circ}{B} e^{\alpha H} + i\alpha \sigma (1 - \alpha H) B e^{\alpha H} \right\} e^{-i\alpha \sigma H T} - \alpha^2 \check{p}_{2,R}, \quad \text{(C.517)}$$

where $\check{p}_{1,R}$ and $\check{p}_{2,R}$ are the rotational contributions. We now have three equations (C.514), (C.515) and (C.517) for three unknowns A, B and C. We can eliminate A and C to get the following differential equation for B:

$$2\alpha \overset{\circ}{B} e^{\alpha H} + i\alpha \sigma (1 - 2\alpha H) B e^{\alpha H} - i\alpha \sigma B e^{-\alpha H}$$
$$= i\alpha \sigma F_2(0, T) e^{-\alpha H} + G(T) e^{i\alpha \sigma H T}, \quad \text{(C.518)}$$

where

$$G = -\alpha \overset{\circ}{F}_2(H, T) + \alpha \overset{\circ}{F}_1(H, T) - \alpha^2 \check{p}_{1,R} + \alpha^2 \check{p}_{2,R}. \quad \text{(C.519)}$$

This equation is of the form

$$a_1 \overset{\circ}{B} + a_2 B = \mathscr{F}(T), \quad \text{(C.520)}$$

where

$$a_1 = 2\alpha e^{\alpha H}, \quad a_2 = i\alpha \sigma (1 - 2\alpha H) e^{\alpha H} - i\alpha \sigma e^{-\alpha H}. \quad \text{(C.521)}$$

This equation can be rewritten using an integrator factor to get

$$\left(e^{(a_2/a_1)T} B \right) = \frac{e^{(a_2/a_1)T}}{a_1} \mathscr{F}(T), \quad \text{(C.522)}$$

which can be integrated to yield the solution

$$B(T) = \frac{e^{-i\Omega T}}{a_1} \int_0^T e^{i\Omega T} \mathscr{F}(T) dT, \quad \text{(C.523)}$$

where

$$\frac{a_2}{a_1} = i\Omega, \quad \Omega = \frac{\sigma}{2} \left[(1 - 2\alpha H) - e^{-2\alpha H} \right]. \quad \text{(C.524)}$$

We see that the solution will be proportional to $e^{-i\Omega T}$, indicating that the disturbance is neutrally stable. Comparing this to the normal mode solution of $e^{-i\omega t}$, we see that $\omega = -\Omega$; i.e., compare Ω with (C.506).

Exercise 8.5 Consider inviscid plane Couette flow. Solve this limiting case by use of the moving coordinate transformation and with initial distribution $\tilde{v}_0(x, y, 0) = \Omega e^{-i(\alpha_0 x + \beta_0 y)}$, which corresponds to a plane wave in the x- and y-directions with wavenumbers α_0 and β_0, respectively.

Solution
We wish to solve

$$\frac{\partial}{\partial T}\Delta\check{v} = 0, \tag{C.525}$$

where

$$\Delta = \frac{\partial^2}{\partial\eta^2} + 2i\alpha\sigma T\frac{\partial}{\partial\eta} - \alpha^2(1+\sigma^2T^2). \tag{C.526}$$

This equation is two-dimensional and in the moving coordinate frame. The equation can be integrated in time to yield

$$\Delta\check{v} = \Delta\check{v}\big|_{T=0}. \tag{C.527}$$

The homogeneous problem is given by

$$\Delta\check{v}_H = 0, \tag{C.528}$$

while the particular (or rotational) problem is given by

$$\Delta\check{v}_R = \Delta\check{v}\big|_{T=0}. \tag{C.529}$$

The homogeneous solution is easily found to be given by

$$\check{v}_H = A(T)e^{-\alpha\eta(1+i\sigma T)} + B(T)e^{\alpha\eta(1-i\sigma T)}. \tag{C.530}$$

Thus the general solution is $\check{v} = \check{v}_H + \check{v}_R$. The boundary conditions are: at $\eta = H$, $\check{v} = 0$ and at $\eta = -H$, $\check{v} = 0$. This gives

$$\check{v}_H(H,T) + \check{v}_R(H,T) = 0,$$

$$\check{v}_H(-H,T) + \check{v}_R(-H,T) = 0$$

or

$$\left[A(T)e^{-\alpha H} + B(T)e^{\alpha H}\right]e^{-i\alpha\sigma TH} = -\check{v}_R(H,T),$$

$$\left[A(T)e^{\alpha H} + B(T)e^{-\alpha H}\right]e^{i\alpha\sigma TH} = -\check{v}_R(-H,T).$$

In matrix form we have

$$\begin{bmatrix} e^{-\alpha H} & e^{\alpha H} \\ e^{\alpha H} & e^{-\alpha H} \end{bmatrix}\begin{bmatrix} A \\ B \end{bmatrix} = \begin{bmatrix} -\check{v}_R(H,T)e^{i\alpha\sigma TH} \\ -\check{v}_R(-H,T)e^{-i\alpha\sigma TH} \end{bmatrix}. \tag{C.531}$$

The determinant is $J = e^{-2\alpha H} - e^{2\alpha H} \equiv -2\sinh q$, where $q = 2\alpha H$. Solving, we get

$$A = \frac{\check{v}_R(H,T)e^{-\alpha H + i\alpha\sigma TH} - \check{v}_R(-H,T)e^{\alpha H - i\alpha\sigma TH}}{2\sinh q}, \tag{C.532}$$

$$B = \frac{\check{v}_R(-H,T)e^{-\alpha H - i\alpha\sigma TH} - \check{v}_R(H,T)e^{\alpha H + i\alpha\sigma TH}}{2\sinh q}. \tag{C.533}$$

The only thing left is to determine the rotational solution $\check{v}_R$.

We assume an initial perturbation of the form

$$\tilde{v}_0(x,y,0) = \Omega e^{-i(\alpha_0 x + \beta_0 y)},$$

which corresponds to a plane wave in x,y with wavenumbers α_0 and β_0, respectively. In the moving coordinate system, the initial disturbance is

$$\tilde{v}_0(\xi,\eta,0) = \Omega e^{-i(\alpha_0\xi + \beta_0\eta)}.$$

Taking the Fourier transform in (ξ,η), we get

$$\widehat{v}_0(\alpha,\beta,0) = \Omega\delta(\alpha - \alpha_0)(\beta - \beta_0).$$

Now taking the Fourier transform in (ξ,η) of the rotational equation

$$\kappa^2\widehat{v}_R = \kappa_0^2\widehat{v}_0, \tag{C.534}$$

where

$$\kappa^2 = \alpha^2 + (\beta - \alpha\sigma T)^2, \quad \kappa_0^2 = \alpha^2 + \beta^2 \tag{C.535}$$

so that

$$\widehat{v}_R = \frac{\kappa_0^2\widehat{v}_0}{\kappa^2} \equiv \frac{\kappa_0^2\Omega\delta(\alpha - \alpha_0)\delta(\beta - \beta_0)}{\kappa^2}. \tag{C.536}$$

We now take the inverse transform in η,

$$\check{v}_R = \frac{\Omega(\alpha^2 + \beta_0^2)e^{-i\beta_0\eta}}{\alpha^2 + (\beta_0 - \alpha\sigma T)^2}\delta(\alpha - \alpha_0). \tag{C.537}$$

Thus, at the boundaries,

$$\check{v}_R(H,T) = \frac{\Omega(\alpha^2 + \beta_0^2)e^{-i\beta_0 H}}{\alpha^2 + (\beta_0 - \alpha\sigma T)^2}\delta(\alpha - \alpha_0),$$

$$\check{v}_R(-H,T) = \frac{\Omega(\alpha^2 + \beta_0^2)e^{i\beta_0 H}}{\alpha^2 + (\beta_0 - \alpha\sigma T)^2}\delta(\alpha - \alpha_0). \tag{C.538}$$

This gives for A and B

$$A = \frac{\Omega(\alpha^2 + \beta_0^2)\delta(\alpha - \alpha_0)}{\alpha^2 + (\beta_0 - \alpha\sigma T)^2}\left[\frac{e^{-i\beta_0 H - \alpha H + i\alpha\sigma TH} - e^{i\beta_0 H + \alpha H - i\alpha\sigma TH}}{2\sinh q}\right],$$

$$B = \frac{\Omega(\alpha^2 + \beta_0^2)\delta(\alpha - \alpha_0)}{\alpha^2 + (\beta_0 - \alpha\sigma T)^2}\left[\frac{e^{i\beta_0 H - \alpha H - i\alpha\sigma TH} - e^{-i\beta_0 H + \alpha H + i\alpha\sigma TH}}{2\sinh q}\right]$$

$$\tag{C.539}$$

so that the final solution is given by

$$
\check{v} = \frac{\Omega(\alpha^2 + \beta_0^2)\delta(\alpha - \alpha_0)}{\alpha^2 + (\beta_0 - \alpha\sigma T)^2}\left[e^{-i\beta_0\eta} + \frac{\sinh[\alpha(\eta - H)]}{\sinh q}e^{i\beta_0 H - i\alpha\sigma T(\eta + H)} \right.
$$

$$
\left. - \frac{\sinh[\alpha(\eta + H)]}{\sinh q}e^{-i\beta_0 H - i\alpha\sigma T(\eta - H)} \right]. \quad \text{(C.540)}
$$

Note that the boundary conditions are satisfied.

We make one more comment here. The initial condition

$$
\check{v}_0(x, y, 0)\Omega e^{-i(\alpha_0 x + \beta_0 y)}
$$

does not satisfy the boundary conditions at $T = 0$. To remedy this, we can assume symmetric or asymmetric solutions. For symmetric solutions, we set

$$
\beta_0 H = \frac{(2n + 1)\pi}{2}, \quad n = 0, 1, 2, \cdots
$$

so that the symmetric rotational solution can be written as

$$
\check{v}_R = \frac{\Omega}{2}\left(\alpha + \frac{(2n+1)\pi}{2H} \right)^2 \delta(\alpha - \alpha_0)\left\{ \frac{e^{i(2n+1)\pi\eta/(2H)}}{\alpha^2 + \left(\dfrac{(2n+1)\pi}{2H} + \alpha\sigma T \right)^2} \right.
$$

$$
\left. + \frac{e^{-i(2n+1)\pi\eta/(2H)}}{\alpha^2 + \left(\dfrac{(2n+1)\pi}{2H} - \alpha\sigma T \right)^2} \right\}. \quad \text{(C.541)}
$$

For the asymmetric solution we set

$$
\beta_0 H = n\pi, \quad n = 0, 1, 2, \cdots
$$

so that the asymmetric rotational solution can be written as

$$
\check{v}_R = \frac{\Omega}{2}\left(\alpha + \frac{n\pi}{H} \right)^2 \delta(\alpha - \alpha_0)\left\{ \frac{e^{in\pi\eta/H}}{\alpha^2 + \left(\dfrac{n\pi}{H} + \alpha\sigma T \right)^2} \right.
$$

$$
\left. + \frac{e^{-in\pi\eta/H}}{\alpha^2 + \left(\dfrac{n\pi}{H} - \alpha\sigma T \right)^2} \right\}, \quad \text{(C.542)}
$$

with either solution requiring a sum over all discrete modes n.

Finally, it should be noted that the solution $\check{v}(x, y, t)$ in physical space can easily be found via the delta-function dependence in the wavenumber α.

C.8 Solutions for Chapter 9

Exercise 9.1 Derive (9.3) and state all necessary conditions and assumptions.

Solution

The kinetic energy of a two-dimensional disturbance is given by

$$e = \frac{1}{2}(u^2 + v^2). \tag{C.543}$$

Begin with the two-dimensional momentum equations

$$u_t + U u_x + v U' = -p_x + Re^{-1}(u_{xx} + u_{yy}), \tag{C.544}$$

$$v_t + U v_x = -p_y + Re^{-1}(v_{xx} + v_{yy}). \tag{C.545}$$

Multiply the *x*-momentum equation (C.544) by u and the *y*-momentum equation (C.545) by v to find

$$\frac{1}{2}u_t^2 + \frac{1}{2}U u_x^2 + uvU' = -up_x + uRe^{-1}(u_{xx} + u_{yy}), \tag{C.546}$$

$$\frac{1}{2}v_t^2 + \frac{1}{2}U v_x^2 = -vp_y + vRe^{-1}(v_{xx} + v_{yy}). \tag{C.547}$$

Add the resulting two equations (C.546) and (C.547) and use the definition of kinetic energy (C.543) to get

$$e_t + U e_x + uvU' = -up_x - vp_y + Re^{-1}(u\nabla^2 u + v\nabla^2 v). \tag{C.548}$$

If we use the continuity equation, $u_x = -v_y$, the final terms of equations (C.546) and (C.547) can be rearranged to give

$$u(u_{xx} + u_{yy}) = -u(v_{xy} - u_{yy})$$

$$= \frac{\partial}{\partial y}[-u(v_x - u_y)] - u_y^2 + v_x u_y - (v_x^2 - u_y^2)^2, \tag{C.549}$$

$$v(v_{xx} + v_{yy}) = -v(v_{xx} - u_{xy}) = \frac{\partial}{\partial x}[v(v_x - u_y)] - v_x^2 + v_x u_y. \tag{C.550}$$

The final terms of (C.548) becomes

$$Re^{-1}(u\nabla^2 u + v\nabla^2 v) = Re^{-1}\frac{\partial}{\partial x}[v(v_x - u_y)]$$

$$-Re^{-1}\frac{\partial}{\partial y}[u(v_x - u_y)] - \frac{1}{Re}(v_x - u_y)^2. \tag{C.551}$$

or, in terms of vorticity, we have

$$Re^{-1}[u\nabla^2 u + v\nabla^2 v] = Re^{-1}\left[\frac{\partial}{\partial x}(v\omega_z) - \frac{\partial}{\partial y}(u\omega_z) - \omega_z^2\right]. \qquad \text{(C.552)}$$

Then equation (C.548) becomes

$$e_t + Ue_x + uvU' = -up_x - vp_y + Re^{-1}\left[\frac{\partial}{\partial x}(v\omega_z) - \frac{\partial}{\partial y}(u\omega_z) - \omega_z^2\right]. \quad \text{(C.553)}$$

If we integrate (C.553) over a single wavelength, many terms drop out because amplification or decay is in time and not space, and we are left with

$$E_t = \int_a^b e_t = \int_a^b -\langle uv\rangle U' dy - Re^{-1}\int_a^b \langle\omega_z^2\rangle dy. \qquad \text{(C.554)}$$

Recall that we have production and dissipation terms over a wavelength. For amplification, the production term must overcome the viscous dissipation term, otherwise, the disturbance will be neutral (production = dissipation) or decaying (dissipation overcomes production).

Exercise 9.2 Derive (9.4) and state all necessary conditions and assumptions.

Solution
Begin with (C.552):

$$e_t + Ue_x + uvU' = -up_x - vp_y + Re^{-1}\left[\frac{\partial}{\partial x}(v\omega_z) - \frac{\partial}{\partial y}(u\omega_z) - \omega_z^2\right]. \quad \text{(C.555)}$$

If we assume a parallel flow and average over one period in time, we find

$$\frac{d}{dx}\int_0^\infty \langle e\rangle U dy = \int_0^\infty -\langle uv\rangle U' dy - \frac{d}{dx}\int_0^\infty \langle pu\rangle dy$$

$$-Re^{-1}\left[\langle\omega_z^2\rangle - \frac{d}{dx}\langle v\omega_z\rangle\right] dy. \qquad \text{(C.556)}$$

But since vorticity is defined as $\omega_z = v_x - u_y$, then we have

$$\frac{d}{dx}\int_0^\infty \langle e\rangle U dy = \frac{d}{dx}\int_0^\infty -\langle uv\rangle U' dy - \frac{d}{dx}\int_0^\infty \langle pu\rangle dy$$

$$-Re^{-1}\left[\langle(v_x - u_y)^2\rangle - \frac{d}{dx}\langle v(v_x - u_y)\rangle\right] dy. \quad \text{(C.557)}$$

Exercise 9.3 Derive the secondary instability theory equations (9.14) and (9.15) using normal modes (9.18) of velocity (v_3) and vorticity (ω_3).

(a) Substitute the secondary disturbance form (9.18) into the continuity equation (9.8) and definition of vorticity (9.11).

(b) Substitute the secondary disturbance form (9.18) into the secondary instability equations (9.14) and (9.15).

(c) Use the normal mode equations obtained from the normal mode continuity
equation and definitions of vorticity (part (a)) to reduce the equations (part
(b)) to normal velocity and vorticity only.

(d) If $A = 0$ in the secondary instability equations, how do the terms that remain compare with the Orr–Sommerfeld and Squire equations?

Solution
The instantaneous velocity and pressure are given by

$$\{\underline{v}, p\} = \{\underline{v}_2, p_2\} + \{\underline{v}_3, p_3\}. \tag{C.558}$$

Similar to linear stability theory, secondary instability decomposes the velocity
and pressure fields into a basic state velocity $\underline{v}_2 = (u_2, v_2, w_2)$ and pressure p_2
and secondary disturbance velocity $\underline{v}_3 = (u_3, v_3, w_3)$ and pressure p_3. The basic
flow is given by

$$\{\underline{v}_2, p_2\}(x, y, z, t) = \{U, W\}(y) + A\{\underline{v}_1, p_1\}(x, y, z, t), \tag{C.559}$$

where the primary mode is given by $\{\underline{v}_1, p_1\}$ and is conventionally determined
by the Orr–Sommerfeld and Squire equations and the base flow may be the
Blasius, Falkner–Scan, channel or other flows. The primary wave is of a wave
form and periodic in x, z.

The wave is therefore traveling with the phase velocity

$$c_x = \omega_r / \alpha_r \sin \phi \quad \text{and} \quad c_z = \omega_r / \alpha_r \cos \phi, \tag{C.560}$$

where ω_r is the wave frequency, α_r is the wavenumber and ϕ is the angle of
the wave. For a two-dimensional primary wave, $\phi = 0$.

The secondary disturbance for velocity and vorticity takes the form (9.18),
or

$$\{\underline{v}_3, \underline{\omega}_3\} = Be^{\kappa} \sum_{n=-\infty}^{\infty} \{\underline{v}_n, \underline{\omega}_n\} e^{i(n/2)\alpha_r(x\cos\phi + z\sin\phi)}, \tag{C.561}$$

where $\kappa = \sigma t + i\beta(z\cos\phi - x\sin\phi) + \gamma(x\cos\phi + z\sin\phi)$. The secondary in-
stability velocity and vorticity are $\underline{v}_3 = (u_3, v_3, w_3)$ and $\underline{\omega}_3 = (\xi_3, \eta_3, \zeta_3)$; $\underline{v}_n =
(u_n, v_n, w_n)$ and $\underline{\omega}_n = (\xi_n, \eta_n, \zeta_n)$ are eigenfunctions; $\beta = 2\pi/\lambda_z$ is a specified
wavenumber; $\sigma = \sigma_r + i\sigma_i$ is a complex temporal eigenvalue (or is a specified
real number for spatial analysis); $\gamma = \gamma_r + i\gamma_i$ is the characteristic exponent in
Floquet theory; B is the amplitude of the secondary instability mode; and α_r is
the primary wavenumber.

(a) From (C.561), we have the following relationships, which will be used in the momentum and continuity equations:

$$\overline{\alpha}_n = \frac{\partial}{\partial x} = -i\beta \sin\phi + \gamma\cos\phi + \frac{in}{2}\alpha_r\cos\phi,$$

$$\frac{\partial}{\partial y} = ()',$$

(C.562)

$$\overline{\beta}_n = \frac{\partial}{\partial z} = -i\beta\cos\phi + \gamma\sin\phi + \frac{in}{2}\alpha_r\sin\phi.$$

Equation (C.558) is substituted into the secondary instability continuity equation (9.8) to yield

$$\overline{\alpha}_n u_n + v'_n + \overline{\beta}_n w_n = 0.$$

(C.563)

For the secondary instability vorticity equation we begin with the relationships for the three-dimensional vorticity (9.11):

$$\left. \begin{array}{l} \xi_n = w_y - v_z = w'_n - \overline{\beta}_n v_n, \\ \eta_n = u_z - w_x = \overline{\alpha}_n u_n - \overline{\beta}_n w_n, \\ \zeta_n = v_x - u_y = \overline{\alpha}_n v_n - u'_n. \end{array} \right\}$$

(C.564)

For the primary instability, we have

$$\xi_1 = w'_1 - i\alpha\sin\phi v_1,$$
$$\eta_1 = i\alpha\sin\phi u_1 - i\alpha\cos\phi w_1,$$
$$\zeta_1 = i\alpha\cos\phi v_1 - u'_1.$$

(b) Substitute the secondary disturbance form (9.18) into the secondary instability equation (9.14):

$$Re^{-1}\eta''_n + \left[Re^{-1}(\overline{\alpha}_n^2 + \overline{\beta}_n^2) - \sigma - (U_o - c_x)\overline{\alpha}_n - (W_o - c_z)\overline{\beta}_n\right]\omega_n$$

$$+ (W'_o\overline{\alpha}_n - U'_o\overline{\beta}_n)v_n + A\left[-u_1\overline{\alpha}_n\eta_n - v_1\eta'_n - w_1\overline{\beta}_n\eta_n\right.$$

$$- i\alpha\cos\phi\eta_1 u_n - \eta'_1 v_n - i\alpha\sin\phi\eta_1 w_n + (\xi_1 + i\alpha\cos\phi v_1)\overline{\alpha}_n v_n + \xi_1 v'_n$$

$$\left. + (\zeta_1 - i\alpha\cos\phi v_1)\overline{\beta}_n v_n - i\alpha\sin\phi v_1 u'_n + i\alpha\cos\phi v_1 w'_n\right]$$

$$= 0.$$

(C.565)

A similar substitution into (9.15) will yield:

$$Re^{-1}v_n^{iv} + \left[2Re^{-1}(\overline{\alpha}_n^2 + \overline{\beta}_n^2) - \sigma - (U_o - c_x)\overline{\alpha}_n - (W_o - c_z)\overline{\beta}_n\right]v_n''$$

$$+ \left[Re^{-1}(\overline{\alpha}_n^2 + \overline{\beta}_n^2) - \sigma - (U_o - c_x)\overline{\alpha}_n - (W_o - c_z)\overline{\beta}_n\right](\overline{\alpha}_n^2 + \overline{\beta}_n^2)v_n$$

$$+ (W_o''\overline{\alpha}_n - U_o''\overline{\beta}_n)v_n + A\left[-(u_1\overline{\alpha}_n + w_1\overline{\beta}_n)(\overline{\alpha}_n^2 + \overline{\beta}_n^2)v_n\right.$$

$$- v_1(\overline{\alpha}_n^2 + \overline{\beta}_n^2)v_n' - (u_1\overline{\alpha}_n + w_1\overline{\beta}_n)v_n'' - v_1 v_n''' + \alpha^2 v_1' v_n - v_1''' v_n$$

$$+ (\alpha^2 u_1 + u_1'' - 2i\alpha\cos\phi v_1')\overline{\alpha}_n v_n + \{\alpha^2(\cos^2\phi - \sin^2\phi)\}v_1$$

$$+ 2i\alpha\cos\phi u_1'\}v_n' + \{\alpha^2(\cos^2\phi - \sin^2\phi)\}w_1 - i\alpha\sin\phi v_1'$$

$$- \alpha^2\sin\phi\cos\phi u_1\}\overline{\beta}_n v_n + (v_1' + 2i\alpha\cos\phi u_1)[v_n'' + (\overline{\beta}_n^2 - \overline{\alpha}_n^2)v_n]$$

$$- 2i\alpha\cos\phi v_1\overline{\alpha}_n v_n' - 2i\alpha(\sin\phi u_1 - \cos\phi w_1)\overline{\alpha}_n\overline{\beta}_n v_n$$

$$- i\alpha\sin\phi v_1\overline{\beta}_n v_n' + i\alpha\cos\phi(\alpha^2 v_1 - v_1'')u_n - 2[\alpha^2(\cos^2\phi - \sin^2\phi)v_1$$

$$- v_1'' - 2i\alpha\cos\phi u_1']\overline{\alpha}_n u_n + 2i\alpha\cos\phi(w_1' - i\alpha\sin\phi v_1)\overline{\beta}_n u_n$$

$$- i\alpha\cos\phi v_1[(\overline{\alpha}_n^2 + \overline{\beta}_n^2)u_n + u_n''] + 2(v_1' + 2i\alpha\cos\phi u_1)\overline{\alpha}_n u_n'$$

$$+ 2i\alpha\cos\phi w_1\overline{\beta}_n u_n' + i\alpha\sin\phi v_1\overline{\alpha}_n\overline{\beta}_n u_n$$

$$+ i\alpha\sin\phi(\alpha^2 v_1 - v_1'')w_n + 2i\alpha\sin\phi(u_1' - i\alpha\cos\phi v_1)\overline{\alpha}_n w_n$$

$$\left. - i\alpha\sin\phi v_1(\overline{\alpha}_n^2 w_n - w_n'') + 2i\alpha\sin\phi u_1\overline{\alpha}_n w_n'\right] = 0. \qquad \text{(C.566)}$$

(c) Use the normal mode equations obtained from the normal mode continuity equations and definitions of vorticity (part (a)) to reduce the equations (part (b)) to normal velocity and vorticity only. First note that to get a real solution for the primary wave, the disturbance must take the form

$$\underline{v} = \underline{v}_1 e^{i\alpha_r(x\cos\phi + z\sin\phi)} + \underline{v}_{-1}e^{-i\alpha_r(x\cos\phi + z\sin\phi)}, \qquad \text{(C.567)}$$

where $\underline{v}_1 = (u_1, v_1, w_1)$ are complex conjugates of $\underline{v}_{-1} = (u_{-1}, v_{-1}, w_{-1})$. To aid in the derivation and to support an equation that can be coded, we introduce the form $\underline{v}_m$ with $m \pm 1$. Then the resulting equations become

$$\left[Re^{-1}\eta_n'' + \left\{Re^{-1}\Delta_n - \sigma - (U_{\mathrm{o}} - c_x)\overline{\alpha}_n - (W_{\mathrm{o}} - c_z)\overline{\beta}_n\right\}\eta_n\right.$$

$$\left. + (W_{\mathrm{o}}'\overline{\alpha}_n + U_{\mathrm{o}}'\overline{\beta}_n)v_n\right]e^N + A\left\{-\left[1 + im\alpha\sin\phi\,\frac{\overline{\beta}_n}{\Delta_n}\right.\right.$$

$$\left. + im\alpha\cos\phi\,\frac{\overline{\alpha}_n}{\Delta_n}\right]v_m\eta_n' + \left[(-\overline{\alpha}_n + m^2\alpha^2\cos\phi\sin\phi\,\frac{\overline{\beta}_n}{\Delta_n}\right.$$

$$\left. - m^2\alpha^2\frac{\overline{\alpha}_n}{\Delta_n})u_m + (1 + im\alpha\cos\phi\,\frac{\overline{\alpha}_n}{\Delta_n})v_m' - (\overline{\beta}_n + m^2\cos^2\phi\,\frac{\overline{\beta}_n}{\Delta_n})w_m\right]\eta_n$$

$$+ im\alpha\left[\sin\phi\,\frac{\overline{\alpha}_n}{\Delta_n} - \cos\phi\,\frac{\overline{\beta}_n}{\Delta_n}\right]v_m v_n'' - \left[(m^2\alpha^2\cos\phi\sin\phi\,\frac{\overline{\alpha}_n}{\Delta_n} + m^2\alpha^2\frac{\overline{\beta}_n}{\Delta_n}\right.$$

$$\left. - im\alpha\sin\phi)u_m - im\alpha\cos\phi\,\frac{\overline{\beta}_n}{\Delta_n}v_m' - (m^2\alpha^2\cos^2\phi\,\frac{\overline{\alpha}_n}{\Delta_n} + im\alpha\cos\phi)w_m\right]v_n'$$

$$\left. - \left[im\alpha\sin\phi\,u_m' - im\alpha\cos\phi\,w_m' - \overline{\alpha}_n w_m' + \overline{\beta}_n u_m'\right]v_n\right\}e^{N+M} = 0 \quad \text{(C.568)}$$

and

$$Re^{-1}v_n^{iv}e^N + \left\{2\Delta_n Re^{-1} - \sigma - (U_{\mathrm{o}} - c_x)\overline{\alpha}_n - (W_{\mathrm{o}} - c_z)\overline{\beta}_n\right\}v_n''e^N$$

$$+ \left\{\Delta_n^2 Re^{-1} - \Delta\sigma - (U_{\mathrm{o}} - c_x)\overline{\alpha}_n\Delta_n - (W_{\mathrm{o}} - c_z)\overline{\beta}_n\Delta_n + W_{\mathrm{o}}''\overline{\beta}_n - U_{\mathrm{o}}''\overline{\alpha}_n\right\}v_n e^N$$

$$+ A\left\{-\left[\frac{\overline{\alpha}_n}{\Delta_n}im\alpha\cos\phi + 1 + \frac{\overline{\beta}_n}{\Delta_n}im\alpha\sin\phi\right]v_m v_n'''\right.$$

$$- \left[(\frac{\overline{\alpha}_n}{\Delta_n}4im\alpha\cos\phi + 1) - 2im\alpha\cos\phi + \frac{\overline{\beta}_n}{\Delta_n}2im\alpha\sin\phi\,\overline{\alpha}_n u_m\right.$$

$$\left. + (\frac{\overline{\alpha}_n}{\Delta_n}2\overline{\alpha}_n - 1)v_m' + (\frac{\overline{\alpha}_n}{\Delta_n}2im\alpha\cos\phi + 1)\overline{\beta}_n w_m\right]v_n''$$

$$+ \left[-\frac{\overline{\alpha}_n}{\Delta_n}\left\{(im^3\alpha^3\cos\phi + 2m^2\alpha^2\cos2\phi\,\overline{\alpha}_n + 2m^2\alpha^2\cos\phi\sin\phi\,\overline{\beta}_n\right.\right.$$

$$\left. - im\alpha\cos\phi\Delta_n + im\alpha\sin\phi\,\overline{\alpha}_n\overline{\beta}_n)v_m - (im\alpha\cos\phi - 2\overline{\alpha}_n)v_m''\right.$$

$$\left.\left. + 4im\alpha\cos\phi\,\overline{\alpha}_n u_m' + 2im\alpha\cos\phi\,\overline{\beta}_n w_m'\right\} - (\Delta_n - m^2\alpha^2\cos2\phi\right.$$

$$+2im\alpha\cos\phi\overline{\alpha}_n + im\alpha\sin\phi\overline{\beta}_n)v_m + v_m'' + 2im\alpha\cos\phi u_m'$$

$$-\frac{\overline{\beta}_n}{\Delta_n}\{im\alpha\sin\phi(m^2\alpha^2 - 2im\alpha\cos\phi\overline{\alpha}_n - \overline{\alpha}_n^2)v_m - im\alpha\sin\phi v_m''$$

$$+2im\alpha\sin\phi\overline{\alpha}_n u_m'\}\Big]v_n' + \Big[\{-\overline{\alpha}_n\Delta_n + m^2\alpha^2\overline{\alpha}_n - m^2\alpha^2\cos\phi\sin\phi\overline{\beta}_n$$

$$+2im\alpha\cos\phi(\overline{\beta}_n^2 - \overline{\alpha}_n^2) - 2im\alpha\sin\phi\overline{\alpha}_n\overline{\beta}_n)\}u_m + (m^2\alpha^2$$

$$-2im\alpha\cos\phi\overline{\alpha}_n - im\alpha\sin\phi\overline{\beta}_n + \overline{\beta}_n^2 - \overline{\alpha}_n^2)v_m' + \overline{\alpha}_n u_m''$$

$$-(\overline{\beta}_n\Delta_n - m^2\alpha^2\cos^2\phi\overline{\beta}_n + 2im\alpha\cos\phi\overline{\alpha}_n\overline{\beta}_n)w_m - v_m'''$$

$$+\overline{\beta}_n w_m''\Big]v_n + im\alpha\Big[\frac{\overline{\beta}_n}{\Delta_n}\cos\phi - \frac{\overline{\beta}_n}{\Delta_n}\sin\phi\Big]v_m\eta_n'' + \Big[\frac{\overline{\beta}_n}{\Delta_n}(2\overline{\alpha}_n v_m'$$

$$+4im\alpha\cos\phi\overline{\alpha}_n u_m + 2im\alpha\cos\phi\overline{\beta}_n w_m) - \frac{\overline{\alpha}_n}{\Delta_n}(2im\alpha\sin\phi\overline{\alpha}_n u_m)\Big]\eta_n'$$

$$+\Big[\frac{\overline{\beta}_n}{\Delta_n}\{(im^3\alpha^3\cos\phi + 2m^2\alpha^2\cos2\phi\overline{\alpha}_n + 2m^2\alpha^2\cos\phi\sin\phi\overline{\beta}_n$$

$$-im\alpha\cos\phi\Delta_n + im\alpha\sin\phi\overline{\alpha}_n\overline{\beta}_n)v_m - (im\alpha\cos\phi - 2\overline{\alpha}_n)v_m''$$

$$+4im\alpha\cos\phi\overline{\alpha}_n u_m' + 2im\alpha\cos\phi\overline{\beta}_n w_m'\} - \frac{\overline{\alpha}_n}{\Delta_n}\Big(im\alpha\sin\phi(m^2\alpha^2$$

$$-2im\alpha\cos\phi\overline{\alpha}_n - \overline{\alpha}_n^2)v_m - im\alpha\sin\phi v_m''$$

$$+2im\alpha\sin\phi\overline{\alpha}_n u_m'\Big)\Big]\eta_n\Big\}e^{N+M} = 0,$$

where

$$\Delta_n = (\overline{\alpha}_n^2 + \overline{\beta}_n^2), \quad e^N = e^{(in/2)\alpha_r(x\cos\phi + z\sin\phi)}$$

and

$$e^M = e^{im\alpha_r(x\cos\phi + z\sin\phi)}.$$

(d) If $A = 0$ in the secondary instability equations, how do the terms that remain compare with the Orr–Sommerfeld and Squire equations? For $A = 0$, (C.568) and (C.569) reduce to

$$\left[Re^{-1}\eta_n'' + \left\{ \Delta_n Re^{-1} - \sigma - (U_o - c_x)\overline{\alpha}_n - (W_o - c_z)\overline{\beta}_n \right\} \eta_n \right.$$

$$\left. + (W_o'\overline{\alpha}_n + U_o'\overline{\beta}_n)v_n \right] e^N = 0, \tag{C.569}$$

and

$$Re^{-1}v_n^{iv}e^N + \left\{ 2\Delta_n Re^{-1} - \sigma - (U_o - c_x)\overline{\alpha}_n - (W_o - c_z)\overline{\beta}_n \right\} v_n'' e^N$$

$$+ \left\{ \Delta_n^2 Re^{-1} - \Delta_n\sigma - (U_o - c_x)\overline{\alpha}_n\Delta_n - (W_o - c_z)\overline{\beta}_n\Delta_n \right.$$

$$\left. + W_o''\overline{\beta}_n - U_o''\overline{\alpha}_n \right\} v_n e^N = 0. \tag{C.570}$$

The equations with $A = 0$ are simply the Squire and Orr–Sommerfeld equations in a reference frame moving with the primary wave.

Exercise 9.4 In the discussion of the secondary instability results of Section 9.3, why are the frequency $F = \omega Re^{-1} \times 10^6$ and spanwise wavenumber $b = \beta Re^{-1} \times 10^3$ used?

Solution
Recall that these parameters are nondimensional numbers, where

$$F = \omega Re^{-1} \times 10^6 = \frac{f^*\delta/U}{U\delta/v} \times 10^6,$$

and

$$b = \beta Re^{-1} \times 10^3 = \frac{\beta^*/U}{U\delta/v} \times 10^3.$$

Therefore, when we integrate the amplification or decay (9.23) of the disturbance as it propagates downstream, the boundary layer thickness (or displacement thickness) grows; however, a constant dimensional frequency (f^*) and spanwise wavenumber (β^*) are required for a distinct instability mode.

Exercise 9.5 By referring to equation (9.22) for the subharmonic profiles, derive the relationship between disturbance profile and eigenfunction for the fundamental mode profiles. (Note, use the $n = 0, \pm 2$ modes.)

Solution
The fundamental mode relationship between the disturbance and the eigenfunctions is given by

$$u_3 = Bu_o + B\cos(\beta z \sqrt{u_2 u_2^* - u_{-2}u_{-2}^*}), \tag{C.571}$$

where $*$ indicates a complex conjugate.

Exercise 9.6 From Section 9.4, use equation (9.46) to show that you can obtain the Orr–Sommerfeld and Squire equations.

Solution
Begin with (9.46) for a single instability mode. We then have the following system of equations:

$$i\alpha u + v' + i\beta w = 0,$$

$$i(\alpha U - \omega)u + vU' + i\alpha p = Re^{-1}\left[u'' - (\alpha^2 + \beta^2)u\right],$$

$$i(\alpha U - \omega)v + p' = Re^{-1}\left[v'' - (\alpha^2 + \beta^2)v\right],$$

$$i(\alpha U - \omega)w + vU' + i\beta p = Re^{-1}\left[w'' - (\alpha^2 + \beta^2)w\right].$$

These equations are nothing more than (2.45)–(2.48) that are used to obtain the Squire and Orr–Sommerfeld equations.

Exercise 9.7 Substituting equations (9.45) and (9.47) into the second-order equations (9.40)–(9.43) leads to equations (9.48). Determine the form of the right-hand sides (a_n, b_n, c_n, d_n).

Solution
With this substitution, (9.48) has coefficients

$$
\begin{aligned}
a_n &= i\alpha_n\tilde{u}_n + \tilde{v}'_n + i\beta_n\tilde{w}_n,\\
b_n &= i(\alpha_n U_o - \omega_n)\tilde{u}_n + \tilde{v}_n U'_o + i\alpha_n\tilde{p}_n - Re^{-1}\left[\tilde{u}''_n - (\alpha^2 - \beta^2)\tilde{u}_n\right],\\
c_n &= i(\alpha_n U_o - \omega_n)\tilde{v}_n + \tilde{p}'_n - Re^{-1}\left[\tilde{v}''_n - (\alpha^2 - \beta^2)\tilde{v}_n\right],\\
d_n &= i(\alpha_n U_o - \omega_n)\tilde{w}_n + i\beta_n\tilde{p}_n - Re^{-1}\left[\tilde{w}''_n - (\alpha^2 - \beta^2)\tilde{w}_n\right].
\end{aligned}
$$

$$(C.572)$$

Exercise 9.8 Derive the PSE equations for two-dimensional disturbances evolving in a two-dimensional basic flow.

Solution
By this point, the reader should be well versed in deriving equations such as the PSE equations, so we will simply say that Section 9.5 clearly shows the process of substituting (9.54) and (9.55) into the Navier–Stokes equations in disturbance form to yield (9.56).

Exercise 9.12 Discuss the process you would follow using secondary instability theory to duplicate the results in Fig. 9.4.

Solution
Figure 9.4 plots the primary and secondary disturbance amplitudes with downstream amplification (or nondimensionally increasing the Reynolds number),

which are determined from equation (9.23). The exact integral relation is replaced by a summation at discrete locations. Recall that the primary disturbance (Orr–Sommerfeld and Squire equations) and the secondary disturbance (Exercise 9.3) are obtained by solving the ordinary differential equations for a given frequency and Reynolds number.

So, to obtain Fig. 9.4, select initial amplitudes (A_o, B_o), the frequency and Reynolds number and solve the equation for the eigenvalues, which of course by now you know include the growth rates (α_i, σ). As shown in Exercise 9.4, the frequency and Reynolds number are changed for the next station downstream along with the primary amplitude A. Numerically the integrations to obtain Fig. 9.4 are easy, since this process can simply be put into a do-loop.

C.9 Solutions for Chapter 10

By now the reader should be extremely familiar with the linear processes associated with hydrodynamic instability theory. Because this chapter is certainly more complicated and is a culmination of the amplification of linear modes, modal interactions and subsequent breakdown, the exercises will primarily focus on essay types of assignments requiring you to think in a cause and effect manner. This cause and effect hypothesizing is consistent with what is required daily from the research engineer.

Exercise 10.1 Outline all of the potential external factors that may impact or induce instabilities. Characterize which factors induce Tollmien–Schlichting waves versus other modes such as crossflow vortex modes.

Solution
It is now widely accepted that stability and transition phenomena are extremely dependent on the initial conditions. These initial conditions are determined by the incoming flow field. In flight, these initial conditions take the form of turbulence, gusts, cloud particulates, insects and dust during take-off and landing, rain and particulate contaminations that are a function of altitude and geographic region (e.g., volcanic ash). In the laboratory, each wind (and water) tunnel has a unique geometric attributes and mechanical systems driving the flow. As such, the facilities will vary in turbulence levels, flow uniformity (i.e., vorticity) and acoustic spectrum.

Surface roughness, discontinuities such as suction slots and holes, critical Reynolds number and pressure gradients couple to the initial flow conditions to induce transition as shown in Fig. C.19.

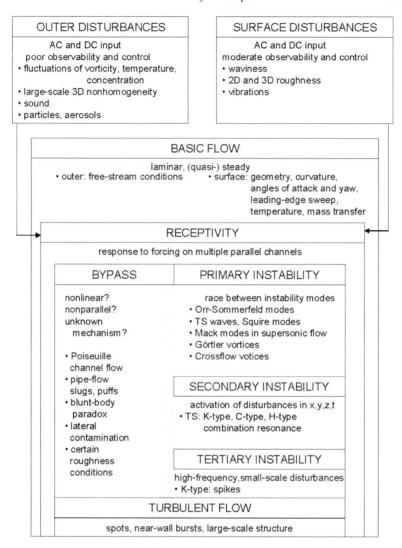

Figure C.19 Pathways to transition (derived from Morkovin, 1985).

At this late point in the text, we know that viscosity is an important factor and destabilizing to the formation and development of Tollmien–Schlichting waves. Also, we know that the Tollmien–Schlichting instability has a predominate direction of propagation close to the freestream direction. We know that if the initial spectrum in the freestream is random, then the initial instability will appear as discrete modes, known as a two-dimensional Tollmien–Schlichting waves. The amplification of Tollmien–Schlichting instabilities is small in regions of favorable pressure gradient and large in regions of adverse pressure gradient.

Crossflow instabilities originate from a base flow that has regions of pressure gradient that cause a curve in the streamlines and a three-dimensional base crossflow velocity profile. The crossflow velocity goes to zero in the freestream and at the wall, so there is an inflection point that is the source of an inviscid instability referred to as the crossflow instability mode. This type of profile can be found in rotating disk flows and in the region near the leading edge of swept wings. The crossflow mode propagates in a direction nearly perpendicular to the local freestream. This mode is characterized by a pair of co-rotating vortices (unlike other centrifugal instabilities that are counter-rotating pairs). The wind-tunnel facilities play an important role in that if the turbulence levels are very low, stationary crossflow modes dominate whereas if the turbulence levels are high traveling crossflow modes dominate. Much more discussion could be given on these instabilities but it is sufficient to say concerning receptivity that the initial conditions, receptivity mechanisms and facilities play a key role into what is observed in the transition process.

Exercise 10.2 Describe the similarities and differences between the Görtler, Dean and Taylor vortices.

Solution

In Chapters 6 and 10, we discussed centrifugal instabilities that occur for shear flows with concave curvature. The Görtler instability occurs in open flows with surfaces that have concave curvature; Dean instabilities occur in curved channel flows; and Taylor instabilities occur in flows between co-rotating cylinders. These names derive, of course, from the researchers who published journal articles on the different flow configurations. Unlike the Tollmien–Schlichting instabilities, which are driven by viscosity, these modes are similar in that counterrotating vortex pairs are observed in the flows and driven by the flows created by the surface curvature. Because of these similarities, we often see a merging of the modes in the literature as in Taylor–Görtler or Taylor–Dean instabilities. Similar to Tollmien–Schlichting instabilities that in themselves do not lead to transition but rather lead to secondary instabilities and subsequent transition, these centrifugal modes do not cause transition, but rather a nonlinear overturning of low- and high-speed momentum. This overturning leads to inflectional profiles and subsequent unstable shear-layer instabilities.

Exercise 10.3 Describe the similarities and differences between the Görtler and crossflow breakdown processes.

Solution

As we just described, both Görtler and crossflow modes are characterized by pairs of rotating vortices with Görtler being counterrotating modes and crossflow being co-rotating modes. In the transition process, both modes experience nonlinear amplification, become asymmetrical and have highly distorted and inflectional mean velocity profiles. These inflectional profiles are unstable and lead to shear layer instabilities that amplify and are the source of transition.

Exercise 10.4 Generate a main routine to loop through frequency and Reynolds numbers so that you can make e^N calculations. For $Re_{\delta^*} = 2{,}240$, calculate the range of N values for each frequency over the neutral curve. What value of N do you use to assess whether transition will occur? This is in fact tracking/predicting a wave of a given physical frequency as it either decays or amplifies downstream.

Solution

The discussion in this chapter has focused on the more general three-dimensional transition problem. In every engineering problem, one must make assumptions to form a solution because often not all information is readily available or specifically stated. As such, here we simplify the problem for this exercise and assume that the question is related to a two-dimensional flat-plate boundary layer (i.e., the Blasius profile solutions hold for simplicity) and transition is dominated by two-dimensional Tollmien–Schlichting waves during the linear process. Looping through frequency and Reynolds number regimes implies that we are solving the spatial stability equations, which means that the routine will solve the Orr–Sommerfeld equation for a sequence of frequency–Reynolds number combinations. This routine will build a loop around the code developed in Exercise 4.4. The outer loop will make use of equation (10.12), where A_o is an arbitrary small number (e.g., 0.0001), and S_0 and S_1 are determined from the solution of the Orr–Sommerfeld equation where $\alpha_i \simeq 0$ for some frequency–Reynolds number combination. Recall that we normally make use of nondimensional quantities in reality when we track the amplification of a wave/instabilities, physical quantities that are tracking the evolution of a dimensional wave evolving downstream. So, the marching process must track single frequency waves as they amplify or decay downstream. The marching/iteration process starts upstream with a small nondimensional frequency and Reynolds number, which are linked to an individual wave ω^*. Then, using the following nondimensional relationships, the evolution will be used to evolve the instability

$$\omega = \frac{\omega^* \delta^*}{U} \quad \text{and} \quad Re = \frac{\rho U \delta^*}{\mu}. \quad \text{(C.573)}$$

Recall that $\delta^* \simeq 1.72(Ux/\mu)^{-1/2}$, so marching from smaller δ^* to larger values implies marching downstream in the boundary layer. Note also that the relationships between physical and nondimensional quantities in equation (C.573) are linear. Then, to begin to solve the Orr–Sommerfeld equation, start with a small value of ω and Re and incrementally increase both where the slope of the increase depends on the instability frequency as graphically shown in Fig. 10.9. As you march from upstream to downstream in x, it is numerically important to make the change in $x(\delta^*)$ small enough so that the integration process in equation (10.12) will have a small error. This is a trial and error process. Also, for a sufficiently small marching increment, the neutral curve location for S_0 and S_1 will become more accurate. The outer-loop changes in frequency and Reynolds number will continue until branch II of the neutral curve is reached, which numerically can be determined when $\alpha_i > 0$. The exercise will provide a series of solutions to the problem $f(\alpha_i, \omega, Re)$, which can be used to integrate from A_0 to A_1 (maximum value after branch II of the neutral curve) for all/any waves with frequency ω as shown in Fig. 10.10. This integration results in the N-values. Make note again that these N-values are empirical correlations that provide an indication of where to expect transition based on a linear theory.

C.10 Solutions for Chapter 11

The exercises contain both derivations and coding. The chapter describes in significant detail how to derive the equations. These derivations are posed as exercise problems because it is important to be able to actually go through the process and derive by pen to understand the nuances associated each one. For the coding exercises, the development of the various routines and their validation are sufficient to provide solutions.

C.11 Solutions for Chapter 12

Begin with the baseline solutions that you obtained from DNS and linear stability theory for Reynolds number based on displacement thickness ($R = 900$) and nondimensional frequency ($F = 86 = \omega/Re \times 10^6$) and the profiles from the Orr–Sommerfeld equation and Blasius basic flows for the initial conditions.

Exercise 12.1 Introduce an oscillatory suction and blowing condition downstream of the wave-forcing location (control actuator). Do a parameter analysis on the amplitude and frequency of the actuator holding the forcing conditions fixed. Discuss your results. Did you observe any wave cancellation or suppression downstream of the actuator? Why or why not?

Solution

Wave cancellation, suppression or amplification will be determined from the summation of the initial disturbance frequency and amplitude linearly combined with the control amplitude and frequency. If the control amplitude exactly matches the disturbance and the frequency is 180 degrees out of phase with the disturbance, then there will be exact disturbance cancellation. However, with numerical issues and in experimental facilities, it is unlikely that exact wave cancellation will ever occur in practice. So, either wave amplitude suppression or amplification should be observed in this exercise.

Exercise 12.2 Write a simple routine to represent the feed-forward strategy discussion in Section 12.4. Implement this strategy as discussion in the section using two grid points upstream of the actuator but downstream of the forcing wave generator. With two sensors, is it easy or difficult to cancel the wave? Why or why not?

Solution

In practice, two or more sensors and actuators can be implemented in numerical simulations, experiments and applications to provide a reasonable control strategy. Using two or more sensors/actuators can never lead to exact cancellation of disturbances similar to the issues described in discussion of Exercise 12.1.

Exercise 12.3 Discuss the process you might use to introduce optimal flow control theory into your DNS code. Would you see these changes to the code as being extremely difficult, somewhat difficult or easy to implement? Explain the rationale for your answer.

Solution

In short, the introduction of optimal flow control theory coupled to DNS code is very difficult to implement and obtain reasonable solutions because of numerical nuances associated with coding. Clearly DNS/optimal flow control coupling is possible as highlighted by the references listed in this chapter.

We just consider DNS, which has many challenges, but which has become more commonplace since the first publication of TCHS. A piece of DNS code should minimize numerical dissipation (i.e., higher-order numerical methods) and should be time-resolved (i.e., higher-order time advancement). The boundary conditions should reflect the true physics of the problem, although with numeric this is not entirely possible. For example, if a mode is introduced by suction/blowing too close to the inflow conditions, erroneous solutions will result in the simulation. The outflow conditions are designed to allow modes to propagate out of the computational domain. This too is challenging and any incorrect implementation can lead to erroneous results at the inflow and corners of the domain because the equations are elliptic. Although instabilities such as the Tollmien–Schlichting mode should be generated by a receptivity mechanism, the numerics involved with discrete Navier–Stokes equations alone can generate the viscous instability. This is generally avoided by inducing the instabilities either through a strong inflow condition (as discussed in Chapter 11) or through direct means such as suction and blowing or through a thermal boundary condition.

Conceptually, it should be trivial to implement the adjoint equations because they have terms very similar to the Navier–Stokes equations. However, if we carefully look at these equations with understanding of the challenges in a numerical implementation, we find that obtaining physically realistic solutions can be challenging. For the physical flat-plate boundary layer problem, the flow has an inflow on the left boundary condition, an outflow on the right boundary condition and the time-evolution of the flow and disturbances travel from left to right in the domain. As discussed earlier, the outflow treatment must permit the propagation of disturbances out of the domain without reflection. For the adjoint equations, the propagation and flow are in the opposite direction in the same domain, with the downstream sensor becomes the disturbance generator in this system. The original inflow now becomes the outflow for the adjoint system, so the numerics in the coupled problem must be able to accommodate this dual boundary condition. Finally, we note that the coefficients of the adjoint system (12.31) and (12.36) have velocities that couple to Navier–Stokes (DNS) equations. They may seem somewhat trivial but for a three-dimensional problem, the equations will require velocities (u, v, w) in four-dimensional space (x, y, z, t). Practical solutions to the coupled problem then require constant read/write velocity vectors to disk and a coupled looping between solving the Navier–Stokes equations and the adjoint system.

References

Acheson, D.J. (1990). *Elementary Fluid Dynamics*. Oxford University Press.

Ames, W.F. (1977). *Numerical Methods for Partial Differential Equations*. Academic Press.

Anders, S. & Fischer, M. (1999). F-16XL-2 supersonic laminar flow control flight test experiment. NASA TP-1999–209683.

Arnal, D. (1994). Boundary layer transition: Predictions based on linear theory. AGARD-R-793.

Arnal, D., Habiballah, M. & Coustols, E. (1984). Laminar instability theory and transition criteria in two- and three-dimensional flow. La Recherche Aérospatiale 1984–2.

Arnal, D., Casalis, G. & Juillen, J.C. (1990). Experimental and theoretical analysis of natural transition on infinite swept wing. In *IUTAM Symposium on Laminar–Turbulent Transition*, D. Arnal & R. Michel (eds). Springer, 311–326.

Arnal, D., Juillen, J.C. & Casalis, G. (1991). The effects of wall suction on laminar–turbulent transition in three-dimensional flow. *ASME FED* **114**, 155–162.

Artola, M. & Majda, A.J. (1987). Nonlinear development of instabilities in supersonic vortex sheets. *Physica D* **28**, 253–281.

Ashpis, D.E. & Erlebacher, G. (1990). On the continuous spectra of the compressible boundary layer stability equations. In *Instability and Transition II*, M.Y. Hussaini & R.G. Voigt (eds). Springer, 145–159.

Baek, P. & Fuglsang, P. (2009). Experimental detection of transition on wind turbine airfoils. European Wind Energy Conference, March, 16–19, 2009.

Balachandar, S., Streett, C.L. & Malik, M.R. (1990). Secondary instability in rotating disk flows. AIAA Paper 90–1527.

Balakumar, P. & Malik, M.R. (1992). Discrete modes and continuous spectra in supersonic boundary layers. *J. Fluid Mech.* **239**, 631–656.

Balsa, T.F. & Goldstein, M.E. (1990). On the instabilities of supersonic mixing layers: A high Mach number asymptotic theory. *J. Fluid Mech.* **216**, 585–611.

Banks, D.W., van Dam, C.P., Shiu, H.J. & Miller, G.M. (2000). Visualization of in-flight flow phenomena using infrared thermography. NASA TM-2000–209027.

517

Barston, F.M. (1980). A circle theorem for inviscid steady flows. *Int. J. Eng. Sci.* **18**, 477–489.

Batchelor, G.K. (1964). Axial flow in trailing line vortices. *J. Fluid Mech.* **20**, 645–658.

Batchelor, G.K. (1967). *An Introduction to Fluid Dynamics*. Cambridge University Press.

Batchelor, G.K. & Gill, A.E. (1962). Analysis of the stability of axisymmetric jets. *J. Fluid Mech.* **14**, 529–551.

Bayly, B.J., Orszag, S.A. & Herbert, T. (1988). Instability mechanisms in shear flow transition. *Ann. Rev. Fluid Mech.* **20**, 359–391.

Benney, D.J. (1961). A non-linear theory for oscillations in a parallel flow. *J. Fluid Mech.* **10**, 209–236.

Benney, D.J. (1964). Finite amplitude effects in an unstable laminar boundary layer. *Phys. Fluids* **7**, 319–326.

Benney, D.J. & Lin, C.C. (1960). On the secondary motion induced by oscillations in a shear flow. *Phys. Fluids* **3**(4), 656–657.

Benney, D.J. & Gustavsson, L.H. (1981). A new mechanism for linear and nonlinear hydrodynamic instability. *Studies in Applied Mathematics* **64**, 185–209.

Bertolotti, F.P. (1985). Temporal and spatial growth of subharmonic disturbances in Falkner–Skan flows. M.S. Thesis, Virginia Polytechnic Institute and State University.

Bertolotti, F.P. (1992). Linear and nonlinear stability of boundary layers with streamwise varying properties. PhD Thesis, The Ohio State University.

Bertolotti, F.P. & Crouch, J.D. (1992). Simulation of boundary layer transition: Receptivity to spike stage. NASA CR-191413.

Bertolotti, F.P. & Joslin, R.D. (1995). The effect of far-field boundary conditions on boundary-layer transition. *J. Comput. Phys.* **118**, May, 392–395.

Bestek, H., Thumm, A. & Fasel, H.F. (1992). Numerical investigation of later stages of transition in transonic boundary layers. First European Forum on Laminar Flow Technology, Hamburg, Germany, March 16–18, 1992.

Betchov, R. & Szewczyk, A. (1963). Stability of a shear layer between parallel streams. *Phys. Fluids* **6**(10), 1391–1396.

Betchov, R. & Criminale, W.O. (1966). Spatial instability of the inviscid jet and wake. *Phys. Fluids* **9**, 359–362.

Betchov, R. & Criminale, W.O. (1967). *Stability of Parallel Flows*. Academic Press.

Bewley, T, Moin, P. & Temam, R. (1996). A method for optimizing feedback control rules for wall bounded turbulent flows based on control theory. Forum on Control of Transition and Turbulent Flows, ASME Fluids Engineering Conference, San Diego.

Biringen, S. (1984). Active control of transition by periodic suction-blowing. *Phys. Fluids* **27**(6), 1345–1347.

Blackaby, N., Cowley, S.J. & Hall, P. (1993). On the instability of hypersonic flow past a flat plate. *J. Fluid Mech.* **247**, 369–416.

Blasius, H. (1908). Grenzschichten in Flüssigkeiten mit kleiner Reibung. *Zeitschrift für Angewandte Mathematik und Physik* **56**, 1. (Translation: Boundary layers in fluids of small viscosity, NACA TM-1256, Feb. 1950).

Blumen, W. (1970). Shear layer instability of an inviscid compressible fluid. *J. Fluid Mech.* **40**, 769–781.

Blumen, W., Drazin, P.G. & Billings, D.F. (1975). Shear layer instability of an inviscid compressible fluid. Part 2. *J. Fluid Mech.* **71**, 305–316.

Bogdanoff, D.W. (1983). Compressibility effects in turbulent shear layers. *AIAA J.* **21**(6), 926–927.

Borggaard, J., Burkardt, J., Gunzburger, M. & Peterson, J. (1995). *Optimal Design and Control.* Birkhauser.

Bower, W.W., Kegelman, J.T., Pal, A. & Meyer, G.H. (1987). A numerical study of two-dimensional instability-wave control based on the Orr–Sommerfeld equation. *Phys. Fluids* **30**(4), 998–1004.

Boyce, W.E. & DiPrima, R.C. (1986). *Elementary Differential Equations and Boundary Value Problems*, 4th edition, John Wiley & Sons.

Breuer, K.S. & Haritonidis, J.H. (1990). The evolution of a localized disturbance in a laminar boundary layer. Part I. Weak disturbances. *J. Fluid Mech.* **220**, 569–594.

Breuer, K.S. & Kuraishi, T. (1994). Transient growth in two- and three-dimensional boundary layers. *Phys. Fluids* **6**, 1983–1993.

Briggs, R.J. (1964). *Electron-Stream Interaction with Plasmas.* MIT Press.

Brown, W.B. (1959). Numerical calculation of the stability of cross flow profiles in laminar boundary layers on a rotating disc and on a swept back wing and an exact calculation of the stability of the Blasius velocity profile. Northrop Aircraft, Inc., Rep. NAI 59–5.

Brown, W.B. (1961a). A stability criterion for three-dimensional laminar boundary layers. In *Boundary Layer and Flow Control*, G.V. Lachmann (ed). Pergamon, Vol. 2, 913–923.

Brown, W.B. (1961b). Exact solution of the stability equations for laminar boundary layers in compressible flow. In *Boundary Layer and Flow Control*, G.V. Lachmann (ed). Pergamon, Vol. 2, 1033–1048.

Brown, W.B. (1962). Exact numerical solutions of the complete linearized equations for the stability of compressible boundary layers. Northrop Aircraft Inc., NORAIR Division Rep. NOR–62–15.

Brown, W.B. (1965). Stability of compressible boundary layers including the effects of two-dimensional linear flows and three-dimensional disturbances. Northrop Aircraft Inc., NORAIR Division Rep.

Brown, W.B. & Roshko, A. (1974). On density effects and large structure in turbulent mixing layers. *J. Fluid Mech.* **64**, 775–816.

Bun, Y. & Criminale, W.O. (1994). Early period dynamics of an incompressible mixing layer. *J. Fluid Mech.* **273**, 31–82.

Burden, R.L. & Faires, J.D. (1985). *Numerical Analysis*, 3rd edition. Prindle, Weber & Schmidt.

Bushnell, D.M. (1984). NASA research on viscous drag reduction II. *Laminar-Turbulent Boundary Layers* **11**, 93–98.

Bushnell, D.M., Hefner, J.N. & Ash, R.L. (1977). Effect of compliant wall motion on turbulent boundary layers. *Phys. Fluids* **20**, S31–S48.

Butler, K.M. & Farrell, B.F. (1992). Three dimensional optimal perturbations in viscous shear flow. *Phys. Fluids A* **4**, 1637–1650.

Canuto, C., Hussaini, M.Y., Quarteroni, A. & Zang, T.A. (1988). *Spectral Methods in Fluid Dynamics*. Springer.

Carpenter, M.H., Gottlieb, D. & Abarbanel, S. (1993). The stability of numerical boundary treatments for compact high-order finite difference schemes. *J. Comp. Phys.* **108**(2), 272–295.

Carpenter, P.W. (1990). Status of transition delay using compliant walls. In *Viscous Drag Reduction in Boundary Layers*, D.M. Bushnell & J.N. Hefner (eds). AIAA, 123, 79–113,

Carpenter, P.W. & Garrad, A.D. (1985). The hydrodynamic stability of flow over Kramer-type compliant surfaces. Part 1. Tollmien–Schlichting instabilities. *J. Fluid Mech.* **155**, 465–510.

Carpenter, P.W. & Garrad, A.D. (1986). The hydrodynamic stability of flow over Kramer-type compliant surfaces. Part 2. Flow-induced surface instabilities. *J. Fluid Mech.* **170**, 199–232.

Carpenter, P.W. & Morris, P.J. (1989). Growth of 3-D instabilities in flow over compliant walls. *4th Asian Congress of Fluid Mechanics*, Hong Kong.

Carpenter, P.W. & Morris, P.J. (1990). The effects of anisotropic wall compliance on boundary-layer stability and transition. *J. Fluid Mech.* **218**, 171–223.

Case, K.M. (1960a). Stability of inviscid plane Couette flow. *Phys. Fluids* **3**(2), 143–148.

Case, K.M. (1960b). Stability of an idealized atmosphere. I. Discussion of results. *Phys. Fluids* **3**, 149–154.

Case, K.M. (1961). Hydrodynamic stability and the inviscid limit. *J. Fluid Mech.* **10**(3), 420–429.

Cebeci, T. & Stewartson, K. (1980). On stability and transition in three-dimensional flows. *AIAA J.* **18**(4), 398–405.

Chandrasekhar, S. (1981). *Hydrodynamic and Hydromagnetic Stability*. Dover Publications.

Chang, C.L. & Malik, M.R. (1992). Oblique mode breakdown in a supersonic boundary layer using nonlinear PSE. In *Instability, Transition, and Turbulence*, M.Y. Hussaini, A. Kumar & C.L. Streett (eds). Springer, 231–241.

Charney, J.G. (1947). The dynamics of long waves in a baroclinic westerly current. *J. Meteor.* **4**, 135–162.

Chen, J.H., Cantwell, B.J. & Mansour, N.N. (1989). Direct numerical simulation of a plane compressible wake: Stability, vorticity dynamics, and topology. PhD Thesis, Stanford University, Thermosciences Division Report No. TF-46.

Chen, J.H., Cantwell, B.J. & Mansour, N.N. (1990). The effect of Mach number on the stability of a plane supersonic wake. *Phys. Fluids A* **2**, 984–1004.

Chimonas, G. (1970). The extension of the Miles–Howard theorem to compressible fluids. *J. Fluid Mech.* **43**, 833–836.

Chinzei, N., Masuya, G., Komuro, T., Murakami, A. & Kudou, D. (1986). Spreading of two-stream supersonic turbulent mixing layers. *Phys. Fluids* **29**, 1345–1347.

Choi, H., Temam, R., Moin, P. & Kim, J. (1993). Feedback control of unsteady flow and its application to the stochastic Burgers equation. *J. Fluid Mech.* **253**, 509–543.

Choudhari, M. (1994). Roughness induced generation of crossflow vortices in three-dimensional boundary layers. *Theor. & Comput. Fluid Dyn.* **5**, 1–31.

Choudhari, M. & Streett, C.L. (1994). Theoretical prediction of boundary-layer receptivity. *25th AIAA Fluid Dynamics Conf.*, June 20–23, 1994, AIAA Paper 94–2223.

Clauser, F.H. & Clauser, M.U. (1937). The effect of curvature on the transition from laminar to turbulent boundary layer. NACA TN-613.

Coles, D. (1965). Transition in circular Couette flow. *J. Fluid Mech.* **21**(3), 385–425.

Cooper, A.J. & Carpenter, P.W. (1997). The stability of the rotating-disc boundary layer flow over a compliant wall. Type I and Type II instabilities. *J. Fluid Mech.* **350**, 231–259.

Cousteix, J. (1992). Basic concepts on boundary layers. *Special Course on Skin Friction Drag Reduction*, AGARD-R-786.

Cowley, S.J. & Hall, P. (1990). On the instability of hypersonic flow past a wedge. *J. Fluid Mech.* **214**, 17–42.

Craik, A.D.D. (1971). Nonlinear resonant instability in boundary layers. *J. Fluid Mech.* **50**(2), 393–413.

Craik, A.D.D. & Criminale, W.O. (1986). Evolution of wavelike disturbances in shear flows: A class of exact solutions of the Navier–Stokes equations. *Proc. R. Soc. London Ser. A* **406**(1830), 13–26.

Crawford, B.K., Duncan, G.T., West, D.E. & Saric, W.S. (2013) Laminar-turbulent boundary layer transition imaging using IR thermography. *Optics and Photonics J.* **3**, 233–239.

Criminale, W.O. (1960). Three-dimensional laminar instability, AGARD-R-266.

Criminale, W.O. (1991). Initial-value problems and stability in shear flows. *Int. Symp. on Nonlinear Problems in Eng. and Sci.*, Beijing, China, 43–63.

Criminale, W.O. & Kovasznay, L.S.G. (1962). The growth of localized disturbances in a laminar boundary layer. *J. Fluid Mech.* **14**, 59–80.

Criminale, W.O. & Drazin, P.G. (1990). The evolution of linearized perturbations of parallel flows. *Studies Appl. Math.* **83**, 123–157.

Criminale, W.O. & Drazin, P.G. (2000). The initial-value problem for a modeled boundary layer. *Phys. Fluids A* **12**, 366–374.

Criminale, W.O. & Lasseigne, D.G. (2002). Use of multiple scales, multiple time in shear flow stability analysis. Personal notes.

Criminale, W.O., Long, B. & Zhu, M. (1991). General three-dimensional disturbances to inviscid Couette flow. *Studies in Appl. Math.* **86**, 249–267.

Criminale, W.O., Jackson, T.L. & Lasseigne, D.G. (1995). Towards enhancing and delaying disturbances in free shear flows. *J. Fluid Mech.* **294**, 283–300.

Criminale, W.O., Jackson, T.L., Lasseigne, D.G. & Joslin, R.D. (1997). Perturbation dynamics in viscous channel flows. *J. Fluid Mech.* **339**, 55–75.

Crouch, J.D. (1994). Receptivity of boundary layers. AIAA Paper 94–2224.

Dagenhart, J.R. & Saric, W.S. (1999). Crossflow stability and transition experiments in a swept wing flow. NASA TP-1999–209344.

Dagenhart, J.R., Saric, W.S., Mousseux, M.C. & Stack, J.P. (1989). Crossflow vortex instability and transition on a 45-degree swept wing. AIAA Paper 89–1892.

Danabasoglu, G., Biringen, S. & Streett, C.L. (1990). Numerical simulation of spatially-evolving instability control in plane channel flow. AIAA Paper 90–1530.

Danabasoglu, G., Biringen, S. & Streett, C.L. (1991). Spatial simulation of instability control by periodic suction blowing. *Phys. Fluids A* **3**(9), 2138–2147.

Davey, A. (1980). On the numerical solution of difficult boundary-value problems. *J. Comput. Phys.* **35**, 36–47.

Davey, A. (1982). A difficult numerical calculation concerning the stability of the Blasius boundary layer. In *Stability in the Mechanics of Continua*, 2nd Symposium, Nümbrecht, Germany, Aug. 31–Sept. 4, 1981, F.H. Schroeder (ed). Springer, 365–372.

Davey, A. & Drazin, P.G. (1969). The stability of Poiseuille flow in a pipe. *J. Fluid Mech.* **36**, 209–218.

Day, M.J., Reynolds, W.C. & Mansour, N.N. (1998a). The structure of the compressible reacting mixing layer: Insights from linear stability analysis. *Phys. Fluids* **10**(4), 993–1007.

Day, M.J., Reynolds, W.C. & Mansour, N.N. (1998b). Parameterizing the growth rate influence of the velocity ratio in compressible reacting mixing layers. *Phys. Fluids* **10**(10), 2686–2688.

Dean, W.R. (1928). Fluid motion in a curved channel. *Proc. R. Soc. London Ser. A* **15**, 623–631.

Deardorff, J.W. (1963). On the stability of viscous plane Couette flow. *J. Fluid Mech.* **15**, 623–631.

Demetriades, A. (1958). An experimental investigation of the stability of the hypersonic laminar boundary layer. California Institute of Technology, Guggenheim Aeronautical Laboratory, Hypersonic Research Project, Memo. No. 43.

Dhawan, S. & Narasimha, R. (1958). Some properties of boundary layer flow during transition from laminar to turbulent motion. *J. Fluid Mech.* **3**(4), 418–436.

Dikii, L.A. (1960). On the stability of plane parallel flows of an inhomogeneous fluid. (in Russian) *Prikl. i Mekh.* **24**, 249–257 (Translation: *J. Appl. Math. Mech.* **24**, 357–369).

DiPrima, R.C. (1959). The stability of viscous flow between rotating concentric cylinders with a pressure gradient acting around the cylinders. *J. Fluid Mech.* **6**, 462–468.

DiPrima, R.C. (1961). Stability of nonrotationally symmetric disturbances for viscous flow between rotating cylinders. *Phys. Fluids* **4**, 751–755.

DiPrima, R.C. & Habetler, G.J. (1969). A completeness theorem for non-selfadjoint eigenvalue problems in hydrodynamic stability. *Arch. Rat. Mech. Anal.* **34**(3), 218–227.

Djordjevic, V.D. & Redekopp, L.G. (1988). Linear stability analysis of nonhomentropic, inviscid compressible flows. *Phys. Fluids* **31**(11), 3239–3245.

Drazin, P.G. (1978). Variations on a theme of Eady. In *Rotating Fluids in Geophysics*, P. H. Roberts & A. M. Soward (eds). 139–169.

Drazin, P.G. & Howard, L.N. (1962). Shear layer instability of an inviscid compressible fluid. Part 2. *J. Fluid Mech.* **71**, 305–316.

Drazin, P.G. & Howard, L.N. (1966). Hydrodynamic stability of parallel flow of inviscid flow. *Adv. Appl. Mech.* **9**, 1–89.

Drazin, P.G. & Davey, A. (1977). Shear layer instability of an inviscid compressible fluid. Part 3. *J. Fluid Mech.* **82**, 255–260.

Drazin, P.G. & Reid, W.H. (1984). *Hydrodynamic stability.* Cambridge University Press.

Drazin, P.G. & Reid, W.H. (2004). *Hydrodynamic stability*, 2nd edition. Cambridge University Press.

Duck, P.W. & Foster, M.R. (1980). The inviscid stability of a trailing line vortex. *J. App. Math. and Phys. (ZAMP)* **31**, 524–532.

Duck, P.W. & Khorrami, M.R. (1992). A note on the effects of viscosity on the stability of a trailing-line vortex. *J. Fluid Mech.* **245**, 175–189.

Duck, P.W., Erlebacher, G. & Hussaini, M.Y. (1994). On the linear stability of compressible plane Couette flow. *J. Fluid Mech.* **258**, 131–165.

Dunn, D.W. (1960). Stability of laminar flows. DME/NAE Quarterly Bulletin No. 1960 (3), National Research Council of Canada, Ottawa, Oct., 15–58.

Dunn, D.W. & Lin, C.C. (1955). On the stability of the laminar boundary layer in a compressible fluid. *J. Aero. Sci.* **22**, 455–477.

Eady, E.A. (1949). Long waves and cyclone waves. *Tellus* **1**, 33–52.

Eckhaus, W. (1962a). Problémes non linéaires dan la théorie de la Stabilité. *J. de Mecanique* **1**, 49–77.

Eckhaus, W. (1962b). Problémes non linéaires de stabilité dans un espace a deux dimensions. I. Solutions péridoques. *J. de Mecanique* **1**, 413–438.

Eckhaus, W. (1963). Problémes non linéaires de stabilité dans un espace a deux dimenions. II. Stabilité des solutions périodques. *J. de Mecanique* **2**, 153–172.

Eckhaus, W. (1965). *Studies in Non-Linear Stability Theory.* Springer.

Ekman, V.W. (1905). On the influence of the Earth's rotation on ocean currents. *Arkiv fŏr Matematik, Astronomi, och Fysik* **2**(11), 1–53.

Eliassen, A., Høiland, E. & Riis, E. (1953). Two-dimensional perturbations of a flow with constant shear of a stratified fluid. Inst. Weather Climate Res., Oslo, Publ. No. 1., Institute of Theoretical Astrophysics, 58 pgs.

Esch, R.E. (1957). The instability of a shear layer between two parallel streams. *J. Fluid Mech.* **3**, 289–303.

Falco, R.E. (1977). Coherent motions in the outer region of turbulent boundary layers. *Phys. Fluids* **20**(10), 5124.

Faller, A.J. & Kaylor, R.E. (1967). Instability of the Ekman spiral with applications to the planetary boundary layer. *Phys. Fluids* **10**, S212–S219.

Fasel, H. (1976). Investigation of the stability of boundary layers by a finite-difference model of the Navier–Stokes equations. *J. Fluid Mech.* **78**, 355–383.

Fasel, H. (1990). Numerical simulation of instability and transition in boundary layer flows. In *IUTAM Symposium on Laminar–Turbulent Transition*, D. Arnal & R. Michel (eds). Springer, 587–597.

Fasel, H. & Thumm, A. (1991). Numerical simulation of three-dimensional boundary layer transition. *Bull. Am. Phys. Soc.* **36**, 2701.

Fedorov, A.V. & Khokhlov, A.P. (1993). Excitation and evolution of unstable disturbances in supersonic boundary layer. In *Transitional and Turbulent Compressible Flows*. ASME FED, Vol. 151.

Fjørtoft, R. (1950). Application of integral theorems in deriving criteria of stability of laminar flow and for the baroclinic circular vortex. *Geofysiske Publikasjoner* **17**, 1–52.

Floquét, G. (1883). Sur les équations differéntielles linéaires á coefficients périodiques. *Annales Scientifiques Ecole Normale Superieure* **2**(12), 47–89.

Fromm, J.E. & Harlow, F.H., (1963). Numerical solution of the problem of vortex street development. *Phys. Fluids* **6**, 975–982.

Fursikov, A.V., Gunzburger, M. & Hou, L. (1996). Boundary value problems and optimal boundary control of the Navier–Stokes system: The two-dimensional case. *SIAM Journal on Control and Optimization* **36**(3), 852–894.

Gad-el-Hak, M. (2000). *Flow Control: Passive, Active, and Reactive Flow Management.* Cambridge University Press.

Gad-el-Hak, M., Pollard, A. & Bonnet, J.P. (eds) (1998). *Flow Control: Fundamentals and Practices.* Springer.

Gaster, M. (1962). A note on the relation between temporally-growing and spatially-growing disturbances in hydrodynamic stability. *J. Fluid Mech.* **14**, 222–224.

Gaster, M. (1965a). The role of spatially growing waves in the theory of hydrodynamic stability. *Prog. Aeron. Sci.* **6**, 251–270.

Gaster, M. (1965b). On the generation of spatially growing waves in a boundary layer. *J. Fluid Mech.* **22**, 433–441.

Gaster, M. (1965c). A simple device for preventing turbulent contamination on swept leading edges. *J. R. Aero. Soc.* **69**, 788–789.

Gaster, M. (1968). Growth of disturbances in both space and time. *Phys. Fluids* **11**(4), 723–727.

Gaster, M. (1974). On the effects of boundary-layer growth on flow stability. *J. Fluid Mech.* **66**, 465–480.

Gaster, M. (1983). The development of a two-dimensional wavepacket in a growing boundary layer. *Proc. R. Soc. London Ser. A.* **384**, 317–332.

Gaster, M. (1988). Is the dolphin a red herring? In *Turbulence Management and Relaminarisation*, H.W. Liepmann and R. Narasimha (eds). Bangalore: IUTAM, 285–304.

Gaster, M. & Davey, A. (1968). The development of three dimensional wave packets in unbounded parallel flows. *J. Fluid Mech.* **32**, 801–808.

Gaster, M. & Grant, I. (1975). An experimental investigation of the formation and development of a wave packet in a laminar boundary layer. *Proc. R. Soc. London Ser. A* **347**, 253–269.

Gear, C.W. (1978). *Applications and Algorithms in Science and Engineering.* Science Research Associates, Inc.

Germano, M., Piomelli, U., Moin, P. & Cabot, W.H. (1991). A dynamic subgrid-scale eddy viscosity model. *Phys. Fluids A* **3**, 1760–1765.

Girard, J.J. (1988). Study of the stability of compressible Couette flow. PhD Thesis, Washington State University.

Glatzel, W. (1988). Sonic instability in supersonic shear flows. *Mon. Not. R. Astron. Soc.* **231**, 795–821.

Glatzel, W. (1989). The linear stability of viscous compressible plane Couette flow. *J. Fluid Mech.* **202**, 515–541.

Gold, H. (1963). Stability of laminar wakes. PhD Thesis, California Institute of Technology.

Goldstein, S. (1930). Concerning some solutions of the boundary layer equations in hydrodynamics. *Proc. Camb. Phil. Soc.* **26**, 1–30.

Goldstein, M.E. (1983). The evolution of Tollmien–Schlichting waves near a leading edge. *J. Fluid Mech.* **127**, 59–81.

Goldstein, M.E. (1985). Scattering of acoustic waves into Tollmien–Schlichting waves by small streamwise variations in surface geometry. *J. Fluid Mech.* **154**, 509–529.

Goldstein, M.E. (1987). Generation of Tollmien–Schlichting waves on interactive marginally separated flows. *J. Fluid Mech.* **181**, 485–518.

Goldstein, M.E. & Wundrow, D.W. (1990). Spatial evolution of nonlinear acoustic mode instabilities on hypersonic boundary layers. *J. Fluid Mech.* **219**, 585–607.

Görtler, H. (1940a). Über eine dreidimensionale Instabilität laminarer Grenzschichten an konkaven Wänden, *Nachr. Akad. Wiss, Göttingen Math-Physik Kl. IIa, Math-Physik-Chem. Abt.* **2**, 1–26 (Translation: NACA Tech. Memo. 1375, June 1954).

Görtler, H. (1940b). Über den Einfluss der Wandkrümmung auf die Enstehung der Turbulenz. *Z. Angew. Math. Mech.* **20**, 138–147.

Görtler, H. & Witting, H. (1958). Theorie der sekundaren Instabilität der laminaren Grenzschichten. *Int. Union Theor. Appl. Mech.*, Grenzschichtforschung, Freiburg, 110–126.

Gottlieb, D. & Orszag, S.A. (1986). *Numerical Analysis of Spectral Methods: Theory and Applications.* SIAM.

Granville, P.S. (1953). The calculation of the viscous drag of bodies of revolution. David Taylor Model Basin Rep. 849.

Greenough, J., Riley, J., Soestrisno, M. & Eberhardt, D. (1989). The effect of walls on a compressible mixing layer. AIAA Paper 89–0372.

Greenspan, H.D. (1969). *The Theory of Rotating Fluids.* Cambridge University Press.

Gregory, N., Stuart, J.T. & Walker, W.S. (1955). On the stability of three-dimensional boundary layers with application to the flow due to a rotating disk. *Phil. Trans. R. Soc. London Ser. A* **248**, 155–199.

Gropengiesser, H. (1969). On the stability of free shear layers in compressible flows. *Deutsche Luft. und Raumfahrt*, FB 69–25 (Translation: NASA TT F-12,786, 1970).

Grosch, C.E. & Salwen, H. (1978). The continuous spectrum of the Orr–Sommerfeld equation. Part 1. The spectrum and the eigenfunctions. *J. Fluid Mech.* **87**, 33–54.

Grosch, C.E. & Jackson, T.L. (1991). Inviscid spatial stability of a three dimensional mixing layer. *J. Fluid Mech.* **231**, 35–50.

Grosch, C.E., Jackson, T.L., Klein, R., Majda, A. & Papageorgiou, D.T. (1991). Supersonic modes of a compressible mixing layer. Unpublished manuscript.

Grosskreutz, R. (1975). An attempt to control boundary-layer turbulence with non-isotropic compliant walls. *Univ. Sci. J. Dar es Salaam* **1**, 65–73.

Guirguis, R.H. (1988). Mixing enhancement in supersonic shear sayers. Part III. Effect of convective Mach number. AIAA Paper 88–0701.

Gunzburger, M. (1995). *Flow Control.* Springer.

Gustavsson, L.H. (1979). Initial-value problem for boundary layer flows. *Phys. Fluids* **22**(9), 1602–1605

Gustavsson, L.H. (1991). Energy growth of three dimensional disturbances in plane Poiseuille flow. *J. Fluid Mech.* **224**, 241–260.

Gustavsson, L.H. & Hultgren, L.S. (1980). A resonance mechanism in plane Couette flow. *J. Fluid Mech.* **98**, 149–159.

Haberman, R. (1987). *Elementary Applied Partial Differential Equations*, 2nd edition. Prentice-Hall, Inc.

Hagan, G. (1855). Über den einfluss der temperatur auf die bewegung des wassers in rohren. *Math. Abh. Akad. Wiss.* (aus dem Jahr 1854), 17–98.

Hains, F.D. (1967). Stability of plane Couette–Poiseuille flow. *Phys. Fluids* **10**, 2079–2080.

Hall, P. & Malik, M.R. (1986). On the instability of a three-dimensional attachment-line boundary layer: Weakly nonlinear theory and a numerical simulation. *J. Fluid Mech.* **163**, 257–282.

Hall, P. & Smith, F.T. (1991). On strongly nonlinear vortex/wave interactions in boundary layer transition. *J. Fluid Mech.* **227**, 641–666.

Hall, P., Malik, M.R. & Poll, D.I.A. (1984). On the stability of an infinite swept attachment line boundary layer. *Proc. Roy. Soc. London Ser. A* **395**, 229–245.

Hama, F.R., Williams, D.R. & Fasel, H. (1979). Flow field and energy balance according to the spatial linear stability theory of the Blasius boundary layer. In *Laminar-Turbulent Transition*, E. Eppler and H. Fasel (eds). Stuttgart, Germany: IUTAM, September 16–22, 1979, 73–85.

Hämmerlin, G. (1955). Über das Eigenwertproblem der dreidimensionalen Instabilität laminarer Grenzschichten an konkaven Wänden. *J. Rat. Mech. Anal.* **4**, 279–321.

Harris, J.E., Iyer, V. & Radwan, S. (1987). Numerical solutions of the compressible 3-D boundary layer equations for aerospace configurations with emphasis on LFC. In *Research in Natural Laminar Flow and Laminar Flow Control*, J.N. Hefner & F.E. Sabo (eds.) March 16–19, 1987. NASA Langley Research Center. NASA CP-2487, 517–545.

Haynes, T.S. & Reed, H.L. (1996). Computations in nonlinear saturation of stationary crossflow vortices in a swept-wing boundary layer. AIAA Paper 96–0182.

Hazel, P. (1972). Numerical studies of the stability of inviscid stratified shear flows. *J. Fluid Mech.* **51**, 39–61.

Healey, J.J. (1995). On the neutral curve of the flat plate boundary layer: Comparison between experiment, Orr–Sommerfeld theory and asymptotic theory. *J. Fluid Mech.* **288**, 59–73.

Hefner, J.N. & Bushnell, D.M. (1980). Status of linear boundary-layer stability theory and the e^N method, with emphasis on swept-wing applications. NASA TP 1645.

Heisenberg, W. (1924). Uber Stabilität und Turbulenz von Flussigkeitsstromen. *Ann. Physik* **74**, 577–627. (Translation: On stability and turbulence of fluid flows. NACA TM-1291, 1951.)

Helmholtz, H. (1868). Über discontinuirliche flüssigkeits-bewegungen. *Akad. Wiss., Berlin, Monatsber.* **23**, 215–228. (Translated by F. Guthrie: On discontinuous movements of fluids. *Phil. Mag.* **36**(4), 337–346, 1868.)

Herbert, Th. (1983). Secondary instability of plane channel flow to subharmonic three-dimensional disturbances. *Phys. Fluids* **26**(4), 871–874.

Herbert, Th. (1984). Secondary instability of shear flows. AGARD-R-709.

Herbert, Th. (1988). Secondary instability of boundary layers. *Ann. Rev. Fluid Mech.* **20**, 487–526.

Herbert, Th. (1991). Boundary-layer transition analysis and prediction revisited. AIAA Paper 91–0737.

Herbert, Th. (1997). Parabolized stability equations. *Ann. Rev. Fluid Mech.* **29**, 245–283.

Herbert, Th., Bertolotti, F.P. & Santos, G.R. (1987). Flóquet analysis of secondary instability in shear flows. In *Stability of Time Dependent and Spatially Varying Flows*, D.L. Dwoyer & M.Y. Hussaini (eds). Springer, 43–57.

Herron, I.H. (1987). The Orr–Sommerfeld equation on infinite intervals. *SIAM Review* **29**(4), 597–620.

Hiemenz, K. (1911). Die grenzschicht an einem in den gleichförmigen flüssigkeitsstrom eingetauchten geraden kreiszylinder. Thesis, Göttingen, *Dingl. Polytechn. J.* **326**, 321.

Hill, D.C. (1995). Adjoint systems and their role in the receptivity problem for boundary layers. *J. Fluid Mech.* **292**, 183–204.

Hocking, L.M. (1975). Non-linear instability of the asymptotic suction velocity profile. *Quart. J. Mech. Appl. Math.* **28**, 341–353.

Høiland, E. (1953). On two-dimensional perturbations of linear flow. *Geofysiske Publikasjoner*, **18**, 1–12

Holmes, B.J., Obara, C.J., Gregorek, G.M., Hoffman, M.J. & Freuhler, R.J. (1983). Flight investigation of natural laminar flow on the Bellanca Skyrocket II. SAE Paper 830717.

Hosder, S. & Simpson, R.L. (2001). Unsteady turbulent skin friction and separation location measurements on a maneuvering undersea vehicle. *39th AIAA Aerospace Sciences Meeting & Exhibit*, January 8–11, 2001, Reno, NV. AIAA Paper 2001–1000.

Howard, L.N. (1961). Note on a paper of John W. Miles. *J. Fluid Mech.* **10**, 509–512.

Howard, L.N. & Gupta, A.A. (1962). On the hydrodynamic and hydromagnetic stability of swirling flows. *J. Fluid Mech.* **14**, 463–476.

Hu, F. Q., Jackson, T.L., Lasseigne, D.G. & Grosch, C.E. (1993). Absolute-convective instabilities and their associated wave packets in a compressible reacting mixing layer. *Phys. Fluids A* **5**(4), 901–915.

Huai, X., Joslin, R.D. & Piomelli, U. (1997). Large-eddy simulation of spatial development of transition to turbulence in a two-dimensional boundary layer. *Theor. Comput. Fluid Dyn.* **9**, 149–163.

Huai, X., Joslin, R.D. & Piomelli, U. (1999). Large-eddy simulation of boundary-layer transition on a swept wedge. *J. Fluid Mech.* **381**, 357–380.

Huerre, P. & Monkewitz, P.A. (1985). Absolute and convective instabilities in free shear layers. *J. Fluid Mech.* **159**, 151–168.

Hughes, T.H. & Reid, W.H. (1965a). On the stability of the asymptotic suction boundary layer profile. *J. Fluid Mech.* **23**, 715–735.

Hughes, T.H. & Reid, W.H. (1965b). The stability of laminar boundary layers at separation. *J. Fluid Mech.* **23**, 737–747.

Hultgren, L.S. & Gustavsson, L.H. (1980). Algebraic growth of disturbances in a laminar boundary layer. *Phys. Fluids* **24**, 1000–1004.

Hultgren, L.S. & Aggarwal, A.K. (1987). Absolute instability of the Gaussian wake profile. *Phys. Fluids* **30**(11), 3383–3387.

Itoh, N. (1996). Simple cases of the streamline-curvature instability in three-dimensional boundary layers. *J. Fluid Mech.* **317**, 129–154.

Iyer, V. (1990). Computation of three-dimensional compressible boundary layers to fourth-order accuracy on wings and fuselages. NASA CR-4269.

Iyer, V. (1993). Three-dimensional boundary layer program (BL3D) for swept subsonic or supersonic wings with application to laminar flow control. NASA CR-4531.

Iyer, V. (1995). Computer program BL2D for solving two-dimensional and axisymmetric boundary layers. NASA CR-4668.

Jackson, T.L. & Grosch, C.E. (1989). Inviscid spatial stability of a compressible mixing layer. *J. Fluid Mech.* **208**, 609–637.

Jackson, T.L. & Grosch, C.E. (1990a). Inviscid spatial stability of a compressible mixing layer. Part 2. The flame sheet model. *J. Fluid Mech.* **217**, 391–420.

Jackson, T.L. & Grosch, C.E. (1990b). Absolute/convective instabilities and the convective Mach number in a compressible mixing layer. *Phys. Fluids A* **2**(6), 949–954.

Jackson, T.L. & Grosch, C.E. (1990c). On the classification of unstable modes in bounded compressible mixing layers. In *Instability and Transition*, M.Y. Hussaini & R.G. Voigt (eds). Springer, Vol. II, 187–198.

Jackson, T.L. & Grosch, C.E. (1991). Inviscid spatial stability of a compressible mixing layer. Part 3. Effect of thermodynamics. *J. Fluid Mech.* **224**, 159–175.

Jackson, T.L. & Grosch, C.E. (1994). Structure and stability of a laminar diffusion flame in a compressible, three dimensional mixing layer. *Theor. Comput. Fluid Dyn.* **6**, 89–112.

Joseph, L.A., Borgoltz, A. & Devenport, W. (2014). Transition detection for low speed wind tunnel testing using infrared thermography. *30th AIAA Aerodynamic Measurement Technology and Ground Testing Conference*, June 16–20, 2014, Atlanta, GA. AIAA Paper 2014–2939.

Joslin, R.D. (1990). The effect of compliant walls on three-dimensional primary and secondary instabilities in boundary layer transition. PhD Thesis, The Pennsylvania State University.

Joslin, R.D. (1995a). Evolution of stationary crossflow vortices in boundary layers on swept wings. *AIAA J.* **33**(7), 1279–1285.

Joslin, R.D. (1995b). Direct simulation of evolution and control of three-dimensional instabilities in attachment-line boundary layers. *J. Fluid Mech.* **291**, 369–392.

Joslin, R.D. (1997). Direct numerical simulation of evolution and control of linear and nonlinear disturbances in three-dimensional attachment-line boundary layers. NASA TP-3623.

Joslin, R.D. (1998). Overview of laminar flow control. NASA TP 1998–208705.

Joslin, R.D. & Morris, P.J. (1992). Effect of compliant walls on secondary instabilities in boundary-layer transition. *AIAA J.* **30**(2), 332–339.

Joslin, R.D. & Streett, C.L. (1994). The role of stationary crossflow vortices in boundary-Layer transition on swept wings. *Phys. Fluids* **6**(10), 3442–3453.

Joslin, R.D., Morris, P.J. & Carpenter, P.W. (1991). The role of three-dimensional instabilities in compliant wall boundary-layer transition. *AIAA J.* **29**(10), 1603–1610.

Joslin, R.D., Streett, C.L. & Chang, C.L. (1992). Validation of three-dimensional incompressible spatial direct numerical simulation code: A comparison with linear stability and parabolic stability equation theories for boundary-layer transition on a flat plate. NASA TP-3205, July.

Joslin, R.D., Streett, C.L. & Chang, C.L. (1993). Spatial direct numerical simulation of boundary-layer transition mechanisms: Validation of PSE theory. *Theor. Comput. Fluid Dyn.* **4**(6), 271–288.

Joslin, R.D., Nicolaides, R.A., Erlebacher, G., Hussaini, M.Y. & Gunzburger, M. (1995). Active control of boundary-layer instabilities: Use of sensors and spectral controller. *AIAA J.* **33**(8), 1521–1523.

Joslin, R.D., Erlebacher, G. & Hussaini, M.Y. (1996). Active control of instabilities in laminar boundary-layer Flow. An overview. *J. Fluids Eng.* **118**, 494–497.

Joslin, R.D., Gunzburger, M.D., Nicolaides, R.A., Erlebacher, G. & Hussaini, M.Y. (1997). Self-contained, automated methodology for optimal flow control. *AIAA J.* **35**(5), 816–824.

Joslin, R.D., Kunz, R.F. & Stinebring, D.R. (2000). Flow control technology readiness: Aerodynamic versus hydrodynamic. *18th Applied Aerodynamics Conference & Exhibit*, August 14–17, 2000, Denver, CO. AIAA Paper 2000–4412.

Kachanov, Y.S. (1994). Physical mechanisms of laminar boundary layer transition. *Ann. Rev. Fluid Mech.* **26**, 411–482.

Kachanov, Y.S. & Levchenko, V.Y. (1984). The resonant interaction of disturbances at laminar–turbulent transition in a boundary layer. *J. Fluid Mech.* **138**, 209–247.

Kachanov, Y.S. & Tararykin, O.I. (1990). The experimental investigation of stability and receptivity of a swept wing flow. In *IUTAM Symposium on Laminar–Turbulent Transition*, D. Arnal & R. Michel (eds). Springer, 499–509.

Kachanov, Y.S., Kozlov, V.V. & Levchenko, V.Y. (1979). Experiments on nonlinear interaction of waves in boundary layers. In *IUTAM Symposium on Laminar–Turbulent Transition*, D. Arnal & R. Michel (eds). Springer, 135–152.

Kaplan, R.E. (1964). The stability of laminar incompressible boundary layers in the presence of compliant boundaries. Massachusetts Institute of Technology, Aero-Elastic and Structures Research Laboratory, ASRL-TR 116–1.

Karniadakis, G.E. & Triantafyllou, G.S. (1989). Frequency selection and asymptotic states in laminar wakes. *J. Fluid Mech.* **199**, 441–469.

Kelvin, Lord (1871). Hydrokinetic solutions and observations. *Phil. Mag.* **4**(42), 362–377. (Reprinted in *Mathematical and Physical Papers*, **4**, 152–165, 1910.)

Kelvin, Lord (1880). On a disturbance in Lord Rayleigh's solution for waves in a plane vortex stratum. *Nature* **23**, 45–46. (Reprinted in *Mathematical and Physical Papers* **4**, 186–187, 1910.)

Kelvin, Lord (1887a). Rectilinear motion of a viscous fluid between parallel plates. In *Mathematical and Physical Papers* **4**, 321–330, 1910.

Kelvin, Lord (1887b). Broad river flowing down an inclined plane bed. In *Mathematical and Physical Papers* **4**, 330–337, 1910.

Kendall, J.M. (1966). Supersonic boundary layer stability experiments. *Bull. Am. Phys. Soc.*

Kennedy, C.A. & Chen, J.H. (1998). Mean flow effects on the linear stability of compressible planar jets. *Phys. Fluids* **10**(3), 615–626.

Kerschen, E.J. (1987). Boundary layer receptivity and laminar flow airfoil design. In *Research in Natural Laminar Flow and Laminar-Flow Control*, J.N. Hefner & F.E. Sabo (eds). March 16–19, 1987. NASA Langley Research Center. NASA CP-2487, 273–287.

Kerschen, E.J. (1989). Boundary-layer receptivity. *AIAA 12th Aeroacoustics Conference*, April 10–12, 1989, San Antonio, TX. AIAA Paper 89–1109.

Khorrami, M.R. (1991). On the viscous modes of instability of a trailing line vortex. *J. Fluid Mech.* **255**, 197–212.

King, R.A. & Breuer, K.S. (2001). Acoustic receptivity and evolution of two-dimensional and oblique disturbances in a Blasius boundary layer. *J. Fluid Mech.* **432**, 69–90.

Kirchgässner, K. (1961). Die instabilität der Strömung zwischen zwei rotierenden Zylindern gegenuber Taylor–Wirbeln fur beliebige Spaltbreiten. *Z. Angew. Math. Phys.* **12**, 14–30.

Klebanoff, P.S., Tidstrom, K.D. & Sargent, L.M. (1962). The three-dimensional nature of boundary layer instability. *J. Fluid Mech.* **12**, 1–34.

Kleiser, L. & Zang, T.A. (1991). Numerical simulation of transition in wall bounded shear flows. *Ann. Rev. Fluid Mech.* **23**, 495–537.

Klemp, J.B. & Acrivos, A. (1972). A note on the laminar mixing of two uniform parallel semi-infinite streams. *J. Fluid Mech.* **55**, 25–30.

Kloker, M. & Fasel, H. (1990). Numerical simulation of two- and three-dimensional instability waves in two-dimensional boundary layers with streamwise pressure gradients. In *IUTAM Symposium on Laminar–Turbulent Transition*, D. Arnal & R. Michel (eds). Springer, 681–686.

Kobayashi, R., Kohama, Y. & Kurosawa, M. (1983). Boundary-layer transition on a rotating cone in axial flow, *J. Fluid Mech.* **127**, 341–52.

Koch, W. (1986). Direct resonances in Orr–Sommerfeld problems. *Acta Mech.* **58**, 11–29.

Kohama, Y., Saric, W.S. & Hoos, J.A. (1991). A high frequency, secondary instability of crossflow vortices that leads to transition. In *Royal Aeronautical Society Conference on Boundary Layer Transition and Control*, Cambridge University.

Koochesfahani, M.M. & Frieler, C.E. (1989). Instability of nonuniform density free shear layers with a wake profile. *AIAA J.* **27**(12), 1735–1740.

Kozusko, F., Lasseigne, D.G., Grosch, C.E. & Jackson, T.L. (1996). The stability of compressible mixing layers in binary gases. *Phys. Fluids* **8**(7), 1954–1963.

Kramer, M.O. (1957). Boundary-layer stabilization by distributed damping. *J. Aero. Sci.* **24**(6), 459–460.

Kramer, M.O. (1965). Hydrodynamics of the dolphin. *Advances in Hydroscience* **2**, 111–130.

Kuo, A.L. (1949). Dynamic instability of two-dimensional nondivergent flow in a barotropic atmosphere. *J. Meteorology* **6**, 105–122.

Kuramoto, Y. (1980). Instability and turbulence of wave fronts in reaction-diffusion systems. *Progr. Theor. Phys.* **63**(6), 1885–1903.

Ladd, D.M. (1990). Control of natural laminar instability waves on an axisymmetric body. *AIAA J.* **28**(2), 367–369.

Ladd, D.M. & Hendricks, E.W. (1988). Active control of 2-D instability waves on an axisymmetric body. *Exp. Fluids* **6**, 69–70.

Lamb, H. (1945). *Hydrodynamics.* Cambridge University Press. Reprinted by Dover.

Landahl, M.T. (1980). A note on an algebraic instability of inviscid parallel shear flows. *J. Fluid Mech.* **98**, 243–251.

Landau, L.D. (1944). On the problem of turbulence. *Akademiya Nauk SSSR. Doklady* **44**, 311–314.

Lasseigne, D.G., Joslin, R.D., Jackson, T.L. & Criminale, W.O. (1999). The transient period for boundary layer disturbances. *J. Fluid Mech.* **381**, 89–119.

Laufer, J. & Vrebalovich, T. (1957). Experiments on the instability of a supersonic boundary layer. *9th Int. Cong. Appl. Mech.* **4**, 121–131.

Laufer, J. & Vrebalovich, T. (1958). Stability of a supersonic laminar boundary layer on a flat plate. California Institute of Technology, Jet Propulsion Laboratory Rep. 20–116.

Laufer, J. & Vrebalovich, T. (1960). Stability and transition of a supersonic laminar boundary layer on an insulated flat plate. *J. Fluid Mech.* **9**, 257–299.

Lecointe, Y. & Piquet, J. (1984). On the use of several compact methods for the study of unsteady incompressible viscous flow round a circular cylinder. *Computers & Fluids* **12**(4), 255–280.

Leehey, P. & Shapiro, P.J. (1980). Leading edge effect in laminar boundary layer excitation by sound. In *IUTAM Symposium on Laminar–Turbulent Transition*, D. Arnal & R. Michel (eds). Springer, 321–331.

Lees, L. (1947). The stability of the laminar boundary layer in a compressible fluid. NACA TR-876.

Lees, L. & Lin, C.C. (1946). Investigation of the stability of the laminar boundary layer in a compressible fluid. NACA TN-1115.

Lees, L. & Reshotko, E. (1962). Stability of the compressible boundary layer. *J. Fluid Mech.* **12**, 555–590.

Lees, L. & Gold, H. (1964). Stability of laminar boundary layers and wakes at hypersonic speeds. I. Stability of laminar wakes. In *Fundamental Phenomena in Hypersonic Flow*, J.G. Hall (ed.). Cornell University Press, 310–339.

Lele, S.K. (1989). Direct numerical simulation of compressible free shear layer flows. AIAA Paper 89-0374.

Lele, S.K. (1992). Compact finite difference schemes with spectral-like resolution. *J. Comput. Phys.* **103**, 16–42.

Lessen, M. (1950). On stability of free laminar boundary layer between parallel streams. NACA R-979.

Lessen, M., Fox, J.A. & Zien, H.M. (1965). On the inviscid stability of the laminar mixing of two parallel streams of a compressible fluid. *J. Fluid Mech.* **23**, 355–367.

Lessen, M., Fox, J.A. & Zien, H.M. (1966). Stability of the laminar mixing of two parallel streams with respect to supersonic disturbances. *J. Fluid Mech.* **25**, 737–742.

Lessen, M., Singh, P.J. & Paillet, F. (1974). The stability of a trailing line vortex. Part 1. Inviscid theory. *J. Fluid Mech.* **63**, 753–763.

Lessen, M. & Paillet, F. (1974). The stability of a trailing line vortex. Part 2. Viscous theory. *J. Fluid Mech.* **65**, 769–779.

Liepmann, H.W. (1943). Investigation of laminar boundary layer stability and transition on curved boundaries. NACA Advisory Conf. Rep. 31730.

Liepmann, H.W. & Nosenchuck, D.M. (1982a). Active control of laminar–turbulent transition. *J. Fluid Mech.* **118**, 201–204.

Liepmann, H.W. & Nosenchuck, D.M. (1982b). Control of laminar instability waves using a new technique. *J. Fluid Mech.* **118**, 187–200.

Ligrani, P.M., Longest, J.E., Kendall, M.R. & Fields, W.A. (1994). Splitting, merging, and spanwise wavenumber selection of Dean vortex pairs. *Exp. Fluids* **18**(1/2), 41–58.

Lilly, D.K. (1966). On the instability of Ekman boundary flow. *J. Atmos. Sci.* **23**, 481–494.

Lin, C.C. (1944). On the stability of two-dimensional parallel flows. *National Academy of Science, US* **30**, 316–323.

Lin, C.C. (1945). On the stability of two-dimensional parallel flows. Parts I, II, III. *Quart. Appl. Math.* **3**, 117–142, 218–234, 277–301.

Lin, C.C. (1954). Hydrodynamic Stability. In *13th Symp. Appl. Math.* Amer. Math. Soc., 1–18

Lin, C.C. (1955). *The Theory of Hydrodynamic Stability*. Cambridge University Press.

Lin, C.C. (1961). Some mathematical problems in the theory of the stability of parallel flows. *J. Fluid Mech.* **10**, 430–438.

Lin, R.S. & Malik, M.R. (1994). The stability of incompressible attachment-line boundary layers: A 2D eigenvalue approach. AIAA Paper 94–2372.

Lin, R.S. & Malik, M.R. (1995). Stability and transition in compressible attachment line boundary layer flow. *Aerotech '95*, Sept 18–21, 1995. Los Angeles, CA. SAE Paper 952041.

Lin, R.S. & Malik, M.R. (1996). On the stability of attachment line boundary layers. Part 1. The incompressible swept Hiemenz flow. *J. Fluid Mech.* **311**, 239–255.

Lingwood, R.J. (1995). Absolute instability of the boundary layer on a rotating-disk boundary layer flow. *J. Fluid Mech.* **299**, 17–33.

Lingwood, R.J. (1996). An experimental study of absolute instability of the rotating-disk boundary-layer flow. *J. Fluid Mech.* **314**, 373–405.

Lingwood, R.J. (1997). On the impulse response for swept boundary-layer flows. *J. Fluid Mech.* **344**, 317–334.

Liou, W.W. & Shih, T.H. (1997). Bypass transitional flow calculations using a Navier–Stokes solver and two equation models. AIAA Paper 97–2738.

Liu, C. (1998). Multigrid methods for steady and time-dependent flow. In *Computational Fluid Dynamics Review*, M. Hafez & K. Oshima (eds). World Scientific Publishers Vol. 1, 512–535.

Liu, C. & Liu, Z. (1997). Advances in DNS/LES. *First AFOSR International Conference on DNS/LES*, August 4–8, 1997. Greyden Press.

Lock, R.C. (1951). The velocity distribution in the laminar boundary layer between parallel streams. *Quart. J. Mech. Appl. Math.* **4**, 42–62.

Lu, G. & Lele, S.K. (1993). Inviscid instability of a skewed compressible mixing layer. *J. Fluid Mech.* **249**, 441–463.

Lu, G. & Lele, S.K. (1994). On the density ratio effect on the growth rate of a compressible mixing layer. *Phys. Fluids* **6**(2), 1073–1075.

Luther, H.A. (1966). Further explicit fifth-order Runge–Kutta formulas. *SIAM Review* **8**(3), 374–380.

Lynch, R.E., Rice, J.R. & Thomas, D.H. (1964). Direct solution of partial difference equations by tensor product methods. *Num. Math.* **6**, 185–199.

Macaraeg, M.G. (1990). Bounded free shear flows: linear and nonlinear growth. In *Instability and Transition*, M.Y. Hussaini & R.G. Voigt (eds). Springer, Vol. II, 177–186.

Macaraeg, M.G. (1991). Investigation of supersonic modes and three-dimensionality in bounded, free shear flows. Comput. *Phys. Commun.* **65**, 201–208.

Macaraeg, M.G., Streett, C.L. & Hussaini, M.Y. (1988). A spectral collocation solution to the compressible stability eigenvalue problem. NASA TP-2858.

Macaraeg, M.G. & Streett, C.L. (1989). New instability modes for bounded, free shear flows. *Phys. Fluids A* **1**(8), 1305–1307.

Macaraeg, M.G. & Streett, C.L. (1991). Linear stability of high-speed mixing layers. *Appl. Numerical Math.* **7**, 93–127.

Mack, L.M. (1960). Numerical calculation of the stability of the compressible, laminar boundary layer. California Institute of Technology, Jet Propulsion Laboratory Rep. 20–122.

Mack, L.M. (1965a). Computation of the stability of the laminar compressible boundary layer. In *Methods in Computational Physics*, B. Alder, S. Fernbach & M. Rotenberg (eds). Academic Press, Vol. 4, 247–299.

Mack, L.M. (1965b). Stability of the compressible laminar boundary layer according to a direct numerical solution. *Recent Developments In Boundary Layer Research*, AGARDograph 97, Part 1, 329–362.

Mack, L.M. (1966). Viscous and inviscid amplification rates of two and three-dimensional disturbances in a compressible boundary layer. *Space Prog. Summary*, 42, IV, November.

Mack, L.M. (1975). Linear stability theory and the problem of supersonic boundary layer transition. *AIAA J.* **13**(3), 278–289.

Mack, L.M. (1976). A numerical study of the temporal eigenvalue spectrum of the Blasius boundary layer. *J. Fluid Mech.* **73**(3), 497–520.

Mack, L.M. (1984). Boundary layer linear stability theory. *Special Course on Stability and Transition of Laminar Flow*, AGARD-R-709.

Mack, L.M. (1987). Review of compressible stability theory. *Stability of Time Dependent and Spatially Varying Flows*, D.L. Dwoyer & M.Y. Hussaini (eds). Springer, 164–187.

Mack, L.M. (1988). Stability of three-dimensional boundary layers on swept wings at transonic speeds. In *Transsonicum III*, IUTAM, J. Zierep & H. Oertel (eds). Springer.

Mack, L.M. (1990). On the inviscid acoustic mode instability of supersonic shear flows. Part I. Two dimensional waves. *Theor. Comput. Fluid Dyn.* **2**, 97–123.

Maddalon, D.V., Collier, F.S., Jr., Montoya, L.C. & Putnam, R.J. (1990). Transition flight experiments on a swept wing with suction. In *IUTAM Symposium on Laminar–Turbulent Transition*, D. Arnal & R. Michel (eds). Springer, 53–62.

Malik, M.R. (1982). COSAL-A black box compressible stability analysis code for transition prediction in three-dimensional boundary layers. NASA CR-165925.

Malik, M.R. (1986). The neutral curve for stationary disturbances in rotating disk flow. *J. Fluid Mech.* **164**, 275–287.

Malik, M.R. (1987). Stability theory applications to laminar-flow control. NASA CP-2487, 219–244.

Malik, M.R. (1990). Numerical methods for hypersonic boundary layer stability. *J. Comp. Phys.* **86**, 376–413.

Malik, M.R., Wilkinson, S.P. & Orszag, S.A. (1981). Instability and transition in rotating disk flow. *AIAA J.* **19**, 1131–1138.

Mattingly, G.E. & Criminale, W.O. (1972). The stability of an incompressible two-dimensional wake. *J. Fluid Mech.* **51**(2), 233–272.

Mayer, E.W. & Powell, K.G. (1992). Instabilities of a trailing vortex. *J. Fluid Mech.* **245**, 91–114.

Meister, B. (1962). Das Taylor-Deansche Stabilitätsproblem für- beliebige Spaltbreiten. *Z. Angew. Math. Phys.* **13**, 83–91.

Meksyn, D. (1950). Stability of viscous flow over concave cylindrical surfaces. *Proc. R. Soc. London Ser. A* **203**, 253–265.

Meksyn, D. & Stuart, J.T. (1951). Stability of viscous motion between parallel flows for finite disturbances. *Proc. R. Soc. London Ser. A* **208**, 517–526.

Michalke, A. (1964). On the inviscid instability of the hyperbolic-tangent velocity profile. *J. Fluid Mech.* **19**, 543–556.

Michalke, A. (1965). On spatially growing disturbances in an inviscid shear layer. *J. Fluid Mech.* **23**, 521–544.

Miklavčič, M. (1983). Eigenvalues of the Orr–Sommerfeld equation in an unbounded domain. *Arch. Rational Mech. Anal.* **83**, 221–228.

Miklavčič, M. & Williams, M. (1982). Stability of mean flows over an infinite flat plate. *Arch. Rational Mech. Anal.* **80**, 57–69.

Miles, J.W. (1958). On the disturbed motion of a plane vortex sheet. *J. Fluid Mech.* **4**, 538–552.

Milling, R.W. (1981). Tollmien–Schlichting wave cancellation. *Phys. Fluids* **24**(5), 979–981.

Mittal, R. & Balachandar, S. (1995). Effect of three-dimensionality on the lift and drag of nominally two-dimensional cylinders. *Phys. Fluids* **7**(8), 1841–1865.

Monkewitz, P.A. & Huerre, P. (1982). Influence of the velocity ratio on the spatial instability of mixing layers. *Phys. Fluids* **25**(7), 1137–1143.

Morkovin, M.V. (1969). On the many faces of transition. In *Viscous Drag Reduction*, C.S. Wells (ed). Plenum Press, 1–31.

Morkovin, M.V. (1985). Bypass transition to turbulence and research desiderata. Transition in Turbines, NASA Conference Publication 2386, 161–204. https://ntrs.nasa.gov/archive/nasa/casi.ntrs.nasa.gov/19850023129.pdf

Morris, P.J. & Giridharan, M.G. (1991). The effect of walls on instability waves in supersonic shear layers. *Phys. Fluids A* **3**(2), 356–358.

Mukunda, H.S., Sekar, B., Carpenter, M., Drummond, J.P. & Kumar, A. (1992). Direct simulation of high speed mixing layers. NASA TP-3186.

Müller, B. & Bippes, H. (1988). Experimental study of instability modes in a three-dimensional boundary layer. AGARD-CP-438.

Murdock, J.W. (1977). A numerical study of nonlinear effects on boundary-layer stability. *AIAA J.* **15**, 1167–1173.

Nayfeh, A. H (1980). Stability of three-dimensional boundary layers. *AIAA J.* **18**, 406–416.

Nayfeh, A.H. (1987). Nonlinear stability of boundary layers, *AIAA 25th Aerospace Sciences Meeting*, January 12–15, 1987, Reno Nevada. AIAA Paper 87–0044.

Nayfeh, A.H. & Bozatli, A.N. (1979). Nonlinear wave interactions in boundary layers. AIAA Paper 79–1496.

Nayfeh, A.H. & Bozatli, A.N. (1980). Nonlinear wave interactions of two waves in boundary layers flows. *Phys. Fluids* **23**(3), 448–459.

Ng, B.S. & Reid, W.H. (1979). An initial value method for eigenvalue problems using compound matrices. *J. Comput. Phys.* **30**(1), 125–136.

Ng, B.S. & Reid, W.H. (1980). On the numerical solution of the Orr–Sommerfeld problem: Asymptotic initial conditions for shooting methods. *J. Comput. Phys.* **38**, 275–293.

Obremski, H.T., Morkovin, M.V. & Landahl, M.T. (1969). A portfolio of stability characteristics of incompressible boundary layers. AGARDograph 134, NATO, Paris.

Orr, W.McF. (1907a). Lord Kelvin's investigations, especially the case of a stream which is shearing uniformly. *Roy. Irish Academy* **A27**, 69–138.

Orr, W.McF. (1907b). The stability or instability of the steady motions of a perfect liquid and of a viscous liquid. *Roy. Irish Academy* **A27**, 9–68.

Orszag, S.A. (1971). Accurate solution of the Orr–Sommerfeld stability equation. *J. Fluid Mech.* **50**, 689–703.

Orszag, S.A. & Patera, A.T. (1980). Subcritical transition to turbulence in plane channel flows. *Phys. Rev. Let.*, **45**, 989–993.

Orszag, S.A. & Patera, A.T. (1981). Subcritical transition to turbulence in plane shear flows. In *Transition and Turbulence*, R.E. Meyer (ed). Academic Press, 127–146.

Pal, A., Bower, W.W. & Meyer, G.H. (1991). Numerical simulations of multi-frequency instability-wave growth and suppression in the Blasius boundary layer. *Phys. Fluids A* **3**(2), 328–340.

Papageorgiou, D.T. (1990a). Linear instability of the supersonic wake behind a flat plate aligned with a uniform stream. *Theoret. Comput. Fluid Dyn.* **1**, 327–348.

Papageorgiou, D.T. (1990b). The stability of two-dimensional wakes and shear layers at high Mach numbers. ICASE Rep. 90–39.

Papageorgiou, D.T. (1990c). Accurate calculation and instability of supersonic wake flows. In *Instability and Transition*, M.Y. Hussaini & R.G. Voigt (eds). Springer, Volume II, 216–229.

Papageorgiou, D.T. & Smith, F.T. (1988). Nonlinear instability of the wake behind a flat plate placed parallel to a uniform stream. *Proc. R. Soc. London Ser. A* **419**, 1–28.

Papageorgiou, D.T. & Smith, F.T. (1989). Linear instability of the wake behind a flat plate placed parallel to a uniform stream. *J. Fluid Mech.* **208**, 67–89.

Papamoschou, D. & Roshko, A. (1986). Observations of supersonic free-shear layers. AIAA Paper 86–0162.

Papamoschou, D. & Roshko, A. (1988). The compressible turbulent shear layer: an experimental study. *J. Fluid Mech.* **197**, 453–477.

Papas, P., Monkewitz, P.A. & Tomboulides, A.G. (1999). New instability modes of a diffusion flame near extinction. *Phys. Fluids* **11**(10), 2818–2820.

Pavithran, S. & Redekopp, L.G. (1989). The absolute-convective transition in subsonic mixing layers. *Phys. Fluids A* **1**(10), 1736–1739.

Peerhossaini, H. (1987). L'Instabilité d'une couche limité sur une paroi concave (les tourbilons de Görtler). Thèse de Doctorat, Univ. Pierre et Marie Curie.

Peroomian, O. & Kelly, R.E. (1994). Absolute and convective instabilities in compressible confined mixing layers. *Phys. Fluids* **6**(9), 3192–3194.

Pfenninger, W. (1965). Flow phenomena at the leading edge of swept wings. *Recent Developments in Boundary Layer Research*, AGARDograph 97.

Piomelli, U. & Zang, T.A. (1991). Large-eddy simulation of transitional channel flow. *Comput. Phys. Commun.* **65**, 224–230.

Piomelli, U. & Liu, J. (1995). Large-eddy simulation of rotating channel flows using a localized dynamic model. *Phys. Fluids* **23**, 839–848.

Piomelli, U., Zang, T.A., Speziale, C.G. & Hussaini, M.Y. (1990). On the large eddy simulation of transitional wall-bounded flows. *Phys. Fluids A* **2**, 257–265.

Planche, O.H. & Reynolds, W.C. (1991). Compressibility effect on the supersonic reacting mixing layer. AIAA Paper 91–0739.

Potter, M.C. (1966). Stability of plane Couette-Poiseuille flow. *J. Fluid Mech.* **24**, 609–619.

Prandtl, L. (1921–1926). Bemerkungen über die entstehung der turbulenz. *Zeitschrift für Angewandte Mathematik und Mechanik* **1**, 431–436; *Physik. Z.* **23**, 1922, 19–23. Discussion after Solberg's paper, 1924; and with F. Noether, *Zeitschrift für Angewandte Mathematik und Mechanik* **6**, 1926, 339, 428.

Prandtl, L. (1930). Einfluss stabilisierender kräfte auf die turbulenz. *Vorträge aus dem Gebiete det Aerodynamik und Verwandter Gebiete*, Aachen. Springer, 1–17.

Prandtl, L. (1935). *Aerodynamic Theory*, W.F. Durand (ed). Springer, Vol 3, 178–190.

Press, W.H., Teukolsky, S.A., Vetterling, W.T. & Flannery, B.P. (1992). *Numerical Recipes in Fortran*, 2nd edition. Cambridge University Press.

Pruett, C.D. & Chang, C.L. (1995). Spatial direct numerical simulation of high-speed boundary-layer flows. Part II: Transition on a cone in Mach 8 flow. *Theor. & Comput. Fluid Dyn.* **7**(5), 397–424.

Pupator, P. & Saric, W. (1989). Control of random disturbances in a boundary layer. AIAA Paper 89–1007.

Raetz, G.S. (1964). Calculation of precise proper solutions for the resonance theory of transition. I. Theoretical investigations. Contract AF 33–657–11618, Final Rept., Document Rept. ASD-TDR, Northrop Aircraft Inc., Norair Division, Hawthorne, CA.

Ragab, S.A. (1988). Instabilities in the wake mixing-layer region of a splitter plate separating two supersonic streams. AIAA Paper 88–3677.

Ragab, S.A. & Wu, J.L. (1989). Linear instabilities in two dimensional compressible mixing layers. *Phys. Fluids A* **1**(6), 957–966.

Rai, M.M. & Moin, P. (1991a). Direct numerical simulation of transition and turbulence in a spatially-evolving boundary layer. AIAA Paper 91–1607.

Rai, M.M. & Moin, P. (1991b). Direct numerical simulation of turbulent flow using finite-difference schemes. *J. Comput. Phys.* **96**, 15–53.

Rayleigh, Lord (1879). On the in stability of jets. *Proc. London Math. Soc.* **10**, 4–13. (Reprinted in *Scientific Papers*, J.W. Strutt (ed). Cambridge University Press, Vol. 1, 361–371, 1899).

Rayleigh, Lord (1880). On the stability or instability of certain fluid motions. *Proc. London Math. Soc.* **11**, 57–70. (Reprinted in *Scientific Papers*, J.W. Strutt (ed). Cambridge University Press, Vol. 1, 474–487, 1899.)

Rayleigh, Lord (1883). Investigation of the character of the equilibrium of an incompressible heavy fluid of variable density. *Proc. London Math. Soc.* **14**, 170–177. (Reprinted in *Scientific Papers*, J.W. Strutt (ed). Cambridge University Press, Vol. 2, 200–207, 1900.)

Rayleigh, Lord (1887). On the stability or instability of certain fluid motions. Part II. *Scientific Papers*, J.W. Strutt (ed). Cambridge University Press, Vol. 3, 17–23.

Rayleigh, Lord (1892a). On the question of the stability of the flow of fluids. *Phil. Mag.* **34**, 59–70. (Reprinted in *Scientific Papers*, J.W. Strutt (ed). Cambridge University Press, Vol. 3, 575–584, 1902.)

Rayleigh, Lord (1892b). On the stability of a cylinder of viscous liquid under capillary force. *Scientific Papers*, J.W. Strutt (ed). Cambridge University Press, Vol. 3, 2–23, 1899.

Rayleigh, Lord (1892c). On the instability of cylindrical fluid surfaces. *Phil. Mag.*, 34, 177–180. (Reprinted *Scientific Papers*, J.W. Strutt (ed). Cambridge University Press, Vol. 3, 594–596, 1902.)

Rayleigh, Lord (1894). *The Theory of Sound*, 2nd edition. Macmillan.

Rayleigh, Lord (1895). On the stability or instability of certain fluid motions. III. *Scientific Papers*, J.W. Strutt (ed). Cambridge University Press, Vol. 4, 203–209, 1899.

Rayleigh, Lord (1911). Hydrodynamical notes. *Phil. Mag.* **21**, 177–195.

Rayleigh, Lord (1913). On the stability of the laminar motion of an inviscid fluid. *Scientific Papers*, J.W. Strutt (ed). Cambridge University Press, Vol. 6, 197–204.

Rayleigh, Lord (1914). Further remarks on the stability of viscous fluid motion. *Phil. Mag.* **28**, 609–619. (Reprinted in *Scientific Papers*, J.W. Strutt (ed). Cambridge University Press, Vol. 6, 266–275, 1920.)

Rayleigh, Lord (1915). On the stability of the simple shearing motion of a viscous incompressible fluid. *Scientific Papers*, J.W. Strutt (ed). Cambridge University Press, Vol. 6, 341–349.

Rayleigh, Lord (1916a). On convection currents in a horizontal layer of fluid when the higher temperature is on the other side. *Phil. Mag.* **32**, 529–546. (Reprinted in *Scientific Papers*, J.W. Strutt (ed). Cambridge University Press, Vol. 6, 432–446, 1920.)

Rayleigh, Lord (1916b). On the dynamics of revolving fluids. *Proc. Roy. Soc. London Ser. A* **93**, 148–154. (Reprinted in *Scientific Papers*, J.W. Strutt (ed). Cambridge University Press, Vol. 6, 447–453, 1920.)

Reddy, S.C. & Henningson, D.S. (1993). Energy growth in viscous channel flows. *J. Fluid Mech.* **252**, 209–238.

Reed, H.L. & Saric, W.S. (1998). Stability of three-dimensional boundary layers. *Ann. Rev. Fluid Mech.* **21**, 235.

Reibert, M.S., Saric, W.S., Carrillo, R.B., Jr. & Chapman, K.L. (1996). Experiments in nonlinear saturation of stationary crossflow vortices in a swept-wing boundary layer. *34th Aerospace Sciences Meeting & Exhibit*, January 15–18, 1996, Reno, NV. AIAA Paper 96–0184.

Reischman, M.M. (1984). A review of compliant coating drag reduction research at ONR. *Laminar-Turbulent Boundary Layers* **11**, 99–105.

Reshotko, E. (1960). Stability of the compressible laminar boundary layer. California Institute of Technology, Guggenheim Aeronautical Laboratory, GALCIT Memo. No. 52.

Reshotko, E. (1976). Boundary-layer stability and transition. *Ann. Rev. Fluid Mech.* **8**, 311–349.

Reshotko, E. (1984). Environment and receptivity. AGARD-R-709.

Reynolds, O. (1883). An experimental investigation of the circumstances which determine whether the motion of water shall be direct or sinuous, and of the law of resistance in parallel channels. *Phil. Trans. R. Soc. London* **34**, 84–99. (Reprinted in *Scientific Papers*, J.W. Strutt (ed). Cambridge University Press, Vol. 2, 51–105.)

Reynolds, W.C. & Potter, M.C. (1967). Finite amplitude instability of parallel shear flows. *J. Fluid Mech.* **27**, 465–492.

Romanov, V.A. (1973). Stability of plane-parallel Couette flow. *Funkcional Anal. i Prolozen* **7**(2), 62–73. (Translation: *Functional Anal. & Its Applications* **7**, 137–146, 1973.)

Rozendaal, R.A. (1986). Natural laminar flow flight experiments on a swept wing business jet: Boundary layer stability analyses. NASA CR-3975.

Salwen, H. & Grosch, C.E. (1981). The continuous spectrum of the Orr–Sommerfeld equation. Part 2. Eigenfunction expansions. *J. Fluid Mech.* **104**, 445–465.

Sandham, N.D. & Reynolds, W.C. (1990). Compressible mixing layer: Linear theory and direct simulation. *AIAA J.* **28**(4), 618–624.

Saric, W.S. (1994a). Görtler vortices. *Ann. Rev. Fluid Mech.* **26**, 379–409.

Saric, W.S. (1994b) Low speed boundary layer transition experiments. In *Transition: Experiments, Theory, and Computations*, T.C. Corke, G. Erlebacher & M.Y. Hussaini (eds). Oxford University Press.

Saric, W.S. (2008). Flight experiments on local and global effects of surface roughness on 2-D and 3-D boundary-layer stability and transition. Final Report AFOSR Grant FA9550–05–1–0044.

Saric, W.S., Hoos, J.A. & Radeztsky, R.H. (1991). Boundary layer receptivity of sound with roughness. In *Boundary Layer Stability and Transition to Turbulence*, D.C. Reda, H.L. Reed & R.K. Kobayashi (eds). *ASME FED* **114**, 17–22.

Schlichting, H. (1932). Über die stabilität der Couette-strömung. *Ann. Physik* (Leipzig) **14**, 905–936.

Schlichting, H. (1933a). Zur entstehung der turbulenz bei der Plattenströmung, *Mathematisch–Physikalische Klasse*, Gessellschaft der Wissenschaften, Göttingen, 181–208.

Schlichting, H. (1933b). Berechnung der anfachung kleiner Störungen bei der Plattenströmung. *Zeitschrift für Angewandte Mathematik und Mechanik* **13**(3), 171–174.

Schlichting, H. (1933c). Laminar spread of a jet. *Zeitschrift für Angewandte Mathematik und Mechanik* **13**(4), 260–263.

Schlichting, H. (1934). Neuere untersuchungen über die turbulenzenstehung. *Naturwiss.* **22**, 376–381.

Schlichting, H. (1935). Amplitudenverteilung und Energiebilanz der kleinen Störungen bei der Plattengrenzschicht. Gesellschaft der Wissenschatten. Göttingen. *Mathematisch-Naturwissenschattliche Klasse* **1**, 47–78.

Schmid, P.J. & Henningson, D.S. (1992a). Channel flow transition induced by a pair of oblique waves. In *Instability, Transition, and Turbulence*, M.Y. Hussaini, A. Kumar & C.L. Streett (eds). Springer, 356–366.

Schmid, P.J. & Henningson, D.S. (1992b). A new mechanism for rapid transition involving a pair of oblique waves. *Phys. Fluids A* **4**(9), 1986–1989.

Schrauf, G., Bieler, H. & Thiede, P. (1992). Transition prediction – the Deutsche Airbus view. In *First European Forum on Laminar Flow Technology*, Hamburg, Germany, March 16–18, 1992, 73–81.

Schreivogel, P. (2010). Detection of laminar-turbulent transition in a free-flight experiment using thermography and hot-film anemometry. In *21st International Congress of the Aeronautical Sciences*, ICAS2010, 1–8.

Schubauer, G.B. & Skramstad, H.K. (1943). Laminar boundary layer oscillations and transition on a flat plate. NACA TR-909.

Schubauer, G.B. & Skramstad, H.K. (1947). Laminar boundary layer oscillations and stability of laminar flow. *J. Aeron. Sci.* **14**(2), 69–78.

Schubauer, G.B. & Klebanoff, P.S. (1955). Contributions on the mechanics of boundary-layer transition. NACA TN-3489.

Schubauer, G.B. & Klebanoff, P.S. (1956). Contributions on the mechanics of boundary-layer transition. NACA Rep. 1289.

Shanthini, R. (1989). Degeneracies of the temporal Orr–Sommerfeld eigenmodes in plane Poiseuille flow. *J. Fluid Mech.* **201**, 13–34.

Shen, S.F. (1952). On the boundary layer equations in hypersonic flow. *J. Aeron. Sci.* **19**, 500–501.

Shin, D.S. & Ferziger, J.H. (1991). Linear stability of the reacting mixing layer. *AIAA J.* **29**(10), 1634–1642.

Shin, D.S. & Ferziger, J.H. (1993). Linear stability of the confined compressible reacting mixing layer. *AIAA J.* **31**(3), 571–577.

Shivamoggi, B.K. (1977). Inviscid theory of stability of parallel compressible flows. *J. de Mecanique* **16**(2), 227–255.

Shivamoggi, B.K. (1979). Effects of compressibility upon the stability characteristics of a free shear layer. *Z. Angew. Math. Mech.* **59**, 405–415.

Shivamoggi, B.K. & Rollins, D.K. (2001). Linear stability theory of zonal shear flows with a free surface. *Geophys. Astrophys. Fluid Dyn.* **95**(1–2), 31–53.

Singer, B.A., Choudhari, M. & Li, F. (1995). Weakly nonparallel and curvature effects on stationary crossflow instability: Comparison of results from multiple scales analysis and parabolized stability theory. NASA CR-198200.

Sivashinsky, G. (1977). Nonlinear analysis of hydrodynamic instability in laminar flames. Part I. Derivation of basic equations. *Acta Astronaut.* **4**, 1117–1206.

Smagorinsky, J. (1963). General circulation experiments with the primitive equations. I. The basic experiment. *Mon. Weather Rev.* **91**, 99–164.

Smith, A.M.O. (1953). Review of research on laminar boundary layer control at the Douglas Aircraft Company El Segundo Division. Douglas Aircraft Co. Rep. No. ES-19475, June.

Smith, A.M.O. (1955). On the growth of Taylor–Görtler vortices along highly concave walls. *Quart. Appl. Math.* **13**, 233–262.

Smith, A.M.O. (1956). Transition, pressure gradient, and stability theory. *International Congress on Applied Mechanics*, Brussels, Belgium, 234–244.

Smith, A.M.O. & Gamberoni, N. (1956). Transition, pressure gradient, and stability theory. Douglas Aircraft Company Rep. ES-26388.

Smith, B.A. (1996). Laminar flow data evaluated. *Aviation Week & Space Tech.*, October 7, 1996.

Smith, F.T. & Brown, S.N. (1990). The inviscid instability of a Blasius boundary layer at large values of the Mach number. *J. Fluid Mech.* **219**, 499–518.

Smol'yakov, A.V. & Tkachenko, V.M. (1983). *The Measurement of Turbulent Fluctuations. An Introduction to Hot-Wire Anemometry and Related Transducers.* Springer.

Sommerfeld, A. (1908). Ein beitraz zur hydrodynamischen erklaerung der turbulenten fluessigkeitsbewegungen. In *Proc. Fourth Inter. Congr. Mathematicians*, Rome, Italy, 116–124.

Spalart, P.R. (1989). Direct numerical study of leading-edge contamination. In *Fluid Dynamics of Three-Dimensional Turbulent Shear Flows and Transition.* AGARD-CP-438, 5.1–5.13.

Spalart, P.R. (1990). Direct numerical study of cross-flow instability. In *IUTAM Symposium on Laminar–Turbulent Transition*, D. Arnal & R. Michel (eds). Springer, 621–630.

Spalart, P.R. & Yang, K.S. (1987). Numerical study of ribbon-induced transition in Blasius flow. *J. Fluid Mech.* **178**, 345–365.

Spooner, G.F. (1980). Fluctuations in geophysical boundary layers. PhD Dissertation, Department of Oceanography, University of Washington.

Spooner, G.F. & Criminale, W.O. (1982). The evolution of disturbances in an Ekman boundary layer. *J. Fluid Mech.* **115**, 327–346.

Squire, H.B. (1933). On the stability for three-dimensional disturbances of viscous fluid flow between parallel walls. *Proc. R. Soc. London Ser. A.* **142**, 621–628.

Srokowski, A.J. & Orszag, S.A. (1977). Mass flow requirements for LFC wing design. AIAA Paper 77-1222.

Streett, C.L. & Macaraeg, M.G. (1989). Spectral multi-domain for large-scale fluid dynamics simulations. *Appl. Num. Math.* **6**, 123–140.

Streett, C.L. & Hussaini, M.Y. (1991). A numerical simulation of the appearance of chaos in finite-length Taylor–Couette flow. *Appl. Num. Math.* **7**, 41–71.

Stuart, J.T. (1960). On the non-linear mechanics of wave disturbances in stable and unstable parallel flows. Part 1. *J. Fluid Mech.* **9**, 353–370.

Swearingen, J.D. & Blackwelder, R.F. (1986). Spacing of streamwise vortices on concave walls. *AIAA J.* **24**, 1706–1709.

Swearingen, J.D. & Blackwelder, R.F. (1987). The growth and breakdown of stream-wise vortices in the presence of a wall. *J. Fluid Mech.* **182**, 255–290.

Synge, J.L. (1938). Hydrodynamic stability. *Semicentennial Publ. Amer. Math. Soc.* **2**, 227–269.

Tadjfar, M. & Bodonyi, R.J. (1992). Receptivity of a laminar boundary layer to the interaction of a three-dimensional roughness element with time-harmonic free-stream disturbances. *J. Fluid Mech.* **242**, 701–720.

Tam, C.K.W. & Hu, F.Q. (1989). The instability and acoustic wave modes of supersonic mixing layers inside a rectangular channel. *J. Fluid Mech.* **203**, 51–76.

Tatsumi, T. (1952). Stability of the laminar inlet flow prior to the formation of Poiseuille region. *J. Phys. Soc. Japan* **7**, 489–502.

Tatsumi, T. & Kakutani, T. (1958). The stability of a two dimensional jet. *J. Fluid Mech.* **4**, 261–275.

Taylor, G.I. (1921). Experiments with rotating fluids. *Proc. Camb. Phil. Soc.* **20**, 326–329.

Taylor, G.I. (1923). Stability of a viscous liquid contained between two rotating cylinders. *Phil. Trans. R. Soc. London Ser. A* **223**, 289–343.

Theofilis, V., Fedorov, A., Obrist, D. & Dallmann, U. (2003). The extended Görtler–Hammerlin model for linear instability of three-dimensional incompressible swept attachment-line boundary layer flow. *J. Fluid Mech.* **487**, 271–313.

Thomas, A.S.W. (1983). The control of boundary-layer transition using a wave-superposition principle. *J. Fluid Mech.* **137**, 233–250.

Thomas, R.H., Choudhari, M.M. & Joslin, R.D. (2002). Flow and noise control: Review and assessment of future directions. NASA TM-2002–211631.

Tietjens, O. (1925). Beiträge zur entsehung der turbulenz. *Zeitschrift für Angewandte Mathematik und Mechanik* **5**, 200–217.

Ting, L. (1959). On the mixing of two parallel streams. *J. Math. Phys.* **28**, 153–165.

Tollmien, W. (1929). Über die Entstehung der Turbulenz. *Mathematisch–Naturwissenschaftliche Klasse. Nachrichten*, Gesellschaft der Wissenschaften, Göttingen, 21–44. (Translation: The production of turbulence. NACA TM-609, 1931.)

Tollmien, W. (1935). Ein allgemeines Kriterium der Instabilität laminarer Geschwindigkeitsverteilungen. *Nachr. Wiss. Fachgruppe, Göttingen, Math. phys.* **1**, 79–114. (Translation: General instability criterion of laminar velocity disturbances. NACA TM 792, 1936.)

Tung, K.K. (1981). Barotropic instability of zonal flows. *J. Atmospheric Sci.* **38**, 308–321.

Vallikivi, M., Hultmark, M., Bailey, S.C.C. & Smits, A.J. (2011). Turbulence measurements in pipe flow using a nano-scale thermal anemometry probe. *Exp. Fluids* **51**(6), 1521–1527.

van Dyke, M. (1975). *Perturbation Methods in Fluid Mechanics*. Annotated edition. The Parabolic Press.

van Dyke, M. (1982). *An Album of Fluid Motion*. The Parabolic Press.

van Ingen, J.L. (1956). A suggested semi-empirical method for the calculation of the boundary layer transition region. University of Delft Rep. VTH-74, Delft, The Netherlands.

Vasilyev, O.V. (2000). High order finite difference schemes on non-uniform meshes with good conservation properties. *J. Comp. Phys.* **157**, 746–761.

Vijgen, P.M.H.W., Dodbele, S.S., Holmes, B.J. & van Dam, C.P. (1986). Effects of compressibility on design of subsonic natural laminar flow fuselages. *AIAA 4th Applied Aerodynamics Conference*, June 9–11, 1986, San Diego, CA. AIAA Paper 86–1825CP.

von Karman, T. (1921). Über laminare und turbulente Reibung. *Zeitschrift für Angewandte Mathematik und Mechanik* **1**, 233–252.

Von Mises, R. & Friedrichs, K.O. (1971). *Fluid Dynamics*. Springer.

Warren, E.S. & Hassan, H.A. (1996). An alternative to the e^N method for determining onset of transition. *AIAA 35th Aerospace Sciences Meeting & Exhibit*, January 6–10, 1997, Reno, NV. AIAA Paper 96–0825.

Warren, E.S. & Hassan, H.A. (1997). A transition closure model for predicting transition onset. *SAE/AIAA World Aviation Congress & Exposition 97*, Anaheim, CA. Paper 97WAC-121.

Watson, J. (1960). On the non-linear mechanics of wave disturbances in stable and unstable parallel flows. Part 2. *J. Fluid Mech.* **9**, 371–389.

Wazzan, A.R., Okamura, T. & Smith, A.M.O. (1966). Spatial stability study of some Falkner–Skan similarity profiles. *Fifth U.S. National Congress on Applied Mechanics*, ASME University of Minnesota, 836.

Wazzan, A.R., Okamura, T. & Smith, A.M.O. (1968). Spatial and temporal stability charts for the Falkner–Skan boundary layer profiles. Report No. DAC-67086, McDonnell–Douglas Aircraft Co., Long Beach, CA.

Werlé, H. (1974). Le tunnel hydrodynamique au service de la recherche Aérospatiale. ONERA No. 156.

White, F.M. (1974). *Viscous Fluid Flow*. McGraw-Hill.

Wiegel, M. & Wlezien, R.W. (1993). Acoustic receptivity of laminar boundary layers over wavy walls. AIAA Paper 93–3280.

Wilkinson, S.P. & Malik, M.R. (1985) Stability experiments in the flow over a rotating disk. *AIAA J.* **21**, 588–595.

Williamson, C.H.K. (1996). Vortex dynamics in the cylinder wake. *Ann. Rev. Fluid Mech.* **28**, 477–539.

Williamson, J.H. (1980). Low storage Runge–Kutta schemes. *J. Comput. Phys.* **35**(1), 48–56.

Willis, G.J.K. (1986). Hydrodynamic stability of boundary layers over compliant surfaces. PhD Thesis, University of Exeter.

Winoto, S.H., Zhang, D.H. & Chew, Y.T. (2000). Transition in boundary layers on a concave surface. *J. Prop. & Power* **16**(4), 653–660.

Witting, H. (1958). Über den Einfluss der Stromlinienkrummung auf die Stabilität laminarer Stromungen. *Arch. Rat. Mech. Anal.* **2**, 243–283.

Wray, A. & Hussaini, M.Y. (1984). Numerical experiments in boundary layer stability. *Proc. R. Soc. London Ser. A* **392**, 373–389.

Yeo, K.S. (1986). The stability of flow over flexible surfaces. PhD Thesis, University of Cambridge.

Zang, T.A. (1991). On the rotation and skew-symmetric forms for incompressible flow simulations. *Appl. Num. Math.* **7**, 27–40.

Zang, T.A. & Hussaini, M.Y. (1987). Numerical simulation of nonlinear interactions in channel and boundary layer transition. In *Nonlinear Wave Interactions in Fluids*, R.W. Miksad, T.R. Akylas, T. Herbert (eds). New York: ASME, AMD-87, 131–145.

Zang, T.A. & Hussaini, M.Y. (1990). Multiple paths to subharmonic laminar breakdown in a boundary layer. *Phys. Rev. Lett.* **64**, 641–644.

Zhuang, M., Kubota, T. & Dimotakis, P.E. (1988). On the instability of inviscid, compressible free shear layers. AIAA Paper 88–3538.

Zhuang, M., Kubota, T. & Dimotakis, P.E. (1990a). Instability of inviscid, compressible free shear layers. *AIAA J.* **28**(10), 1728–1733.

Zhuang, M., Kubota, T. & Dimotakis, P.E. (1990b). The effect of walls on a spatially growing supersonic shear layer. *Phys. Fluids A* **2**(4), 599–604.

Author Index

Subject Index